Geophysical Monograph Series

Including

IUGG Volumes
Maurice Ewing Volumes
Mineral Physics Volumes

Geophysical Monograph 132

Mountain Building in the Uralides
Pangea to the Present

Dennis Brown
Christopher Juhlin
Victor Puchkov
Editors

American Geophysical Union
Washington, DC

Library of Congress Cataloging-in-Publication Data
Mountain building in the Uralides: Pangea to the present /Dennis Brown, Christopher Juhlin, Victor Puchkov, editors.
    p. cm -- (Geophysical monograph ; 132)
  ISBN 0-87590-991-4
    1. Orogeny--Russia (Federation)--Ural Mountains. 2. Geology, Structural--Russia (Federation)--Ural Mountains. I. Brown, Dennis, 1958- II. Juhlin, Christopher. III. Puchkov, Victor Nikolaevich. IV. Series.

QE621.5.R8 M68 2002
551.8'2'094743--dc21   2002034153

ISSN 0065-8448
ISBN 0-87590-991-4

CONTENTS

CONTENTS

Section 4. Late- to Post-Orogenic History and the Uplift of the Ural Mountains

PREFACE

Extending for more than 2000 kilometers from the islands of Novaya Zemlya in the north to the Aral Sea in the south, the Uralide orogen forms the geographical and geological divide between Europe and Asia. For more than a century the Uralides have been one of the key areas of geological research in Russia, and have provided much of its mineral and petroleum wealth for the last 50 years. Nevertheless, the geology and tectonic evolution of the Uralide orogen were relatively unknown in the international literature until recently, when EUROPROBE and GEODE (European Science Foundation scientific programmes) brought together Russian, European, and American earth scientists to work in the Uralides project and the Urals Mineral Province project, respectively. Much of the recent research has focused around two deep seismic surveys, Europrobe's Seismic Reflection Profiling in the Urals (ESRU) survey in the Middle Urals and the multicomponent Urals Seismic Experiment and Integrated Studies (URSEIS) survey in the South Urals. These experiments were accompanied by a large number of geological, geochemical, geochronological, and geophysical studies.

This volume brings together many of the results of these projects, presenting data that will provide the basis for an enhanced understanding of the Uralide orogen. It is designed to address the large-scale orogenic processes that were active in the formation and preservation of the Uralides. This is achieved by developing four broad themes; (1) the Uralide crustal structure, with focus on the ESRU and URSEIS seismic transects that provide information on formative tectonic processes; (2) the well-preserved arc-continent collision zone, with geochemical and structural analyses of the island arc and oceanic system; (3) the timing and processes of subduction-related and late orogenic granitoid melt generation and emplacement in the internal part of the orogen using geochemical and geochronological constraints; and (4) the still unanswered question of the age of Ural Mountain topography, which gains new clarity in fission track studies and rheological modeling.

In the following introductory section we briefly present an overview of the Uralides, outlining the tectonic and geological background upon which this volume builds. The geological and geographic nomenclatures applied to the Uralides, and used throughout this volume, are also presented.

We would like to express our gratitude to the many people who have contributed manuscripts to this volume, and to those who have provided manuscript reviews. Many of the ideas expressed in the papers in this volume were discussed and developed in EUROPROBE and GEODE workshops. The numerous colleauges who praticipated in these discussions are also thanked.

It is hoped that this volume will be of interest to geoscientists, researchers, and students working on orogenic processes worldwide and in the Uralides in particular.

Dennis Brown
Christopher Juhlin
Victor Puchkov

Introduction

Dennis Brown

Institute Jaume Almera, Barcelona, Spain

Christopher Juhlin

Department of Earth Sciences, Uppsala University, Uppsala, Sweden

Victor Puchkov

Institute of Geology, Ufa, Russia

INTRODUCTION

The Uralide orogen of Russia and the Variscide-Appalachian orogenic system of Europe and North America are the main orogenic edifices built during the Paleozoic assembly of Pangaea (Figure 1). Unlike the Variscide-Appalachian orogenic system, which was largely rifted apart by the opening of the Atlantic ocean or extensively overprinted by post orogenic processes, the Uralides have been preserved within a tectonic plate (Eurasia), providing an opportunity to study an intact Paleozoic orogen. Most of what is known about the Uralide orogen is confined to the Ural Mountains, a narrow range of low to moderate topography extending for nearly 2500 kilometers from near the Aral Sea in the south to the islands of Novaya Zemlya in the Arctic Ocean (Plate 1a). To the south and east of the Ural Mountains exposure is poor and much of the orogen is buried beneath Mesozoic and Cenozoic sediments and has not been extensively studied (Plate 1b). The full extent of the Uralides is best seen by its roughly north-south oriented, short wavelength magnetic signature,

which abruptly interupts the long wavelength signature of the Baltica, Siberia and Kazakhstan plates (Plate 1c). For descriptive purposes the Uralides have traditionally been divided into a number of longitudinal zones that are largely based on the ages and paleogeography of the dominant rocks within them [e.g., *Ivanov* et al., 1975; *Khain*, 1985; *Puchkov*, 1997]. From west to east these zones are; the Pre-Uralian zone, the West Uralian zone, the Central Uralian zone, the Magnitogorsk-Tagil zone, the East Uralian zone, and the Trans-Uralian zone (Figure 2). Additionally, the Uralides have been divided geographically into the South, Middle, North, Cis-Polar and Polar Urals (Figure 2).

The Pre-Uralian, West Uralian and Central Uralian zones contain syntectonic Late Carboniferous to Early Triassic sediments of the foreland basin, Paleozoic platform and slope rocks, and Archean and Proterozoic rocks of the East European Craton (part of Baltica). These three zones were affected by Uralide deformation and make up the foreland thrust and fold belt [e.g., *Kamaletdinov*, 1974; *Brown* et al., 1996]. The Magnitogorsk-Tagil zone consists of Silurian to Devonian intra-oceanic island arc volcanic rocks and overlying volcanoclastic sediments. The Magnitogorsk-Tagil zone is sutured to the former continental margin of Baltica along the Main Uralian fault. The East Uralian zone is composed predominantly of deformed and metamorphosed volcanic arc fragments with minor amounts of Precambrian and Paleozoic rocks thought to represent continental crust [e.g., *Puchkov*,

Mountain Building in the Uralides: Pangea to the Present
Geophysical Monograph 132
Copyright 2002 by the American Geophysical Union
10.1029/132GM01

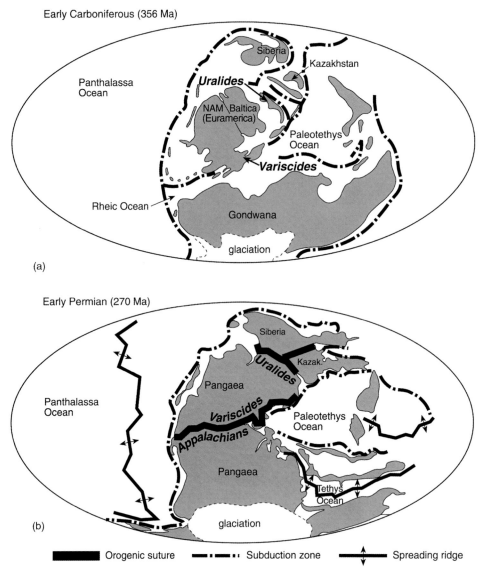

Figure 1. Paleogeographic maps for, (a) the Early Carboniferous, and (b) the Early Permian showing the development of the Uralides. (redrafted from C. R. *Scotese*, PALEOMAP Project, www.scotese.com [*Scotese*, 2001]).

1997, 2000; *Friberg*, 2000]. The East Uralian zone was extensively intruded by Carboniferous and Permian granitoids [*Fershtater* et al., 1997; *Bea* et al., 1997], forming the "main granite axis" of the Uralides. The East Uralian zone is juxtaposed against the Magnitogorsk-Tagil zone along the East Magnitogorsk – Serov-Mauk fault system. The Trans-Uralian zone is composed of Devonian and Carboniferous volcano-plutonic complexes overlain by terrigenous red-beds and evaporites [e.g., *Puchkov*, 1997, 2000]. Ophiolitic material and high pressure rocks have also been reported [e.g., *Puchkov*, 2000]. The contact between the East Uralian and Trans-

Uralian zones is only exposed in the Southern Urals, where it is a serpentinite mélange. Rocks that unequivocally belong to either the Kazakhstan or Siberia plates do not outcrop in the Uralides.

The tectonic evolution of the Uralides began during the Late Paleozoic as the continental margin of Baltica (part of Euroamerica in Figure 1) was subducted eastward (current coordinates) beneath the Magnitogorsk and Tagil island arcs during the Late Devonian and Early Carboniferous (Figure 1a). This was followed by the accretion of volcanic arcs, oceanic crust, and possible continental fragments along the margin of the growing

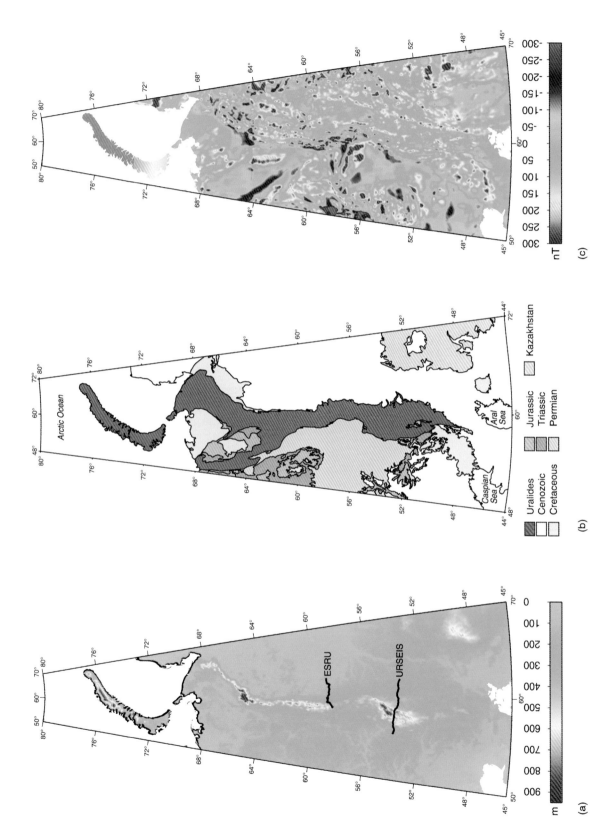

Plate 1. (a) Shaded relief map of the topography of the Ural Mountains. (b) Map showing the extent of the outcropping Uralides. (c) Shaded reliefmap of the magnetic anomolies of the Uralides.

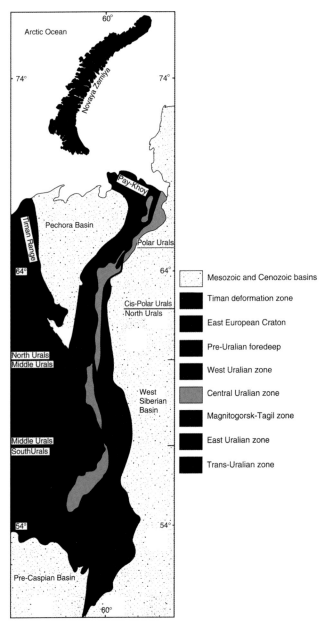

☐	Mesozoic and Cenozoic basins
■	Timan deformation zone
■	East European Craton
■	Pre-Uralian foredeep
■	West Uralian zone
▨	Central Uralian zone
■	Magnitogorsk-Tagil zone
■	East Uralian zone
■	Trans-Uralian zone

Figure 2. Map of the Uralides showing the location of the various zones and geographic areas referred to in the book. In this map the area affected by the Old Cimmerian deformation and Pay-Khoy-Novozemelian foldbelt has been included in the Uralides.

belt and foreland basin of the Uralides developed [*Kamaletdinov*, 1974; *Brown* et al., 1997; *Puchkov*, 1997], while widespread strike-slip faulting accompanied by melt generation and granitoid emplacement took place in the interior part of the orogen [*Bea* et al., 2002]. Following an episode of extensive Early Triassic trapp volcanism and a later deformation hiatus, the mid-Jurassic intraplate Old Cimmerian deformation event led to the uplift of the Timan Range and formation of the Pay-Khoy-Novozemelian foldbelt in the northernmost Uralides, and resulted in localised deformation southward [e.g., *Puchkov*, 1997] (see Figure 2). *Sengor* et al. [1993] suggested that the dominant orogen-forming processes in the Uralides were related to subduction and accretion, forming what they termed a "Turkic-type" orogen.

The volume deals mostly with the South and Middle Urals where much of the recent research was carried out. It is divided into 4 sections. The first section deals predominantly with the Uralide crustal structure and the tectonic processes that went into its formation. In order to understand either of these it is necessary to know how and when the continental margin of Baltica evolved and what its distribution in the Uralides is. In this section, Puchkov presents the Paleozoic margin of Baltica and interprets its stratigraphy in terms of rifting and opening of the ocean and development of a passive margin. The ESRU and URSEIS seismic surveys have shown the orogen to be bivergent [e.g., *Ecthler* et al., 1996; *Knapp* et al., 1996; *Steer* et al., 1998; *Juhlin* et al., 1998; *Friberg*, 2000; *Tryggvason* et al., 2001], and have confirmed many previous Russian interpretations [e.g., *Ryzhiy* et al., 1992] of its crustal architecture (e.g., a crustal root beneath the Magnitogorsk-Tagil zone [*Thouvenot* et al., 1995; *Carbonell* et al., 1996; 1998; *Juhlin* et al., 1998]). Here, *Brown* et al. compare and contrast the URSEIS, ESRU and Alapaev reflection seismic sections, suggesting how the crustal architecture of the South and Middle Urals might be related. Using gravity and magnetic data, Kimbell et al. model these sections, and suggest that the potential field data can discriminate zones in which either subduction or intracontinental plutonism processes dominated, further strengthening the interpretation of geological features from the South to the Middle Urals. One of the key questions surrounding the Uralides is the age of its root and the role it has played in the evolution of the orogen. Is it, for example, a preserved Paleozoic feature, or is it much younger and somehow related to the formation of the current topography of the Ural Mountains? With the new images provided by the geophysical experiments it is now possible to begin searching for the answers to these questions. In this

orogen and, finally, the collision of the Kazakhstan and Siberian plates as the Uralian ocean basin closed by the Late Carboniferous to Early Permian (Figure 1b) [e.g., *Hamilton*, 1970; *Zonenshain* et al., 1984, 1990; *Puchkov*, 1997, 2000]. From the Early Permian to the Early Triassic the western foreland thrust and fold section, Diaconescu and Knapp use a variety of geo-physical data to investigate the role of a phase-change Moho in the preservation of the orogenic root in the South Urals. The crustal architecture of the Uralides formed mainly by subduction and accretion processes, whereas that of the Applachian-Variscide system formed mostly by continent-continent collision and it has been extensively affected by post-orogenic processes. In the final paper in this section, Alvarez-Marron compares the tectonic evolution of the Uralides in the South Urals with that of the Variscides of Germany.

The second section looks at aspects of the exception-ally well preserved island arc and oceanic system, and the arc-continent collision zone that is developed along much of the length of the Uralides. In the South Urals, studies on the geochemistry of the arc volcanic rocks [*Spadea* et al., 1998], on the structure of the accretionary complex and fore arc [*Alvarez-Marron* et al., 2000; *Brown* et al., 1998, 2001], on high pressure rocks along the suture zone [e.g., *Beane and Connolly*, 2000; *Hetzel*, 1999; *Brown* et al., 2000], and on ophiolitic, mafic and ultra-mafic material [*Savelieva* et al., 1997; *Scarrow* et al., 1999] have shown that the crustal structure of the arc-continent collision, and the Paleozoic tectonic processes that went into its formation, can be favorably compared with those in currently active settings such as the west Pacific [*Puchkov*, 1997; *Brown* et al., 1998; *Brown and Spadea*, 1999]. In this section, *Spadea* et al. present geo-chemical and isotope data from the Magnitogorsk fore arc region, comparing them to those from the Izu-Bonin arc. The extensive ophiolite and mafic-ultramafic belt in the Uralides record processes that affected the oceanic crust, mantle, and suprasubduction zone during arc-continent collision [*Savelieva*, 1987; *Savelieva and Nesbitt*, 1996; *Edwards and Wasserburg*, 1985; *Spadea* et al., 1998; *Scarrow* et al., 1999]. Here, *Savelieva* et al. discuss various aspects of the age, generation, deformation and emplacement of the ophiolite and mafic-ultramafic mas-sifs throughout the Uralides. The results of the research into arc-continent collision in the Uralides are aiding studies into the processes of formation of its large volcanic hosted massive sulfide field. Located mostly in the island arc volcanic rocks and allochthons in the accretionary complex, these deposits are mainly unde-formed and unmetamorphosed; a number of them are exceptionally large in size, and others contain fossilized vent fauna assemblages [*Zaykov* et al., 1996; *Little* et al., 1997]. In this section, *Herrington* et al., present new geochemical and isotope data from massive sulfide deposits in the Magnitogorsk arc that help further develop a model for their formation during arc-continent collision. Another important feature of the arc-continent

collision zone in the Uralides is the presence of high pressure rocks along the suture zone and its immediate footwall. For example, the Maksutovo Complex in the South Urals has been extensively studied, and the age (380–375 Ma) [*Matte* et al., 1993; *Shatsky* et al., 1996; *Beane and Connelly*, 2000] and pressure and temperature conditions (20 ± 4 kbar and 550 ± 50°C) [*Beane* et al., 1995; *Hetzel* et al., 1998; *Schulte and Blümel*, 1999] of high pressure metamorphism are well known. Fission-track thermochronological studies [*Leech* et al., 2000] have helped to understand the low temperature part of its exhumation history. In the final paper in this section, Willner et al. present data on detrital grains from the syncollisional forearc and accretionary complex sedi-ments that indicate that high pressure rocks similar to those in the Maksutovo Complex were at the surface and supplying material to basins in these areas early in the arc-continent collision history.

The third section deals mostly with the internal part of the orogen where Devonian to Permian granitoids record subduction-related and late orogenic intraconti-nental processes [*Fershtater* et al., 1997; *Bea* et al., 1997; *Gerdes* et al., 2001; *Montero* et al., 2000]. Geochronology has shown that these granitoids developed first in the South Urals and later in the Middle Urals [*Bea* et al., 1997; *Montero* et al., 2000]. In this section, *Bea* et al. present the age, the mineralogy, and the geochemical and isotope composition of twelve granitoids, which they use to discuss the implications for timing, nature of the source rock, imput of energy for melting, and the polarity of successive magmatic arcs in the Uralides. Geochronology has proven to be a powerful tool in helping understand the processes that occured in the metamorphic and plutonic rocks of Uralides. Here, *Scarrow* et al. present a compilation of geochronological data collected in the Uralides over the past forty years.

The fourth section deals with one of the teasing, but still unanswered questions about the Uralides: what is the age of the topography of the Ural Mountains (Plate 1a)? There is widespread evidence that the Uralides were largely peneplaned during the Jurassic [*Borisevich*, 1992; *Puchkov*, 1997], so why do they have a topographic expression today? *Borisevich* [1992], and *Piwowar* [1997] have suggested that the Ural Mountains are the result of Late Cenozoic uplift of the orogen, although preliminary fission-track data [*Seward* et al., 1997] were not able to detect any post-Jurassic move-ment. In this section, *Seward* et al. present a much larger fission track data set that largely confirms the absence of any post-Jurassic movement that can be measured by this method, although disturbed age-altitude relation-ships in the South Urals foreland thrust and fold belt (where the relief is greater) may indicate recent uplift. *Piwowar* [1997] suggested that the Ural Mountains form part of a Neogene-aged topographic bulge that extends northwestward from Tien Shan. Could the topography of the Ural Mountains then be the result of Cenozoic far-field intraplate stress caused by the collision of India and Eurasia? Here, Mikhailov et al. present strength and strain modeling that suggests that the uplift is focussed along several crustal scale discontinuities, and may be caused by NW-SE oriented intraplate stress.

Dennis Brown
Christopher Juhlin
Victor Puchkov

REFERENCES

Alvarez-Marron, J., D. Brown, A. Perez-Estaun, V. Puchkov and Y. Gorozhanina, Accretionary complex structure and kinematics during Paleozoic arc-Continent collision in the southern Urals, *Tectonophysics*, 325, 175–191, 2000.

Bea F., G. Fershtater and P. Montero, Variscan granitoids of the Urals: Implications for the evolution of the orogen, this volume, 2002.

Bea, F., G. Fershtater, P. Montero, V. Smirnov and E. Zin'kova, Generation and evolution of subduction-related batholiths from the central Urals: Constraints on the P-T history of the Uralian orogen, *Tectonophysics*, 276, 103–116, 1997.

Beane, R. J. and J. N. Connelly, ^{40}Ar/^{39}Ar, U-Pb and Sm-Nd constraints on the timing of metamorphic events in the Maksyutov Complex, southern Urals, Ural Mountains, *J. Geol. Soc. of London*, 157, 811–822, 2000.

Beane, R. J., J. G. Liou, R. G. Coleman and M. L. Leech, Petrology and retrograde P-T path for eclogites of the Maksyutov Complex, southern Ural Mountains, Russia, *Isl. Arc*, 4, 254–266, 1995.

Borisevich, D. V., Neotectonics of the Urals, *Geotectonics*, 26, 41–47, 1992.

Brown, D., J. Alvarez-Marron, A. Perez-Estaun, V. Puchkov, Y. Gorozhanina and P. Ayarza, Structure and evolution of the Magnitogorsk forearc basin: Identifying into upper crustal processes during arc-continent collision in the southern Urals, *Tectonics*, 20, 364–375, 2001.

Brown, D., R. Hetzel and J. H. Scarrow, Tracking arc-continent collision subduction zone processes from high pressure rocks in the southern Urals, *J. Geol. Soc. of London*, 157, 901–904, 2000.

Brown, D. and P. Spadea, Processes of forearc and accretionary complex formation during arc-continent collision in the southern Ural Mountains, *Geology*, 27, 649–652, 1999.

Brown, D., C. Juhlin, J. Alvarez-Marron, A. Perez-Estaun and A. Oslianski, Crustal-scale structure and evolution of an arc-continent collision zone in the southern Urals, Russia, *Tectonics*, 17, 158–171, 1998.

Brown, D., J. Alvarez-Marron, A. Perez-Estaun, Y. Gorozhanina, V. Baryshev and V. Puchkov, Geometric and kinematic evolution of the foreland thrust and fold belt in the Southern Urals, *Tectonics*, 16, 551–562, 1997.

Brown, D., V. Puchkov, J. Alvarez-Marron and A. Perez-Estaun, The structural architecture of the footwall to the Main Uralian Fault, southern Urals, *Earth–Sci. Rev.*, 40, 125–147, 1996.

Carbonell, R., A. Perez-Estaún, J. Gallart, J. Diaz, S. Kashubin, J. Mechie, R. Stadtlander, A. Schulze, J. H. Knapp and A. Morozov, A crustal root beneath the Urals: Wide-angle seismic evidence, *Science*, 274, 222–224, 1996.

Carbonell, R., D. Lecerf, M. Itzin, J. Gallart and D. Brown, Mapping the Moho beneath the southern Urals, *Geophys. Res. Lett.*, 25, 4229–4233, 1998.

Echtler, H. P., M. Stiller, F. Steinhoff, C. M. Krawczyk, A. Suleimanov, V. Spiridonov, J. H. Knapp, Y. Menshikov, J. Alvarez-Marron and N. Yunusov, Preserved collisional crustal architecture of the southern Urals – Vibroseis CMP-profiling, *Science*, 274, 224–226, 1996.

Edwards, L. R. and G. J. Wasserburg, The age and the emplacement of obducted oceanic crust in the Urals from Sm-Nd and Rb-Sr systematics, *Earth Planet. Sci. Lett.*, 72, 389–404, 1985.

Fershtater, G. B., P. Montero, N. S. Borodina, E. V. Pushkarev, V. Smirnov, E. Zin'kova and F. Bea, Uralian magmatism: An overview, *Tectonophysics*, 276, 87–102, 1997.

Friberg, M., Tectonics of the Middle Urals, Ph.D. Thesis, Uppsala University, 2000.

Gerdes, A., P. Montero, F. Bea, G. Fershtater, N. Borodina, T. Osipova and G. Shardakova, Peraluminous granites frequently with mantle-like isotope compositions: The continental-type Murzinka and Dzhabyk batholiths of the eastern Urals, *Int. J. Earth Sci*, 90, 2001.

Hamilton, W., The Uralides and the motion of the Russian and Siberian platforms, *Geol. Soc. Am. Bull.*, 81, 2553–2576, 1970.

Hetzel, R., Geology and geodynamic evolution of the high-P/low-T Maksyutov Complex, southern Urals, Russia, *Geol. Rundsch.*, 87, 577–588, 1999.

Hetzel, R., H. P. Echtler, W. Seifert, B. A. Schulte and K. S. Ivanov, Subduction- and exhumation-related fabrics in the Paleozoic high-pressure/low-temperature Maksyutov Complex, Antingan area, Southern Urals, Russia, *Geol. Soc. Am. Bull.*, 110, 916–930, 1998.

Ivanov, S. N., A. S. Perfiliev, A. A. Efimov, G. A. Smirnov, V. M. Necheukhin and G. B. Fershtater, Fundamental features in the structure and evolution of the Urals, *Am. J. Sci.*, 275, 107–130, 1975.

Juhlin, C., M. Friberg, H. Echtler, A. G. Green, J. Ansorge, T. Hismatulin and A. Rybalka, Crustal structure of the Middle Urals: Results from the (ESRU) Europrobe Seismic Reflection Profiling in the Urals Experiments, *Tectonics*, 17, 710–725, 1998.

Kamaletdinov, M. A., *The Nappe Structures of the Urals* (in Russian), 228 pp., Nauka, Moscow, 1974.

Khain, V. E., Geology of the USSR. 272 pp., Schweizerbart, Germany, 1985.

Knapp, J. H., D. N. Steer, L. D. Brown, R. Berzin, A. Suleimanov, M. Stiller, E. Lüschen, D. Brown, R. Bulgakov and A.V. Rybalka, A lithosphere-scale image of the Southern Urals from explosion-source seismic reflection profiling in URSEIS '95, *Science*, 274, 226–228, 1996.

Leech, M. L. and D. F. Stockli, The late exhumation history of the ultrahigh-pressure Maksyutov Complex, south Ural Mountains, from new apatite fission-track data, *Tectonics*, 19, 153–167, 2000.

Little, C. T. S., R. J. Herrington, V. V. Maslennikov, N. J. Morris and V. V. Zaykov, Silurian hydrothermal-vernt community from the southern Urals, Russia, *Nature*, 385, 146–148, 1997.

Matte, P., H. Maluski, R. Caby, A. Nicolas, P. Kepezhinskas and S. Sobolev, Geodynamic model and $^{39}Ar/^{40}Ar$ dating for the generation and emplacement of the high pressure (HP) metamorphic rocks in SW Urals, *C.R. Acad. Sci. Ser. II*, 317, 1667–1674, 1993.

Montero, P., F. Bea, A. Gerdes, G. Fershtater, E. Zin'kova, N. Borodina, T. Osipova and V. Smirnov, Single-zircon evaporation ages and Rb-Sr dating of four major Variscan batholiths of the Urals. A perspective on the timing of deformation and granite generation, *Tectonophysics*, 317, 93–108, 2000.

Piwowar, T. J., Long-wavelength Neogene flexural uplift of the Southern Urals and central Eurasia, unpublished M.Sc., Cornell University, U.S.A., 74 pp., 1997.

Puchkov, V. N., Paleogeodynamics of the central and southern Urals (in Russian), Ufa Dauria, 145 pp., 2000.

Puchkov, V. N., Structure and geodynamics of the Uralian orogen, in *Orogeny Through Time*, edited by J.-P. Burg and M. Ford, Geol. Soc. of London, *Spec. Publ.*, 121, 201–236, London, 1997.

Ryzhiy, B. P., V. S. Druzhinin, F. F. Yunusov and I. V. Ananyin, Deep structure of the Urals region and its seismicity, *Phys. Earth Planet. Inter.*, 75, 185–191, 1992.

Savelieva, G. N., A. Y. Sharaskin, A. A. Saveliev, P. Spadea and L. Gaggero, Ophiolites of the southern Uralides adjacent to the East European continental margin, *Tectonophysics*, 276, 117–138, 1997.

Savelieva, G. N. and R.W. Nesbitt, A synthesis of the stratigraphic and tectonic setting of the Uralian ophiolites, *J. Geol. Soc.*, London, 153, 525–537, 1996.

Savelieva G. N., *Gabbro-ultrabasite Assemblages of Uralian Ophiolites and Their Analogues in the Recent Oceanic Crust* (in Russian), Nauka, Moscow, 242 pp., 1987.

Scarrow, J. H., G. N. Savelieva, J. Glodny, P. Montero, A. N. Pertsev, L. Cortesogno and L. Gaggero, The Mindyak Palaeozoic lherzolite ophiolite, Southern Urals: Geochemistry and geochronology, *Ofioliti*, 24, 239–246, 1999.

Scotese, C. R., *Atlas of Earth History, PALEOMAP Project*, Arlington, Texas, 52 pp., 2001.

Schulte, B. A. and P. Blümel, Prograde metamorphic reactions in the high-pressure Maksyutov Complex, Urals, *Geol. Rundsch.*, 87, 561–576, 1999.

Sengor, A. M. C., B. A. Natal'lin and V. S. Burtman, Evolution of the Altaid tectonic collage and Palaeozoic crustal growth in Eurasia, *Nature*, 364, 299–307, 1993.

Seward, D., A. Perez-Estaun and V. Puchkov, Preliminary fission-track results from the Southern Urals: Sterlitamak to Magnitogorsk, *Tectonophysics*, 276, 281–290, 1997.

Shatsky, V. S., Jagoutz, E. and O. A. Kozemenko, New evidence for Upper Devonian age of high pressure metamorphism of Maksyutov Complex, *Eos Transactions*, 77, 767, 1996.

Spadea, P., L. Y. Kabanova and J.H. Scarrow, Petrology, geochemistry and geodynamic significance of Mid-Devonian boninitic rocks from the Baimak-Buribai area [Magnitogorsk Zone, southern Urals), *Ofioliti*, 23, 17–36, 1998.

Steer, D. N., J. H. Knapp, L. D. Brown, H. P. Echtler, D. L. Brown and R. Berzin, Deep structure of the continental lithosphere in an unextended orogen: An explosive-source seismic reflection profile in the Urals (Urals Seismic Experiment and Integrated Studies (URSEIS 1995)), *Tectonics*, 17, 143–157, 1998.

Thouvenot, F., S. N. Kashubin, G. Poupinet, V. V. Makovskiy, T. V. Kashubina, Ph., Matte and L. Jenatton, The root of the Urals: Evidence from wide-angle reflection seimics, *Tectonophysics*, 250, 1–13, 1995.

Tryggvason, A., D. Brown and A. Perez-Estaun, Crustal architecture of the southern Uralides from true amplitude processing of the URSEIS vibroseis profile, *Tectonics*, 20, 1040–1052, 2001.

Zaykov, V. V., V. V. Maslennikov, E. V. Zaykova and R. J. Herrington, Hydrothermal activity and segmentation in the Magnitogotsk-west Mugodjarian zone on the margins of the Urals palaeo-ocean, in *Tectonic, Magmatic, Hydrothermal and Biological Segmentation of Mid-Ocean Ridges*, edited by C. J. MacLeod, P. A. Tyler, and C. L. Walker, *Geol. Soc of London, Spec. Publ.*, 118, 199–210, 1996.

Zonenshain, L. P., M. I. Kuzmin and L. M. Natapov, Uralian Foldbelt, in *Geology of the USSR: A Plate-Tectonic Synthesis*, edited by B. M. Page, Geodyn. Ser. vol. 21, pp. 27–54, AGU, Washington, D.C., 1990.

Zonenshain, L. P., V. G. Korinevsky, V. G. Kazmin, D. M. Pechersky, V. V. Khain and V. V. Mateveenkov, Plate tectonic model of the south Urals development, *Tectonophysics*, 109, 95–135, 1984.

Paleozoic Evolution of the East European Continental Margin Involved in the Uralide Orogeny

V. Puchkov

Institute of Geology, Ufa, Russia

The Paleozoic formations exposed to the West of the Main Uralian fault belonged to a passive margin of Balica. The margin formed in the Early Paleozoic as a result of break-up of a bigger continent. The rugged outline of the margin, with promontories and recesses, predetermined many features of the future Uralide foldbelt. Three stages of development of the margin are established: a rift stage; a passive stage; and a collisional stage. Late Cambrian to Early Ordovician rifting was followed by seafloor spreading and subsidence of the newly formed margin. The margin developed a typical topographic profile with a shelf, continental slope, and continental rise, whose facial zones can be recognised in the rocks of the Uralides. The passive margin stage was accompanied by weak tectonic activity, with rare, localised episodes of rifting. The accumulation of sediments was governed by subsidence and a strong dependence on sea-level changes. The change to the collisional stage can be recognised in the accumulation of greywacke flysch, which filled a deep-water basin and was deposited over the continental shelf. These collision-related processes were strongly diachronous, starting in the Late Devonian in the Southern Urals, and propagating northward, were they ended in Pay-Khoy as late as the Jurassic.

1. INTRODUCTION

Before the advent of the theory of plate tectonics, the Paleozoic passive margin of Baltica involved in the Uralides was traditionally treated as a miogeosyncline [e.g., *Kheraskov*, 1967]. Many other terms from the pre-plate tectonics era could also be applied to what are now considered the deep water bathyal formations of the

Mountain Building in the Uralides: Pangea to the Present
Geophysical Monograph 132
Copyright 2002 by the American Geophysical Union
10.1029/132GM02

margin, such as leptogeosyncline, terrigenous geosyncline, and cryptogeosyncline [*Trümpy*, 1960; *Zonenshain*, 1972; *Knipper*, 1963]. Nevertheless, a large volume of knowledge accumulated in the frame of the geosynclinal theory allowed a series of paleofacial schemes to be created, including the territory of the continental margin [*Nalivkin*, 1960]. Even then, within this "fixist" scheme, one could notice and point out that there was no clear boundary between "platform" and "miogeosynclinal" shelf facies [e.g., *Chermnykh and Puchkov*, 1972; *Puchkov*, 1975]. Unfortunately this scheme followed a misleading idea that the Central Uralian zone (a structurally uplifted zone east of the current position of bathyal sediments, Figure 1) was the provenance area for Ordovician, Devonian and Lower Carboniferous quartz sandstones

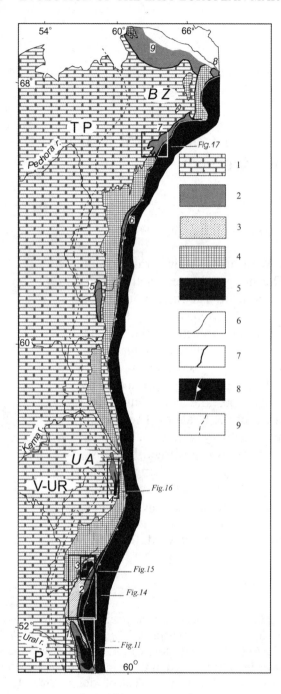

Figure 1. Major structural elements and complexes of Baltica Paleozoic passive margin involved into the Urals. Position of detailed maps is given by frames. Symbols in boxes: 1. Paleozoic rift and shelf complexes. 2. Paleozoic rift, bathyal and abyssal complexes. 3. Upper Devonian Zilair flysch of the Southern Urals. Its analogues in the northern areas are shown undivided from bathyal complexes. 4. Proterozoic crystalline basement of Baltica, exhumed in the Central Uralian zone. 5. Paleozoic oceanic and island-arc complexes. 6–9. Geological boundaries: 6. Normal stratigraphic or weakly displaced. 7. Thrusts. 8. Main Uralian fault. 9. Western boundary of the foreland thrust and fold belt. Letters and numbers in the scheme: P — Pricaspian basin, V-Ur — Volgo-Uralian basin, TP — Timano-Pechora basin, *UA* — Ufimian amphitheatre, *BZ* — Bolshezemelian promontory; 1 — Sakmara synform, 2 — Suvanyak complex in the Uraltau antiform, 3 — Kraka allochthon, 4 — Nizhnie Sergi allochthon, 5 — Malaya Pechora allochthon, 6 — Polya tectonic zone, 7 — Lemva allochthon, 8 — Baydarata tectonic zone, 9 — Kara allochthon.

the Sakmara-Lemva zone (see section 2) were shallow water sediments [*Evseev*, 1960; *Kondiain*, 1963]. This idea stemmed from the long lived Beloussov's theory [e.g., *Beloussov*, 1975] that states that the thickness of sediments is a measure of subsidence and therefore these thin, condensed facies ought to have been deposited in shallow water. Only by investigating the nature of deep water sediments along modern continental margins and their comparison with ancient analogues did it become evident that the sediments of the "miogeosyncline" were partly deposited in deep water conditions [*Puchkov*, 1973; *Khvorova* et al., 1978]. It is worth noting, however, that the much criticised but true interpretation of the Lemva facies in the Polar Urals as allochthonous deep water sediments came as early as in 1940's [*Voynovsky-Kriger*, 1945]. With the re-introduction of nappe tectonics in the Uralides by *Kamaletdinov* [1974] it was finally understood that these "miogeosyncline" sediments were not in their original setting, but were tectonically juxtaposed against the platform sediments. And, finally, the discovery and systematic study of conodonts provided a powerful tool for resolving the stratigraphy of the bathyal sediments that were involved in these nappe complexes, allowing them to be correlated reliably along the Uralides (Figure 2), [*Puchkov*, 1973, 1979].

With these new ideas firmly entrenched it became possible to understand how and when the passive margin of Baltica evolved. After the Late Vendian orogeny, an intense rifting grading to drifting took place in the Late Cambrian to Early Ordovician. The position of the nascent margin only partially coincided with the trend of the underlying crystalline basement, thus implying that

found along the western slope of the Urals. The idea was popular until criticised by *Smirnov* [1957], and finally discarded some ten years later.

Ideas began to change more quickly when a reappraisal of the Cambro-Ordovician volcano-terrigenous facies that underlie the platform shelf and slope led to their interpretation as rift complexes [*Goldin and Puchkov*, 1974]. Despite this, it remained difficult to overcome the idea that the cherty-terrigenous condensed deposits of

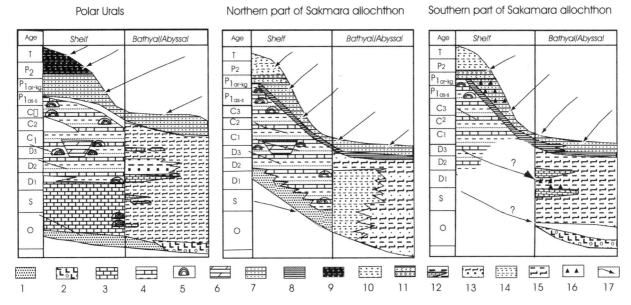

Figure 2. The correlation scheme of the Paleozoic geological formations belonging to the passive margin of Baltica continent. A — in the North of the Urals, B, C — in the Southern Urals (B — Uraltau antiform, C — Sakmara allochthon). Symbols in the boxes: Typical lithological complexes (formations in the Russian sense) of the Uralian passive margin of the Paleozoic continent. 1. basal terrigenous oligomyctic sandstone, siltstone and shale, and graben molassoid; 2. graben terrigenous-volcanogenic; 3. layered shallow water limestone; 4. layered shallow water limestone and terrigenous oligomyctic, undifferentiated; 5. biohermal and reefal; 6. domanik; 7. greywacke flysch; 8. preflysch condensed; 9. coal-bearing molasse; 10. variegated molasse; 11. evaporitic; 12. cherty-carbonate (with knotty limestones); 13, 14. flysch of the passive margin: 13. cherty-terrigenous oligomyctic; 14. terrigenous oligomyctic; 15. cherty-shale; 16. olistostrome. 17. Directions of terrigenous provenance.

a great deal of a continent was split off and drifted away (Figure 3). One of the results of this process was the formation of a paleo-Uralian ocean and a Baltica passive margin facing it.

By the Middle Devonian the margin of Baltica in the Southern Uralides began to change from a passive setting to an active one. From the Devonian to the Early Carboniferous its leading edge was subducted beneath the Magnitogorsk volcanic arc and syntectonic clastic sediments were deposited across the margin and incorporated into an accretionary complex [*Puchkov*, 1996, 1997; *Brown* et al., 1998; *Brown and Spadea*, 1999; *Alvarez-Marron* et al., 2000]. Meanwhile, westward of this area of activity, the deposition of shallow water platform sediments continued uninterrupted. In the Early Carboniferous tectonic activity along the former margin had stopped, and platform sedimentation along with rift volcanism took place within the accreted arc. In the Middle Carboniferous the Uralian ocean basin closed and the Kazakhstanian continent collided with Baltica [*Puchkov*, 2000].

Taking into account that the early collisional history in the Uralides has been the subject of many publications,

this paper will be restricted mainly to the development of the Uralian passive margin, before it was involved in collision.

Three major structural elements of the Uralides are situated to the west of the Main Uralian fault; the Preuralian foredeep, the West Uralian and Central Uralian zones. The rocks within these zones initially belonged to the margin of Baltica, including the Precambrian crystalline basement upon which the Paleozoic platform was built. One must keep in mind that the name of Baltica has two slightly different meanings. The name is usually applied to a continent which is thought to have existed since the break up of Rodinia in the Neoproterozoic and through the most of Paleozoic [e.g., *Torsvik* et al., 1996]. However, the margin under discussion formed only in the Early Ordovician, as a result of the break up of a bigger continent [*Puchkov*, 1979a, 2001]. The margin existed during most of the Paleozoic, evolving until the Middle to Late Devonian when it collided in the Southern to Middle Urals with the Magnitogorsk and Tagil island arcs and was incorporated into the Uralide orogen. The Uralide orogeny appears to have migrated from South to North

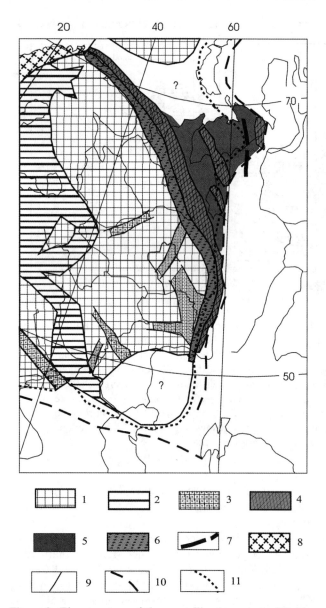

Figure 3. The structure of the crystalline basement of Baltica continent. 1–3. Craton: 1. Archean foldbelts. 2. Mostly Paleoproterozoic foldbelts. 3. Meso- and Neoproterozoic aulacogens. 4–7. Timanides/Preuralides (Late Vendian foldbelt): 4. Anticlinoria. 5. Synclinoria. 6. Foredeep filled with molasse. 7. Ophiolitic suture zone (Late Riphean). 8. Caledonides. 9–11. Geological boundaries: 9. Faults. 10. Main Uralian fault. 11. The western boundary of Uralides.

(in modern coordinates), affecting the margin at different times [*Puchkov*, 1996, 2000].

The crystalline basement upon which the platform was built is composed predominately of strongly folded and metamorphosed rocks ranging in age from Archean to Vendian. This basement, which was exhumed in the Central Uralian zone during the Uralide orogeny, shows evidence of Pre-Uralide deformation events (Figure 3). Two major areas can be differentiated; an Archean to Paleoproterozoic craton consolidated before 1.6 Ga (the Karelian orogeny) and a foldbelt at the periphery of the craton, formed as a result of a Late Vendian (600–550 Ma) orogeny (Pre-Uralides, or Timanides) that has recently been correlated with the Cadomian orogeny of Central and Western Europe [*Puchkov*, 1997; *Giese* et al., 1999].

The sedimentary cover of the platform is many kilometers thick and tends to grow eastwards, close to the margin (Figure 4). But the structural style of the cover differs strongly depending on the age of the basement (compare Figures 3, 4 and 5). The older part of the Archean to Neoproterozoic craton experienced several episodes of rifting resulting in the formation of deep grabens, or aulacogens. These structures ceased to exist in the Late Vendian time, by the onset of the Cadomian orogeny. The basement rocks, as well as Proterozoic sediments covering them, are beyond the scope of this paper, and will not be discussed farther. The paper will mainly describe the Paleozoic evolution of the margin of Baltica involved in the Uralide orogen, giving a brief description of the characteristics of the sediments and their depositional environment and tectonic setting from the time of rifting to the start of collision.

2. THE PALEOZOIC MARGIN

2.1. Early Paleozoic Rift Stage

A well studied representative of the rift stage facies is found in the Sakmara allochthon of the Southern Urals (Figures 2, 6, and 11) [*Ruzhentsev*, 1976; *Khvorova* et al., 1978; *Ivanov and Puchkov*, 1984a]. Here, the Upper Cambrian(?) to Ordovician (Tremadoc) Kidryasovo and (Tremadoc to Early Arenig) Kuagash formations are composed of coarse-grained, poorly sorted conglomerate, sandstone, and siltstone that were derived from a continental basement. These formations also contain layers of trachybasalt, rhyolite tuffs, some limestone, and jasper. The facies and thickness (up to 1000 m) are very variable; they are moderately folded and practically unmetamorphosed [*Khvorova* et al., 1978]. Also in the Sakmara allochthon (mostly in the northern part), are the Lower Cambrian Terekla and Mednogorsk formations [*Russian Stratigraphic Committee*, 1993]. These formations are composed of sandstone and conglomerate with tuff and subalkaline basalt layers and large blocks of limestone with Lower Cambrian algae and archaeocyathids, giving the formation an appearance of an olistostrome. The occurrence of Ordovician (Tremadoc)

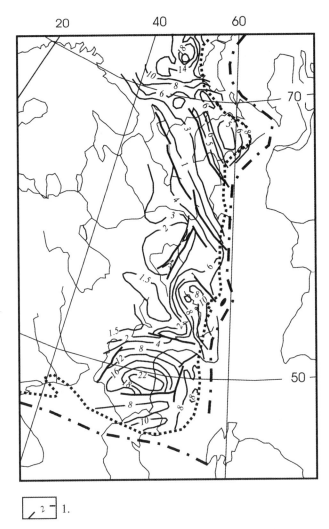

Figure 4. The depth to the crystalline basement at the eastern margin of Baltica continent. 1. Isopleths of depth to the crystalline basement, km. The other symbols are in the Figure 3.

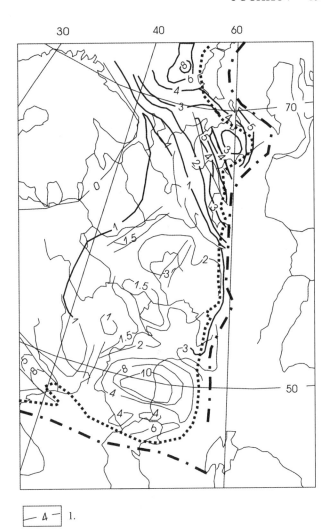

Figure 5. Thickness of the Paleozoic (Ordovician to Lower Permian) pre-molasse sediments at the eastern margin of Baltica. 1. Isopleths of thicknesses of the Paleozoic (Ordovician to Lower Permian) pre-molasse sediments. The other symbols are in the Figure 3.

acritarchs in the Terekla formation [*Chibrikova and Olli*, 1999] suggests that it might be a lateral equivalent of the Kidryasovo and Kuagash formations. West of the Sakmara allochthon, the Lower to Middle Ordovician terrigenous sediments are present in a latitudinal graben-like depression situated close to Orenburg (Figure 6) [*Lukinykh and Belyayev*, 1998].

Farther North, in the Middle Urals, the rift facies is Lower Ordovician in age, and developed as a narrow band immediately west of the Main Uralian fault (Figures 6 and 16). They are represented by the Kozino and Kolpakovo formations, which are composed of 1000 to 3000 meters of conglomerate, quartzitic sandstone, some marble, layers of tuff, basalt and trachybasalt, and some rhyolite [*Russian Stratigraphic Committee*, 1993].

In the lower thrust sheet of the Bardym allochthon the Ordovician is represented by the Nizhnie Sergi formation of tuffs, basalt, limy sandstone, limestone and shale of Caradoc age [*Puchkov and Ivanov*, 1982]. These formations are deformed and slightly metamorphosed.

Late Cambrian to Lower Ordovician rift processes in the Cis-Polar and Polar Urals are reflected by thick quartzitic polymictic conglomerates of the Telpos, Mani-tanyrd and Pogurei formations and various terrigenous-volcanic, volcanic, subvolcanic and intrusive complexes underlying and partially substituting them [*Puchkov*, 1975; *Goldin* et al., 1999; *Dembovsky* et al., 1990]. The volcanic units are aligned subparallel to the main structural grain of the Uralides (Figures 2, 6, and 17).

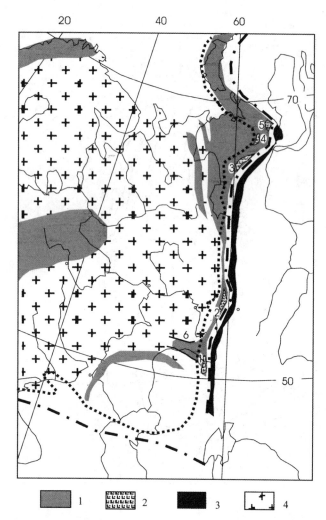

Figure 6. The Ordovician complexes at the margin of Baltica. The symbols in the boxes: 1. The areas of modern development of the Ordovician shelf sediments. 2. The areas of development of the Ordovician rift complexes. 3. The Ordovician ophiolites. 4. The areas where the Ordovician is absent. The numbers in the Figure 1–5: The areas of development of rift complexes. 1 – Sakmara allochthon. 2 — Nizhnie Sergi area. 3 — Lemva zone. 4 — Manitanyrd area. 5 — Baydarata area. 6 — Orenburg graben-like depression. The other symbols are in the Figure 3.

In the northernmost Polar Urals (Baydarata area, Figure 6) rift volcanism is represented by the Ordovician (Upper Tremadoc to Lower Llanvirn) basalt and rhyolite of the Kharapeshor formation [*Dembovsky* et al., 1988].

The origin of all the above mentioned formations was connected with continental to shallow water conditions, all of them ensialic, and all situated to the west of the Main Uralian fault.

East of the Main Uralian fault, or in the uppermost thrust sheets of allochthons to the west of it,

the Ordovician formations are quite different. They are represented by basalt with layers of chert or jasper, indicating at more open sea, deep water environment. By chemical composition they vary from MORB and related types of ophiolite basalt to subalkaline olivine basalt of probably Afar or Red Sea type [*Seravkin* et al., 1992; *Savelieva* et al., 1998; *Tevelev* et al., 2000]. According to paleontological and geochronological data, they are never older than the Early Ordovician (Late Arenig), thus setting a probable time limit for the onset of seafloor spreading and opening of paleo-Uralian ocean [*Ivanov* et al., 1986; *Ronkin* et al., 1997].

2.2. The Passive Margin Stage: The Continental Shelf

The passive margin stage of the development of the area was characterized by steady submergence [*Puchkov*, 1975, 1979a], the general absence of volcanism, and only weak, localized tectonics (except of a special episode of Devonian dissipated rifting accompanied by a basaltic volcanism which developed not only in the margin, but spread over the whole Baltica) [*Ivanov and Puchkov*, 1984b; *Puchkov*, 1988; *Fokin* et al., 2001] (Figures 4, 5, 6, 7, and 8). This stage was accompanied by the formation of a typical topography for a passive margin, consisting of shelf, continental slope and continental rise. The URSEIS seismic sections show that the margin was subducted to at least 50 km underneath the Magnitogorsk arc [e.g., *Tryggvason* et al., 2001]. Nevertheless, fragments of the sedimentary complexes of the margin are present in thrust sheets at the surface, permitting us to restore some general features of this "topography".

Along the western slope of the Ural Mountains, the Ordovician to Carboniferous continental shelf of Baltica margin involved in the Uralides is developed in a territory called Belsk-Yelets zone (after the rivers Belaya in the south and Yelets in the north). Information on the stratigraphy and lithology of these sediments is summed up in the last edition of the stratigraphic charts of the Uralides [*Russian Stratigraphic Committee*, 1993]. The shelf facies is characterized by the predominance of shallow water layered limestone with thick-walled fauna, subordinate bioherms and reefs, dolomite, shale, siltstone and quartz sandstone (Figure 2).

It is worth noting that all the paleomagnetic data [e.g., *Didenko* et al., 2001] indicate that from the Middle Ordovician until the Carboniferous the margin of Baltica discussed here was situated in a tropical to equatorial zone, so the conditions were favourable for the growth of carbonate reefs. Ordovician to Lower Carboniferous (Visean) reefs occur in the Polar Urals, and Silurian to Early Devonian reefs occur along the whole length of the Western slope of the Ural Mountains [*Antoshkina*, 1994;

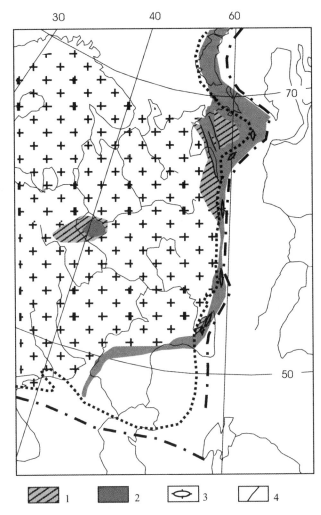

30   40   60

70

50

| 1 | 2 | 3 | 4 |

Figure 7. The areas of modern development of the Silurian and Lower Devonian (pre-Emsian) shelf sediments. 1. Areas of modern development of Silurian shelf sediments. 2. Areas of modern development of Silurian and Devonian shelf sediments. 3. Barrier reefs. 4. Devonian (post-Pragian) normal faults of the overprinted Pechora-Kolva graben system. The other symbols are in the Figure 3.

Nikonov et al., 2000] (Figure 7). Another indication for the low latitudinal position of the margin are evaporites, which appear periodically in the Timan-Pechora province (e.g., Middle to Late Ordovician salt and anhydrites, and the Early and Late Devonian (Frasnian) and Early Carboniferous (Serpukhovian) anhydrites) [*Nikonov* et al., 2000]. Another climatic indicator, bauxite, is also typical for the Devonian along the whole length of the margin. Early Carboniferous (Lower to Middle Visean) coals also developed in the south of the Timan-Pechora basin and in most of the Volga-Urals basin [*Nalivkin*, 1960; *Maksimov*, 1970; *Nikonov* et al., 2000].

Subsidence of the platform margin in the Ordovician to Early Carboniferous is recorded in two large scale sedimentary cycles that can be traced along the whole length of the Uralides. The beginning of transgression events is marked in most places by quartz-rich, terrigenous, partly alluvial sediments and lagoonal dolomite. Widespread open sea limestone marks the peak of the transgressions, and red coloured sandstone and dolomite with some sulphate appear close to the end of the cycles.

Subsidence of the platform appears to have varied depending on the age of the basement. For example, from the Ordovician (Figure 6) to Early Devonian (Figure 7) the central part of the continent (Volga-Urals province), having the oldest (Archean to Early Proterozoic) basement, was strongly uplifted compared to the Timan-Pechora province and Pricaspian depression, which have (or are thought to have) a Cadomian age basement. The shelf deposits which cover the basement in the Southern and Middle Urals are generally coarse-grained and of shallow water, near shore type, whereas in the Timan-Pechora basin in the north or the Pricaspian depression in the south, the deposits are thinner, and are often of open sea type. Moreover, the pre-existing structures in the basement appear to have influenced the architecture of the basins developed on the passive margin.

The first major transgression that occurred after rifting started in the Early Ordovician (Llandeilo) and peaked in the Late Silurian. Lower Devonian redbed and dolomitic near shore sedimentation indicates regression during the Early Devonian (Lohkovian and Pragian). This was a time of a maximum development of barrier reefs, which mark the edge of the continental shelf (Figure 7).

A new, bigger transgression cycle started in the late Early Devonian (Emsian), with deposition of the Takata sandstone. The Takata sandstone was deposited over a wide area, much wider than the underlying regressive sediments, so in some places the stratigraphic break under them is very large. But neither here, nor in other parts of the Lower to Middle Paleozoic sections of the western slope of the Ural Mountains have any angular unconformities that would be suggestive of Caledonian deformation been found. In the easternmost sections, the Takata sandstone is substituted by limestone, and finally disappears. In these sections no stratigraphic unconformity is recorded at the base of the Takata level.

Terrigenous horizons of quartz sandstone appear several times higher up in the Devonian sections, marking smaller scale transgressions inside the larger cycle.

A number of these small scale cycles have been recognized, and some sandstone layers are petroleum targets in the Volga-Urals and Timan-Pechora provinces, and the Preuralian foredeep [*Maksimov*, 1970; *Nikonov* et al., 2000]. The near shore origin of these deposits is suggested by the presence of sandy bar, deltaic and shoestring riverbed facies. These facies have been traced by seismic profiling and extensively intersected in some places by drilling [e.g., *Gataulin*, 1999]. More open sea conditions are reflected in the development of pure shallow water limestone and bioherms.

The late Early Devonian (Emsian) to Early Carboniferous (Early Visean) cycle of sedimentation is marked by the appearance of the deep water Domanik facies, represented by marl, oil shale, and chert [*Mirchink*, 1965; *Maksimov*, 1970; *Parasyna* et al., 1989]. The first appearance of the Domanik facies (Infradomanik) is in the Middle Devonian (Upper Eifelian) of the Southern Urals and the adjoining Russian platform [*Russian Stratigraphic Committee*, 1993]. The second appearance was in the Late Devonian (Frasnian), when it was much more widely developed and marked the axes of branching basins in the vast area of the eastern part of the East European Craton (the Kama-Kinel trough system). During the Late Devonian and Early Carboniferous (Tournaisian and Early Visean) troughs were developed across much of the continental margin (Figure 8). The shoulders of the Kama-Kinel trough system are marked by chains of bioherms and small reefs, and extensional faults which would have controlled the development of the trough system have not been recognized. The total thickness of the Frasnian to Lower Visean sediments in the troughs and on their shoulders are the same, suggesting they are not simple grabens. The reason for this is thought to be an absence of terrigenous influx combined with rapid subsidence of the axial parts of the basins, which did not permit reefs to grow there. Another possible factor governing the location of the troughs could have been a gentle warping of the platform, connected either with rifting or with an arc-continent collision; these possibilities will be discussed later, in Section 3.

From the Middle to Late Devonian Baltica underwent rifting, basaltic (Trap) volcanism and formation (or reactivation of some Proterozoic) of grabens, such as the Donetzk-Pripyat and Pechora-Kolva systems (Figure 9). The western part of the Pechora-Kolva graben system was involved in the Uralide deformation at the lower reaches of Schugor river [*Nikonov* et al., 2000]. Recent data suggest that volcanism connected with this phase of rifting had ended by the Tournaisian [*Wilson* et al., 1999].

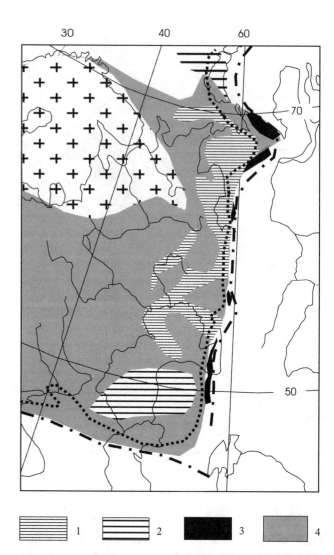

Figure 8. The types of deep water sediments at the eastern margin of Baltica (1–3): 1. The Domanik facies (Frasman to Toumaisian). 2. The facies of the Barentz/Novaya Zemlya and Pricaspian depressions (Upper Devonian to Lower Permian). 3. Sakmara-Lemva facies (Ordovician to Upper Devonian/ Lower Permian). 4. Shelf facies. The other symbols are in the Figures 3 and 9.

The final stage of the sedimentary cycle under discussion took place during the Early Carboniferous (Tournaisian to Early Visean). The Tournaisian sedimentation was restricted mostly to the territory of the western slope of the Ural Mountains, where shallow water limestone and dolomite were accumulated, and to the Kama-Kinel trough system. The troughs were rapidly filled by limestone, marl, chert, siltstone, quartz sandstone, and, locally, coal seams. By the end of the Early Visean they were completely filled and do not show up in the sedimentary record after that.

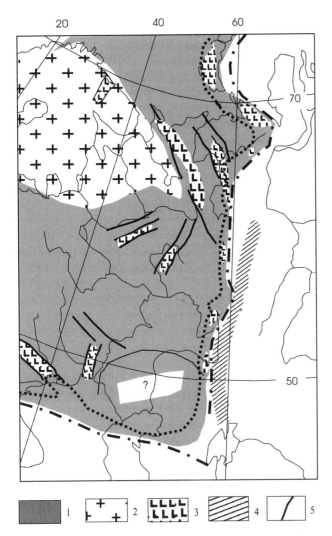

Figure 9. Devonian grabens and graben-related volcanics in the eastern half of Baltica [*Puchkov*, 1988; *Fokin* et al., 2001]. 1. The area of maximal Devonian transgression. 2. The area of absence of the Upper Devonian sediments. 3. The area of development of Devonian rift (mostly trapp) volcanics. 4. The Magnitogorsk island arc. 5. Normal faults connected with rifts. The other symbols are in the Figure 3.

A vast transgression of a new sedimentary cycle started in the Early Carboniferous (Middle Visean). In the northern and southern parts of the margin carbonate sediments were predominant (there are also evaporites in the northeast of the Timan-Pechora basin); but in the Middle, North and Cis-Polar Urals the transgression started with accumulation of a so-called "coal-bearing formation" [*Nalivkin*, 1960] represented by quartz sandstone, siltstone, and shale with coal seams. The transgressive cycle developed throughout the Carboniferous and Early Permian. This part of the history is relevant

only to the northern part of the margin, which ceased to exist by the Late Devonian in the Southern Urals and by the Early Permian in the north.

Generally speaking, the margin underwent subsidence (in part modified by global sealevel changes) and the accumulation of sediments. Within this setting, extension in the Devonian led to the development of grabens and widespread basaltic magmatism. Indications of local compressional deformation have also been described [e.g., *Fokin* et al., 2001]. It is expressed in uplifts, thrusts and the inversion of some grabens and depressions of the Volga-Uralian Province in the Late Devonian (Frasnian to Early Famennian) and the Early Carboniferous (Early Visean). In the Timan-Pechora province, analogous events took place later, in the Late Carboniferous to Early Permian and post-Triassic time [*Chermnykh and Puchkov*, 1972; *Timonin*, 1998; *Nikonov* et al., 2000] (Figure 10). Therefore, the diachroneity of the orogenic events is reflected in a diachroneity of local movements along the margin.

It is very important to note that there was no predetermined boundary between the Uralides and the platform: the outer boundary of the foreland deformation crosses the facial zones and earlier platform structures at different angles (Figures 3 to 9).

2.3. Passive Margin Stage: The Continental Slope and Rise

The sediments that can be attributed to the continental slope and continental rise of the margin are found in the West Uralian zone as separate, thrust bound units. Altogether they belong to Sakmara-Lemva (or Zilair-Lemva) structural/facial zone, called after the names of the corresponding rivers in the north and south of the Uralides [*Russian Stratigraphic Committee*, 1993].

The tectonic position together with a short stratigraphic and lithological description of the deep water sediments of the continental margin preserved along the western slope of the Ural Mountains is needed because a considerable amount of data have been compiled since the last review made by the author (*Puchkov*, 1979a). The concise description is given below, unit after unit in a northward direction (Figures 1, 2, 11, 15, 16, and 17).

2.3.1. Sakmara Allochthon (Synform). In the southernmost part of the Urals, the Sakmara allochthon (Figure 11) structurally overlies the parautochthonous Upper Devonian turbidites of the Zilair formation, which overlies the Devonian shelf deposits. The allochthon consists of a series of deformed thrust sheets with a crude "nappe stratigraphy". The lowermost position is occupied by thrust sheets composed of deep water facies

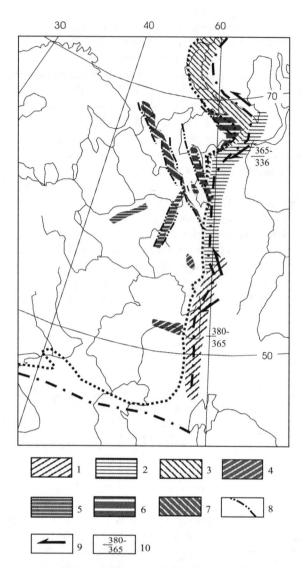

Figure 10. The correlation of orogenic and platform structures. 1–3. The different parts of the Uralian Qrogen: 1. The flysch of eastern provenance appeared in the Late Frasnian. The main phase of collision took place in the Middle Carboniferous to Early Permian. 2. The flysch of eastern provenance appeared in the Early Visean. The main phase of collision was in the Middle Carboniferous to Late Permian. 3. The flysch of eastern provenance appeared in the Late Permian (Artinskian). The main phase of collision was in the Early Jurassic. 4–7. Platform anticlinal structures; inverted aulacogens among them: 4. Formed in the Late Devonian. 5. Formed in the Early Carboniferous. 6. Formed in the Late Carboniferous/Early Permian. 7. Formed after the Early Triassic (in the Early Jurassic?). 8. The thrusts formed after the Early Triassic (in the Early Jurassic?). 9. The general directions of strike-slip movements at the salients of the margin. 10. The interval of isotopic age determinations for the HP-LT metamorphics (Ma). The other symbols are in the Figure 3.

sediments of the margin, grading upward into an olistostrome and the Zilair formation. These sediments are overlain tectonically by a serpentinite melange, pillow lava and chert, tuff turbidites, polymictic olistostromes and other formations of oceanic and island arc nature. At the top, large blocks (i.e., the ultramafic massifs of Kempirsay and Khabarny) outcrop. The root of these thrust sheets is thought to be in the Main Uralian fault, and material from its hanging wall is represented here by the Voznesensk-Prisakmara melange zone [*Ruzhentsev*, 1976; *Khvorova* et al., 1978; *Ivanov and Puchkov*, 1984a; *Puchkov*, 2000].

The units situated at the base of the Sakmara allochthon are partially represented by volcanics and an olistostrome containing blocks of Cambrian limestone. Besides these, coarse terrigenous sediments with flows of subalkaline basalt, and layers of rhyolite tuff, chert, limestone of the Kidryasovo and Kuagash formations are found. These deposits are interpreted to have been deposited in grabens [*Khvorova* et al., 1978]. Higher in the section the Lower to Middle Ordovician Kuragan formation is composed of shale, siltstone, tuffite, with minor arkosic and quartz sandstone. These are thought to have been deposited in deep water conditions, with a distal western provenance [*Khvorova* et al., 1978]. They are replaced upward by the Middle to Upper Ordovician reddish chert and shale of the Pismenskaya unit [*Puchkov* et al., 1990]. From the Upper Ordovician to the Upper Devonian (Frasian), the bathyal sediments in the Sakmara allochthon are composed of condensed, cherty sediments, with rare interbeds of sandstone. The Silurian and Lower Devonian belong to the Sakmara formation, which is composed of graptolitic carbonaceous shale, phtanite with rare layers of uneven-bedded, knotty limestone (orthoceratid-bearing in Silurian and tentaculitic in Pragian), and minor cannibalistic sandstone and conglomerate. Lower Devonian alkaline volcanics of the Chanchar formation (probably a rift complex) are reported from the middle part of Sakmara zone, but its age and tectonic position are still not clear. The Lower and Middle Devonian Kyzilflot and Aitpaika formations resemble the Sakmara formation, with the exception that they locally contain blocks of shallow water limestone enveloped in a matrix of chert with slump structures, suggesting that these formations are oligomictic olistostromes [*Puchkov*, 1979a, 2000]. The Frasnian Eginda formation is composed of chert and shale and is overlain by the Zilair formation.

The above description is very generalized. Due to intense deformation no complete sections are preserved. Moreover, there are some important differences between the sections. For example, after the latest data

57°30'

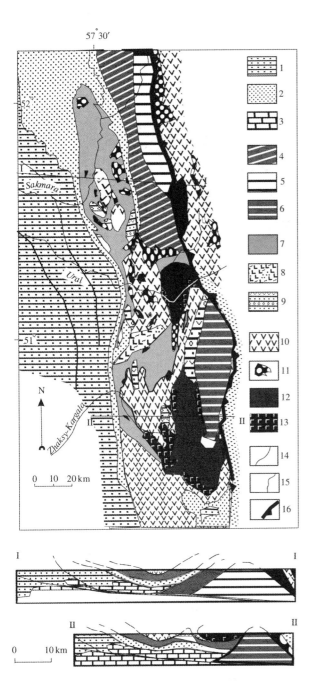

Figure 11. The tectonic scheme of the Sakmara allochthon and surrounding structures. 1–3. Paraautochthon: 1. Lower Carboniferous-Permian flysch and molasse. 2. Famennian Zilair flysch (involved in allochthons in the east). 3. Paleozoic Ordovician to Devonian shelf deposits (shown only in the cros sections). 4–13. Allochthons: 4–6. The Lower tectonic unit (Uraltau complex): 4. Suvanyak greenschist complex of quartzite, shale and chert (mostly Paleozoic deep water sediments). 5. Maksyutovo HP-LT metamorphic complex (mostly Paleozoic oceanic sediments and volcanics). 6. Suvanyak and Maksytovo complexes, undifferentiated. 7–9. The Middle tectonic unit: 7. Lower Ordovician to Devonian bathyal complex (shale, chert, limestone, olistostrome). 8. Devonian alkaline basalts, tuff, sandstone. 9. Cambrian?-Lower Ordovician rift complex (subalkaline basalts, conglomerate, sandstone). 10–13. The Upper tectonic unit: 10. Ordovician to Middle Devonian oceanic and island arc complexes. 11. Serpentinitic melange. 12. Ultramafic rocks.13. Gabbro. 14–16. Geological boundaries: 14. Stratigraphic contacts. 15. Tectonic contacts. 16. Main Uralian fault.

to Devonian (Frasnian inclusive) chert, Cambrian and Devonian organogenic limestone, undated quartz sandstone, crystalline schists and basalt. Such a variety of fragments in the olistostrome suggests involvement of the continental margin in the processes of collision and formation of an accretionary prism. The upper tectonic unit (Rysayevo) is composed of condensed Devonian (Lohkovian to Frasnian) chert overlain by Famennian (Zilair) greywacke, which has olistostromes containing blocks of Frasnian chert at the base. We think that the Late Frasnian to Lower Famennian complexes of these two sections correspond to debris flow channels and inter channel areas at the slope of Uraltau ridge, which had started to grow by this time (Figure 12).

According to *Chibrikova* [1997], the deposition of the Zilair formation in the Sakmara allochthon started in the Late Frasnian, unlike the other parts of the Southern Urals where deposition began in the Famennian. Diachroneity of the lower boundary of the Zilair formation and its general tendency of becoming younger to the west is supported by conodont data [*Puchkov*, 2000] (Figures 13 and 14). The general situation is very typical for the development of flysch troughs [e.g., *Sinclair*, 1997]. The eastern provenance of the Zilair formation is supported by lithological and mineralogical studies [e.g., *Willner* et al., 2002].

2.3.2. Suvanyak Complex. The major part of the Ural-Tau antiform is composed of the Suvanyak complex (Figure 14), a thick, polydeformed greenschist facies metamorphic unit consisting mostly of quartz sandstone,

[*Ryazantsev* et al., 2000; *Aristov* et al., 2000], in the western part of the Sakmara zone, close to the valleys of the Sakmara and Ural rivers, two types of bathyal sections can be established. We suggest that they be called the Shaitantau and Rysayevo types. The lower tectonic unit (Shaitantau type) is represented by the Ordovician and Silurian rift to bathyal sediments described above, which are overlain by an Upper Devonian polymictic olistostrome. The olistostrome is several hundreds of metres thick, and consists of Ordovician

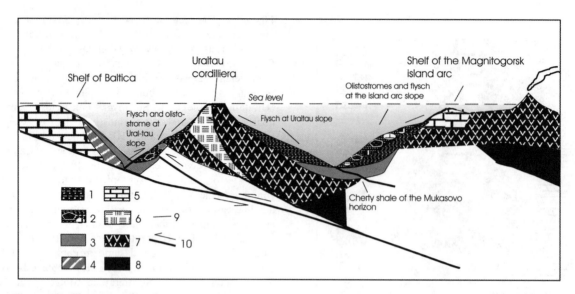

Figure 12. Character of sedimentation in the Zilair flysch troughs, Frasnian and Famennian [modified from *Gorozhanin and Puchkov*, 2001]. 1. Flysch. 2. Flysch and olistostrome. 3. Cherty shale. 4. Quartz sandstone, chert, shale. 5. Limestone and other shallow water deposits. 6. HP-LT metamorphics. 7. Island arc volcanics. 8. Ophiolite mafic-ultramafics. 9. Provenance of terrigenous material. 10. Thrusts and their directions.

siltstone and shale, with some chert (where recognisable) [e.g., *Alvarez-Marron* et al., 2000]. It is important to note that no syngenetic limestone is present. For many years rocks in the Suvanyak complex were thought to be

Proterozoic in age [*Russian Stratigraphic Committee*, 1993], although Paleozoic fauna had been reported from it [*Krinitsky and Krinitskaya*, 1965; *Puchkov*, 1979b, 2000; *Puchkov* et al., 1997, 1998; *Chibrikova and Olli*, 1997].

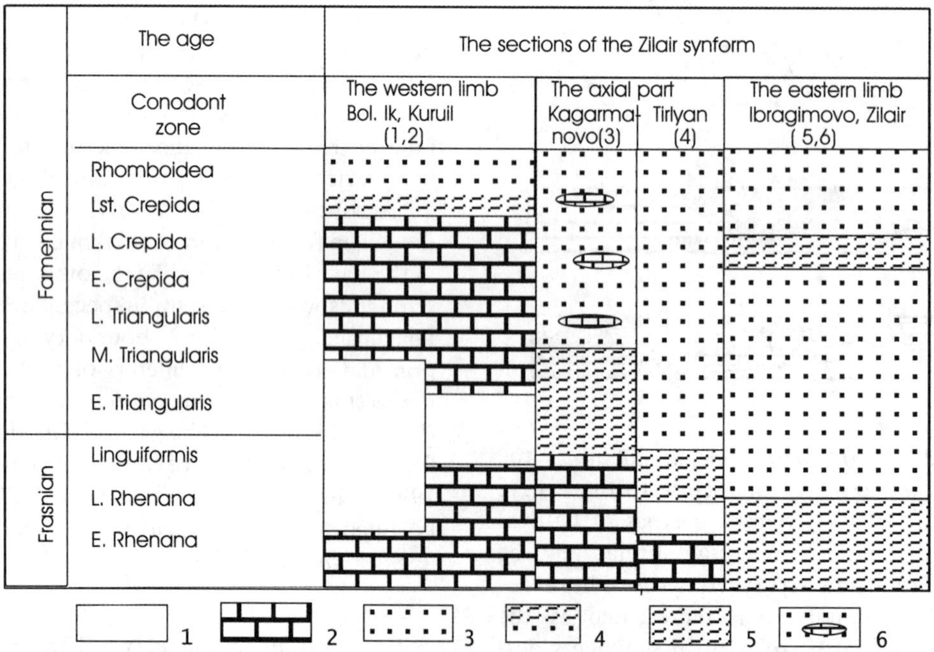

The age		The sections of the Zilair synform		
	Conodont zone	The western limb Bol. Ik, Kuruil (1,2)	The axial part Kagarma-novo(3) Tirlyan (4)	The eastern limb Ibragimovo, Zilair (5,6)
Famennian	Rhomboidea			
	Lst. Crepida			
	L. Crepida			
	E. Crepida			
	L. Triangularis			
	M. Triangularis			
	E. Triangularis			
Frasnian	Linguiformis			
	L. Rhenana			
	E. Rhenana			

1 2 3 4 5 6

Figure 13. Correlation of sections at the base of the Upper Devonian Zilair series. 1. Information is absent (a fault or no outcrops). 2. Limestone. 3. Zilair greywacke flysch. 4. Greywacke with layers of chert. 5. Chert. 6. Shale and siltstone with lenses of marl. Positions of the sections characterized by conodonts (points 1 to 6) see in Figure 14.

gneiss and granite clasts indicate the possibility of a crystalline basement provenance. The overlying Frasnian deposits are represented by a condensed unit of chert and subordinate shale (Ibragimovo horizon), with layers of greywacke sandstones in its upper part, grading upwards into the Famennian Zilair formation turbidites.

2.3.3. The Uzyan Nappe. The Uzyan nappe, which tectonically underlies the Kraka ultramafic massifs (Figure 15), consists of an imbricated series of sedimentary rocks devoid of carbonate and dated solely by pelagic fauna (conodonts, graptolites and chitinozoans) of Ordovician, Silurian and Devonian age. The Uzyan nappe tectonically overlies the Zilair formation [*Puchkov*, 1995; *Puchkov* et al., 1998; *Brown* et al., 1996]. Sediments in the Uzyan nappe are composed of quartz sandstone, siltstone, shale, with layers of chert predominating in the top of the succession. There are some basalt flows in the area, but neither their type, nor their stratigraphic position is known for certain. The sediments of the Uzyan nappe resemble the Paleozoic rocks of the Suvanyak complex, but being less deformed and metamorphosed. The absence of limestone argues deposition below the level of the carbonate compensation depth, perhaps on the abyssal plain.

2.3.4. Nizhnie Sergi. The Nizhnie Sergi thrust sheet of the Bardym allochthon [*Puchkov and Ivanov*, 1982; *Zhivkovich and Cekhovich*, 1984] is situated in the centre of the Ufimian amphitheatre, in the Middle Urals (Figure 16). The allochthon consists of two thrust sheets: Nizhnie Sergi and Nyazepetrovsky. The Nyazepetrovsky thrust sheet consists of Silurian to Early Devonian island arc complexes, with tectonised serpentinites at the contact between the two thrust sheets. The allochthonous nature of the Nizhnie Sergi thrust sheet was demonstrated by five boreholes which penetrated Ordovician to Devonian bathyal sediments and entered into Devonian shelf limestone at an average depth of 1000 m. The oldest rocks of the Nizhnie Sergi thrust sheet are Ordovician (Caradocian) sandstone, limestone, volcanic tuff and basalt with shallow water fauna. These are overlain by a condensed section of graptolitic carbonaceous shale, phtanite, light coloured chert and rare nodular limestone. In the Lower Devonian there is another, quite different facies (the Aptechnogorskian formation), represented by a thick unit of turbiditic quartz sandstone, siltstone and shale, evidently of western provenance, with some subalkaline basalt [*Puchkov and Ivanov*, 1982]. The section ends with a Late Devonian (Frasnian) condensed unit, the Korsakov formation, composed of chert and subordinate shale. The Nizhnie Sergi lithology

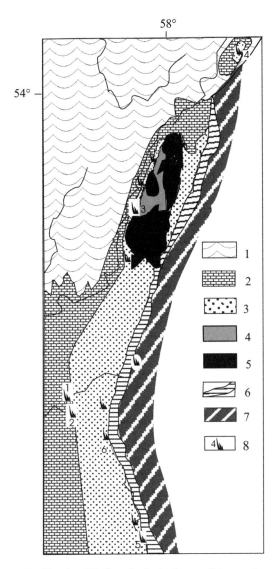

Figure 14. The simplified geological scheme of the northern and middle part of the Zilair synform. 1. Precambrian basement. 2. The Paleozoic shelf complexes (Ordovician to Upper Devonian). 3. Zilair (Famennian) flysch. 4. Abyssal Ordovician to Devonian sediments of the Uzyan-nappe. 5. Ultramafic bodies and serpentinitic melange of the Kraka allochthon. 6. Cherts of the Ibragimovo horizon (Frasnian). 7. Metamorphic deep water sediments of the Suvanyak complex (Ordovician to Middle Devonian). 8. The sections close to the base of Zilair flysch, characterized by conodonts (see Figure 13).

The Suvanyak complex stratigraphy is now interpreted to represent a thick (3 to 5 km) series of deep water sediments, composed of quartzitic terrigenous deposits of Ordovician, Silurian and Devonian age, with intercalations of chert in the upper part [*Puchkov*, 2000]. Enigmatic, rare, thin beds of conglomerate containing

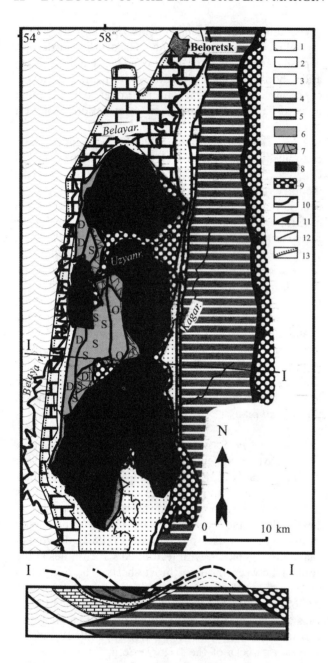

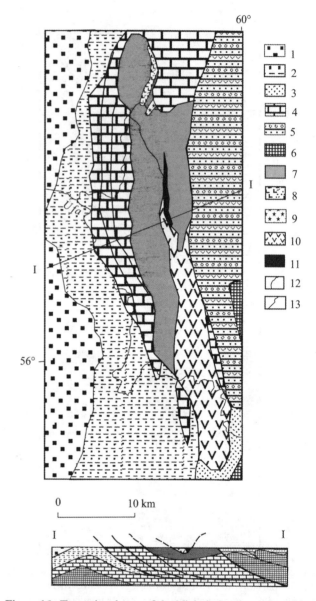

Figure 15. The tectonic scheme of the Kraka area. 1–3. Para-autochthon: 1. Precambrian basement. 2. Ordovician to Frasnian shelf deposits. 3. Famennian (Zilair) greywacke. 4–9. Allochthons: 4–5. Lower unit: 4. Ordovician to Givetian Suvanyak abyssal complex. 5. Frasnian Ibragimovo cherts. 6–7. Middle unit: 6. Sub-Kraka abyssal complex (Uzyan nappe). Fauna occurences: O — Ordovician, S — Silurian, D — Devonian. 7. Imbricated zones within the Uzyan nappe. 8–9. The Upper unit: 8. Kraka ultramafic massifs. 9. Serpentinitic melange. 10–13. Geological boundaries: 10. Thrusts. 11. Main Uralian fault. 12. Normal stratigraphic contact. 13. Contact with angular unconformity.

Figure 16. Tectonic scheme of the Nizhnie Sergi area [modified from *Puchkov and Ivanov*, 1982]. 1–6: Paraautochthon. 1. Molasse (Lower Permian), 2. Late flysch (Middle Carboniferous to Lower Permian). 3. Zilair flysch (Famennian). 4. Shelf sediments, predominately carbonates (Upper Ordovician to Lower Carboniferous). 5. Rift complexes: coarse grained terrigenous sediments and volcanics (Lower Ordovician). 6. Metamorphic complexes (Precambrian). 7–11. Bardym allochthon: 7–9. The Lower (Nizhnie Sergi) unit. 7. Bathyal sediments, Middle Ordovician to Frasnian. 8. Rift sandstones, basalts and rhyolite tuffs (Lower Devonian). 9. Rift sandstones, limestones and basaltoids (Middle Ordovician). 10–11. The Upper (Nyazepetrovsk) unit: 10. The oceanic and predominately island arc volcanics and intrusions (Silurian to Lower Devonian). 11. Serpentinitic melange. 10–11. Geological boundaries: 10. Stratigraphic contacts. 11. Tectonic contacts.

is typical for the deep water facies along the whole Uralide passive margin.

2.3.5. The Polya Zone. The Polya zone (Figure 1) consists of a narrow strip of sandstone and shale, attributed to the Polya series, immediately to the west of the Main Uralian fault in the Northern and Cis-Polar Urals [*Puchkov*, 2000]. In the stratotype of the series, the Sarankhapner formation is comprised of conglomerate, quartzitic sandstone and volcanics with associated diabase dikes of rift affinity, followed upward by quartzitic sandstone, siltstone and shale (partially carbonaceous and containing Ordovician graptolites and chitinozoans) of the Khomasya and Polya formations. The presence of chitinozoans suggests that these rocks may be also of Silurian age [*Karsten* et al., 1989]. Locally, the series contains olistostromes with blocks of Ordovician limestones and carbonaceous shales with Frasnian conodonts in the matrix [*Petrov and Puchkov*, 1994].

The Polya zone has generally been thought to be Ordovician, and to be overlain conformably by the Upper Ordovician to Silurian island arc volcanics of the Tagil-Magnitgorsk zone [*Russian Stratigraphic Committee*, 1993]. Later it had been shown, however, that Polya and Tagil-Magnitogorsk zones are separated by the Main Uralian fault.

2.3.6. The Malaya Pechora Allochthon. The Malaya Pechora allochthon covers a large area in the sub-Polar Urals, in the upper reaches of the Pechora and Kisunya rivers [*Puchkov*, 1979b] (Figure 1). The Malaya Pechora section starts with Lower Ordovician variegated sandstone and siltstone, containing Lower Ordovician brachiopods that resemble those found in the Pogurey formation of the Lemva area (see section 2.3.7.). Higher up in the section, the Kisunya formation is composed of reddish and greenish siltstone. It grades upward into the Baskaya formation, which is composed of grey-greenish siltstone and shale with thin layers and lenses of pelitic limestone containing Middle Ordovician fauna. The Silurian is poorly represented. Lower Devonian, darkish siltstone and shale with some limestone (including a Pragian age nodular tentaculitic limestone) are typical of some sections of the Malaya Pechora allochthon. The overlying Lower to Middle Devonian Gorevskaya formation is represented by green-greyish quartz sandstone, siltstone and shale with rare lenses of detrital limestone. Upward in the section, the Evtropinskaya formation consists of a condensed section of chert with subordinate shale and interbeds of limestone containing abundant Frasnian and Lower Famennian conodonts. The Malaya Pechora section is completed by a unit of turbiditic greywacke sandstone, siltstone and shale with lenses of detrital limestone containing Visean, Serpukhovian and Upper Carboniferous foraminifera [*Puchkov*, 1979a].

2.3.7. The Lemva Allochthon. The Lemva allochthon in the Polar Urals is the biggest (and the classic) area of deep water sediments in the western slope of the Ural Mountains [*Puchkov*, 1979a] (Figure 17). The Upper Cambrian to Lower Ordovician Pogurey formation, which is composed of conglomerate and sandstone, magmatic Sivyaga-Lemva trachybasalt, trachyrhyolite, alaskitic and Vangiryu-Sedyiu-Pogurei alkaline basalt complexes, are thought to represent the rift phase of the margin of Baltica [*Goldin* et al., 1999]. These are overlain by a thick series of reddish siltstone and shale of the Lower Ordovician Grubeyu formation, an analogue of the Kuragan formation of the Sakmara allochthon (see section 2.3.1.). The younger Ordovician, Silurian and Lower Devonian sediments are developed in two types of sections. The first (the western Lemva or Kharota type) section is represented by the Khaima, Yunkovozh, and Pravogrubeyu formations (which were attributed earlier to the Ordovician Kechpel formation), and the Kharota formation (Silurian to Lower Devonian Pragian) which is composed of shale (partly carbonaceous) and carbonates (including the lower level of nodular Middle Ordovician limestone). The second section (Central Lemva), consists of the Moliudshor and Kharota formations, which are composed of Middle Ordovician to Lower Devonian graptolite and conodont-bearing phtanites, carbonaceous chert and shale, and with no carbonate. In most of the sections the Lower to Middle Devonian is represented by turbidites with an alternation of quartz sandstone, siltstone, shale and chert, and barite mineralisation. The Upper Devonian and Lower Carboniferous (Tournaisian) in both types of sections is composed mostly of chert, with or without layers of limestone [*Puchkov*, 1979a; *Shishkin*, 1989; *Didenko* et al., 2001]. In the Paga river valley a Devonian trachybasalt formation is present. In the author's opinion, which is not generally accepted, it can be correlated with the Devonian rift formations of the platform [*Goldin* et al., 1999].

In the southeastern part of the Lemva zone, *Ruzhentsev and Aristov* [1998] establish two more types of section within two subzones; the Grubeshor and East Lemva subzones. But, unlike the other sections, only Ordovician sediments have been reliably identified. Both subzones contain volcanics, and the East Lemva subzone also contains abundant dikes. These manifestations of volcanism may be attributed to the Ordovician rift process.

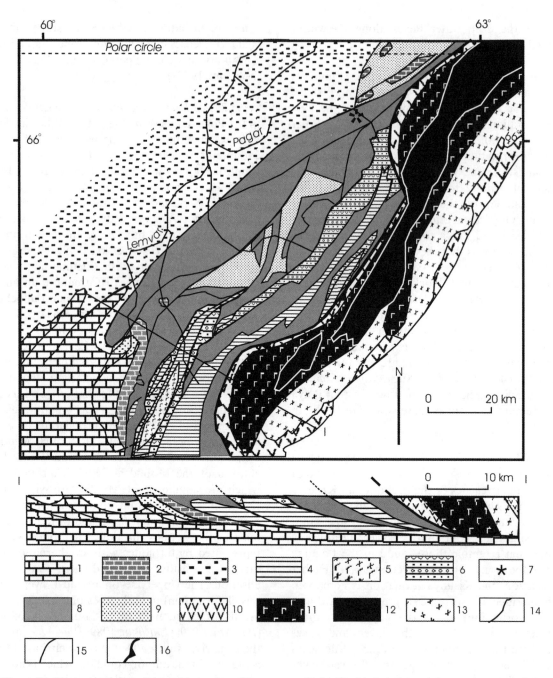

Figure 17. Tectonic scheme of the southern part of Lemva area (Polar Urals). 1–3. Autochthon and parautochthon: 1. Ordovician to Carboniferous shelf deposits (mainly limestone, subordinate sandstone). 2. Ordovican to Carboniferous outer shelf to bathyal deposits: shale, limestone, sandstone, chert (Western Lemva facial type). 3. Lower Permian flysch and molasse of the Preuralian foredeep. 4–9. The lower allochthons: 4. Precambrian volcanic and metamorphic rocks. 5. Lemva gabbro-granite massif. 6. Upper Cambrian to Lower Ordovician rift sandstone, conglomerate and volcanics. 7. Upper Devonian trachybasalts (Paga complex). 8. Bathyal to abyssal Lower Ordovician to Lower Carboniferous deposits (Central Lemva, Grubeshor and Eastern Lemva facial types). 9. Visean to Lower Permian greywacke flysch of the residual deep water trough in place of Kharota and Central Lemva type facies. 10–13. The Upper allochthon: 10. Ordovician to Devonian riftogenic, ophiolitic and island arc volcanic and sedimentary complexes. 11. Gabbro. 12. Ultramafics. 13 Silurian tonalites. 14–16. Geological boundaries: 14. Stratigraphic and intrusive contacts. 15. Faults. 16. Main Uralian fault.

The Early Carboniferous (Early Visean) Ray-Iz formation sandstone and conglomerate with fragments of magmatic, metamorphic and sedimentary rocks is suggestive of an eastern provenance [*Saldin*, 1999]. More widespread is the Visean and younger greywacke turbidites of the Jayu formation. The Jayu formation grades westward into a condensed section of grey and black chert and shale which is in turn overlain by distal turbidites of the Upper Carboniferous? to Lower Permian Kechpel formation. By the Early Permian (Artinskian), turbidites quickly propagated over the shelf to the west and the Preuralian foredeep started to form [*Puchkov*, 1979a; *Saldin*, 1996, 1998; *Nikonov* et al., 2000].

2.3.8. Baydarata.   In the Baydarata area (Figure 1) of the northernmost Polar Urals, a Lemva-type succession begins with the Cambrian Oyukha formation, which consists of terrigenous pyroclastic sediments represented mainly by tuffs of andesitic basalt, andesite, and dacite composition, with lenses of tuff conglomerate and rarelayers of phyllite, sandstone and silty limestone [*Dembovsky* et al., 1988]. The unit is overlain conformably by the Cambrian(?) to Lower Ordovician (Tremadoc) Talota formation, which is composed of quartz and feldspar sandstone, siltstone and shale. The overlying Upper Tremadoc to Lower Llanvirn Kharapeshor formation is highly variable in lithology, being composed of different combinations of limestone, shale, quartz siltstone, basalt and rare trachybasalt and rhyolite. Upward in the section, the Salepeyakha formation is composed of shale, cherty shale (weakly carbonaceous), with layers of silty limestone, characterized by rare Ordovician conodonts. The Salepeyakha formation is conformably overlain by carbonaceous cherty shale of the Kharota formation with Lower Silurian (Llandovery) graptolites.

2.3.9. Kara Tectonic Zone of Pay-Khoy.   In the opinion of some geologists [*Timonin and Yudin*, 1999] the Kara tectonic zone of Pay-Khoy does not belong to the Uralides, but to Paykhoides. Nevertheless, this depends only on how we agree upon the exact meaning of the term Uralides. In the author's opinion, it is possible to also include Old Cimmerian thrust and fold structures of the Pay-Khoy and Novaya Zemlya into the Uralides, because the deformation of this age can be traced far to the south in the Uralides as well. For completeness, a description of the Kara zone is given here.

Paleozoic facies of the Pay-Khoy and the western slope of the Ural Mountains are so similar that a former continuity between them is evident; they certainly belonged to the same continental margin of Baltica. The

data on the geology of the Kara deep water sediments were published in quite a few papers [*Ustritsky*, 1961; *Yenokyan*, 1970; *Belyakov* et al., 1981; *Elisseev* et al., 1984; *Puchkov and Ivanov*, 1985; *Yudina and Puchkov*, 1987; *Yudovich* et al., 1998; *Rogov* et al., 1988; *Timonin and Yudin*, 1999]. The Kara zone sediments are composed of up to 2500 m of Ordovician, predomiantely terrigenous sediments, but with an increase in cherty and carbonate components upward in the section. The section starts with a basal unit (Uppermost Cambrian?) of polymictic conglomerate, sandstone and limy sandstone, overlying with angular unconformity [*Dedeev and Getsen*, 1987] different horizons of the Upper Riphean and Vendian. The basal unit is overlain by Lower Ordovician (Tremadoc to Llandeilo) shale, which is in turn overlain by Caradoc to Ashgil cherty-limy shale and limestone. Comparatively thin (several hundreds of meters) Silurian and Lower Devonian sediments are represented by chert, cherty and limy shale, shaly and nodular limestone. No fauna except graptolites and conodonts is found in this part of the section. The Lower (Emsian) to Middle Devonian is composed of quartz sandstone, cherty and carbonate shale. There are no volcanics in the Devonian of Pay-Khoy, but numerous sills of gabbro-diabase, located mostly in the Ordovician rocks, are dated the Middle Devonian [*Ostaschenko*, 1979] and therefore are penecontemporaneous with the basalts of the East European Craton (Figure 9). A condensed Middle Frasnian to Tournaisian section of shale and coloured and black chert also occurs. The rest of this fairly condensed part of the section, corresponding to a large period of time in which 700 m of sediments were deposited during 60 Ma or more (Early Carboniferous (Visean) to Lower Permian time) is represented by carbonaceous shale, cherty shale, detrital limestone and limestone breccias with graded bedding. The upper 100 m of this black shale section, dated by goniatites, is Early Permian (Asselian and Sakmarian) in age. The shale grades upward to Artinskian flysch.

The resemblance between the Kara and Lemva facies is striking. The marker horizons of nodular limestone (counterparts of *ammonitico rosso* of Alps or *Tentakulitenknollenkalk* of Saxo-Thüringia), which have been traced from the Southern to the Polar Urals, are also present in Kara [*Puchkov and Ivanov*, 1985]. However, the differences are important, particularly a very late (Artinskian) change from deep water shaly facies to turbidites of the Uralian provenance in the Pay-Khoy.

The outer margin of the bathyal facies (its contact with the shelf facies) was regarded by *Ustritsky* [1961] as gradational, but later on it was revisited by many geologists and its overthrust nature, with a displacement of not less than 30 km, was accepted [*Timonin and Yudin*, 1999]. Its inner boundary is not exposed.

3. DISCUSSION

From the Ordovician to the Late Devonian in the Southern Urals, or until the Early Carboniferous and even later in the northern, the continental margin evolved from a rift-drift stage to a passive margin, and finally an active stage as island arcs accreted to it. This margin is now preserved in western slope of the Ural Mountains, in the footwall of the Main Uralian fault. The rift facies are characterized by Late Cambrian? to Ordovician coarse-grained terrigenous sediments with subalkaline volcanics overlying Precambrian (essentially Cadomian) basement. The rift stage of development of the margin was followed by a passive margin stage (with only minor tectonic activity) characterized by a general subsidence and accumulation of thick sedimentary successions. These sediments record the topography of the margin. By analogy with the modern continental margins, shelf, continental slope and continental rise (shallow-water to abyssal) facies can be recognised.

Shelf sediments are characterized by the presence of continental (lacustrine, alluvial, etc.), lagoonal, littoral, and neritic sediments, the latter being characterized by the presence of thick-walled fauna, development of bioherms and reefs. Only in the periods of the highest stand of the sea level were relatively deep water sediments of the Domanik type deposited. Generally speaking, the sedimentation in the shelf was influenced by interaction between subsidence of the margin and eustatic oscillations of the sea level, with the participation of local, minor tectonic activity. Processes at the margin resulted in the formation of three major oil producing sedimentary basins adjoining it (Timan-Pechora, Volga-Urals, and Pricaspian). The architecture of these basins was strongly influenced by the different anisotropy and rigidity of their respective basements.

The continental slope (bathyal) facies are characterized by the alternation of thick terrigenous sediments of continental provenance and thin, condensed (starved) shale, chert and limestone; among the latter, marker horizons of nodular limestones are traced from the south to the north of the Uralides. The alternation of sediments depended largely on the oscillations of the sea level: the terrigenous series and marginal barrier reefs formed during the high stands of the platform, while the starved series appear when the platform source areas were separated by the widespread carbonate shelf seas.

Among the deep water sediments there are sections completely devoid of autigenic carbonates (Suvanyak complex, Uzyan nappe, central sections of the Lemva allochthon). They could have originated at still greater depths, below the level of carbonate compensation, on the continental rise.

The development of a passive continental margin finished with the onset of arc-continent collision in the Early Devonian. This is reflected in the sedimentary record by a change from terrigenous provenance from the platform to a provenance predominantly from the growing orogen [e.g., *Willner* et al., 2002] (Figure 2). But this event did not start simultaneously along the whole margin. The diachroneity of the greywacke flysch is revealed both in the transversal (Figures 13 and 14) and along-strike directions of the margin (Figures 2 and 10) and suggests a diachroneity of collision from north to south. This diachroneity is also indicated by the age of HT-LP metamorphism, which is Middle Devonian in the Southern Urals and Lower Carboniferous in the Cis-Polar and Polar Urals [*Shacky* et al., 1997, 1999; *Ivanov* et al., 2000]. During arc-continent collision the bathyal to abyssal facies were scraped off their basement and thrust over the Zilair formation and/or over the underlying shelf rocks (Figures 11, 12, 15, 16, and 17). Fragments of transitional neritic to bathyal facies are rare (like the Kharota facial type in Lemva zone) and probably are also allochthonous.

Less evident is the answer to the question if the margin was really passive (Atlantic-type) or belonged to Japanese-type back arc sea, as it stated by some researchers [*Nikishin* et al., 1996; *Fokin* et al., 2001].

The following arguments can be given in support of a passive Atlantic-type character of the margin of Baltica as opposed to the idea of its back arc position during the Devonian. The volcanic successions in both the Tagil and Magnitogorsk arcs have been clearly shown to be of intra-oceanic island arc affinity [*Seravkin* et al., 1992; *Spadea* et al., 2002], whereas there is no evidence to support the presence of a Japanese or Kuril type of the margin. The island arc volcanism developed outboard of the continental margin and has geochemical features indicating intraoceanic formation as well as the eastward dip of the subduction zone [*Bobokhov*, 1991; *Seravkin* et al., 1992; *Yazeva and Bochkarev*, 1998]. There is clear evidence that the arc collided with the passive margin in the last half of the Devonian, in a tectonic setting similar to that which appears now in the southwest Pacific where island arcs are colliding with the Australian and Eurasian continents [e.g., *Puchkov*, 1996, 1997; *Brown* et al., 1998, 2001; *Brown and Spadea*, 1999]. Conversely, the Famennian events at the continent-island arc boundary, such as an exhumation of a HP-LT metamorphic complex, and formation of two flysch basins divided by the Uraltau ridge (Figure 12) [*Willner* et al., 2002] cannot be adequately explained by a process involving back arc compression as suggested by *Fokin* et al. [2001]. Such a process is not supported by the data. Finally, many lithological and faunal features of the bathyal sections throughout the 2000-km long continental margin are so persistent that it is difficult to presume that there were any barriers along it, as is the case with the Japanese-like margin. All these arguments taken together do not permit us to consent with the opinion [e.g., *Nikishin* et al., 1996; *Fokin*, 2001] that the Devonian rifting manifested along the passive margin under discussion was a result of back arc extension above a west dipping subduction zone in the Uralian ocean.

Yet, Devonian rifting certainly played an important role in the development of the margin. Some of the grabens formed by this event can be traced from the platform into the foreland fold and thrust belt (e.g., Pechora-Kozhva inverted graben in the Pechora basin) [*Chermnykh and Puchkov*, 1972; *Nikonov* et al., 2000]. Extension may also have governed the locatisation of the Kama-Kinel trough system, but definitive evidence for this has not been found. In places rifting was more extensive and the Ordovician to Middle Devonian shelf deposits grade upward to deep water deposits that extend into the Permian, signifying considerable reworking of the passive margin. This is most clearly expressed in Novaya Zemlya [*Korago* et al., 1989; *Timonin*, 1998], where Middle to Late Devonian rifting was accompanied by extensive basaltic volcanism. Rifting has also been suggested as one of several alternatives for the origin of the deep water Pricaspian basin, where sedimentation at a probable depth of 1500 m also started in the Late Devonian, and ended by the Permian (Kungurian) when the basin had been filled by a large thickness of evaporites [*Zonenshain* et al., 1990; *Brunet* et al., 1999] (Figures 8 and 9).

Based on the distribution of sediments and volcanics, it is difficult to relate the subsidence of the margin in the Lower Devonian (Emsian) and the formation of the Kama-Kinel trough system to the arc-continent collision that was taking place at that time. This collision affected only Southern to Middle Urals part of the margin, while marine transgression, subsidence, basaltic volcanism, and formation of basins took place along the whole length of the margin, as well as deep within the craton (Figures 8 and 9). In general, Late Devonian and Early Carboniferous (Visean) compressional reactivation of some structures of the Volga-Uralian province mentioned by *Fokin* et al. [2001] may be related with this process of collision, with the later reactivation and inversion of grabens and depressions taking place (Figure 10).

It is also important to clarify the question of the outline of the margin as it was during the time between rifting and collision, and the influence that this may have had on the subsequent collisional structure. It is well known that the paleo continental and paleo oceanic complexes in the structure of the Uralides are juxtaposed at the surface by the Main Uralian fault. However, the continental crust of Baltica can be traced in the subsurface some 25 to 50 km farther east, beneath the Magnitogorsk and Tagil arcs [*Echtler* et al., 1996; *Juhlin* et al., 1998; *Tryggvason* et al., 2001]. Nevertheless, the margin had its embayments and promontories, which are complementary now to salients and recesses of the orogen, and which played an important structural role. The idea itself is not new. *Karpinsky* [1919] suggested that the Ufimian and Bolshezemelian uplifts, formed by rigid blocks of the East European Craton, played an important role in development of the architecture of the Uralides. The same idea in much more detail was developed in the structural analysis of the Appalachians [*Thomas*, 1977]. An analogy can be made between the Appalachians/Ouachita and Ural/Pay-Khoy relationships. In both cases a sharp change in the trend of the margin had a significant influence on its collisional history and the resulting architecture of the orogen. For example, in the Uralides structures developed around the Ufimian and Bolshezemelian uplifts can be related to this phenomenon. As can the structures of Pay-Khoy and the Polar Urals (Figure 10).

Acknowledgments. D. Brown (as a co-editor), P. Ziegler and A. Nikishin (as reviewers) are thanked for their very helpful comments on the manuscript. This work has been supported by the MinUrals project (N1CA2-CT-2000-10011) This is a EUROPROBE Uralides and Variscides Project publication.

REFERENCES

Alvarez-Marrón, J., D. Brown, A. Pérez-Estaún, V. Puchkov and Y. Gorozhanina, Accretionary complex structure and kinematics during Paleozoic arc-continent collision in the Southern Urals, *Tectonophysics*, 325, 175–191, 2000.

Antoshkina, A. I., *Paleozoic Reefs of the Pechora Urals* (in Russian), 154 pp., Nauka, St. Petersburg, 1994.

Aristov, V. A., S. V. Ruzhentsev, K. E. Degtyarev, D. V. Borisenok and I. V. Latysheva, The Devonian stratigraphy of the Sakmara and Sakmara-Voznesensk zones of the Southern Urals (in Russian), in *General and Regional Questions of Geology*, 2, pp. 46–58, Geological Institute, Moscow, 2000.

Beloussov, V. V., *The Fundamental Principles of Geotectonics* (in Russian), 257 pp., Nedra, Moscow, 1975.

Belyakov, L. N., V. S. Yenokyan and V. A. Chermnykh, The Carboniferous sediments of Pay-Khoy and Vaigatch island (in Russian), in *Paleozoic Stratigraphy of the North-East of the European Part of the USSR*, pp. 33–52, Inst. of Geol., Syktyvkar, 1981.

Bobokhov, A. S. *The Endogenic Dynamic System of the Southern Uralian Island Arc* (in Russian), 181 pp., Nauka, Moscow, 1991.

Brown, D., V. Puchkov, J. Alvarez-Marrón and A. Pérez-Estaún, The structural architecture of the footwall to the Main Uralian fault, Southern Urals, *Earth Sci. Rev.*, 40, 125–147, 1996.

Brown, D., J. Alvarez-Marrón, A. Pérez-Estaún, Ye. Gorozhanina, V. Baryshev and V. Puchkov, Geometric and kinematic evolution of the foreland thrust and fold belt in the Southern Urals, *Tectonics*, 16, 551–562, 1997.

Brown, D., C. Juhlin, J. Alvarez-Marrón, A. Pérez-Estaún and A. Oslianski, Crustal-scale structure and evolution of an arc-continent collision zone in the Southern Urals, Russia, *Tectonics*, 17, 158–171, 1998.

Brown, D. and P. Spadea, Process of forearc and accretionary complex formation during arc-continent collision in the southern Ural Mountains, *Geology*, 27, 649–652, 1999.

Brunet, M.-F., Yu. A. Volozh, M. P. Antipov and L. I. Lobkovsky, The geodynamic evolution of the Precaspian Basin (Kazakhstan) along a north-south section, *Tectonophysics*, 313, 85–106, 1999.

Chermnykh, V. A. and V. N. Puchkov, Conclusion [in Russian), in *Atlas of the Lithological-Paleogeographic Maps of the Paleozoic and Mesozoic of the Northern Preuralian Region*, edited by V. A. Chermnykh, pp. 143–152, Nauka, Leningrad, 1972.

Chibrikova, V., Age and stratigraphy of the Zilair deposits of the Urals (in Russian), *Otechestvennaya Geologiya*, 12, 31–35, 1997.

Chibrikova, V. and V. A. Olli, The first occurrences of acritarchs in the metamorphic complex of Ural-Tau (Southern Urals) (in Russian), *Izvestiya of the Earth Sci. and Ecol. Branch*, Ac. Sci. Bashk., Ufa, 1, 42–48, 1997.

Chibrikova, V. and V. A. Olli, Exotic blocks in the Paleozoic sections of the Southern Urals and Northern Caucasus (in Russian), *Ezhegodnik-1997*, pp. 21–25, Inst. of Geol., Ufa, 1999.

Dedeev, V. A. and V. G. Getsen, *Riphean and Vendian of the European North of the USSR* (in Russian), 124 pp., Inst. Geol., Syktyvkar, 1987.

Didenko, A. N., S. A. Kurenkov and S. V. Ruzhentsev, *The Tectonic History of the Polar Urals* (in Russian), 191 pp., Nauka, Moscow, 2001.

Dembovsky, B. Ya., Z. P. Dembovskaya, M. L. Klyuzhina, A. S. Miklyayev, V. A. Nassedkina, V. N. Puchkov and V. V. Tereshko, *New Data on the Precambrian and Lower Paleozoic Stratigraphy of the Western Slope of the Northern Urals* (in Russian), edited by V. Puchkov and B. Chuvashov, 63 pp., Institute of Geol. and Geochem., Sverdlovsk, 1988.

Dembovsky, B. Ya., V. A. Nassedkina and M. L. Klyuzhina, *The Ordovician of the Cis-Polar Urals. Geology, Lithology, Stratigraphy* (in Russian), edited by V. Puchkov, 196 pp., Uralian Branch of the Ac. Sci. USSR, Sverdlovsk, 1990.

Echtler, H. P., M. Stiller, F. Steinhoff, C. M. Krawczyk, A. Suleimanov, V. Spiridonov, J. H. Knapp, Y. Menshikov, J. Alvarez-Marron and N. Yunusov, Preserved collisional crustal architecture of the Southern Urals — Vibroseis CMP-profiling, *Science*, 274, 224–226, 1996.

Elisseyev, A. I., Ya. E. Yudovich, A. A. Belyayev and G. F. Semenov, *Sedimentary Formations of Pay-Khoy and Prospects for New Raw Material* (in Russian), 50 pp., Komi Branch of Ac. Sci. USSR, Syktyvkar, 1984.

Evseev, K. P., The Paleozoic stratigraphy and facies of the Lemva structural zone of the Polar Urals (in Russian), *Vsesoyuz. Geol. Inst. Bull.*, 2, 24–39, 1960.

Fokin, P. A., A. M. Nikishin and P. A. Ziegler, Peri-Uralian and Peri-Paleo-Tethyan rift sections of the East-European craton, in *Peri-Tethys Memoir 6: Peri-Tethyan Rift/Wrench Basins and Passive Margins, Mem. Mus. Nat. Hist.*, 186, 347–368, 2001.

Gataulin, R. M., The exploration of small complex oil traps in Bashkortostan (in Russian), Thesis of Cand. of Sci. Referate, 20 pp., BashNIPINeft, Ufa, 1999.

Giese, U., U. Glasmacher, V. I. Kozlov, I. Matenaar, V. N. Puchkov, L. Stroink, W. Bauer, S. Ladage and R. Walter, Structural framework of the Bashkirian anticlinorium, SW Urals, *Geol. Rundsch.*, 87, 526–544, 1999.

Glasmacher, U. A., P. Reynolds, A. A. Alekseev, V. N. Puchkov, K. Taylor, V. Gorozhanin and R. Walter, ^{40}Ar/^{39}Ar Thermochronology west of the Main Uralian fault, southern Urals Russia, *Geologische Rundschau*, 87, 515–525, 1999.

Goldin, B. A., E. P. Kalinin and V. N. Puchkov, *Magmatic Formations of the Western Slope of the Northern Urals and Their Minerageny* (in Russian), edited by N. Yushkin, 214 pp., Inst. of Geol., Syktyvkar, 1999.

Gorozhanina, Ye. and V. Puchkov, The model of sedimentation in the forearc area of the Magnitogorsk island arc in the Late Devonian (in Russian), in *The Geology and Prospects of the Raw Material Base of the Bashkortostan*, pp. 5–12, Inst. of Geology, Ufa, 2001.

Ivanov, K. S., L. A. Karsten and G. Maluski, The first data on the age of the subductional (eclogite-glaucophane schist) metamorphism in the Cis-Polar Urals (in Russian), in *Subduction Paleozones: Tectonics, Magmatism, Metamorphism and Sedimentation*, pp. 121–128, Inst. of Geol. and Geochem., Ekaterinburg, 2000.

Ivanov, K. S. and V. N. Puchkov, *Geology of the Sakmara Zone of the Urals (New Data)* (in Russian), 86 pp., Sverdlovsk, 1984a.

Ivanov, K. S. and V. N. Puchkov, The Devonian basaltoid volcanism of the Zilair-Lemva zone of the Urals (in Russian), in *The Geology of the Conjunction Zone Between the Urals and the East-European Platform*, pp. 41–50, Inst. of Geol. and Geochem., Sverdlovsk, 1984b.

Ivanov, S. N., V. N. Puchkov, K. S. Ivanov, G. I. Samarkin, I. V. Semenov, A. I. Pumpyansky, A. M. Dymkin, Yu. A. Poltavets, A. I. Rusin and A. A. Krasnobayev, *The Formation of the Earth's Crust of the Urals* (in Russian), 248 pp., Nauka, Moscow, 1996.

Juhlin, C., M. Friberg, H. Echtler, A. G. Green, J. Ansorge, T. Hismatulin and A. Rybalka, Crustal structure of the Middle Urals: Results from the (ESRU) Europrobe Seismic Reflection Profiling in the Urals Experiments, *Tectonics*, 17, 710–725, 1998.

Karpinsky, A. P., On the tectonics of the European Russia (in Russian), *Izvestia of Russian Acad. of Sci.*, 573–590, 1919.

Karsten, L. A., V. N. Puchkov and N. M. Zaslavskaya, The Geology of the Main Uralian fault in the Cis-Polar Urals (in Russian), *Izvestiya of the USSR Acad. of Sci, Ser. Geol.*, 4, 133–136, 1989.

Kamaletdinov, M. A., *The Nappe Structures of the Urals* (in Russian), 230 pp., Nauka, Moscow, 1974.

Khain, V. E. and Yu. G. Leonov (Editors), International Tectonic map of Europe, 3rd Edition, VSEGEI, St. Petersbourg, 1998.

Kheraskov, N. P., *Tectonics and Formations* (in Russian), 255 pp., Nauka, Moscow, 1967.

Khvorova, I. V., T. A. Voznesenskaya, B. P. Zolotarev, M. N. Ilyinskaya and S. V. Ruzhentsev, *Formations of the Sakmara Allochthon* (in Russian), 232 pp., Nauka, Moscow, 1978.

Knipper, A. L., *The Tectonics of the Baikonur Synclinorium* (in Russian), 205 pp., Nauka, Moscow, 1963.

Kondiain, O. A., Structural-facial character of the Northern Urals at the Middle Devonian and Late Devonian-Tournaisian Stages of Development (in Russian), *Proceedings of the Vsesoyuz. Geol. Inst.*, 92, 22–34, 1963.

Korago, Ye. A., G. N. Kovaleva and G. N. Trufanov, Formations, tectonics and geological history of the Novozemelian Cimmerides (in Russian), *Geotektonika*, 6, 40–61, 1989.

Krinitsky, D. D. and V. M. Krinitskaya, On the discovery of the Silurian deposits among the ancient formations of the western slope of Ural-Tau Range, Southern Bashkiria (in Russian), *Materials on the Geology and Deposits of the Southern Urals*, 4, 54–65, 1965.

Maksimov, S. P. (Ed.), *Geology and Oil and Gas Deposits of the Volga-Urals Province* (in Russian), 800 pp., Nedra, Moscow, 1970.

Mirchink, M. F., Oil prospects and direction of geological exploration in the Kama-Kinel system of depression, *Petroleum Geol.*, 9, 419–494, 1965.

Nalivkin, D. V. (Ed.), *Atlas of the Lithological-Paleogeographic Maps of the Russian Platform, Proterozoic and Paleozoic* (in Russian), Moscow, 1960.

Nikishin, A. M., P. A. Ziegler, R. A. Stephenson, S. A. P. L. Cloetingh, A. V. Furne, P. A. Fokin, A. V. Ershov, S. N. Bolotov, M. V. Korotaev, A. S. Alekseev, V. I. Gorbachev, A. Lankreuer, E. Yu. Bembinova and V. I. Shalimov, Late Precambrian to Triassic history of the East-European craton: dynamics of sedimentary basin evolution, *Tectonophysics*, 268, 23–63, 1996.

Nikonov, N. I., V. I. Bogatski, A. V. Martynov, Z. V. Larionova, V. M. Laskin, L. V. Galkina, E. G. Dovhikova, O. L. Ermakova, P. K. Kostygova, T. I. Kuranova, K. A. Moskalenko, Yu. A. Pankratov, E. L. Petrenko, E. V. Popova, A. I. Surina and G. A. Shabanova, *Timan-Pechora Sedimentary Basin. Atlas of Geological Maps* (in Russian), 64 sheets, Timan-Pechora Research Center, Ukhta, 2000.

Ostaschenko, B. A., *Petrology and Ores of the Central Pay-Khoy Basaltoid Complex* (in Russian), 112 pp., Nauka, Leningrad, 1979.

Petrov, G. A. and V. N. Puchkov, The Main Uralian fault in the Northern Urals (in Russian), *Geotektonika*, 1, 25–37, 1994.

Puchkov, V. N., Paleotectonics of the Lemva zone (in Russian), *Geotectonics*, 6, 342–345, 1973.

Puchkov, V. N., *Structural Connections Between the Cis-Polar Urals and Adjacent Part of the East-European Platform* (in Russian), 208 pp., Nauka, Leningrad, 1975.

Puchkov, V. N., *Bathyal Complexes of the Passive Margins of Geosynclines* (in Russian), 260 pp., Nauka, Moscow, 1979.

Puchkov, V. N., The occurrences of Devonian conodonts at the western slope of the Urals and their importance for the stratigraphy of the Lemva-type Paleozoic (in Russian), in *Conodonts of the Urals and Their Stratigraphic Significance*, edited by G. Papulov and V. Puchkov, pp. 33–52, Inst. Geol. and Geochem., Sverdlovsk, 1979b.

Puchkov, V. N., The place of intraplate events in the history of foldbelts (in Russian), in *Intraplate Events in the Earth's Crust*, pp. 167–175, Nauka, Moscow, 1988.

Puchkov, V. N., New data on the geology of the Sub-Kraka allochthonous complexes (Southern Urals) (in Russian), in *Ezhegodnik-1994*, pp. 3–9, Inst. of Geol., Ufa, 1995.

Puchkov, V. N., The origin of the Ural-Novozemelian foldbelt as a result of an uneven, oblique collision of continents (in Russian), *Geotektonika*, 5, 66–75, 1996.

Puchkov, V. N., Structure and geodynamics of the Uralian orogen, in *Orogeny Through Time*, edited by J.-P. Burg and M. Ford, *Geo. Soc. Spec. Pub.*, 121, 201–234, London, 1997.

Puchkov, V. N., *Paleogeodynamics of the Southern and Middle Urals* (in Russian), 146 pp., Dauria, Ufa, 2000.

Puchkov, V. N., Did the Late Vendian supercontinent exist?, in *Assembly and Break-up of Rodinia Supercontinent*, edited by E. Sklyarov, pp. 172–178, Inst. of Earth's Crust, Irkutsk, 2001.

Puchkov, V. N., V. N. Baryshev and V. N. Pazukhin, New data on the stratigraphy of the terrigenous cherty Devonian at the western slope of the Bashkirian Urals (in Russian), *Ezhegodnik-1996*, pp. 24–31, Inst. of Geol., Ufa, 1998.

Puchkov, V. N. and K. S. Ivanov, *Geology of Allochthonous Bathyal Complexes of the Ufimian Amphitheatre* (in Russian), 36 pp., Inst. Geol. and Geochem., Sverdlovsk, 1982.

Puchkov, V. N. and K. S. Ivanov, Pelagic "knotty" limestones at the western slope of the Urals (in Russian), *Bull. Mosc. Soc. Nat. Researchers*, 2, 59–68, 1985.

Puchkov, V. N. and K. S. Ivanov, On the stratigraphy of the Upper Devonian and Lower Carboniferous sediments of the Sakmara zone (in Russian), in *The New Nata on the Geology of the Urals*, edited by V. Puchkov, pp. 84–93, Inst. Geol. and Geophys., Sverdlovsk, 1987.

Puchkov, V. N., K. S. Ivanov and V. A. Nassedkina, The first data on the Ordovician cherty formations in the western slope of the Urals (in Russian), in *New Data on the Geology of the Urals, West Siberia and Kazakhstan*, pp. 16–20, Institute of Geology and Geochemistry, Sverdlovsk, 1990.

Puchkov, V. N., A. Pérez-Estaún, D. Brown and J. Alvarez-Marron, The marginal fold and thrust belt of an orogen: structure and origin (at the example of the Bashkirian Urals) (in Russian). *Herald of Geol., Geochem., Geophys. and Mining Branch of RAS*, 1(3), 70–99, 1998b.

Rogov, V. S., E. I. Galitskaya, V. I. Davydov and A. V. Popov, New data on the stratigraphy of manganese-bearing sediments of the Permian and Carboniferous of Pay-Khoy (in Russian), *Soviet Geol.*, 6, 59–68, 1988.

Ronkin, Yu. L., B. A. Kaleganov, E. V. Pushkarev and O. P. Lepikhina, On the problem of isotopic age of the ophiolites of the Urals (in Russian), in *The Geology and Raw Materials of the Western Urals*, pp. 64–66, Ekaterinburg, 1997.

Ruzhentsev, S. V., *The Marginal Ophiolitic Allochthons* (in Russian), 170 pp., Nauka, Moscow, 1976.

Ruzhentsev, S. V. and V. A. Aristov, The new data on the geology of the Polar Urals (in Russian), in *The Urals: Fundamental Problems of Geodynamics and Stratigraphy*, pp. 25–41, Nauka, Moscow, 1998.

Ruzhentsev, S. V., V. A. Aristov and P. M. Kucherina, The Upper Devonian-Carboniferous ophiolites and a bathyal series of the Polar Urals (in Russian), *Doklady Ac. Sci.*, 365, 802–805, 1999.

Ryazantsev, A. V., S. V. Dubinina and D. V. Borisenok, The tectonic approachment of the Paleozoic complexes in the Devonian accretionary structure of the Sakmara zone of the Southern Urals (in Russian), in *General and Regional Questions of Geology*, 2, pp. 25–45, Geol. Inst., Moscow, 2000.

Saldin, V. A., Upper Paleozoic terrigenous formations of the Lemva zone of the Urals (in Russian). Candidate of Sci. thesis, Inst. of Geol., Syktyvkar, 1996.

Saldin, V. A., Stratigraphic position of the Ray-Iz svita in the Polar Urals (in Russian), in *Geology of the European North of Russia*, 3, pp. 29–37, Inst. of Geol., Syktyvkar, 1999.

Savelieva, G. N., A. Ya. Sharaskin, A. A. Savelyev, A. L. Knipper, P. Spadea and L. Gaggero, Ophiolites of the conjugation zone between the southern Uralides and the periphery of the East-European continent (in Russian), in *The Urals: Fundamental Problems of the Geodynamics and Stratigraphy*, pp. 93–118, Nauka, Moscow, 1998.

Seravkin, I. B., A. M. Kosarev and D. N. Salikhov, *Volcanism of the Southern Urals* (in Russian), 197 pp., Nauka, Moscow, 1992.

Shacky, V. S., E. Iagoutz and O. A. Kozmenko, Sm-Nd dating of a high-baric metamorphism of the Maksyutovo complex (Southern Urals) (in Russian), *Doklady of Academy of Sci.*, 352, 812–815, 1997.

Shacky, V. S., V. A. Simonov, E. Iagoutz, O. A. Kozmenko and S. A. Kurenkov, New data on the age and paleogeo-dynamic conditions of origin of the Polar Urals eclogites (in Russian), in *Tectonics, Geodynamics and Processes of Magmatism and Metamorphism*, II, pp. 296–298, GEOS, Moscow, 1999.

Shishkin, M. A., The tectonics of the southern part of the Lemva zone (Polar Urals) (in Russian), *Geotektonika*, 3, 86–95, 1989.

Sinclair, H. D., Tectonostratigraphic model for underfilled peripherial foreland basins: an Alpine perspective, *Bull. Geol. Soc. Am.*, 129, 324–343, 1997.

Smirnov, G. A., *The Materials for the Paleogeography of the Urals. The Visean Stage* (in Russian), 120 pp., Institute of Geology and Mining, Sverdlovsk, 1957.

Russian Stratigraphic Committee, *Stratigraphic Charts of the Urals (Precambrian, Paleozoic)* (in Russian), schemes 151, explanation notes 152 pp., Ekaterinburg, 1993.

Spadea, P., M. D'Antonio, A. Kosarev, Y. Gorozhanina and D. Brown, Arc-Continent Collision in the Southern Urals: Petrogenetic Aspects of the Fore-Arc Complex, this volume.

Tevelev, A. V., I. A. Kosheleva and A. V. Ryazantsev, The composition and structural position of the Ordovician cherty basalt complexes of the Southern Urals (in Russian), in *General and Regional Questions of Geology*, 2, pp. 25–45, Geol. Inst., Moscow, 2000.

Timonin, N. I., *The Pechora Plate: The Geological History in the Phanerozoic* (in Russian), edited by V. Puchkov, 240 pp., Uralian branch of Russian Ac. Sci., Ekaterinburg, 1998.

Torsvik, T. H., M. A. Smethurst, J. G. Meert, R. Van der Voo, W. S. McKerrow, M. D. Brasier, B. A. Sturt and H. J. Walderhaug, Continental break-up and collision in the Neoproterozoic and Paleozoic — a tale of Baltica and Laurentia, *Earth Science Reviews*, 40, 229–258, 1996.

Thomas, W. A., Evolution of Appalachian-Ouachita salients and recesses from reentrants and promontories in the continental margin, *Am. J. of Sci.*, 277, 1233–1278, 1977.

Timonin, N. I. and V. V. Yudin, *The Tectonics of Pay-Khoy* (in Russian), 33 pp., Komi Branch of Russian Ac. Sci., Syktyvkar, 1999.

Trümpy, R., Paleotectonic evolution of the Central and Western Alps, *Bull. Geol. Soc. Am.*, 71, 6, 1960.

Tryggvason, A., D. Brown and A. Perez-Estaun, Crustal architecture of the southern Uralides from true amplitude processing of the URSEIS vibroseis profile, *Tectonics*, 20, 1040–1052, 2001.

Zhivkovich, A. E. and P. A. Chekhovich, *Paleozoic Formations and Tectonics of the Ufimian Amphitheatre* (in Russian), 184 pp., Moscow, Nauka, 1985.

Ustritsky, V. I., On the facial zonation of the Middle Paleozoic sediments of Pay-Khoy and the northern part of the Polar Urals (in Russian), *Transactions of NIIGA*, 123, 41–60, 1961.

Voinovsky-Kriger, K. G., Two complexes of the Paleozoic at the western slope of the Polar Urals (in Russian), *Soviet Geol.*, 6, 27–44, 1945.

Willner, A., T. Ermolaeva, Y. Gorozhanina, V. Puchkov, U. Kramm and R. Walter, Surface signals of an Arc-Continent Collision with Exhumation of High-Pressure Rocks: the Detritus of the Upper Devonian Zilair Formation in the Southern Urals, this volume.

Wilson, M., J. Wijbrans, P. A. Fokin, A. M. Nikishin, V. I. Gorbachev and B. P. Nazarevich, $^{40}Ar/^{39}Ar$ dating, geochemistry and tectonic setting of Early Carboniferous dolerite sills in the Pechora basin, foreland of the Polar Urals, *Tectonophysics*, 313, 107–118, 1999.

Yazeva, R. G. and V. V. Bochkarev, *The Geology and Geodynamics of the Southern Urals* (in Russian), 204 pp., Institute of Geology and Geochemistry, Ekaterinburg, 1998.

Yudina, A. B. and V. N. Puchkov, The occurrences of conodonts in the Devonian and Carboniferous sediments of the shaly zone of Pay-Khoy and their stratigraphic significance (in Russian), in *Stratigraphy and Paleogeography*, pp. 56–57, Inst. of Geology, Syktyvkar, 1987.

Yenokyan, V. S., Silurian and Lower Devonian deposits of the Northwestern part of the Yugor Peninsula and the Petchora Sea Islands (in Russian), *Uchenye Zapiski NIIGA*, 30, pp. 5–25, 1970.

Yudovich, Ya. E., A. A. Belyaev and M. P. Ketris, *Geochemistry and Ore Genesis in the Black Shales of Pay-Khoy* (in Russian), 364 pp., Nauka, Ekaterinburg, 1998.

Zonenshain, L. P., *The Doctrine of Geosynclines and its Application to the Central Asian Folded Belt* (in Russian), 235 pp., Nedra, Moscow, 1972.

Zonenshain, L. P., M. I. Kuzmin and L. M. Natapov, Geology of USSR: a plate-tectonic synthesis, *Am. Geophys. Union, Geodynamic Series*, 21, 242 pp., 1990.

V. Puchkov, Ufimian Geoscience Center, Russian Academy of Sciences, ul. Karl Marx 16/2, Ufa 45000, Bashkiria, Russia (puchkv@anrb.ru)

The Crustal Architecture of the Southern and Middle Urals From the URSEIS, ESRU, and Alapaev Reflection Seismic Surveys

D. Brown[1], C. Juhlin[2], A. Tryggvason[1], D. Steer[3], P. Ayarza[4], M. Beckholmen[2], A. Rybalka[5] and M. Bliznetsov[5]

The Urals Seismic Experiment and Integrated Studies (URSEIS), Euro-probe's Seismic Reflection Profiling in the Urals (ESRU), and reprocessed Russian reflection/refraction seismic surveys have shown the known Uralides to be bivergent, with a crustal root along the central volcanic axis of the orogen. In the Southern (URSEIS) and Middle (ESRU and Alapaev) Urals the East European Craton crust thickens eastward from ~40 km to ~48 km, and is imaged by sub-horizontal to east-dipping reflectivity that can be related to its Paleozoic and older evolution. The suture zone between the East European Craton and the accreted terranes, the Main Uralian fault, is poorly imaged in the URSEIS section, but in the ESRU and Alapaev sections it is imaged as an abrupt change from a zone of east-dipping reflectivity that extends from the surface into the middle crust. East of the Main Uralian fault, the Magnitogorsk (Southern Urals) and the Tagil (Middle Urals) volcanic arcs display moderate to weak upper crustal reflectivity, and diffuse middle to lower crust reflectivity. The Moho beneath both arc complexes is poorly imaged in the reflection data, but based on refraction data is interpreted to be at 50 to 55 km depth. East of the arc complexes, the Uralide structural architecture is dominated by a wide zone of anastomosing strike-slip faulting into which numerous syntectonic Late Carboniferous and Permian granitoids intruded. This area is imaged in the seismic sections as clouds of diffuse reflectivity interspersed with, or cut by sharp, predominantly west-dipping reflections. In the Southern and Middle Urals, west-dipping

[1]Institute Jaume Almera, Barcelona, Spain

[2]Department of Earth Sciences, Uppsala University, Uppsala, Sweden

[3]University of Akron, Akron, US

[4]Department of Geology, University of Salamanca, Salamanca, Spain

[5]Bazhenov Geophysical Expedition, Scheelite, Russia

Mountain Building in the Uralides: Pangea to the Present
Geophysical Monograph 132
Copyright 2002 by the American Geophysical Union
10.1029/132GM03

reflectivity of the Trans-Uralian zone extends from the middle crust into the lower crust where it appears to merge with the Moho. The boundaries and internal faults of the strike-slip fault system are well marked by magnetic anomalies, allowing them to be correlated between the seismic sections.

1. INTRODUCTION

The Uralide orogen developed during the Late Paleozoic as the continental margin of the former East European Craton was subducted eastward (current coordinates) beneath a chain of intra-oceanic island arcs (Magnitogorsk and Tagil) during the Late Devonian and Early Carboniferous [*Puchkov*, 1997; *Brown and Spadea*, 1999]. This was followed by closure of the paleo-Uralian ocean basin to the east and accretion of volcanic arcs and continental crust along the eastern margin of the growing orogen from the Late Carboniferous through to the Late Permian-Early Triassic [e.g., *Zonenshain* et al., 1984, 1990; *Puchkov*, 1997]. In the south, the Uralides began to form during the Middle to Late Devonian as the continental margin of the former East European Craton subducted eastward beneath the Magnitogorsk island arc. As a consequence, an accretionary complex was developed and emplaced over the subducting slab, and by the Early Carboniferous the arc was sutured to the continental margin along the Main Uralian fault zone [e.g., *Puchkov*, 1997; *Brown* et al., 1998; *Brown and Spadea*, 1999]. A deformation hiatus until the Late Carboniferous followed along the eastern margin of the East European Craton. Farther north exposure is poor and tectonic interpretations are heavily reliant on geophysical and borehole data. It is generally accepted, however, that the Tagil volcanic arc collided with the eastward subducting East European Craton in the Early Carboniferous [e.g., *Puchkov*, 1997]. This interpretation is based on the presence of Early Carboniferous (Visean) flysh-type deposits with an eastern provenance overlying bathyal deposits in the upper reaches of the Pechora River [*Puchkov*, this book].

From the Late Carboniferous through to the Late Permian-Early Triassic, the paleo-Uralian ocean basin to the east closed, as island arcs and continental fragments were accreted to the eastern flank of the developing Uralides. It is not clear to what extent terrane amalgamation, deformation and metamorphism occurred within these fragments prior to accretion. Closure of the paleo-Uralian ocean basin was accompanied by westward thrusting of the East European Craton Precambrian basement and its late Paleozoic platform cover to form a foreland thrust and fold belt and a largely Permian-age foreland basin [*Kamaletdinov*, 1974; *Brown* et al., 1997].

During late stages of the collision, extensive wrench or transpressive faulting appears to have dominated along the central axis of the orogen, fragmenting the Tagil arc and juxtaposing metamorphic terranes within the East and Trans-Uralian zones [*Ayarza* et al., 2000a; *Friberg* et al., 2000; *Hetzel and Glodny*, 2002]. This stage of orogenic development was accompanied by widespread crustal and mantle melting and the intrusion of numerous plutons [*Bea* et al., this book, *Gerdes*, 2002] (Plate 1). The Middle Urals was also affected by Mesozoic to Tertiary deformation that resulted in folding and thrusting of Lower Triassic deposits in the Tagil zone [*Puchkov*, 1997] and the development of the West Siberian Basin. The easternmost zones (East and Trans-Uralian zones) are widely covered by Mesozoic and Cenozoic sediments.

The Uralides are preserved, relatively intact, within Eurasia and provide a rare opportunity to investigate the crustal architecture of an entire Paleozoic orogen. With this in mind, nearly 1000 km of new deep reflection seismic data have been acquired in the Southern (URSEIS) and Middle (ESRU) Urals since 1993 under the auspices of EUROPROBE [e.g., *Berzin* et al., 1996; *Ecthler* et al., 1996; *Knapp* et al., 1996; *Steer* et al., 1998; *Juhlin* et al., 1996, 1998], and c. 500 km of shallow Russian data (Alapaev, R114 and R115) have been reprocessed [e.g., *Steer* et al., 1995; *Brown* et al., 1998] (Plate 1). In addition, two refraction/wide-angle reflection profiles (UWARS and URSEIS) have been acquired [*Thouvenot* et al., 1995; *Carbonell* et al., 1996, 2000]. These experiments have shown the Uralide crust to be highly reflective, to have what appears to be a bivergent structural architecture, and to have clear reflection characteristics for each tectonic unit [*Echtler* et al., 1996; *Knapp* et al., 1996; *Steer* et al., 1998; *Juhlin* et al., 1998; *Friberg*, 2000]. The Moho appears to have distinct characteristics from west to east across the orogen that can be related to the different tectonic units and events [*Steer* et al., 1998; *Juhlin* et al., 1998]. Despite similarities, the URSEIS, ESRU and Alapaev sections display crustal reflectivity patterns that suggest the Uralide crust is different in the Southern Urals than in the Middle Urals [e.g., *Steer* et al., 1995, 1998; *Juhlin* et al., 1998; *Ayarza* et al., 2000], indicating that different late- or post-tectonic processes may have been active in the two areas. The aim of this paper is to integrate

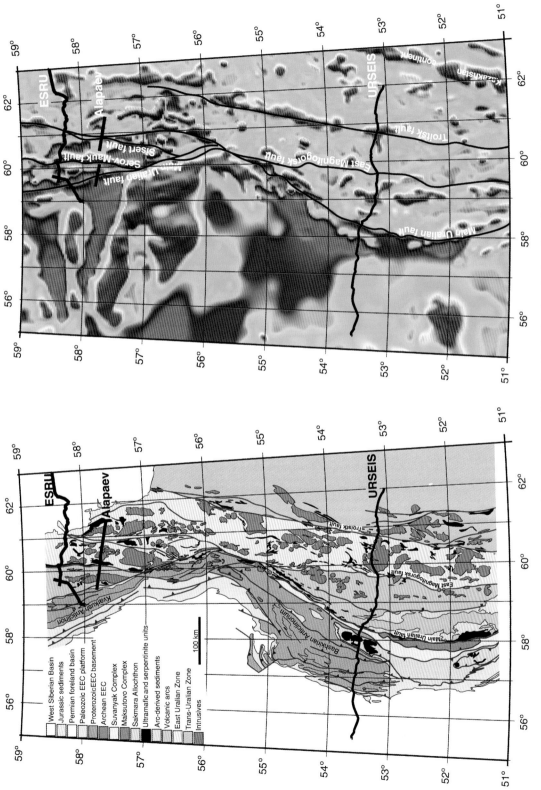

Plate 1. (a) Simplified geological map of the Southern and Middle Urals showing the location of the URSEIS, ESRU and Alapaev reflection seismic surveys. (b) Magnetic map of the Southern and Middle Urals showing the location of the seismic surveys and the main tectonic boundaries that they cross.

the sesimic data with the known surface geology, placing special emphasis on the structures that bound the major tectonic units, and to develop a unified crustal-scale model for the architecture of the Uralides from the URSEIS section in the south and the ESRU section in the north.

2. GEOLOGICAL FRAMEWORK

A generally accepted division of the Uralides involves a number of longitudinal zones that are largely based on the ages and paleogeography of the dominant rocks within them [e.g., *Ivanov* et al., 1975; *Khain*, 1985; *Puchkov*, 1991]. In this paper, however, we use a modified subdivision that includes the recognition of tectonic units. This subdivision consists of a foreland thrust and fold belt composed of East European Craton rocks, the accreted Magnitogorsk and Tagil island arcs, the East Uralian zone, and the Trans-Uralian zone. Each of these tectonic units and their bounding faults is described below. Additionally, the area of the Uralides discussed in this paper has been traditionally divided geographically into the Southern and Middle Urals at about 56° N.

2.1. The Foreland Thrust and Fold Belt

South of approximately 53° N, the foreland thrust and fold belt consists of a narrow (~20 km) zone of upright to west-verging folds and thrusts that record only minor amounts of shortening. Between 53° N and 56° N it forms a ~120 km wide west-verging, basement involved thrust stack that records ~20 km of Uralide shortening [*Brown* et al., 1997; *Perez-Estaun* et al., 1997]. Much of this part of the foreland thrust and fold belt is composed of the gently south-plunging, Precambrian-cored Bashkirian Anticlinorium (Plate 1) that was deformed and metamorphosed during the Late Proterozoic [*Shatsky*, 1963; *Puchkov*, 1993], and which was only mildly deformed and metamorphosed during the Paleozoic [*Glasmacher* et al., 1999; *Geise* et al., 1999]. In the Southern Urals, an accretionary complex, related to the collision of the Magnitogorsk island arc with the East European Craton, overlies the foreland thrust and fold belt (Plate 1) [*Brown* et al., 1998; *Brown and Spadea*, 1998; *Alvarez-Marron* et al., 2000]. North of 56° N, the foreland thrust and fold belt is a narrow, N-S trending, west-verging basement-involved thrust stack measuring ~50 to 75 km in width (Plate 1). It extends northward into the linear thrust belt of the Middle Urals, where it is cored in the east by the Kvarkush Anticlinorium. Rocks in the Kvarkush Anticlinorium were deformed and metamorphosed during the Precambrian, and only mildly

reworked during the Paleozoic. The amount of shortening in the foreland thrust and fold belt has not been calculated for this part of the orogen.

2.2. The Main Uralian Fault Zone

The Main Uralian fault zone (the arc-continent suture) in the Southern Urals is an up to 10 km wide mélange containing material that was tectonically eroded from the volcanic arc, including a number of mantle fragments [e.g., *Savelieva* et al., 1997]. In the northernmost part of the Southern Urals (~55° N), it was intruded by an undeformed phase of the Syrostan batholith (Plate 1), dated at 327 ± 2 [*Montero* et al., 2000]. Intrusion of the batholith marked the end of tectonic activity along the Main Uralian fault in this area. In the Middle Urals, the Main Uralian fault is poorly exposed. *Juhlin* et al. [1998] defined it as a 10 km wide zone containing an underlying thrust stack that contains strongly deformed and metamorphosed sandstones, quartzites, and quartz-mica schists of apparent East European Craton affinity. Here we define it to be the contact between the East European Craton and the Tagil arc, and the metamorphic thrust stack to be in its footwall (see also *Knapp* et al. [1998]). *Ayarza* et al. [2000a] suggest that there are significant differences in the evolution of the Main Uralian fault between the Southern and Middle Urals, and in the latter area it may be extensively reworked.

2.3. The Magnitogorsk and Tagil Volcanic Arcs

In the Southern Urals, the Silurian to Late Devonian Magnitogorsk volcanic arc is composed of a complete island arc volcanic sequence that begins with Emsian boninite-bearing arc-tholeiites in the forearc region, followed by Emsian to Givetian arc-tholeiite to calc-alkaline volcanism typical of a mature arc [*Seravkin* et al., 1992; *Brown and Spadea*, 1998; *Spadea* et al., this book]. These volcanic units form the basement on which up to 5000 m of westward-thickening, Frasnian- to Famennian-age forearc basin volcanoclastic sediments were deposited [*Maslov* et al., 1993; *Brown* et al., 2001]. Lower Carboniferous shallow water carbonates unconformably overlie the arc edifice. Locally, Lower Carboniferous granitoids intrude the arc. Deformation in the Magnitogorsk volcanic arc is low, with only minor, open folding and minor thrusting [*Brown* et al., 2001]. The metamorphic grade barely exceeds seafloor metamorphism.

In the Middle Urals, the Middle Silurian Tagil arc has also been interpreted to be an intra-oceanic island arc [*Yazeva and Bochkarev*, 1996; *Bosch* et al., 1997] with predominantly Silurian andesitic magmatism in the west and Lower Devonian trachytes and volcanoclastics in the east. These volcanic and volcanoclastic rocks are overlain by 2000 meters of Lower and Middle Devonian limestone that in the east is intercalated with calc-alkaline volcanics [*Antsigin* et al., 1994; *Yazeva and Bochkarev*, 1994]. The Tagil arc forms an open synformal structure [e.g., *Bashta* et al., 1990; *Ayarza* et al., 2000b] that has been metamorphosed to lower greenschist facies.

2.4. The East Magnitogorsk–Serov-Mauk Fault System

The Magnitogorsk arc is structurally juxtaposed against the East Uralian zone along a mélange zone that has been named the East Magnitogorsk fault [*Ayarza* et al., 2000a]. It has tentatively been correlated with the Serov-Mauk fault, which sutures the Tagil arc to the East Uralian zone in the Middle Urals. The Serov-Mauk fault is a strike-slip fault zone that can be traced as a prominent magnetic anomaly throughout the Middle Urals (Plate 1). Along the entire length (> 700 km) of the East Magnitogorsk–Serov-Mauk fault system there is a significant jump in metamorphic grade from the volcanic arcs in the west to the East Uralian zone in the east.

2.5. The East Uralian Zone

The East Uralian zone is composed of deformed and metamorphosed Precambrian and Paleozoic continental-type crust and island arc fragments [e.g., *Puchkov*, 1997, 2000; *Friberg*, 2000]. It was intruded by numerous granitoid batholiths and subordinate diorite and gabbro intrusions during the Carboniferous and the Early Permian [*Sobolev* et al., 1964; *Fershtater* et al., 1997; *Bea* et al., 1997, this book]. The regional metamorphic grade ranges from greenschist to granulite facies. In the northern part of the Southern Urals and in the Middle Urals a number of metamorphic complexes have been identified. These include the Salda, Murzinka-Adui, Sisert, Krasnogvardeiskii, Petrokamensk, and Alapaevsk complexes [e.g., *Antsigin* et al., 1994], which are juxtaposed along tectonic contacts that have been extensively reworked by a roughly north-south striking strike-slip fault system. Many of the granitoid batholiths intruded syntectonically into this fault system.

The eastern contact of the East Uralian zone is only known in the Southern Urals, where it is a mélange containing local relics of harzburgite. In the area crossed by the URSEIS section, the mélange is intruded by a late, undeformed phase of the Dzhabyk granite that has been dated at 291 ± 4 Ma [*Montero* et al., 2000]. The late orogenic, dextral strike-slip Troitsk fault lies within the mélange.

2.6. The Trans-Uralian Zone

The Trans-Uralian zone is not well known due to its poor exposure; it only outcrops in the Southern Urals. The best known units are Devonian and Carboniferous calc-alkaline volcano-plutonic complexes which are composed predominately of volcanoclastics and lava flows that are intruded by co-magmatic gabbro-diorite and diorite plutons [e.g., *Puchkov*, 1997, 2000]. Ophiolite units and high pressure rocks have also been reported [*Puchkov*, 2000]. The volcano-plutonic complexes are overlain by terrigenous red-beds and evaporites. Deformation has not been well studied, although it appears that the Devonian and Lower Carboniferous units are affected by open to tight folds.

3. THE SEISMIC PROFILES

The acquistion and processing parameters for the URSEIS section can be found in *Tryggvason* et al. [2001], for the ESRU section in *Juhlin* et al. [1998] and *Friberg* et al. [2000], and for the Alapaev section in *Steer* et al. [1995]. All sections have been presented in earlier papers, except for the ESRU99 segment (the easternmost ~70 km shown in Plate 3) which is unpublished up to now. Important to note for this paper is that the URSEIS data were acquired with two source types, vibroseis [*Echtler* et al., 1996] and explosives [*Steer* et al., 1998]. The vibroseis sections presented in this study image the upper crust much better than the explosive ones (due to the higher fold), but the signal did not penetrate to the Moho along the entire URSEIS section with the vibroseis source. The line drawings shown here (Plates 1 to 3) were generated from coherency filtered stacks with the events picked automatically and then depth migrated. The migration velocities used are based upon refraction/wide-angle data presented in *Carbonell* et al. [1996] for the Southern Urals and *Juhlin* et al. [1996] for the Middle Urals.

3.1. The Foreland Thrust and Fold Belt

In the URSEIS section (Plate 2), from kilometer 0 to about kilometer 50, the foreland thrust and fold belt is characterized by a ~20 km thick zone of continuous, subhorizontal reflectivity that is related to the Riphean to Permian sediments. A thin band of weak, openly concave upward reflectivity at ~20 km depth is interpreted to mark the top of the Archean crystalline basement. Below this, the crust is transparent and the Moho is not imaged. From near kilometer 50 to the Main Uralian fault at about kilometer 146, the foreland thrust and fold belt is characterized by patchy, east and west dipping reflectivity in the upper and middle crust. From kilometer 117 to the

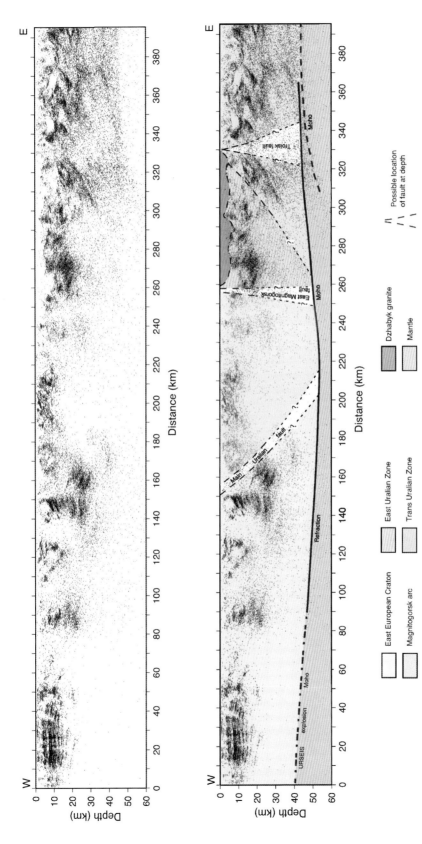

Plate 2. Uninterpreted and interpreted line drawings of the coherency filtered, depth migrated URSEIS vibroseis data. See Figure 1 for location. The colors represent the different tectonic units that make up the Uralide crust. The main suture zones that bind these tectonic units together have been interpreted to end at the Moho. Their exact location at depth cannot be unambiguously interpreted, and have therefore been shown as a zone in which they may possibly occur. The location of the URSEIS explosion source reflection Moho [*Steer* et al., 1998] and the Refraction Moho [*Carbonell* et al., 1998] are shown along with the Moho imaged in this data set.

Main Uralian fault the uppermost ~5 km of the section images the accretionary complex, the base of which is seen as a thin band of reflectivity that dips moderately eastward. Beneath and to the west of the accretionary complex this reflectivity is related to polydeformed and variably metamorphsed Proterozoic rocks within the Bashkirian Anticlinorium. The lower crust is nearly transparent and the Moho is not imaged.

The ESRU (Plate 3) and the Alapaev (Plate 4) sections do not extend westward across the entire foreland thrust and fold belt in the Middle Urals. From kilometer 0 to the Main Uralian fault at about kilometer 40 the ESRU section mostly crosses the polydeformed and variably metamorphosed Proterozoic rocks within the Kvarkush Anticlinorium. At the western end of the section, a band of subhorizontal reflections at ~5 to ~7 km depth may represent the easternmost edge of undeformed Riphean sediments. Below this, the East European Craton is represented by diffuse reflectivity. Eastward, the Kvarkush Anticlinorium in both the ESRU and the Alapaev sections is imaged as a zone of diffuse reflectivity in the upper crust, with strong, moderately east-dipping reflectivity in the middle and lower crust (ESRU only) that appears to be cross cut by moderately west-dipping reflectivity in the lower crust. In both sections, in the immediate footwall to the Main Uralian fault, a narrow band of moderately east-dipping reflectivity in the upper crust is related to strongly deformed Proterozoic and Paleozoic rocks in the eastern limb of the Kvarkush Anticlinorium. Note that in previous interpretations [Steer et al., 1995; Juhlin et al., 1998; Friberg, 2000] this reflection package was included in the Main Uralian fault (see Section 2.2.). The Moho is not clearly imaged beneath the East European Craton in the ESRU data, but by projecting the wide-angle Moho from the UWARS experiment some 60 km to the south [Thouvenot et al., 1995] onto the ESRU section it appears that the crustal thickness reaches ~52 km [see also, Juhlin et al., 1998].

3.2. Main Uralian Fault

Where crossed by the URSEIS section, the Main Uralian fault is approximately 4 km wide, cropping out from about kilometer 152 to kilometer 156 (Figure 1 and Plate 2). The Main Uralian fault is nonreflective in the URSEIS section, but in the upper 30 km it is marked by a change in reflectivity from the East European Craton to the Magnitogorsk arc (Figure 1). Based on this, we interpret the Main Uralian fault to dip steeply eastward beneath the Magnitogorsk arc, and perhaps to extend to the Moho at approximately kilometer 210.

The Main Uralian fault (as defined in Section 2.2.) in the ESRU (at the surface at about kilometer 40) and the Alapaev (at the surface at about kilometer 17) sections is imaged as a sharp change from moderately east-dipping reflectivity associated with deformed East European Craton rocks to gently west-dipping to subhorizontal reflectivity in the Tagil arc (Figure 1 and Plates 3 and 4). On the basis of these observations it can be traced into the middle crust in both sections.

3.3. Magnitogorsk and Tagil Volcanic Arcs

The URSEIS section crosses the Magnitogorsk arc from the Main Uralian fault zone at kilometer 156 to the East Magnitogorsk fault zone at about kilometer 255 (Plate 2). The reflectivity in this region is characterized by patchy, noncoherent to coherent reflections in the upper ~10 km. Below this, reflectivity is diffuse or the arc crust is transparent. The Moho is not imaged.

The ESRU (from about kilometer 41 to kilometer 81) and Alapaev (from near kilometer 20 to kilometer 41) sections (Plates 3 and 4) image the upper crust of the Tagil arc as an open synformal structure (ESRU), or as gently west-dipping reflectivity (Alapaev). The middle crust is characterized in both sections by patchy, generally gently east-dipping reflectivity. From kilometer 23 to kilometer 32 the middle crust is highly reflective in the ESRU section, being characterized by gently east-dipping to subhorizontal reflections.

3.4. East Magnitogorsk–Serov-Mauk Fault System

The East Magnitogorsk fault zone is crossed by the URSEIS section at about kilometer 255 (Plate 2). It is not imaged in the upper ~6 km, but a sharp contrast in reflectivity between the arc and the East Uralian zone from ~6 to ~23 km depth is interpreted to image the location of the fault zone (Figure 2).

The Serov-Mauk fault is crossed by the ESRU section at about kilometer 81, and by the Alapaev section at about kilometer 41, neither of which image it unambiguously (Figure 2 and Plates 3 and 4). Juhlin et al. [1998] interpret steeply west dipping reflections that project up to the surface location of the Serov-Mauk fault in the ESRU section to indicate that it dips steeply westward, truncating the Tagil arc in the middle crust (Figure 2 and Plate 3). Here, we also propose an alternative interpretation in which the Serov-Mauk fault is subvertical to steeply east-dipping, truncating to the west a patch of highly coherent middle and lower crustal reflectivity at about kilometer 80 in ESRU, and extending to the base of the crust at an apparent offset in the Moho at about kilometer 85 (Figure 2).

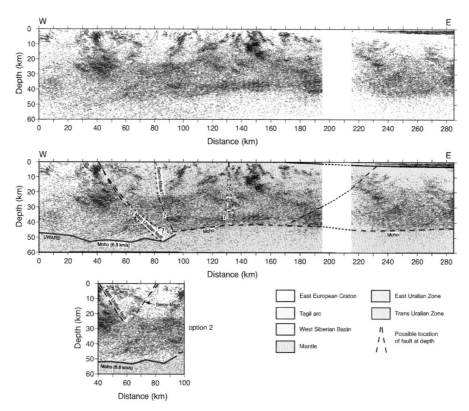

Plate 3. Uninterpreted and interpreted line drawings of the coherency filtered, depth migrated ESRU data. See Figure 1 for location. The colors represent the different tectonic units that make up the Uralide crust. The main suture zones that bind these tectonic units together have been interpreted to end at the Moho. Their exact location at depth cannot be unambiguously interpreted, and have therefore been shown as a zone in which they may possibly occur. The location of the UWARS wide-angle Moho is from *Juhlin* et al. [1998].

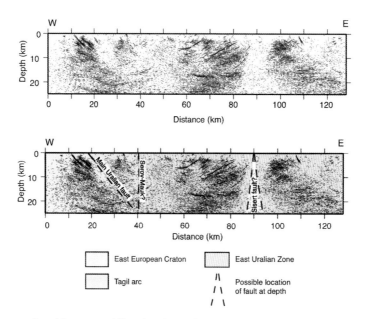

Plate 4. Uninterpreted and interpreted line drawings of the coherency filtered, depth migrated Alapaev data. See Figure 1 for location. The colors represent the different tectonic units that make up the Uralide crust. The main suture zones that bind these tectonic units together have been interpreted to end at the base of the section.

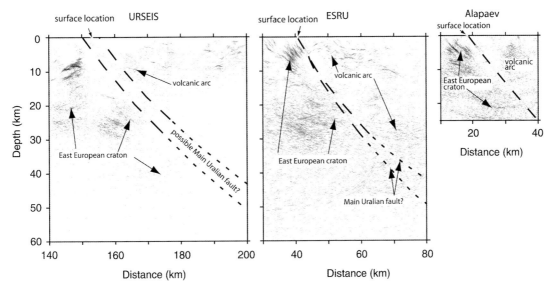

Figure 1. Detailed line drawings of the URSEIS, ESRU and Alapaev sections showing the interpreted subsurface location of the Main Uralian fault. The shorter dash indicates a lesser degree of confidence in the interpretation.

3.5. East Uralian Zone

The URSEIS section crosses the East Uralian zone from about kilometer 255 to near kilometer 330, where the entire surface geology comprises the Dzhabyk granite (Plate 2). In this area, the upper 5 to 6 km of the section is transparent, and the change to moderately coherent reflections has been interpreted to image the base of the granite [*Brown and Tryggvason*, 2001]. Below the Dzhabyk granite, to a depth of ~20 km, the crust is characterized by east-dipping patches of openly undulating reflectivity that, in the east, become shallowly west-dipping. From near kilometer 300 to about kilometer 325, a ~10 km thick, west-dipping band of reflections between 12 to 35 km depth forms the Kartaly reflection sequence of *Echtler* et al. [1996]. On the basis of similarities in reflection character, the Kartaly reflection sequence has recently been interpreted to represent Trans-Uralian zone crust [*Tryggvason* et al., 2001]. Beneath this reflection sequence, the middle and lower crust are characterized by moderately west-dipping reflections that may extend westward to about kilometer 260. The Moho is marked by a sharp decrease in reflectivity. The crust beneath the Kartaly reflection sequence reaches a maximum thickness of ~53 km.

The East Uralian zone in the ESRU (kilometer 81 to kilometer 230) and the Alapaev (kilometer 41 to the end) sections is imaged as a region of complex upper crustal reflectivity with a highly reflective and uniform middle and lower crust (Plates 3 and 4). The Moho deepens

slightly westward from ~42 to ~45 km, with a possible offset near kilometer 85. This step appears in the same position (i.e., beneath the surface expression of the Serov-Mauk fault) in UWARS wide-angle data to the south [*Thouvenot* et al., 1995]. From the Serov-Mauk fault at kilometer 81 to the Sisert fault at about kilometer 132, the upper ~20 km of the East Uralian zone crust in the ESRU section is characterized by moderately to steeply west-dipping reflectivity. Within this region, there is minor shallowly east-dipping reflectivity in the westernmost upper 10 km and in the easternmost 10 km. From kilometer 41 to about kilometer 61 in the Alapaev section the East Uralian zone is imaged as diffuse, moderately east-dipping reflectivity, and from kilometer 61 to the Sisert fault at about kilometer 90 it shows strong, moderately west-dipping reflectivity.

The Sisert fault is crossed by the ESRU section at about kilometer 132, and by the Alapaev section at about kilometer 90 (Figure 3 and Plates 3 and 4). In the upper 10 km of both sections there are minor changes in the dip of reflections across the Sisert fault, but these changes cannot be traced into the middle and lower crust [see also, *Friberg* et al., 2000]. In the Alapaev section, the area beneath the surface location of the Sisert fault is nearly transparent, with the transparent zone truncating prominent west-dipping reflectivity to the west and patchy reflectivity in the upper 10 km to the east.

East of the Sisert fault, from kilometer 132 to about kilometer 230, the upper crust is characterized in the

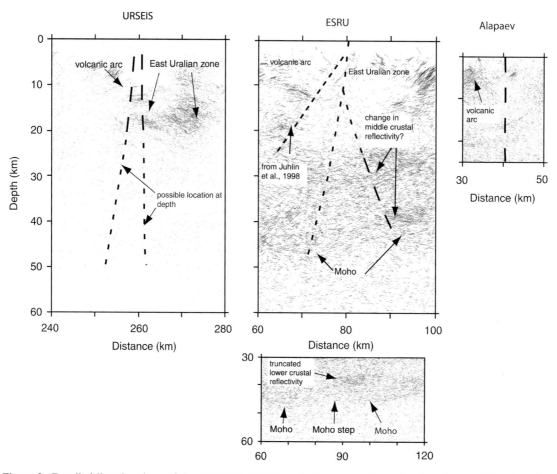

Figure 2. Detailed line drawings of the URSEIS, ESRU and Alapaev sections showing the possible subsurface location of the East Magnitogorsk–Serov-Mauk fault zone. The shorter dash indicates a lesser degree of confidence in the interpretation. The enlarged ESRU panel shows the truncated lower crustal reflectivity and the Moho step that provides the basis for the extrapolation of the Serov-Mauk fault to the Moho.

ESRU section by patchy, gently east and west dipping reflectivity. From about kilometer 170 eastward to the end of the section, the uppermost part of the section images horizontal reflectivity of the eastward-thickening West Siberian Basin. The middle and lower crust is characterized by moderately east-dipping to subhorizontal reflectivity. The Moho beneath the East Uralian zone is imaged as a 42 to 44 km deep feature. In the Alapaev section, from the Sisert fault to the end of the section, the upper 25 km of the crust appears as patchy, predominately east-dipping reflectivity.

3.6. Trans-Uralian Zone

From about kilometer 328 to kilometer 334, the URSEIS section images a zone of nearly transparent crust that corresponds to the boundary between the East

Uralian and the Trans-Uralian zones in the Southern Urals (Figure 4 and Plate 2). The Troitsk fault appears at kilometer 330 (Figure 4). From kilometer 334 to kilometer 420, the upper ~5 km of the crust displays east and west dipping reflectivity. Between ~5 and ~25 km depth, however, the Trans-Uralian zone is characterized by a series of east and west dipping intensely reflective zones that extend westward beneath the East Uralian zone. The lower crust displays thin bands of west-dipping reflectivity that appears to merge with the Moho, which is imaged as a sharp transition from weakly reflective lower crust to transparent upper mantle at ~49 km depth.

The boundary between the East Uralian and Trans-Uralian zones in the ESRU section is interpreted to occur in the subsurface at about kilometer 233, coincident with

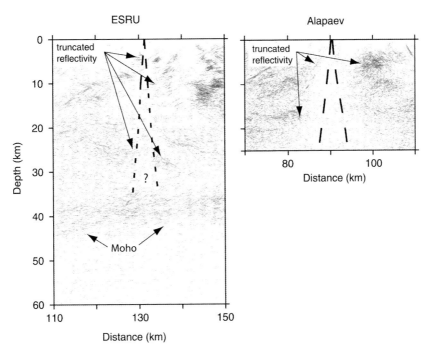

Figure 3. Detailed line drawings of the ESRU and Alapaev sections showing the truncations of upper and middle crustal reflectivity that indicate the possible subsurface location of the Sisert fault. The shorter dash indicates a lesser degree of confidence in the interpretation.

a prominent magnetic lineament that extends to where it is exposed near the URSEIS section (Plate 1). The upper crust of the Trans-Uralian zone is nearly transparent, with only local, weak, west-dipping reflections (Figure 4

and Plate 3). In contrast, the middle and lower crust is characterized by strong moderately west-dipping reflectivity that extends westward beneath the East Uralian zone where it appears to merge with the Moho. The

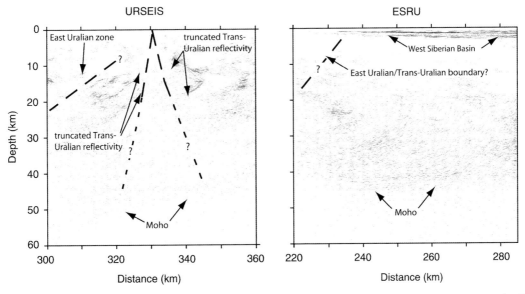

Figure 4. Detailed line drawings of the URSEIS and ESRU sections showing the boundary between the East Uralian and Trans-Uralian zones. The location of the boundary in the ESRU profile is determined from the magnetic data in Plate 1. The Troitsk fault is crossed by the URSEIS section at ~kilometer 330.

Moho is imaged as an openly concave upward structure at ~46 km depth.

4. DISCUSSION

The overall reflection seismic pattern of the Uralide crust as imaged by the URSEIS and ESRU data is bivergent (Plates 2 and 3), perhaps representing the original collision-related crustal architecture [see also, *Echtler* et al., 1996; *Knapp* et al., 1996; *Steer* et al., 1998; *Tryggvason* et al., 2001]. However, the Uralide orogen records a long and complex subduction/accretion history prior to the final collision that gave it its bivergent architecture [e.g., *Zonenshain* et al., 1984, 1990; *Puchkov*, 1997; *Brown and Spadea*, 1999; *Friberg*, 2000]. In addition, it records a complex late-orogenic history that involved extensive wrench faulting accompanied by widespread melt generation and granitoid emplacement in the interior of the orogen [*Ayarza* et al., 2000a; *Friberg*, 2000; *Echtler* et al., 1997; *Fershtater* et al., 1997; *Bea* et al., this book]. Here, we suggest that this late orogenic activity is expressed predominantly in the East Uralian zone, where it significantly overprinted and/or reworked much of the subduction- and accretion-related tectonic fabric, giving this zone its varied and complex reflection seismic character (see below). Below, we argue that the major tectonic boundaries associated with the wrench fault system can be correlated in the seismic data from south to north. Despite problems in imaging these boundaries in the seismic data, we interpret them to affect the entire crust, and perhaps the Moho.

The URSEIS and ESRU sections indicate that crustal thickness and Moho topography change somewhat between the Southern and Middle Urals (Plates 2 and 3), although the crustal root can be seen to extend along the western volcanic axis of the orogen. Crustal thickness can only be estimated in the eastern part of the URSEIS vibroseis section, where it reaches ~43 km and thickens westward (Plate 2). Based on the explosion source reflection data [*Steer* et al., 1998] and the wide-angle reflection data [*Carbonell* et al., 1998] the Uralide crust is shown to be ~41 km thick in the western end of the section, thickening to ~53 km in the central part, beneath the westernmost Magnitogorsk arc (Plate 2). The ESRU section (Plate 3) images a crustal thickness of ~46 km associated with the Trans-Uralian zone, thinning to ~42 km beneath the East Uralian zone. The Moho is not imaged beneath the Tagil arc, but by projecting the results of the UWARS section onto it the crust appears to reach a thickness of ~52 km [*Juhlin* et al., 1998].

An important difference between the URSEIS and the ESRU sections is the middle and lower crustal reflectivity. The origin of this reflectivity is enigmatic, and may be the result of the interaction of a number of processes [e.g., *Juhlin* et al., 1998]. It is noteworthy that this reflectivity (with the exception of the east-dipping reflectivity associated with the East European Craton) occurs beneath areas affected by the late orogenic wrench fault system (i.e., the East Uralian zone and, to a lesser extent, the Tagil arc). The major structures in this region do not appear to penetrate beyond the top of the middle crustal reflectivity. This suggests that the middle and lower crustal reflectivity in the Middle Urals was formed very late in the history of the orogen, overprinting even the wrench fault system. There is no conclusive evidence that the middle and lower crustal reflectivity is related to late- or post-orogenic extension, as suggested by *Knapp* et al. [1998]. *Friberg* et al. [2000] identify east dipping reflections in the ESRU section that they interpret to be related to extension. However, due to poor exposure, field verification of widespread extension is not available. Also, the west-dipping reflectivity of the Trans-Uralian zone in the ESRU99 data (Plate 3) appears to be uninterrupted beneath the West Siberian Basin, where the locus of extension for the West Siberian Basin should be located, suggesting that in this key area the crustal structure has not been affected by any major late- or post-orogenic extension, or by any other process that could have led to the middle and lower crustal reflectivity imaged to the west. Recently, *Koyi* et al. [1999] suggested that the lower crustal reflectivity in the ESRU section could be the result of root rebound and ductile flow in the lower crust. The association of the lower crustal reflectivity and the relatively thinner crust associated with the East Uralian zone lead us to tentatively interpret these features to be the result of lower crustal flow arising from the development of the wrench fault system. The absence of lower crustal reflectivity in the URSEIS section (see also the explosion-source data [*Steer* et al., 1998]) may be related to melt generation and emplacement of the Dzhabyk batholith.

The foreland thrust and fold belt as imaged in the URSEIS and, to a lesser extent, the ESRU and Alapaev sections is that of a west-vergent, basement-involved thrust stack that is flanked along its hinterland by a basement culmination. The reflection seismic fabric in these basement culminations represent the effects of extensive pre-Uralide deformation and metamorphism.

Along the eastern margin of the foreland thrust and fold belt, all three seismic sections image east-dipping upper and middle crustal reflectivity that extends beneath the accreted volcanic arc complexes. This reflectivity is interpreted to be related to the East European Craton. Based on its truncation, it is possible to trace the Main

Uralian fault to dip moderately eastward to at least 30–35 km depth. In the Southern Urals, the high pressure Maksutovo Complex (in the immediate footwall of the Main Uralian fault, but to the south of URSEIS) appears to represent East European Craton rocks that were subducted to a depth of 50–70 km [*Schulte and Blumel*, 1999; *Hetzel*, 1999; *Brown* et al., 2000], suggesting that the East European Craton was subducted to at least an upper mantle level. Since the Main Uralian fault represents the paleo subduction zone along which the East European Craton and the volcanic arcs (Magnitogorsk and Tagil) collided, we interpret it to extend into the lower crust and even to the Moho as indicated on Plates 2 and 3. *Juhlin* et al. [1998] and *Friberg* [2000] have suggested, however, that in the Middle Urals the Main Uralian fault does not extend beyond the top of the middle crustal reflectivity imaged in the ESRU section (alternative II in Plate 3).

The reflection characteristics of the volcanic arc complexes show some variation between the URSEIS (Magnitogorsk) and the ESRU and Alapaev (Tagil) sections. The upper crustal reflectivity is overall similar, imaging mappable structures in the volcanics and volcaniclastic sediments, whereas the Tagil arc middle and lower crust is relatively reflective compared to that of the Magnitogorsk arc.

The East Magnitogorsk fault in the Southern and the Serov-Mauk fault in the Middle Urals form a major tectonic boundary across which there is a significant increase in metamorphic grade and change in amount and style of deformation. This fault system can be traced for about 700 km along strike (Plate 1), and where it outcrops the deformation associated with it is predominately strike-slip, or transpressive [*Echtler* et al., 1997; *Friberg*, 2000; *Hetzel and Glodny*, 2002]. Neither the East Magnitogorsk nor the Serov-Mauk fault is directly imaged on the reflection seismic data, however, making it difficult to fix their location and geometry in the subsurface. In the URSEIS section, the westward truncation of the East Uralian zone reflectivity at about kilometer 260 (Figure 2 and Plate 2) suggests that the East Magnitogorsk fault can be traced from the surface to approximately 23 km depth as a (near) vertical structure. It is not clear in either the ESRU or Alapaev sections what happens to the Serov-Mauk fault zone at depth. Based on reflection seismic and borehole data, *Sokolov* [1988] suggested that the Serov-Mauk fault has a variable east-west dip direction, which suggests that over its length it may be a steep feature. The ESRU section displays a distinct change in Moho reflectivity and a slight Moho offset at about kilometer 85. *Thouvenot* et al. [1995] also image a Moho offset in the UWARS

data which they associated with the Serov-Mauk fault. *Doll* et al. [1996] and *Stern and McBride* [1998] have shown that steeply dipping to vertical strike-slip faults affect the entire crust in a number of Paleozoic and younger orogens. They also show that a Moho offset is characteristic of these faults. On the basis of these arguments we suggest that the Serov-Mauk fault (and also the East Magnitogorsk fault) may affect the entire crust and Moho (Plate 3). This is in contrast to *Juhlin* et al. [1998] and *Friberg* [2000] who interpret the Serov-Mauk fault to be imaged in the upper crust of ESRU as a set of west-dipping reflectors (see option 2 in Plate 3). In their interpretation, this set of reflectors is associated with a fault that outcrops locally along the eastern margin of the Serov-Mauk serpentinite zone [e.g., *Friberg*, 2000], whereas here we interpret the serpentinite zone itself to represent the deformation zone along the boundary between the Tagil arc and the East Uralian zone (option 1 of Plate 3).

The Sisert fault is a major internal structure in the East Uralian zone. It appears to splay off the East Magnitogorsk/Serov-Mauk fault system and can be traced northward for more than 150 km before disappearing beneath the West Siberian Basin (Plate 1). It is not imaged directly in either the ESRU or Alapaev sections, although in the ESRU section the juxtaposition of reflectors with contrasting dips, and patchy middle crustal reflectivity may indicate its presence. In the Alapaev section its surface expression coincides with a several kilometer wide zone of transparent crust across which two zones of contrasting reflectivity are juxtaposed (Figure 3). This can be traced to the base of the Alapaev section at about 23 km depth, suggesting that the Sisert fault is a near vertical structure. Based on the observations outlined above, we tentatively suggest that the Sisert fault may be a near vertical feature that extends into middle and possibly the lower crust (Plate 3).

In the Southern Urals, the boundary between the East Uralian and the Trans-Uralian zone is a wide mélange zone whose magnetic signature can be traced for nearly 500 km northward beneath the West Siberian Basin (Plate 1). In the URSEIS and ESRU sections, this boundary is imaged as a moderately to steeply west-dipping change in reflectivity from generally east-dipping reflectivity of the East Uralian zone to west-dipping reflectivity characteristic of the Trans-Uralian zone. In the URSEIS section, the transparent crust associated with the area affected by the Troitsk fault suggests that it is a steeply dipping to vertical fault that cuts across the Trans-Uralian zone, truncating reflectivity to at least middle crustal levels. It is not clear if it extends into the lower crust, although this possibility has been shown in Plate 2. The Troitsk fault cannot be identified in the ESRU section. In both the URSEIS and ESRU sections, the west-dipping reflectivity in the Trans-Uralian zone appears to extend to the Moho, suggesting that crustal-scale, east-vergent imbrication of the entire crust occurred. Reflections in the Trans-Uralian zone can be interpreted to be related to subduction-accretion processes during accretion and imbrication of volcanic arc terranes, oceanic and continental crust to, and along, the western margin of the Kazakhstan plate [e.g., *Tryggvason* et al., 2001].

5. CONCLUSIONS

The overall reflection seismic pattern of the Uralide crust, as imaged by the URSEIS, ESRU and Alapaev data, is bivergent. This reflection pattern can be interpreted to be the result of a preserved, collision-related crustal architecture. In the interior part of the orogen, however, late-orogenic wrench faulting accompanied by widespread melt generation and granitoid emplacement has significantly overprinted and/or reworked much of the subduction/accretion-related tectonic fabric, giving this zone its varied and complex reflection seismic character. The major tectonic boundaries associated with the wrench fault system can be correlated from south to north between the seismic sections, and are interpreted to have developed on a crustal scale. Perhaps the most important difference between Uralide crust imaged by the URSEIS and the ESRU sections is the middle and lower crustal reflectivity imaged in ESRU. The origin of this reflectivity is enigmatic, but its almost exclusive development within the wrench fault system represented largely by the East Uralian zone suggests that it may be the result of a process, or processes related to the evolution of this wrench fault system. For the first time, the boundary between the East Uralian and Trans-Uralian zones can be confidently correlated with a strong magnetic anomaly that is traced for ~500 km northward from the URSEIS to the ESRU section. There are some differences in crustal thickness and Moho topography between the Southern and the Middle Urals, but in general these are minor. The crustal root can be seen to extend along the western volcanic axis of the orogen from at least the latitude of URSEIS to that of ESRU.

Acknowledgments. This work has been funded by the TMR Network URO (ERBFMRXCT-960009) of the European Union. C. Juhlin was funded, in part, by the Swedish Research Council. A. Green and J. Hall are thanked for their comments on the manuscript. This is a EUROPROBE Uralides Project publication.

REFERENCES

Alvarez-Marron, J., D. Brown, A. Perez-Estaun, V. Puchkov and Y. Gorozhanina, Accretionary complex structure and kinematics during Paleozoic arc-continent collision in the Southern Urals, *Tectonophysics*, 325, 175–191, 2000.

Antsigin, N. I., K. K. Zoloev, M. L. Kluzina, V. A. Nasedkina, B. A. Popov, B. I. Chuvashov, M. V. Shurigina, O. A. Sherbakov and V. M. Iackusev, Descriptions of the stratigraphic schemes of the Urals (in Russian), *Urals Stratigraphic Commitee*, Ekaterinburg, 1994.

Ayarza, P., D. Brown, J. Alvarez-Marrón and C. Juhlin, Contrasting tectonic history of the arc-continent suture in the Southern and Middle Urals: Implications for the evolution of the orogen, *J. Geol. Soc.*, 157, 1065–1076, 2000a.

Ayarza, P., C. Juhlin, D. Brown, M. Beckholmen, G. Kimbell, R. Pechning, L. Pevzner, R. Pevzner, C. Ayala, M. Bliznetsov, A. Glushkov and A. Rybalka, Integrated geological and geophysical studies in the SG4 borehole area (Tagil volcanic arc—Middle Urals): Location of seismic reflectors and source of the reflectivity, *J. Geophys. Res.*, 105, 21, 333–21, 352, 2000b.

Bashta, E. G., V. N. Kushkov, L. N. Shatornaya, A. N. Glushkov and V. A. Sergeev, First results from the drilling and investigations of the Urals superdeep borehole (SG-4) (in Russian), *Min. Geol. SSSR*, 19–30, 1990.

Bea, F., G. Fershtater, P. Montero, V. Smirnov and E. Zin'kova, Generation and evolution of subduction-related batholiths from the central Urals: Constraints on the P-T history of the Uralian orogen, *Tectonophysics*, 276, 103–116, 1997.

Bea F., G. Fershtater and P. Montero, Variscan granitoids of the Urals: Implications for the evolution of the orogen, this volume.

Berzin, R., O. Oncken, J. H. Knapp, A. Perez-Estaun, T. Hismatulin, N. Yunusov and A. Lipilin, Orogenic evolution of the Ural Mountains: Results from an integrated seismic experiment, *Science*, 274, 220–221, 1996.

Bosch, D., A. A. Krasnobayev, A. Efimov, G. Savelieva and F. Boudier, Early Silurian ages for the gabbroic section of the mafic-ultramafic zone from the Urals, in *EUG 9, Terra Nova*, edited by E. R. Oxburgh, Cambridge Pub., Strasbourg, 121, 1997.

Brown, D. and A. Tryggvason, Ascent mechanism of the Dzhabyk batholith, southern Urals: Constraints from URSEIS reflection seismic profiling, *J. Geol. Soc.*, 158, 881–884, 2001.

Brown, D., J. Alvarez-Marron, A. Perez-Estaun, V. Puchkov, P. Ayarza and Y. Gorozhanina, Structure and evolution of the Magnitogorsk forearc basin: Identifying upper crustal processes during arc-continent collision in the Southern Urals, *Tectonics*, 20, 364–375, 2001.

Brown, D., R. Hetzel and J. H. Scarrow, Tracking arc-continent collision subduction zone processes from high pressure rocks in the Southern Urals, *J. Geol. Soc.*, 157, 901–904, 2000.

Brown, D. and P. Spadea, Processes of forearc and accretionary complex formation during arc-continent collision in the Southern Ural Mountains, *Geology*, 27, 649–652, 1999.

Brown, D., C. Juhlin, J. Alvarez-Marron, A. Perez-Estaun and A. Oslianski, Crustal-scale structure and evolution of an arc-continent collision zone in the Southern Urals, Russia, *Tectonics*, 17, 158–171, 1998.

Brown, D., J. Alvarez-Marron, A. Perez-Estaun, Y. Gorozhanina, V. Baryshev and V. Puchkov, Geometric and kinematic evolution of the foreland thrust and fold belt in the Southern Urals, *Tectonics*, 16, 551–562, 1997.

Carbonell, R., A. Perez-Estaún, J. Gallart, J. Diaz, S. Kashubin, J. Mechie, R. Stadtlander, A. Schulze, J. H. Knapp and A. Morozov, A crustal root beneath the Urals: Wide-angle seismic evidence, *Science*, 274, 222–224, 1996.

Carbonell, R., D. Lecerf, M. Itzin, J. Gallart and D. Brown, Mapping the Moho beneath the Southern Urals, *Geophys. Res. Lett.*, 25, 4229–4233, 1998.

Carbonell, R., J. Gallart, A. Perez-Estaun, J. Diaz, S. Kashubin, J. Mechie, F. Wenzel and J. Knapp, Seismic wide-angle constraints on the crust of the southern Urals, *J. Geophys. Res.*, 105, 13,755–13,777, 2000.

Doll, W. E., W. J. Domoracki, J. K. Costain, C. Çoruh, A. Ludman and J. T. Hopeck, Seismic reflection evidence for the evolution of a transcurrent fault system: The Norumbega fault zone, Maine, *Geology*, 24, 251–254, 1996.

Echtler, H. P., K. S. Ivanov, Y. L. Ronkin, L. A. Karsten, R. Hetzel and A. G. Noskov, The tectono-metamorphic evolution of gneiss complexes in the Middle Urals, Russia: A reappraisal, *Tectonophysics*, 276, 229–251, 1997.

Echtler, H. P., M. Stiller, F. Steinhoff, C. M. Krawczyk, A. Suleimanov, V. Spiridonov, J. H. Knapp, Y. Menshikov, J. Alvarez-Marron and N. Yunusov, Preserved collisional crustal architecture of the Southern Urals–Vibroseis CMP-profiling, *Science*, 274, 224–226, 1996.

Fershtater, G. B., P. Montero, N. S. Borodina, E. V. Pushkarev, V. Smirnov, E. Zin'kova and F. Bea, Uralian magmatism: An overview, *Tectonophysics*, 276, 87–102, 1997.

Friberg, M., Tectonics of the Middle Urals, Ph. Thesis, Uppsala University, 2000.

Friberg M., C. Juhlin, A. G. Green, H. Horstmeyer, J. Roth, A. Rybalka and M. Bliznetsov, Europrobe seismic reflection profiling across the eastern Middle Urals and West Siberian Basin, *Terra Nova*, 12, 252–257, 2000.

Gerdes, A., P. Montero, F. Bea, G. Fershtater, N. Borodina, T. Osipova and G. Shardakova, Peraluminous granites frequently with mantle-like isotope compositions: The continental-type Murzinka and Dzhabyk batholiths of the eastern Urals, *Int. J. Earth Sci.*, 91, 1–17, 2002.

Giese, U., U. Glasmacher, V. I. Kozlov, I. Matenaar, V. N. Puchkov, L. Stroink, W. Bauer, S. Ladage and R. Walter, Structural framework of the Bashkirian anticlinorium, SW Urals, *Geols. Rundsch.*, 87, 526–544, 1999.

Glasmacher, U. A., P. Reynolds, A. A. Alekseyev, V. N. Puchkov, K. Taylor, V. Gorozhanin and R. Walter, $^{40}Ar/^{39}Ar$ thermochronology west of the Main Uralian fault, Southern Urals, Russia, *Geols. Rundsch.*, 87, 515–525, 1999.

Hetzel, R., Geology and geodynamic evolution of the high-P/low-T Maksyutov Complex, Southern Urals, Russia, *Geols. Rundsch.*, 87, 577–588, 1999.

Hetzel, R. and J. Glodny, A crustal-scale, orogen parallel strike-slip fault in the Middle Urals: Age, magnitude of displacement and geodynamic significance, *Int. J. Earth Sci.*, 91, 231–245, 2002.

Ivanov, S. N., A. S. Perfilyev, A. A. Yefimov, G. A. Smirnov, V. M. Necheukhin and G. B. Fershtater, Fundamental features in the structure and evolution of the Urals, *Am. J. Sci.*, 275, 107–130, 1975.

Juhlin, C., M. Friberg, H. Echtler, A. G. Green, J. Ansorge, T. Hismatulin and A. Rybalka, Crustal structure of the Middle Urals: Results from the (ESRU) Europrobe Seismic Reflection Profiling in the Urals Experiments, *Tectonics*, 17, 710–725, 1998.

Juhlin, C., J. H. Knapp, S. Kashubin and M. Bliznetsov, Crustal evolution of the Middle Urals based on seismic reflection and refraction data, *Tectonophysics*, 26, 21–34, 1996.

Kamaletdinov, M. A., *The Nappe Structures of the Urals* (in Russian), 228 pp., Nauka, Moscow, 1974.

Khain, V. E., *Geology of the USSR; First part, Old cratons and Paleozoic fold belts*, Gebrüder Borntraeger, Berlin-Stuttgart, 1985.

Knapp, J. H., D. N. Steer, L. D. Brown, R. Berzin, A. Suleimanov, M. Stiller, E. Lüschen, D. Brown, R. Bulgakov and A. V. Rybalka, A lithosphere-scale image of the Southern Urals from explosion-source seismic reflection profiling in URSEIS '95, *Science*, 274, 226–228, 1996.

Knapp, J. H., C. C. Diaconescu, M. A. Bader, V. B. Sokolov, S. N. Kashubin and A. V. Rybalka, Seismic reflection fabrics of continental collision and post-orogenic extension in the Middle Urals, central Russia, *Tectonophysics*, 288, 115–126, 1998.

Koyi, H. A., A. G. Milnes, H. Schmeling, C. J. Talbot, C. Juhlin and H. Zeyen, Numerical models of ductile rebound of crustal roots beneath mountian belts, *Geophys. J. Int.*, 139, 556–562, 1999.

Maslov, V. A., V. L. Cherkasov, V. T. Tischchenko, A. I. Smirnova, O. V. Artyushkova and V. V. Pavlov, On the stratigraphy and correlation of the middle Paleozoic complexes of the main copper-pyritic areas of the Southern Urals (in Russian), *Ufimsky Nauchno Tsentr*, Ufa, Russia, 1993.

Montero, P., F. Bea, A. Gerdes, G. Fershtater, E. Zin'kova, N. Borodina, T. Osipova and V. Smirnov, Single-zircon evaporation ages and Rb-Sr dating of four major Variscan batholiths of the Urals. A perspective on the timing of deformation and granite generation, *Tectonophysics*, 317, 93–108, 2000.

Perez-Estaun, A., J. Alvarez-Marron, D. Brown, V. Puchkov, Y. Gorozhanina and V. Baryshev, Along-strike structural variations in the foreland thrust and fold belt of the southern Urals, *Tectonophysics*, 276, 265–280, 1997.

Puchkov, V. N., Paleozoic evolution of the East European Craton continental margin involved into the Urals, this volume.

Puchkov, V. N., Paleogeodynamics of the central and southern Urals (in Russian), *Ufa Pauria*, 145 pp., 2000.

Puchkov, V. N., Structure and geodynamics of the Uralian orogen, in *Orogeny through time*, edited by J.-P. Burg and M. Ford, M., *Geol. Soc. Spec. Pub.*, 121, 201–236, 1997.

Puchkov, V. N., Paleozoic of the Urals-Mongolian foldbelt. Occasional Publication of ESRI, *University of South Carolina, Nov. Ser.(II)*, 69 pp., 1991.

Puchkov, V. N., Paleo-oceanic structures of the Ural Mountains, *Geotektonika, N3*, 18–33, 1993.

Savelieva, G. N., A. Y. Sharaskin, A. A. Saveliev, P. Spadea and L. Gaggero, Ophiolites of the Southern Uralides adjacent to the East European continental margin, *Tectonophysics*, 276, 117–138, 1997.

Schulte, B. A. and P. Blümel, Prograde metamorphic reactions in the high-pressure Maksyutov Complex, Urals, *Geols Rundsch*, 87, 561–576, 1999.

Seravkin, I., A. M. Kosarev and D. N. Salikhov, *Volcanism of the Southern Urals* (in Russian), 195 pp., Nauka, Moscow, Russia, 1992.

Shatsky N. S., Riphean era and Baikalian epoch of folding (in Russian), *Akademician Shatsky, Selected works. v. 1*, Moscow, AN USSR, 600–619, 1963.

Sobolev, I. D., E. M. Ananieva and M. N. Annenkova, Geological map of the north, middle and east part of south Urals (in Russian), *Uralian Geol. Com.*, Sverdlovsk, 1964.

Sokolov, V. B., Structure and tectonic position of the Serov-Mauk serpentinite belt according to seismic studies, *Geotectonics*, 22, 40–46, 1988.

Spadea, P., M. D'Antonio, A. Kosarev, Y. Gorozhanina and D. Brown, Arc-continent collision in the Southern Urals: Petrogenetic aspects of the fore arc complex, this volume.

Steer, D. N., J. H. Knapp, L. D. Brown, A. V. Rybalka and V. B. Sokolov, Crustal structure of the Middle Urals based on reprocessing of Russian seismic reflection data, *Geophys. J. Int.*, 123, 673–682, 1995.

Steer, D. N., J. H. Knapp, L. D. Brown, H. P. Echtler, D. L. Brown and R. Berzin, Deep structure of the continental lithosphere in an unextended orogen: An explosive-source seismic reflection profile in the Urals (Urals Seismic Experiment and Integrated Studies (URSEIS 1995)), *Tectonics*, 17, 143–157, 1998.

Stern, T. A. and J. H. McBride, Seismic exploration of continental strike-slip zones, *Tectonophysics*, 286, 63–78, 1998.

Thouvenot, F., S. N. Kashubin, G. Poupinet, V. V. Makovskiy, T. V. Kashubina, Ph., Matte and L. Jenatton, The root of the Urals: Evidence from wide-angle reflection seimics, *Tectonophysics*, 250, 1–13, 1995.

Tryggvason, A., D. Brown and A. Perez-Estaun, Crustal architecture of the southern Uralides from true amplitude processing of the URSEIS vibroseis profile, *Tectonics*, 20, 1040–1052, 2001.

Yazeva, R. G. and V. V. Bocharev, Post-collisional Devonian magmatism of the northern Urals, *Geotectonics*, 27, 316–325, 1994.

Yazeva, R. G. and V. V. Bochkarev, Silurian island arc of the Urals: Structure, evolution and geodynamics, *Geotectonics*, 29, 478–489, 1996.

Zonenshain, L. P., M. I. Kuzmin and L. M. Natapov, Uralian foldbelt, in *Geology of the USSR: A Plate-Tectonic Synthesis*, Geodyn. Ser., vol. 21, edited by B. M. Page, pp. 27–54, AGU, Washington, D.C., 1990.

Zonenshain, L. P., V. G. Korinevsky, V. G. Kazmin, D. M. Pechersky, V. V. Khain and V. V. Matveenkov, Plate tectonic model of the South Urals, *Tectonophysics*, 109, 95–135, 1984.

D. Brown, Instituto de Ciencias de la Tierra "Jaume Almera", c/ Lluis Sole i Sabaris s/n, 08028 Barcelona, Spain

C. Juhlin and M. Beckholmen, Department of Earth Sciences, Geophysics, Uppsala University, Villavaegen 16, S-752 36 Uppsala, Sweden

A. Tryggvason, Swedish Geological Survey, Box 670, 75128 Uppsala, Sweden

D. Steer, 316 Crouse Hall, University of Akron, Akron, OH 44325-4101, USA

P. Ayarza, Departamento de Geologia, Universidad de Salamanca, 37008-Salamanca, Spain

A. Rybalka and M. Blitznetsov, Bazhenov Geophysical Expedition, Communarov, Scheelite, Zarechny 624051, Russia

Insights Into the Architecture and Evolution of the Southern and Middle Urals From Gravity and Magnetic Data

G. S. Kimbell[1], C. Ayala[2], A. Gerdes[3], M. K. Kaban[4], V. A. Shapiro[5] and Y. P. Menshikov[6]

Magnetic, Precambrian crystalline basement is inferred to form a geophysically coherent block beneath the southwestern part of the Altaids which is thrust beneath the eastern side of the Uralides. To the west, magnetic basement of the East European Craton is separated from the western Palaeozoic arc terranes of the Uralides by a pre-Uralian terrane. Gravity and magnetic anomalies over the Uralides can be correlated with extensive belts of subduction-related (dense, magnetic) and late orogenic (low density, non-magnetic) plutons. The latter form two sub-parallel belts which coincide with interpreted suture zones, suggesting that transpressional reactivation of these sutures may have contributed to the initiation and distribution of such plutonism. The overall geometry of the plutonic markers suggests that an observed northward decrease in the age of geological events along the Uralide orogen may be due to initial docking of arcs and microcontinents in the south and their subsequent anticlockwise rotation.

1. INTRODUCTION

Potential field (gravity and magnetic) data over the Uralide orogen reveal geological features with contrasting density and/or magnetic properties that can be traced along strike for distances of 1000 km or more. By integrating images and models based on these data with the results of other geophysical surveys, and with geological, geochemical and isotopic data, it is possible to build up a picture of key components of the orogen and the way in which they have been assembled.

Previous analysis of magnetic field variations in the Southern Urals by *Shapiro* et al. [1997] and *Ayala* et al. [2000] has emphasized the importance of deep seated magnetic basement in generating the observed long wavelength anomalies. *Ayala* et al. [2000] have modelled the geometry of the concealed eastern margin of the magnetic crystalline basement of the East European Craton to the west of the Uralides. *Shapiro* et al. [1997] noted that deep magnetic basement was also required further east and suggested that this could be crystalline basement associated with a Kazakhstan continent. Recently published interpretations of gravity data from the region [*Döring* et al., 1997; *Döring and Götze*, 1999; *Scarrow* et al., submitted] have tended to focus on the nature of the crustal root that occurs beneath the orogen,

[1]*British Geological Survey, Keyworth, UK*
[2]*Instituto de Ciencias de la Tierra, Barcelona, Spain*
[3]*NERC Isotope Geosciences Laboratory, Keyworth, UK*
[4]*Institute of Physics of the Earth, Moscow, Russia*
[5]*Institute of Geophysics, Ekaterinburg, Russia*
[6]*Bazhenov Geophysical Expedition, Sheelit, Russia*

Mountain Building in the Uralides: Pangea to the Present
Geophysical Monograph 132
This paper not subject to U.S copyright.
Published in 2002 by the American Geophysical Union
10.1029/132GM04

but the gravity field also contains strong signatures from shallower structures.

In this paper we present images of regional gravity and magnetic data extending across the Southern and Middle Urals. The key features we identify are magnetic, crystalline basement massifs on either side of the orogen and a set of anomalies within the intervening area which we ascribe to orogen-parallel plutonic belts. The nature (subduction-related or late orogenic) and age of plutons within these belts is reviewed using geochemical and isotopic data. This information is brought together within the context of geotectonic models for the evolution of the Uralides and Altaids.

2. TECTONIC MODELS

2.1. The Uralides

The Palaeozoic Uralide Orogen developed as the result of accretion of terranes to the eastern margin of the East European Craton [Hamilton, 1970; Ivanov et al., 1975; Zonenshain et al., 1990]. Puchkov [1997, and references therein] provides a valuable review of the extensive research into this orogen that is reported in the Russian literature. The Main Uralian fault has been interpreted by many authors as the east dipping suture between the East European Craton and island arc terranes represented by the Magnitogorsk zone in the Southern Urals and the Tagil zone in the Middle Urals (Figure 1). The main development of the Magnitogorsk arc was during Devonian times and its accretion occurred during the Late Devonian to Early Carboniferous [Brown et al., 1998]. The Tagil arc is somewhat older (Ordovician–Silurian); the timing of its accretion is difficult to determine but appears most likely to have occurred in the Early Carboniferous [Ayarza et al., 2000].

Subsequent closure of the ocean to the east of these arcs involved the accretion of further oceanic and possibly microcontinental terranes. The East Uralian zone is separated from the Magnitogorsk zone by a serpentinitic mélange zone [Puchkov, 1997], which Ayarza et al. [2000] call the East Magnitogorsk fault (Figure 1). Puchkov [1997] describes the East Uralian zone as being distinguished by the presence of microcontinental complexes, although Friberg et al. [2000] observe that, in the Middle Urals, at least some of these are Palaeozoic subduction-related magmatic complexes. Echtler et al. [1996] placed the boundary between the East Uralian zone and the Trans-Uralian zone in the Southern Urals at a westward dipping seismic reflector, the Kartaly reflection sequence, which projects to surface immediately to the east of the Troitsk (Chelyabinsk) fault (Figure 1).

The Eastern Volcanic Belt (sometimes termed the East Uralian Volcanogenic Belt) lies in the vicinity of the Troitsk fault. It is described by Puchkov [1997] as containing allochthonous Palaeozoic ophiolitic and island arc complexes. Russian geological maps indicate components ranging in age from Ordovician to Carboniferous. Yazeva and Bochkarev [1996] describe this zone as a collage of tectonic lenses and sheets along the suture zone between the East Uralian and Trans-Uralian zones. They note the presence of Silurian components in the region south of around 56° N and argue that these represent along-strike equivalents of the Tagil zone to the north. The oceanwards shift of this southern sector of the Silurian arc is inferred to result from the influence of an intervening (Chelyabinsk–Mugodzhary) microcontinental block during accretion [Yazeva and Bochkarev, 1996].

The Trans-Uralian zone is generally assumed to be predominantly composed of island arc units; however, it is poorly exposed as a result of Mesozoic–Cenozoic cover of the West Siberian Basin, and is the least well understood part of the orogen. In this paper we assume that the Trans-Uralian zone extends westwards to the major structural lineament defined by the Troitsk fault/Kartaly reflection sequence/Eastern Volcanic Belt. Note that Puchkov [1997] defines the Trans-Uralian zone differently, applying this term to the Palaeozoic ensialic terranes to the east of the Denisovka suspect suture zone, which lies about 90 km east of the Troitsk fault.

The final closure of the Uralide orogen occurred during Late Carboniferous to Early Triassic times, and the compression resulting from this convergence is recorded in the development of a foreland thrust and fold belt [Brown et al., 1997]. Shortening within this belt was minor in the Southern Urals [Brown et al., 1997] but has not been quantified in the Middle Urals. The Tagil zone is, however, more strongly deformed than the Magnitogorsk zone. It is likely that shortening was greater in the area that currently lies around 56° N (Figure 1) because the margin of the East European Craton forms an indenter here (the "Ufimian amphitheatre" [Puchkov, 1997]). Ayarza et al. [2000] infer that, in the Late Carboniferous to Permian, the Main Uralian fault was reworked as the western strand of a strike-slip fault system, which included the East Magnitogorsk and Troitsk faults. A further indication of final closure is provided by the widespread intrusion of late orogenic granites of Permian age, in particular the "Main Granite Axis" within the East Uralian zone (see Section 3).

2.2. The Altaids

In contrast to the stable continental area underpinned by the East European Craton, the area to the east

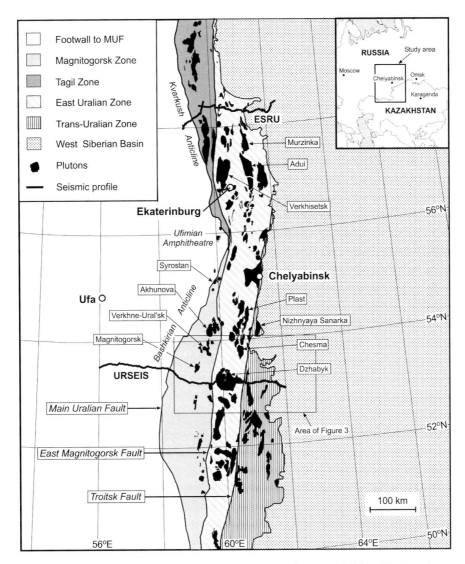

Figure 1. Selected geological elements from the Southern and Middle Urals region.

of the Uralides was one of considerable tectonic activity during the Palaeozoic, when Asia was developing by accretion around a nucleus provided by the Angaran (Siberian) craton. *Sengör* et al. [1993] and *Sengör and Natal'in* [1996] provide a model for the evolution of this vast Altaid accretionary orogen which involves the development of subduction-accretion complexes along a single magmatic arc (the Kipchak Arc) and subsequent reorganisation by strike-slip faulting and oroclinal bending. The Kipchak Arc is inferred to have been initiated along a nearly 7000 km long sliver of Precambrian basement, which rifted from a Russian-Angaran supercontinent. In the Southern and Middle Urals such rifting is correlated by *Sengör* et al. [1993] with the early Palaeozoic opening of the Uralian ocean. As accretionary

complexes grew along this arc, subduction "backsteps" occurred, leading to a series of magmatic fronts younging in the oceanward direction.

Components of the postulated Kipchak Arc are exposed in Kazakhstan immediately to the southeast of the present study area. These include slivers of pre-Altaid basement interspersed with Vendian(?) and Cambrian to Ordovician accretionary complexes and Ordovician magmatic rocks. *Sengör and Natal'in* [1996] suggest that these were dismembered by major strike-faulting in the Ordovician, which was initiated because of collision of the southern tip of the migrating arc with the Mugodzhar microcontinent (a component of the East Uralian zone). In this model, subsequent subduction-accretion led to the development of Devonian and Carboniferous magmatic arcs at the margins of the Kazakhstan collage, and parts of these are now aligned approximately parallel with the Uralide orogen as a result of (clockwise) rotation and oroclinal bending.

The *Sengör* et al. [1993] and *Sengör and Natal'in* [1996] model has provided a powerful stimulus for further research into the Altaids. A key question is the degree to which the accretionary orogen model can explain the complex geology of this large region. Can the observations be reconciled with the single subduction polarity required by this model? What proportion of the Altaid crust is juvenile, and what is the balance between classical collisions (i.e., those involving substantial continental fragments) and accretionary tectonics? Some shortcomings in the model are suggested by detailed analysis of the Karatau fault system, one of main boundaries within the Kazakhstan collage and a structure that may project into the southeastern corner of the present study area. Field observations by *Allen* et al. [2001] suggest that major strike-slip movements along this fault system are pre-Late Riphean in age, rather than Ordovician, as required by the *Sengör and Natal'in* [1996] model. Ordovician deformation is present but is interpreted by *Allen* et al. [2001] to be more likely to have resulted from a compressional event, perhaps associated with a collision to the northeast.

3. PLUTONISM

Numerous small- to medium-sized plutonic complexes are exposed in a ~200 km wide area in the accreted terranes east of the Main Uralian fault (Figure 1) [*Fershtater* et al., 1997]. Most of the available data are for plutons that lie within the Tagil, Magnitogorsk and East Uralian zones, since the widespread Mesozoic cover further east has limited the evidence to bodies inferred from geophysical data and proved by drilling. Tectono-magmatic events usually were not synchronous in the Southern and Middle Urals [*Fershtater* et al., 1997]. Several north-south striking, 200–600 km long belts, each formed by intrusions with a similar origin or source, can be distinguished.

3.1. Plutons in the Tagil and Magnitogorsk Zones

The western island arc zones, Tagil and Magnitogorsk, are both characterized by relatively mafic and dense plutonic bodies. Those in the Tagil zone are predominantly Silurian in age. They include zoned plutonic complexes of the "Platinum Belt" which may be subduction-related intrusions formed as part of the root (feeder) zones of the Tagil zone and subsequently

tectonically juxtaposed with their comagmatic volcanic counterparts during collision [*Savelieva and Nesbitt*, 1996]. Within the Magnitogorsk zone there are a number of Devonian to Early Carboniferous subduction-related granitoids (e.g., Magnitogorsk, Verkhne-Ural'sk, Petropalovsk and Syrostan) which are interpreted to have been produced primarily by melting of oceanic crust and upper mantle [*Ronkin*, 1989; *Fershtater* et al., 1994, 1998; *Salikhov and Mitrofanov*, 1994; *Montero* et al., 2000]. Large granite-granodiorite massifs (e.g., Akhunova) and smaller rift-related gabbro-granite complexes were intruded somewhat later in the Carboniferous [*Fershtater* et al., 1994; *Puchkov*, 1997].

3.2. Plutons in the East Uralian Zone

Numerous, felsic potassium-rich granites and biotite-granodiorites of Late Carboniferous to Late Permian age form the "Main Granite Axes" in the East Uralian zone [*Fershtater* et al., 1994, 1997]. These complexes were generated after ocean closure, predominantly by partial melting of reworked Middle Palaeozoic arc complexes and Proterozoic crust [*Bea* et al., 1997; *Gerdes* et al., 2002]. Representative examples in the Southern Urals are Dzhabyk and Chesma, which were generated about 30 Ma earlier than the main late orogenic complexes, Murzinka and Adui, of the Middle Urals (Figure 1) [*Montero* et al., 2000]. Very similar Nd isotope composition and major and trace element chemistry, together with geophysical evidence (see below), suggest that the Dzhabyk and Chesma granites jointly form a much larger batholith [*Gerdes* et al., 2002; *Shatagin* et al., 2000].

In the Southern Urals, granodiorites and tonalites from several Early Carboniferous complexes (e.g., Chernorechensk, Plast and Chelyabinsk; [*Lozovaya* et al., 1972; *Fershtater* et al., 1994]), which lie immediately to the east of the Main Granite Axis, have a composition very similar to the western subduction-related granitoids [*Bea* et al., 1997; *Shatakin* et al., 2000; *Gerdes* et al., unpublished data]. Inherited zircons in the Chelyabynsk granitoids indicate the presence of Palaeoproterozoic crust [*Krasnobaev* et al., 1997]. Together with Late Devonian to Early Carboniferous volcanosedimentary calc-alkaline sequences south of Chelyabinsk [*Lehmann* et al., 1999] this suggests late Palaeozoic subduction below an active micro-continental margin in the Eastern Urals.

In the Middle Urals, Silurian to Late Carboniferous subduction-related complexes are exposed in the East Uralian zone, to the west of the late orogenic granites [*Fershtater* et al., 1997; *Friberg* et al., 2000a]. The largest of these is the composite Verkhisetsk batholith (Figure 1), which lies about 60 km to the west of the Main Granite Axis [*Fershtater* et al., 1994; *Bea* et al., 1997]. Late Carboniferous granodiorites and tonalites within this batholith have adakite-like chemistry suggesting that they were generated from very young subducted oceanic crust. More felsic intrusions in the inner part of the batholith were generated during the Early Permian by remelting of the older rocks. Granulite-facies gneisses directly west of the late orogenic granites in the Middle Urals have been interpreted as Devonian to Early Carboniferous island arc intrusions [*Friberg* et al., 2000a], although these were previously inferred to be of micro-continental origin [e.g., *Fershtater* et al., 1997].

3.3. Isotopic and Geochemical Data

There are clear differences in major and trace element chemistry and Nd isotope composition between the western subduction-related complexes of the Magnitogorsk zone and the eastern late orogenic batholiths of the East Uralian zone. The main components of the latter usually have lower εNd_t, which varies from $+3$ to -12, compared to values of $+6$ to $+0.5$ in the subduction-related complexes [*Bea* et al., 1997; *Ronkin* et al., 1989; *Shatagin* et al., 2000; *Gerdes* et al., 2002]. Positive εNd values of magmatic rocks usually indicate their derivation from a depleted mantle or oceanic source. Due to fractionation of Sm/Nd during magmatic differentiation and crustal reworking, the εNd of crustal rocks evolves through time to lower values. Thus old continental sources typically have $\varepsilon Nd_t < -5$. The slightly positive εNd_t of 0 to 1.6 in various high-Si late orogenic granites of the Southern Urals is interpreted to reflect the relatively high amount of juvenile Paleozoic arc material in the accreted terranes [*Gerdes* et al., 1999, 2002; *Shatagin* et al., 2000]. Negative εNd_t of -9 to -12 [*Gerdes* et al., 2002] and inherited zircons [*Krasnobaev* et al., 1997] in some granitoids indicate the presence of old Paleoproterozoic crust. The typical observed εNd_t of around $+1$ corresponds to an average late Proterozoic crustal residence (Nd model ages of 1.0–1.2 Ga), suggesting protoliths in the Southern Urals composed of a relatively uniform mixture of juvenile and old crustal components.

The late orogenic granites also have lower average Mg–Fe–Ca contents compared to subduction-related granitoids, reflecting their felsic composition, and are enriched in incompatible elements, such as Rb, Cs and the radiogenic heat producing elements K, Th and U [*Gerdes* et al., 2002]. This indicates that these elements were already enriched in their protoliths due to crustal recycling and differentiation.

4. GEOPHYSICAL DATA SOURCES

4.1. Aeromagnetic Data

The magnetic data used to produce the images shown in Plate 1 have been published on CD-ROM by the US National Geophysical Data Center [*NGDC*, 1997]. This compilation results from digitisation of 1:2,500,000 scale aeromagnetic maps of the former Soviet Union [*Makarova*, 1974] in collaboration with the Ministry of Geology of the USSR. The maps were based on surveys flown between 1951 and 1966 at a variety of line spacings and altitudes. Given the scale of the source maps, some distortion and loss of resolution is inevitable (compared with that possible from original data sources). Nonetheless this compilation provides an invaluable overview of regional magnetic structure [*Zonenshain* et al., 1991]. Plate 1 includes the pseudogravity transform of the aeromagnetic field, i.e. the gravity field that would arise if there was a simple relationship between density and magnetization [*Baranov*, 1957]. This transformation enhances the signal from large, deep-seated magnetic sources and suppresses that from local, near surface bodies.

The more detailed magnetic image of the Southern Urals shown in Plate 2 is based on digitized aeromagnetic survey data provided by the Bazhenov Geophysical Expedition. The flight lines were 2 km apart and oriented E-W; the sensor elevation was 300 m above terrain.

4.2. Gravity Data

The Bouguer gravity anomaly field imaged in Plate 1 is based on values averaged within $10'$ N $\times$ $15'$ E cells (18.5 km NS and 14–18 km EW within the present study area). The observed gravity is referenced to the International Gravity Standardisation Net 1971 and a Bouguer reduction density of 2.67 Mg/m^3 has been used. Terrain corrections have been applied out to a radius of 200 km. Further details of the source data are provided by *Döring* et al. [1997]. The field is heavily smoothed as a result of the averaging process, so cannot be used for resolving local structure, but it does provide a useful regional overview of anomaly patterns.

Plate 2 includes an image of residual Bouguer anomalies (i.e., with a long wavelength component removed) over the Southern Urals. This is based on a data grid with 2 km node spacing provided by the Bazhenov Geophysical Expedition.

4.3. Magnetic Variation Data

Shapiro et al. [1986] and *Shapiro* [1988] report studies in which measurements of temporal variations in the Earth's magnetic field in the Urals region, complemented

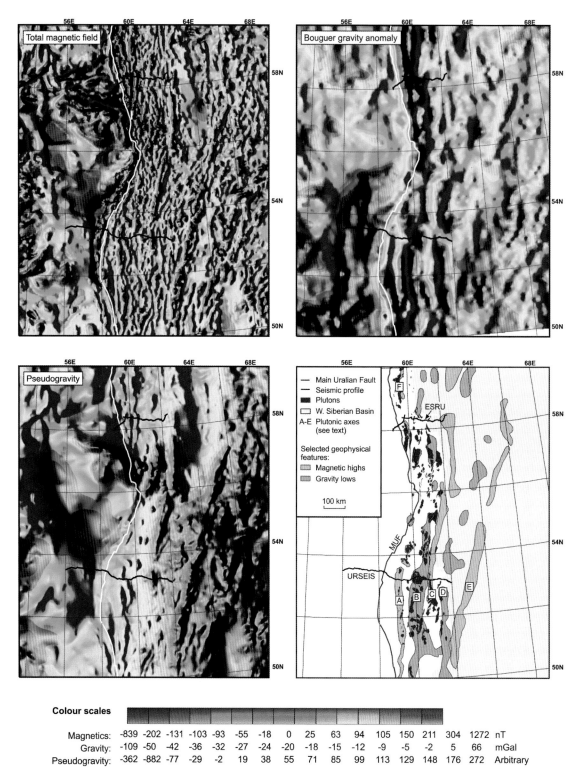

Plate 1. Regional geophysical anomalies in the Southern and Middle Urals. Geophysical images employ equal colour area and shaded-relief with illumination from the west.

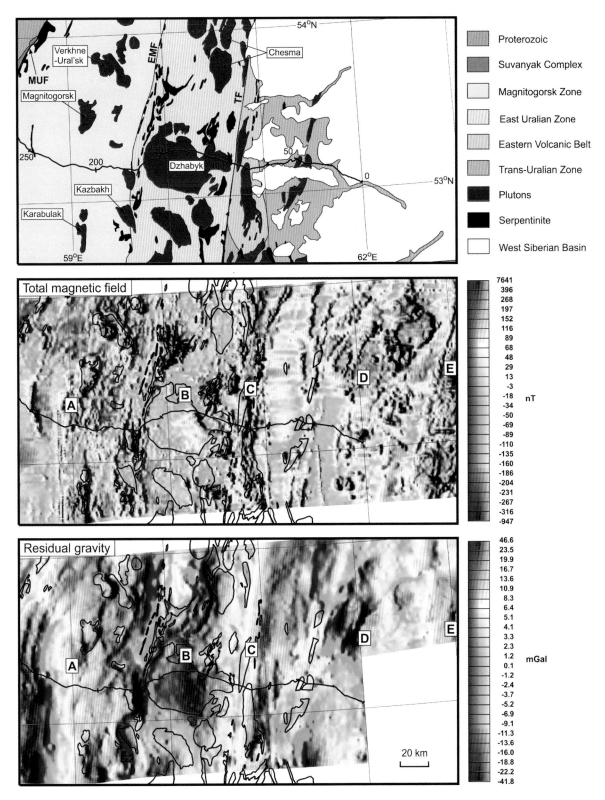

Plate 2. Summary geology, total magnetic field and residual Bouguer gravity anomalies for the Southern Urals. Geophysical images employ equal colour area and shaded-relief with illumination from the west. A–E are geophysical anomaly axes (cf. Plate 1). The thick black line is the URSEIS profile (on the geological map this is annotated with distance from the profile origin in km). Abbreviations: MUF = Main Uralian fault; EMF = East Magnitogorsk fault; TF = Troitsk fault.

by magnetotelluric traversing, have been used to interpret lithospheric conductivity variations. We have correlated the positions of the apparent conductor axes identified by these studies with features mapped using the gravity and (static) magnetic fields.

5. QUALITATIVE ANALYSIS

As discussed by *Shapiro* et al. [1997] and *Ayala* et al. [2000], there is a marked contrast between the magnetic signatures over the East European Craton, to the west of the Main Uralian fault, and those over the area to the east. Over the East European Craton, broad, large amplitude magnetic highs associated with relatively shallow Archean basement are interspersed with lows over the thick sedimentary sequences within Riphean aulocogens. The area east of the Main Uralian fault is dominated by short wavelength N-S oriented magnetic anomalies, but these are superimposed on a long wavelength eastward rise in magnetic field that is resolved very clearly by the pseudogravity transform (Plate 1). Caution is necessary when attempting to resolve long wavelength magnetic variations from compilations of data acquired at different times because of the potential for inaccuracies in the removal and reconciliation of reference fields. However, the eastward rise in magnetic field has been confirmed by a special programme of long aeromagnetic profiles across the Urals designed to avoid such problems [*Shapiro* et al., 1986, 1997].

Shapiro et al. [1997] suggested that the eastward rise in magnetic field was due to concealed magnetic Precambrian crystalline basement of a Kazakhstan continent. Precambrian rocks crop out just to the east of the present study area in the Kokshetau Massif (53° N, 67–72° E) and the Ulutau-Arghanaty Massif (47–51° N, 67° E) [*Glukhan and Serykh*, 1996] and are in places associated with high amplitude magnetic anomalies. These outcrops lie towards the eastern edge of the regional magnetic high; further east, local magnetic anomalies occur but the regional anomaly level is lower [e.g., *Zonenshain* et al., 1991]. Precambrian crystalline basement does thus appear to be the most likely cause for this regional magnetic high, and this qualitative analysis suggests that such basement extends westwards from the known Precambrian outcrops towards the Urals.

On a broad regional scale, the gravity field (Plate 1) is dominated by a positive anomaly over the dense Magnitogorsk and Tagil zones [*Sokolov*, 1992]. This relatively short wavelength high is superimposed on a longer wavelength low due to the crustal root which isostatically compensates the upper crustal load (*Döring* et al. [1997], *Döring and Götze* [1999], *Scarrow* et al.

[submitted]). The gravity field to the east of the Main Uralian fault contains further, short-wavelength N- to NNE-trending features that can be correlated with the magnetic anomaly pattern.

For the present discussion we will focus on a series of anomaly axes defined on the basis of their gravity and magnetic signatures. Axes A–F have been defined in Plate 1, and Plate 2 shows where A–E cross the Southern Urals. The wavelength of the geophysical responses indicates that the sources of these anomalies lie within the upper crust. Of particular interest is the correlation between these axes and plutonic bodies. Compositionally, the subduction-related complexes are rich in ferromagnesian minerals including magnetite, and the late orogenic plutons are quartz-rich and magnetite-poor [*Fershtater* et al., 1994, 1997; *Bea* et al., 1997; *Gerdes* et al., 2002]. It is thus to be expected that, in general, subduction-related plutons will be characterized by positive gravity and magnetic signatures whereas the late orogenic bodies will be characterized by a negative gravity response and lack of associated magnetic anomaly. Thus the geophysical data should aid in tracing plutonic complexes of similar composition and distinguishing their origin. This is very much a simplification, but one that can be tested against the observations. The correlations we note between particular geological units and gravity and magnetic signatures have long been recognized by Russian scientists, who have, for example, used them to extend the mapping of geological structures in areas of poor exposure. Our purpose is to delineate the broad patterns revealed by such correlations and use these as a framework for discussing the geodynamic evolution of the orogen.

5.1. Axis A

Comparison with detailed geological mapping indicates that axis A can be correlated with plutonic bodies within the Magnitogorsk zones (Plate 1). Local magnetic anomalies occur over the Devonian to Early Carboniferous Verkhne-Ural'sk, Magnitogorsk and Karabulak plutons (Plate 2). The data suggests that there may be a concealed link between the Magnitogorsk and Karabulak outcrops and this forms the northern part of an apparently continuous anomaly, which can be traced southwards for about 400 km (Plate 1). The gravity effect of these bodies is either neutral or positive, and bearing in mind the relatively dense nature of their host, this implies an intermediate-basic composition. The physical properties of the plutons inferred from their geophysical signatures are thus compatible with the subduction-related origin inferred from their geochemistry (Section 3.1).

5.2. Axis B

Axis B is a belt of Bouguer gravity anomaly lows within the East Uralian zone, which is associated with the low density, late orogenic plutons of the Main Granite Axis (Plate 1) [*Fershtater* et al., 1994, 1997]. The magnetic response over these features is generally subdued. These geophysical characteristics are well illustrated in the Southern Urals (Plate 2), where there is clear evidence of a subsurface linkage between the large Dzhabyk pluton and the Chesma pluton to the north. The Kazbakh pluton has the signature characteristic of this axis but is offset to the west, implying that it either stitches the East Magnitogorsk fault or has been tectonically emplaced westwards. Conspicuous magnetic anomalies elsewhere in the East Uralian zone are associated with serpentinite outcrops (Plate 2).

5.3. Axis C

Axis C (Plate 1) appears to coincide with the boundary between the East Uralian zone and the Trans-Uralian zone, close to the Troitsk fault and the surface projection of the Kartaly reflection sequence. It is characterized by positive magnetic anomalies and generally by a positive gravity effect. The data from the Southern Urals indicate the close correlation between axis C and the Eastern Volcanic Belt (Plate 2). There is an offset in the geophysical axis at about 56° N, the segment north of this latitude lying further to the east in an area where the source rocks are concealed by the West Siberian Basin (Plate 1). The coincidence of this offset and the northward limit of the Silurian components of the Eastern Volcanic Belt identified by *Yazeva and Bochkarev* [1996] might be an indication that these components are offset to the east and extend further north than these authors suggest. However, it is not possible at this stage to correlate the geophysical responses over this zone with a particular age of source rock. For example, the anomalies over the southern end of the Eastern Volcanic Belt (from the URSEIS profile southwards) are centred over a gabbroic body which is assigned a Carboniferous age in Russian geological maps, although details of the dating procedure used are not known. Regional magnetic images suggest that axis C is obliquely cut by NNW trending linear structures (late faults?) in the Middle Urals.

5.4. Axis D

Axis D delineates a second belt of Bouguer gravity anomaly lows which lies approximately 100 km east of the Main Granite Axis (Plate 1). In the Southern Urals it is represented by the distinct gravity lows which occur to the north and south of the eastern end of the URSEIS

profile (Plate 2). These anomalies are very likely to be due to voluminous granitic intrusions. Adjacent to URSEIS they occur in areas where sedimentary cover of the West Siberian Basin is present, but the outcrop pattern suggests that this cover is too thin to be the cause of the observed gravity effects, and published geological maps indicate that granites of Permian age are exposed along the axis further south where the cover rocks are absent [Geology of the USSR, 1:1,000,000 map series]. There is a close correlation between gravity and magnetic signatures to the north of the eastern end of the URSEIS profile (Plate 2). The gravity anomalies suggest a granitic intrusion with several cupolas; the roof region of each cupola is generally magnetically quiet, but there is typically a short wavelength magnetic anomaly surrounding it, which could be due to magnetization of the country rocks as a result of metamorphism and/or a relatively magnetic outer intrusive phase. There is evidence of skarn-type magnetite deposits locally [*G. B. Fershtater*, personal communication]. The background magnetic levels in this area are relatively high because of the long wavelength effects discussed above. This, together with the fact that the granitic bodies are not associated with magnetic lows, suggests that they do not puncture the underlying magnetic basement. Further short wavelength magnetic anomalies are probably caused by serpentinitic rocks associated with the NNE-trending Denisovka suture [*Puchkov*, 1997], which crosses the eastern end of the URSEIS profile. There is an apparent correlation between Axis C and a conductivity anomaly delineated by *Shapiro* et al. [1988], although the latter is more likely to relate conductive rocks associated with the Denisovka suture zone than to the plutonic bodies, which are generally electrically resistive [*Jones*, 1993].

5.5. Axis E

Axis E (Plate 1) is a prominent, almost linear geophysical feature that can be traced over a SSW-NNE distance of at least 1200 km, extending well to the south of the study area [*Belyankin* et al., 1972 and references therein]. It is characterized by positive magnetic and gravity anomalies. The source is concealed by rocks of the West Siberian Basin, but geophysical surveys and drilling have identified the anomalies to be associated with the Valeryanovka zone, a zone of andesitic-basaltic extrusives and comagmatic intrusive granitoids [*Belyankin* et al., 1972]. *Puchkov* [1997] describes this as a marginal volcano-plutonic belt of mid-Viséan to Serpukhovian age. It is interpreted by *Sengör* et al. [1993] as a Carboniferous arc at the western edge of the Altaid collage.

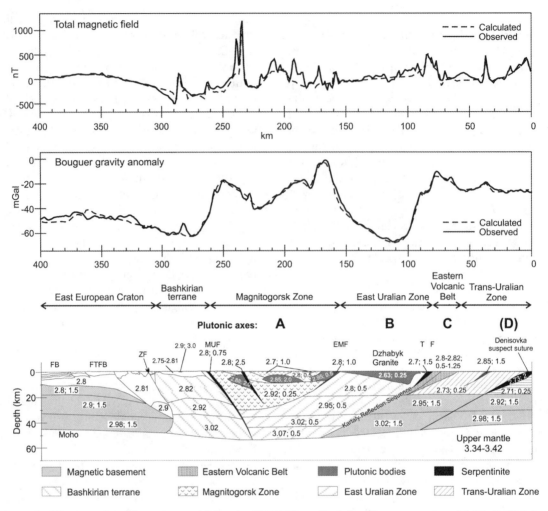

Figure 2. 2D magnetic and gravity model for the URSEIS profile [after *Scarrow* et al., submitted]. Numbers indicate density (Mg/m³); magnetization (A/m). Zero magnetization is assumed unless otherwise indicated. Abbreviations: EMF — East Magnitogorsk fault; FB — foreland basin; FTFB — foreland thrust and fold belt; MUF — Main Uralian fault; TF — Troitsk fault; ZF — Zuratkul fault.

5.6. Axis F

Axis F is a set of pronounced positive gravity and magnetic anomalies associated with the Silurian basic-ultrabasic plutonic complexes of the Platinum Belt within the Tagil zone (Plate 1) [*Sokolov*, 1993].

6. 2D MODELLING

6.1. *Magnetic and Gravity Model for the URSEIS Profile*

A 2D gravity and magnetic model along the URSEIS seismic profile [*Berzin* et al., 1996; *Echtler* et al., 1996; *Knapp* et al., 1996; *Carbonell* et al., 2000] is shown in Figure 2. The observed geophysical profiles employ the high resolution Southern Urals data illustrated in Plate 2, but the gravity field has been modified on the basis of

the low resolution dataset (Plate 1) to restore its long wavelength components. The profile covers the full length of the URSEIS profile (i.e., extending to the west of the area shown in Plate 2) in order to illustrate the inferred magnetic basement configuration. We will only discuss selected aspects of this model here (see *Scarrow* et al. [submitted] and *Ayala* et al. [2000] for further details, and also *Shapiro* et al. [1997] for discussion of the assumptions underlying the magnetic modelling). Distances along the profile are referred to an origin at its eastern end for compatibility with previous publications [e.g., *Berzin* et al., 1996].

The magnetic profile contains large amplitude, short wavelength features (in particular an anomaly over a serpentinite zone at 240 km) which tend to mask the longer

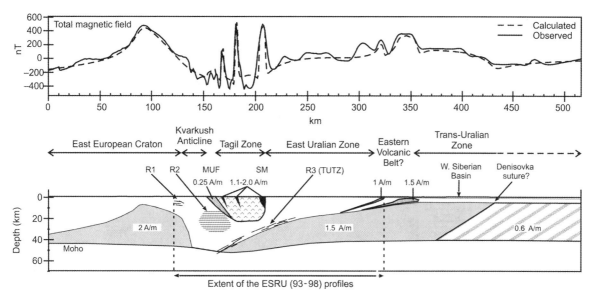

Figure 3. Simple 2D magnetic model along the extended ESRU profile. Assumed magnetizations in A/m are indicated; unlabeled units have zero magnetization. Abbreviations: MUF — Main Uralian fault; R1-R3 — features from ESRU seismic reflection data (see text); SM — Serov Mauk serpentinite zone.

wavelength effects. However, there is a significant long wavelength magnetic low centered over the orogen which has an amplitude of several hundred nanotesla. *Shapiro* et al. [1997] have argued that this probably arises because the crust along the axis of the orogen is on average less magnetic than that underlain by Precambrian continental crystalline basement to the west and east. The magnetic crystalline basement of the East European Craton at the west end of the profile is truncated at around 300 km, about 50 km to the west of the Main Uralian fault. *Ayala* et al. [2000] concluded that the intervening region contained a non-magnetic terrane which was accreted against the East European Craton in Late Vendian times; they noted that the structures which truncate the magnetic basement at middle crustal levels may have been initiated by Early Riphean rifting. The likelihood of a distinct "Bashkirian terrane" had been inferred from geological observations [*Brown* et al., 1996, 1997; *Glasmacher* et al., 1999]. The eastward rise in the magnetic field on the eastern side of the profile is modelled by introducing magnetic basement east of the Kartaly reflection sequence (east of about 150 km). Alternative geometries are possible, but this provides an acceptable fit to observed anomalies and employs a major crustal structure defined seismically at depth and as a probable suture zone from surface observations.

The modelling confirms that upper crustal sources are required to explain the short wavelength anomalies observed over axes A–C. In line with the qualitative assess-

ment, a concealed link between the Magnitogorsk and Karabulak plutons of Axis A is indicated at about 200 km (Figures 3 and 4). The Dzhabyk pluton (Axis B; 100–150 km) thickens eastward, its base dipping in the same direction as reflection fabrics observed in the underlying basement on the URSEIS profile [*Echtler* et al., 1996]. As indicated by surface observations, the Eastern Volcanic Belt (Axis C; 70 km) is shown as an allochthonous slice trapped between the East Uralian zone and the Trans-Uralian zone and projecting down towards the west dipping Kartaly reflection sequence. Plutons of Axis D do not directly underlie the URSEIS profile but it is possible that their off-line effects contribute to the relatively low apparent density for the upper crust required in the modelling [*Scarrow* et al., submitted]. Prominent west dipping reflection fabrics on the URSEIS profile [*Echtler* et al., 1996; *Knapp* et al., 1996] indicate that the Denisovka suture probably dips towards the west and this geometry has been assigned to the magnetic rocks incorporated in this suture zone in the model.

6.2. Magnetic Model for the ESRU Profile

The ESRU (Europrobe Seismic Reflection in the Urals) project has acquired seismic reflection data along a profile across the Middle Urals (Figure 1) through a series of surveys conducted between 1993 and 1999 [*Juhlin* et al., 1995, 1997, 1998; *Friberg* et al., 2000b; *Brown* et al., 2002]. The 2D magnetic model shown in Figure 3 coincides with the ESRU profile in its central

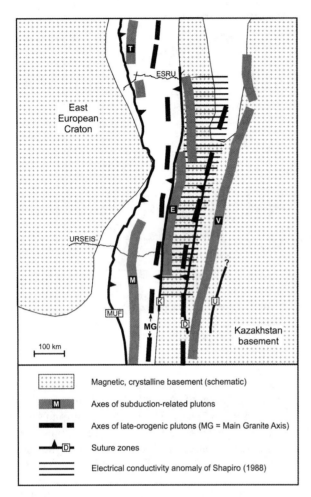

Figure 4. Summary figure showing location of interpreted components of the Uralide Orogen discussed in the text. Subduction-related axes: E — Eastern Volcanic Belt; M — Magnitogorsk Arc; T — Tagil Arc; V — Valeryanovka Arc. Sutures: D — Denisovka; K — Kartaly; MUF — Main Uralian fault; U — ?Urkash. The traces of the Denisovka and Urkash suspect sutures are as mapped by *Puchkov* [1997]. Note that only the easternmost conductivity feature of *Shapiro* [1988] is indicated (further anomalies occur over and to the east of the Main Uralian fault).

part, and is extrapolated to the west and east to provide the regional context. Note that the observed magnetic field profile is based on the low resolution dataset (see Section 4.1) which does not reproduce short wavelength variations accurately. The main aim of this model is to examine possible sources for the longer wavelength magnetic variations and explore the degree to which models of the type developed in the Southern Urals might be applicable in this area.

The magnetic low over the Uralides is well resolved on this profile and centred at 150–200 km (Figure 3). To the

east of this, a high amplitude magnetic anomaly indicates very magnetic rocks extending over a significant proportion of the crustal thickness. These most probably belong to the crystalline Precambrian basement of the East European Craton [*Ayala* et al., 2000]. The 2D assumption is inaccurate here because of the three-dimensional nature of the source (Plate 1). Nonetheless the evidence does suggest that this basement is not truncated at the Main Uralian fault but that there is an intervening non-magnetic zone. Flat lying reflectors at the eastern end of the ESRU profile (R1 in Figure 3) have been interpreted as the signatures of autochthonous Proterozoic strata deposited on the East European Craton [*Juhlin* et al., 1998]; similar strata occur at the western end of URSEIS [*Echtler* et al., 1996]. There is a correlation between the eastward limit of these undeformed strata and that of the underlying magnetic basement. The Kvarkush Anticline, which lies at around 140 km, immediately west of the Main Uralian fault, contains Riphean rocks that were deformed and metamorphosed prior to the Uralide orogeny, and lie in a similar structural position to those of the Bashkirian terrane of the Southern Urals [*Gee* et al., 1996]. This section of the profile appears to be associated with high reflectivity in the middle crust (R2 in Figure 3) [*Juhlin* et al., 1998].

Only a very simplified representation of the magnetic structure of the Tagil zone is included in this model. Anomalies arise from serpentinitic rocks (e.g., in the Serov Mauk serpentinite zone at the eastern margin of the arc) and from magnetic units within the volcanic sequence (see *Ayala* [2000] for further details). Plutonic bodies that are an along-strike equivalent of the sources of anomaly axis F (Plate 1) occur about 10 km south of the ESRU profile and are well resolved by detailed magnetic and gravity data [*Ayala*, 2000].

The magnetic model (Figure 3) incorporates a middle to lower crustal magnetic bock on the eastern side of the orogen which appears directly comparable to the block east of the Kartaly reflection sequence in the URSEIS model (Figure 2). *Juhlin* et al. [1998] identified dipping reflectors on ESRU which they called the Trans-Uralian thrust zone (TUTZ; R3 in Figure 3) and correlated with the Kartaly reflection sequence. The magnetic modelling is compatible with such an interpretation, although the evidence is not conclusive, as the Trans-Uralian thrust zone has a relatively weak seismic signature and a somewhat better replication of observed magnetic anomalies is possible if the deeper part of the magnetic source is truncated at about 200 km, slightly to the east of the postulated thrust. The latter would be more compatible with recent reinterpretation of the ESRU profile in which the Serov-Mauk serpentinite zone is identified as the surface expression of a crustal scale steeply-dipping structure [*Brown* et al., 2002]. Regardless of the nature of its westward truncation, it appears reasonable to conclude that magnetic basement underlies the East Uralian zone in the Middle Urals.

The source of the apparent continuation of the magnetic anomaly axis C (Plate 1) at 320–350 km in the vicinity of the eastern end of the ESRU profile is modelled schematically as allochthonous magnetic units (Eastern Volcanic Belt?) at the boundary between the East Uralian zone and Trans-Uralian zone (Figure 3). Again the parallel with the structure modelled in the Southern Urals (Figure 2) is clear. The interpretation is less secure in the Middle Urals because of the lack of exposure and possibility of contributions from other sources (e.g., Triassic trap basalts).

A distinct eastward drop in magnetic field which occurs at 430 km, approximately 100 km east of the eastern end of ESRU, has been speculatively modelled as the result of crustal magnetization contrasts across the Denisovka suture (Figure 3). The implication is that a less magnetic terrane may be sandwiched between the Denisovka suture and the Kazakhstan basement to the east, a possibility which is reflected in the pseudogravity anomaly patterns (Plate 1). An alternative explanation for this feature is that it is due to vertical offsets in a relatively shallow magnetic basement without a major lateral basement property contrast.

7. DISCUSSION

The potential field analysis and modelling described above has led to the identification of; (i) magnetic, crystalline basement massifs to the west and east of the Uralide orogen; (ii) a series of magmatic axes generated by subduction of Uralian oceanic lithosphere; and (iii) axes of late orogenic plutons generated during the final stages of ocean closure. The geometry of these components is indicated schematically in Figure 4.

The geophysical signature of the Archean magnetic crystalline basement of the East European Craton has been discussed previously by *Ayala* et al. [2000], who identified a non-magnetic pre-Uralian terrane sandwiched between this basement and the Main Uralian fault. The ESRU model described above suggests the possibility of a similar configuration in the Middle Urals, with the non-magnetic terrane corresponding at surface with the Kvarkush Anticline.

The magnetic data indicate that a substantial area to the east of the Southern and Middle Urals is underlain by magnetic crust. This crust appears to extend from the

vicinity of the Precambrian outcrops of the Kokshetau and Ulutau-Arghanaty massifs in the east to the eastern side of the Uralides in the west. Our interpretation is that magnetic, Precambrian basement associated with a Kazakhstan continent is thrust beneath the eastern Uralides, along a major boundary represented by the Kartaly reflection sequence/Eastern Volcanic Belt. Either the crustal blocks to the east and west of the Denisovka suture are founded on similar basement or the magnetic crust extends under this area beneath a mid-crustal detachment (Figure 2). The isotope geochemistry of the Nizhnyaya Sanarka massif, which has been intruded through the magnetic basement about 100 km north of the URSEIS profile (Figure 1) appears to support a Precambrian origin [*Shatagin* et al., 2000]. Negative εNd values and a Nd model age (1.2 Ga) for this massif correlate with those of granites intruded into the Kokchetau Precambrian massif to the east [*Shatagin* et al., 2000]. Further north, the strongly negative εNd values observed in components of the Murzinka batholith (Figure 1; [*Gerdes* et al., 2002]) and inherited zircons in Chelyabinsk granitoids [*Krasnobaev* et al., 1997] provide evidence of the influence of old basement on granites of the East Uralian zone.

Although much of the Altaid collage is interpreted to contain a relatively low proportion of Precambrian crystalline basement [*Sengör* et al., 1993; *Sengör and Natal'in*, 1996], the part that forms the eastern strip of our study area appears to be an exception. It is likely that this basement block was largely consolidated by the Ordovician, although the major movements on some of the structures affecting it may have occurred significantly earlier [*Allen* et al., 2001]. In the *Sengör and Natal'in* [1996] model this part of the basement was derived from the northernmost (Ishim-Stepnyak) part of the Precambrian sliver along which the Kipchak arc developed. To the east of the Kokshetau massif (east of about 70° E), a reduction in long wavelength magnetic field [*NGDC*, 1997] and positive εNd values in plutonic rocks [*Heinhorst* et al., 2000] imply that such basement represents a minor crustal component. The regional magnetic data suggest that the northern limit of the magnetic basement lies close to the northeast corner of our study area. In addition to Ordovician and older structures, it is possible that the northward truncation of this basement is influenced by Permian strike-slip movements on NW-trending shear zones [*Sengör* et al., 1993].

We have traced axes of subduction-related magmatism on the basis of their magnetic (and to a lesser extent gravitational) signatures. Similar regional magnetic anomaly axes have been identified in northeast Japan and Antarctica. In Japan, the magnetic anomalies were ascribed to an 800 km long belt of Early Cretaceous subduction-related plutons (the plutonic roots of the Kitakami arc) intruded into a strike-slip fault system within a pre-Cretaceous accretionary prism [*Finn*, 1994]. In Antartica, the approximately 2000 km long Pacific Margin Anomaly has been correlated with a belt of Early Cretaceous subduction-related plutons intruded into arc-parallel extensional faults in the pre-Cretaceous basement [*Vaughan* et al., 1998; *Johnson*, 1999]. Plutonic sources have been identified for the magnetic anomalies along the axes of the Magnitogorsk and Tagil zones; plutonic components are also likely to play a significant role in generating the anomalies over the Eastern Volcanic Belt and the Valeryanovka arc. As with the examples from Japan and Antartica, an association with faulting is likely in at least some of these cases. The geophysical signatures provide a means of identifying important regional axes, which would not be easily recognized on the basis of field mapping. This is self evident in areas of poor exposure, but also applies in well exposed areas because local geological complexity can mask the underlying broad trends.

The Main Uralian fault is an east dipping arc-continent suture [*Brown* et al., 1998] whereas further east the Kartaly reflection sequence and Denisovka fault appear to mark west dipping suture zones (Figure 4). *Puchkov* [1997] suggests that the Urkash fault, to the east of the Valeryanovka zone, may also be a suture (Figure 4). There is a correlation between some of these suture zones and electrical conductivity anomalies. *Shapiro* et al. [1986] and *Shapiro* [1988] identified meridional conductor axes on either side of the Uralide orogen. Conductive zones to the east can be correlated with the Main Uralian fault and the Vendian terrane boundary further west. A separate conductor on the east side of the Uralides can be correlated with the east dipping Denisovka suture (Figure 4).

A subduction chronology can be tentatively outlined:

(i)   Subduction during the Silurian is recorded in the Tagil zone and also in the Eastern Volcanic Belt [*Fershtater* et al., 1997; *Yazeva and Bochkarev*, 1996]. An eastward polarity is probable in the case of the former and *Yazeva and Bochkarev* [1996] infer the same polarity for the latter.

(ii) During the Devonian to Early Carboniferous, subduction-related magmatism is recorded in the Magnitogorsk zone in the Southern Urals [*Fershtater* et al., 1997]. In the Middle Urals, arc-related rocks of a similar age are found to the east of the Tagil zone in the East Uralian zone; *Friberg*

et al. [in press] suggest that some of these may have been generated because of eastward migration of the Tagil magmatic axis as a result of a change in subduction angle. This part of the Middle Urals has not been investigated in detail within the present study, although initial examination of the low resolution magnetic data suggests that the most conspicuous anomalies here are due to serpentinitic rather than plutonic sources. Another event that may have occurred during this period is westward subduction of the Denisovka ocean, perhaps beneath a continental fragment which now forms the western part of the Trans-Uralian zone.

(iii) Arc-continent collision in the Early Carboniferous could have been the trigger for an eastward jump in subduction to the location now represented by the Kartaly reflection sequence in the Southern Urals. The seismic data indicate a westward polarity (beneath the East Uralian zone) for this subduction, which may have been responsible for the Early Carboniferous granodiorite-tonalite plutons observed in that zone (Section 3.2). The geophysical signatures of these plutons are not, however, distinguished as a separate axis in the analysis described above, probably in part because of overprinting by the effects of more voluminous late orogenic granites. Further east, subduction at the margin of the Kazakhstan accretionary continent generated the Valeryanovka arc during the Early Carboniferous [*Puchkov*, 1997]. The apparent extension of the Kazakhstan magnetic basement beneath the upper crustal expression of this arc suggests either that it was intruded through this basement without substantially altering its magnetic properties, or that it has been thrust across it within a thin skinned unit.

Two belts of voluminous, low density granite plutons have been identified. The western belt is the well known Main Granite Axis [*Fershtater* et al., 1994, 1997] and the second belt lies sub-parallel to this and 100–200 km further east. The Main Granite Axis is known be a late orogenic feature and the eastern belt is assumed to be analogous on the basis of the geophysical signatures and Permian ages indicated for exposed components on published geological maps. The mechanism by which these granites were generated is not well understood. Crustal thickening as a result of convergence across the orogen is an important prerequisite, but other contributing factors such as magmatic underplating [*Bea* et al., 1997] and burial of protoliths with upper crustal heat production [*Gerdes* et al., 2002] have also been invoked.

There is a close spatial correlation between the late orogenic plutonic belts and the Kartaly and Denisovka sutures (Figure 4). This suggests that transpressional reactivation of these sutures during final closure may have played an important role in triggering the generation of the plutons and controlling their geographical distribution. The plutons of the Main Granite Axis appear to lie in the immediate hanging wall of the Kartaly suture, whereas components of the eastern axis may stitch the Denisovka suture. Further precise dating of the eastern belt of late orogenic plutons is required. A northward decrease in the age of the components of the Main Granite Axis has previously been identified. For example, *Montero* et al. [2000] have determined ages of 291 ± 4 Ma for the Dzhabyk pluton in the Southern Urals and 254 ± 5 Ma for the Murzinka pluton in the Middle Urals. Does the second axis show a similar trend, and is there a difference between the ages of plutons in the two belts at similar latitudes? Such dating would help to resolve whether the two axes represent two distinct intrusive events or whether they are both expressions of the same (northward migrating) climactic event combined with the influence of pre-existing structures. The strike-slip component of movement on a transpressively reactivated suture would have been synchronous along its length, but the maximum in the compressive component might have migrated along the orogen and could thus be compatible with the observed age differences.

The northward younging trend is not confined to the late orogenic granites but is reflected in other indicators of Uralide orogenesis. *Puchkov* [1997] describes a "gradual and continuous" shift of geological events northwards along the orogen. The geometries summarized in Figure 4 suggest a possible explanation for this trend. Features on the west side of the orogen are generally oriented N-S whereas those to the east have a NNE-SSW orientation. The migration of events along the margin could be explained by initial docking of arcs and microcontinents in the south and then anticlockwise rotation of these components as convergence proceeded. This inferred rotation direction is opposite to that implied by *Sengör and Natal'in* [1996], but is supported by recent palaeomagnetic data from Kazakhstan [*Weindl* et al., 2000]. Differences in the configuration of Lower and Upper Palaeozoic arcs between the Southern and Middle Urals may be pre-collisional (transform related?) in origin. Further complexity is introduced in the Middle Urals by the more intense deformation resulting from the influence of the indentor formed by the Ufimian amphitheatre, together with strike-slip reworking. The geophysical data, however, provide evidence for the continuity of more easterly features (e.g., the boundary between the East Uralian zone and Trans-Uralian zone; the Valeryanovka arc) between the Southern and Middle Urals.

Acknowledgments. Our research was supported by EC TMR Programme through the URO Network (ERBFMR-XCT960009). This paper is published with the permission of the Executive Director, British Geological Survey (NERC). It is a contribution to the Uralides project of the EUROPROBE programme.

REFERENCES

Allen, M. B., G. I. Alsop and V. G. Zhemchuzhnikov, Dome and basin refolding and transpressive inversion along the Karatau fault system, southern Kazakstan, *J. Geol. Soc., London*, 158, 83–95, 2001.

Ayala, C., Application of potential field methods to geological mapping and determination of upper crustal structure in the vicinity of the SG4 borehole, Middle Urals, Russia. *British Geol. Sur. Technical Report WK/00/09*, pp. 41, +16 figs, 2000.

Ayala, C., G. S. Kimbell, D. Brown, P. Ayarza and Y. P. Menshikov, Magnetic evidence for the geometry and evolution of the eastern margin of the East European Craton in the Southern Urals, *Tectonophysics*, 320, 31–44, 2000.

Ayarza, P., D. Brown, J. Alvarez-Marrón and C. Juhlin, Contrasting tectonic history of the arc-continent suture in the Southern and Middle Urals: implications for the evolution of the orogen, *J. Geol. Soc., London*, 157, 1065–1076, 2000.

Baranov, V., A new method for the interpretation of aeromagnetic maps: pseudogravimetric anomalies, *Geophysics*, 22, 359–383, 1957.

Belyankin, V. I., V. V. Buklin, L. G. Kiryokhin and V. I. Samodurov, South extension of the Valer'yanovka zone of the Transural region (in Russian), *Dok. Akad. Nauk SSSR*, 212, 54–57, 1972.

Bea, F., G. Fershtater, P. Montero, V. Smirnov and E. Zin'kova, Generation and evolution of subduction-related batholiths from the central Urals: Constraints on the P-T history of the Uralian orogen, *Tectonophysics*, 276, 103–116, 1997.

Berzin, R., O. Oncken, J. H. Knapp, A. Pérez-Estaún, T. Hismatulin, N. Yunusov and A. Lipilin, Orogenic evolution of the Ural Mountains: Results from an integrated seismic experiment, *Science*, 274, 220–221, 1996.

Brown, D., J. Alvarez-Marrón and A. Pérez-Estaún, The structural architecture of the footwall to the Main Uralian fault, southern Urals, *Earth Sci. Rev.*, 40, 125–147, 1996.

Brown, D., J. Alvarez-Marrón, A. Pérez-Estaún, Y. Gorozhanina, V. Baryshev and V. Puchkov, Geometric and kinematic evolution of the foreland thrust and fold belt in the southern Urals, *Tectonics*, 16, 551–562, 1997.

Brown, D., C. Juhlin, J. Alvarez-Marrón, A. Pérez-Estaún and A. Oslianski, Crustal-scale structure and evolution of an arc-continent collision zone in the southern Urals, Russia, *Tectonics*, 17, 158–171, 1998.

Brown, D., C. Juhlin, A. Tryggvason, D. Steer, P. Ayarza, M. Beckholmen, A. Rybalka and M. Bliznetsov, The crustal architecture of the Southern and Middle Urals from the URSEIS, ESRU, and Alapaev reflection seismic surveys, this volume.

Carbonell, R., J. Gallart, A. Pérez-Estaún, J. Diaz, S. Kashubin, J. Mechie, F. Wenzel and J. Knapp, Seismic wide-angle constraints on the crust of the southern Urals, *J. Geophys. Res.*, 105B, 13755–13777, 2000.

Döring, J., H. J. Götze and M. Kaban, Preliminary study of the gravity field of the southern Urals along the URSEIS-95 seismic profile, *Tectonophysics*, 276, 49–62, 1997.

Döring, J. and H. J. Götze, The isostatic state of the southern Urals crust, *Geol. Rundsch.*, 87, 500–510, 1999.

Echtler, H. P., M. Stiller, S. Steinhoff, C. M. Krawczyk, A. Suleimanov, V. Spiridonov, J. H. Knapp, Y. Menshikov, J. Alvarez-Marrón and N. Yunusov, Preserved collisional crustal architecture of the southern Urals–Vibroseis CMP-profiling, *Science*, 274, 224–226, 1996.

Fershtater, G. B., N. S. Borodina, M. S. Rapoport, T. A. Osipova, B. H. Smirnov and M. Y. Levin, *Orogenic Granitoid Magmatism of the Urals* (in Russian), Miass, Russ. Acad. Sci. Urals Branch, 1994.

Fershtater, G. B., P. Montero, N. S. Borodina, E. V. Pushkarev, V. N. Smirnov and F. Bea, Uralian magmatism: an overview, *Tectonophysics*, 276, 87–102, 1997.

Fershtater, G. B., F. Bea, N. S. Borodina and P. Montero, Lateral zonation, evolution, and geodynamic interpretation of magmatism of the Urals: New petrological and geo-chemical data, *Petrology*, 6, 409–433, 1998.

Finn, C., Aeromagnetic evidence for a buried Early Cretaceous magmatic arc, northeast Japan, *J. Geophys. Res.*, 99, 22165–22185, 1994.

Friberg, M., A. Larionov, G. A. Petrov and D. Gee, Paleozoic amphibolite-granulite facies magmatic complexes in the hinterland of the Uralide Orogen. *Int. J. Earth Sci.*, 89, 21–39, 2000a.

Friberg, M., C. Juhlin, A. G. Green, H. Horstmeyer, J. Roth, A. Rybalka and M. Bliznetsov, Europrobe seismic reflection profiling across the eastern Middle Urals and West Siberian Basin, *Terra Nova*, 12, 252–257, 2000b.

Friberg, M., C. Juhlin, M. Beckholmen and A. G. Green, Palaeozoic tectonic evolution of the Middle Urals in the light of the ESRU seismic experiments, *J. Geol. Soc., London*, in press.

Gee, D. G., M. Beckholmen, M. Friberg, A. N. Glushkov and V. Sokolov, Baikalian-age structural control of Uralian tectonics - evidence from the Middle Urals, *EUROPROBE Uralides and Variscides Workshop, Granada, 23–29 March 1996, Scientific Programme and Abstracts*, 1996.

Gerdes, A., P. Montero, F. Bea, T. Osipova, N. S. Borodina and G. Fershtater, Late-orogenic continental-type granites of the Urals: Composition and petrogenetic implications, *Acta Universitatis Carolinae, Geologica*, 42, 255–256, 1998.

Gerdes, A., P. Montero, F. Bea, G. Fershtater, N. Borodina, T. Osipova and G. Shardakova, Peraluminous granites frequently with mantle-like isotope compositions: The continental-type Murzinka and Dzhabyk batholiths of the eastern Urals, *Int. J. Earth Sci.*, 91, 1–17, 2002.

Glasmacher, U. A., P. Reynolds, A. A. Alekseyev, V. N. Puchkov, K. Taylor, V. Gorozhanin and R. Walter, $^{40}Ar/^{39}Ar$ Thermochronology west of the Main Uralian fault, southern Urals, *Geol. Rundsch.*, 87, 515–525, 1999.

Glukhan, I. V. and V. I. Serykh, Geology and tectonic evolution of central Kazakhstan (in Russian), in *Granite-related Ore Deposits of Central Kazakhstan and Adjacent Areas*, edited by V. Shatov, R. Seltmann, A. Kremenetsky, B. Lehmann, V. Popov and P. Ermolov, pp. 11–24, Glagol Publishing House, St. Petersburg, 1996.

Hamilton, W., The Uralides and the motion of the Russian and Siberian platforms, *Geol. Soc. Am. Bull.*, 81, 2553–2576, 1970.

Heinhorst, J., B. Lehmann, P. Ermolov, V. Serykh and S. Zhurutin, Paleozoic crustal growth and metallogeny of Central Asia: evidence from magmatic-hydrothermal ore systems of Central Kazakhstan. *Tectonophysics*, 328, 69–87.

Ivanov, S. N., A. S. Perfiliev, A. A. Efimov, G. A. Smirnov, V. M. Necheukhin and G. M. Fershtater, Fundamental features in the structure and evolution of the Urals, *Am. J. Sci.*, 275, 107–130, 1975.

Johnson, A. C., Interpretation of new aeromagnetic data from the central Antartic Peninsula, *J. Geophys. Res.*, 104, 5031–5046, 1999.

Jones, A. G., Electromagnetic images of modern and ancient subduction zones, *Tectonophysics*, 219, 29–54, 1993.

Juhlin, C., S. Kashubin, J. H. Knapp, V. Makovsky and T. Ryberg, Project conducts seismic reflection profiling in the Ural mountains, *EOS, Trans. Am. Geophys. Union*, 76, 197–199, 1995.

Juhlin, C., M. Bliznetsov, L. Pevzner, T. Hismatulin, A. Rybalka and A. Glushkov, Seismic imaging of reflectors in the SG4 borehole, Middle Urals, Russia, *Tectonophysics*, 276, 1–18, 1997.

Juhlin C., M. Friberg, H. P. Echtler, T. Hismatulin, A. Rybalka, A. G. Green and J. Ansorge, Crustal structure of the Middle Urals: Results from the (ESRU) Europrobe seismic reflection profiling in the Urals experiments, *Tectonics*, 17, 710–725, 1998.

Knapp, J., D. Steer, L. Brown, R. Berzin, A. Suleimanov, M. Stiller, E. L. Shen, D. Brown, R. Bulgakov, S. Kashubin and A. Rybalka, Lithosphere-scale seismic image of the southern Urals from explosion-source reflection profiling, *Science*, 274, 226–228, 1996.

Krasnobaev, A. A., V. A. Davydov, G. P. Kuznetsov and N. V. Cherednichenko, Proterozoic zircons in the eastern slope of Urals (in Russian), *Dok. Akad. Nauk*, 355, 246–249, 1997.

Lehmann, B., J. Heinhorst, U. Hein, M. Neumann, J. D. Weisser and V. Fedesejev, The Bereznjakovskoje gold trend, southern Urals, Russia, *Mineralium Deposita*, 34, 241–249, 1999.

Lozovaya, L. S., M. A. Garris and Grevtsova, A. P., The Hercynian cycle of magmatism and metamorphism in the Urals (in Russian), in *Problems of Isotopic Geochronology in the Urals and the Eastern Part of the Russian Platform*, pp. 98–114, Inst. of Geol., Akad. Nauk SSSR, Ufa, 1972.

Makarova, A. A., *Map of the anomalous magnetic field of the territory of the USSR and adjacent marine areas* (18 sheets at 1:2500000 scale), USSR Ministry of Geology, VSEGEI, Leningrad, 1974.

Montero, P., F. Bea, A. Gerdes, G. Fershtater, N. Zin'kova, N. Borodina, T. Osipova and V. Smirnov, Single-zircon stepwise-evaporation $^{207}Pb/^{206}Pb$ and Rb/Sr dating of four major Uralian batholiths. A perspective on the timing of deformation and granite generation, *Tectonophysics*, 317, 93–108, 2000.

NGDC, *Magnetic Anomaly Data for the Former Soviet Union* (on CD-ROM), National Geophysical Data Center, National Oceanic and Atmospheric Administration, Boulder, Colorado, 1997.

Puchkov, V. N., Structure and geodynamics of the Uralian orogen, in *Orogeny Though Time*, edited by J. P. Burg and M. Ford, *Geol. Soc. Spec. Publ.*, 121, 201–236, 1997.

Ronkin, Y. L., Strontium isotopes as indicators of the magmatic evolution of the Urals (in Russian), *Yearbook 1988, Inst. Geol. Geokhim., Urals Nauchn. Centre, Akad. Nauk SSSR*, 107–109, 1989.

Salikhov, D. N. and V. A. Mitrofanov, The Late Devonian-Early Carboniferous intrusive magmatism of the Magnitogorsk Megasyncline, Southern Urals (in Russian), *Ufimskii Nauchn. Tsentr Ross. Akad. Nauk, Ufa*, 1994.

Savelieva, G. N. and R. W. Nesbitt, A synthesis of the stratigraphic and tectonic setting of the Uralian ophiolites, *J. Geol. Soc., London*, 153, 525–537, 1996.

Scarrow, J. H., C. A. Ayala and G. S. Kimbell, Insights into orogenesis: Getting to the root of the continent-ocean-continent collision in the Southern Urals, Russia, submitted to *J. Geol. Soc., London*.

Sengör, A. M. C., B. A. Natal'in and V. S. Burtman. Evolution of the Altaid tectonic collage and Palaeozoic crustal growth in Eurasia, *Nature*, 364, 299–307, 1993.

Sengör, A. M. C. and B. A. Natal'in, Paleotectonics of Asia: Fragments of a synthesis, in *The Tectonic Evolution of Asia*, edited by A. Yin and M. Harrison, pp. 486–640, Cambridge Uni. Press, 1996.

Shapiro, V. A., The Urals-T'Yan Shan electrical conductivity anomaly (in Russian), *Dokl. Akad. Nauk SSSR*, 299, 598–602, 1988.

Shapiro, V. A., A. V. Tsirulsky, N. V. Fedorova, F. I. Nikonova, A. G. Dyakonova, A. V. Chursin and L. O. Tutmina, The anomalous magnetic field and its dynamics used to study the deep structure and modern geodynamic processes of the Urals, *J. Geodynamics*, 5, 221–235, 1986.

Shapiro, V. A., N. V. Fedorova, F. I. Nikonova, A. V. Chursin, Y. P. Menshikov and G. S. Kimbell, Preliminary investigation of the crustal structure of the southern Urals by geomagnetic methods, *Tectonophysics*, 276, 35–47, 1997.

Shatagin, K. N., O. V. Astrakhantsev, K. E. Degtyarev and M. V. Luchitskaya, The heterogeneity of the continental crust in the Eastern Urals: The results of an isotope-geochemical study of Palaeozoic granitoids, *Geotectonics*, 34, 380–396, 2000.

Sokolov, V. B., Crustal structure of the Urals, *Geotectonics*, 26, 357–366, 1992.

Vaughan, A. P. M., C. D. Wareham, A. C. Johnson and S. P. Kelley, A Lower Cretaceous, syn-extensional magmatic source for a linear belt of positive magnetic anomalies: The Pacific Margin Anomaly (PMA), western Palmer Land, Antartica, *Earth Planet. Sci. Lett.*, 158, 143–155, 1998.

Weindl, R., V. Bachtadse, D. V. Alexeiev, A. Zwing and H. Echtler, Palaeomagnetism of Middle to Upper Palaeozoic sediments from the Karatau range, southern Kazakhstan. *INTAS EUROPROBE Timpebar-Uralides Workshop, Abstract Volume*, 38–39, 2000.

Yazeva, R. G. and V. V. Bochkarev, Silurian island arc of the Urals: Structure, evolution and geodynamics, *Geotectonics*, 29, 478–489, 1996.

Zonenshain, L. P., M. I. Kuzmin and L. M. Natapov, *Geology of the USSR: A Plate-Tectonic Synthesis*, Geodyn. Ser., Vol. 21, AGU, Washington D.C., 1990.

Zonenshain, L. P., J. Verhoef, R. Macnab and H. Myers, Magnetic imprints of continental accretion in the U.S.S.R., *EOS, Trans. Am. Geophys. Union*, 72, 305+310, 1991.

Conxi Ayala, Instituto de Ciencias de la Tierra "Jaume Almera", CSIC, 08028 Barcelona, Spain

Axel Gerdes, NERC Isotope Geosciences Laboratory, Keyworth, Nottingham, NG12 5GG, UK

Mikhail Kaban, Institute of Physics of the Earth, Bolshaya Gruzinskaya 10, 123810, Moscow, Russia

Geoff Kimbell, British Geological Survey, Keyworth, Nottingham, NG12 5GG, UK

Seva Shapiro, Institute of Geophysics, 100 Amundsen Str., Ekaterinburg, 620016, Russia

Yuri Menshikov, Bazhenov Geophysical Expedition, Sheelit, Sverdlovsk 624051, Russia

Role of a Phase: Change Moho in Stabilization and Preservation of the Southern Uralide Orogen, Russia

Camelia C. Diaconescu[1,2,3] and James H. Knapp[1,2]

[1]University of South Carolina, South Carolina, USA
[2]Cornell University, Ithaca, New York, USA
[3]National Institute for Earth Physics, Bucharest-Magurele, Romania

Geophysical (URSEIS experiment) and geological data from the Southern Uralides of central Russia provide the basis for a geodynamic model involving eclogitization of the Uralian crustal root in Late Triassic to Early Jurassic time as a mechanism for stabilization and preservation of this Paleozoic orogen. The crustal structure of the orogen implies eastward subduction of the East European continental crust, and balanced restoration implies a significant volume of crust (comprised of ~70% European crust, and ~30% accreted terranes) was carried to sub-Moho depths of up to 70 km. The lack of a clearly defined near-vertical incidence reflection Moho corroborated by coincident wide-angle reflection data suggest that the Moho is a sub-horizontal gradational boundary at ~50–53 km depth beneath the axis of the Southern Uralides. Previous modeling of a subdued (−50 mgal) regional Bouguer gravity minimum across the orogen suggests a subsurface load that is interpreted here as substantiation for a metamorphic phase-change of the lower crust to mantle-like eclogite facies rocks. Timing of eclogitization appears to be constrained by (1) superposition of a nearly flat Moho across the Paleozoic Uralian orogenic fabric, and (2) zircon and apatite fission-track minimum ages of 180–200 Ma, marking an upper age limit to cooling of rocks exposed at the surface, and, implicitly, to significant uplift and erosion in the Southern Uralides. The proposed eclogitization of the Southern Uralian root zone may have led to an isostatically balanced system with subdued topography, and thereby presumably served to stabilize and preserve the orogenic structure.

1. INTRODUCTION

Seismological investigations in orogenic settings in recent years have led to a revised integration of the two classic views of isostatic compensation of mountain belts (Airy vs. Pratt equilibrium). New studies [*Jones* et al., 1994; *Wernicke* et al., 1996] indicate that orogenic loads can in large part be supported by density heterogeneities

Mountain Building in the Uralides: Pangea to the Present
Geophysical Monograph 132
Copyright 2002 by the American Geophysical Union
10.1029/132GM05

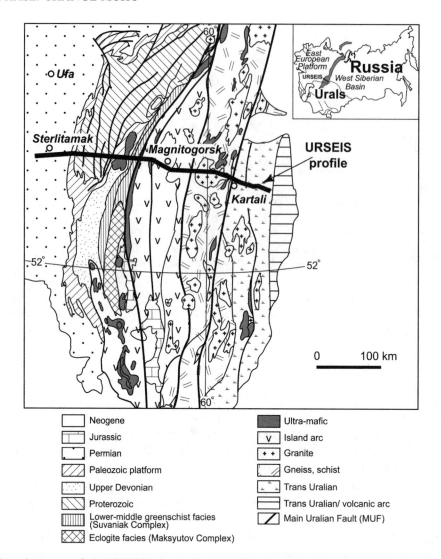

Figure 1. Location map of the URSEIS deep seismic profile showing the principal tectonic elements of the Southern Uralides (modified after *Berzin* et al., 1996).

in the lower crust or upper mantle rather than exclusively by thickening of the crust. While results from young mountain belts, such as the Sierra Nevada, suggest that high elevations may be supported by low density bodies in the upper mantle [e.g., *Ducea and Saleeby*, 1996; *Jones and Phinney*, 1998], analyses of older mountain belts document various mechanisms for isostatic compensation. Seismic profiles from the Archean age Baltic [*BABEL Working Group*, 1990] and Canadian [*Henstock* et al., 1998] shields suggested that they have been stable for over 1.5 Ga and still preserve their crustal roots as revealed by depressed Moho boundaries with significant relief. In contrast, Paleozoic orogens such as the Appalachians, Caledonides, and Variscides underwent

post-orogenic collapse and extension as indicated by relatively flat and shallow Mohos and the wide Atlantic Ocean in between [*Cook* et al., 1979; *Meissner* et al., 1987; *Nelson* et al., 1987; *Andersen* et al., 1991; *Boundy* et al., 1992; *Austrheim* et al., 1997].

A notable exception to the extended Paleozoic orogens is the Southern Uralide orogen of Central Russia (Figure 1), which still preserves its collisional architecture [*Hamilton*, 1970; *Druzhinin* et al., 1988; *Berzin* et al., 1996; *Carbonell* et al., 1996; *Echtler* et al., 1996; *Knapp* et al., 1996]. A regional Bouguer gravity minimum (~ -50 mgal) and the lack of significant topographic relief across the axis of the orogen make the Southern Uralides yet another example of an orogenic belt where the thickening of the crust does not exclusively support the mountain load [*Druzhinin* et al., 1988, 1990; *Kruse and McNutt,* 1988]. However, while the Southern Uralides seem to have preserved their orogenic structure for over 250 Ma without undergoing orogenic collapse and post-orogenic extension, there is evidence that the Middle Uralides were affected by early Mesozoic extension as indicated by *Knapp* et al. [1998].

Earlier geophysical investigations of the Southern Uralides indicated that a pronounced crustal root (10–15 km thick) underlies the orogen [*Druzhinin* et al., 1988, 1990; *Thouvenot* et al., 1995; *Berzin* et al., 1996; *Carbonell* et al., 1996; *Echtler* et al., 1996; *Juhlin* et al., 1996; *Knapp* et al., 1996]. The presence of a Uralian crustal root has long been a subject of controversy since the crust appears to be much thicker than required for the compensation of the subdued topography [*Kruise and McNutt*, 1988; *Döring and Götze*, 1999]. The URSEIS (Urals Seismic Experiment and Integrated Studies) deep seismic profile across the Southern Uralides displays a highly reflective subhorizontal Moho reflection at ~42–45 km depth beneath the Uralian foreland and hinterland (Figure 2a). While the subhorizontal Moho on both sides of the orogen deepens gently toward the central part of the orogen, it loses the pronounced reflective character and cannot be clearly identified on the seismic reflection profile. Previous interpretations of this relationship involved projection of the Moho boundary to depths of ~60 km [*Berzin* et al., 1996; *Carbonell* et al., 1996, 1998]. More recent analysis of the velocity structure of the crustal root suggests it is characterized by high P-wave velocity (7.7–8.0 km/s) [*Druzhinin* et al., 1988; *Thouvenot* et al., 1995; *Carbonell* et al., 1998, 2000], and it was interpreted as either remnant of the Paleozoic collision [*Kruise and McNutt*, 1988] or interlayered sequences of eclogites and peridotites [*Carbonell* et al., 2000].

An increasing number of multidisciplinary studies of collisional zones and an abundance of geophysical data in the past years suggest that the composition and structure of the continental lower crust may play a critical role in the geodynamic development of mountain belts [*Laubscher*, 1990; *Andersen* et al., 1991; *Dewey* et al., 1993; *Platt and England*, 1994; *Baird* et al., 1996; *Wernicke* et al., 1996; *Austrheim* et al., 1997; *Le Pichon* et al., 1997]. Of particular interest lately has been the metamorphic phase-change of the orogenic lower crust to eclogite facies rocks, as this process is being considered responsible for gravitational destabilization of orogenic belts. Partial or full metamorphic phase change of the thickened lower crust from granulite to eclogite facies assemblages was proposed in a number of orogenic belts, such as the Norwegian Caledonides, Variscides, Alps, Himalayas, and Trans-Hudson orogen [*Laubscher*, 1990; *Austrheim*, 1991; *Andersen* et al., 1991; *Dewey* et al., 1993; *Baird* et al., 1995]. Eclogitization of orogenic roots was suggested as a mechanism of triggering delamination of the lower crust and uppermost mantle [*Laubscher*, 1990; *Bousquet* et al., 1997], delamination followed by collapse and post-orogenic extension [*Austrheim*, 1990; *Austrheim* et al., 1997], or subsidence of the overlying upper crust and subsequent formation of sedimentary basins [*Baird* et al., 1995, 1996]. Conversely, retrogression of eclogite to granulite facies rocks was proposed as a mechanism for large-scale uplift without surface shortening [*Le Pichon* et al., 1997]. Since high pressure rocks are only exposed in a few orogenic sections worldwide, deep seismic profiling and mass balance techniques have been used lately to remotely study deep orogenic roots [e.g., *Laubscher*, 1990].

Here we present a model for post-orogenic eclogitization of the Southern Uralide crustal root that rests on a series of geophysical (seismic, gravity, thermal) and geological (crustal restoration, fission track, surface geology) data. Furthermore, we compare the Southern Uralides with other orogens of different ages that were proposed to have experienced eclogitization of the crustal roots, and discuss possible scenarios in support of long-lived stability of orogenic systems and mechanisms for isostatic compensation unrelated to crustal thickness.

2. GEOLOGIC FRAMEWORK

The Urals of Central Russia form the modern geographic boundary between Europe and Asia, and resulted from the Late Paleozoic collision between the East European and Siberian cratons through a collage of island arcs and microcontinental terranes in between the two cratons [*Sengör* et al., 1993]. The Urals together with the Appalachians, the Caledonides, and the Variscides comprise the major zones of continental convergence that contributed to the edifice of the Late Paleozoic Pangea supercontinent [*Hamilton*, 1970; *Sengör* et al., 1993].

Formation of the Uralides began with rifting and development of a passive continental margin on the East European platform in Late Cambrian to early Ordovician time [*Hamilton*, 1970; *Zonenshain* et al., 1984]. The subsequent tectonic evolution of the Uralides involved amalgamation of various lithospheric elements during the Permian or early Triassic time, with formation of island arcs, back-arc basins, and oceanic crust by successive convergence of the East European platform, Siberian

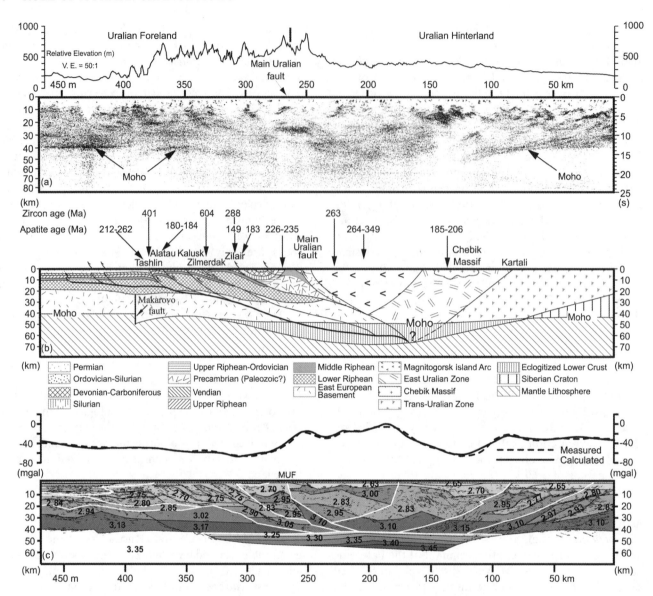

Figure 2. (a) Migrated time section of the URSEIS profile (modified after *Knapp* et al., 1996). (b) Crustal-scale cross-section of the URSEIS profile suggesting that the majority of the Uralian crustal root at ~53–70 km depth (horizontal pattern) is of East European affinity. Moho was picked as PmP arrivals on stacked versions of the wide-angle profile (*Carbonell* et al., 1998). Numbers on the top of cross-section represent cooling ages from zircon and apatite fission-track data after *Seward* et al. (1997); (c) Gravity model along the URSEIS profile showing crustal densities and their calculated versus measured Bouguer gravity effect (adapted from *Döring and Götze*, 1999).

craton, and Kazakhstan [*Hamilton*, 1970; *Zonenshain* et al., 1984; *Zonenshain* et al., 1990; *Puchkov*, 1996].

Earlier studies suggested that the Uralides exhibit several superficial geometric similarities with other orogens of Paleozoic age such as the Appalachians, Ouachitas, Variscides or Caledonides including a highly imbricated transition zone from the foreland basin to the hinterland [e.g., *Rodgers*, 1990]. However, fundamental differences were recognized including a thick-skinned foreland fold and thrust belt and reduced shortening for the Uralides [*Brown* et al., 1996] as well as the presence of a pronounced Uralide crustal root [*Druzhinin* et al., 1988; *Carbonell* et al., 1996; *Juhlin* et al., 1996; *Knapp* et al., 1996; *Steer* et al., 1998].

The foreland fold and thrust belt of the Southern Uralides forms a west-vergent thrust system west of the Main Uralian fault (the inferred Paleozoic suture between the East European craton to the west and accreted terranes to the east) involving both Paleozoic and Precambrian strata in the deformation [*Zonenshain* et al., 1990; *Brown* et al., 1996, 1997]. The Riphean and Vendian sections attain thicknesses in excess of 19 km and were extensively deformed during the Late Paleozoic time with predominantly west vergent thrusting [*Skripiy and Yunusov*, 1989; *Brown* et al., 1997]. A wide zone of deformation, in which Permian strata have been folded into ramp anticlines cored by blind thrusts (Figure 1) [*Skripiy and Yunusov*, 1989; *Brown* et al., 1997; *Diaconescu* et al., 1998], marks the transition from the foreland basin to the foreland fold and thrust belt.

An early phase of eclogitization in the Southern Uralides is clearly recorded in the Maksyutov Complex, a 15 × 200 km body in the footwall of the Main Uralian fault (Figure 1). *Lennykh* et al. [1995], *Hetzel* et al. [1998], *Dobretsov* et al. [1996], and *Beane* [1997] suggested that this complex consists of three main rock types including high-pressure eclogite facies rocks, metasandstones (blueschist-facies), and a metamorphosed mafic-ultramafic melange (greenschist facies). The protolith and the metamorphic age of the rocks forming the Maksyutov Complex remain a subject of controversy [e.g., *Zakharov and Puchkov*, 1994; Hetzel, 1999; Leech and Stockli, 2000]. However, metamorphosed mafic and quartz-rich rocks exposed in the Maksyutov Complex preserve evidence for a Paleozoic high-pressure metamorphic event during the assembly of the Southern Uralides [*Matte* et al., 1993; *Beane* et al., 1995, 1997; *Hetzel* et al., 1998; *Beane and Connelly*, 2000; *Leech and Stockli*, 2000].

The Uralian hinterland, east of the Main Uralian fault (Figure 1), consists of several island arc assemblages, microcontinents, and ophiolite suites that were obducted onto the East European craton throughout the late Paleozoic until Early Carboniferous time. The island arcs were interpreted to be Devonian and Early Carboniferous in age and were amalgamated east of the inferred east-dipping subduction zone [*Zonenshain* et al., 1990; *Berzin* et al., 1996].

3. CRUSTAL-SCALE RESTORATION

Balanced-cross sections have proven to be a powerful technique for understanding the deformation style in foreland fold and thrust belts [e.g., *Dahlstrom*, 1970; *Allmendinger* et al., 1990]. While this technique was initiated through structural interpretations of orogenic systems from surface geologic information [*Dahlstrom*, 1970], it was subsequently developed to constrain crustal-scale interpretations from deep seismic reflection profiles including structural and lithologic boundaries, main detachments and/or the base of the crust [*Allmendinger* et al., 1990].

A recently acquired ~500-km dynamite and vibroseis near-vertical and wide-angle incidence deep seismic reflection profile (URSEIS) across the Southern Uralides provides an excellent means for investigating the crustal architecture and composition of this orogen through use of crustal-scale balanced sections [*Berzin* et al., 1996; *Carbonell* et al., 1996; *Echtler* et al., 1996; *Knapp* et al., 1996] (Figure 1). The Southern Uralides, as shown by the URSEIS profile, constitute a bivergent orogen with highly reflective structures within the crust, both in the foreland basin and hinterland (Figure 1) [e.g., *Berzin* et al., 1996]. A clear image of the Moho boundary was obtained in both the Uralian foreland to (the west) and hinterland (to the east) at approximately 42–45 km (Figure 2a), as indicated by an abrupt downward change in reflectivity. This well-defined Moho reflection dies out toward the central part of the orogen that is dominated by a zone of diffuse reflectivity (175–300 km distance in Figure 2a). However, the Moho was previously projected to a depth of 60 km and interpreted to represent the base of the crustal root from initial processing of the wide-angle data and the downward diminution of the zone of diffuse reflectivity beneath the axis of the orogen [*Carbonell* et al., 1996; *Knapp* et al., 1996; *Steer* et al., 1998].

Several interpretations of the URSEIS near-vertical incidence vibroseis and dynamite seismic sections have already been published by *Berzin* et al. [1996], *Echtler* et al. [1996], *Diaconescu* et al. [1998], and *Döring and Götze* [1999]. Here we attempt to reinterpret the combined URSEIS vibroseis (upper 7 s/20 km) and dynamite (down to 25s/~80 km) sections based on (1) reflection character throughout the crust, (2) surface geology, and (3) crustal-scale restoration of the Southern Uralide fold and thrust belt west of the Zilair fault. While there have been recent efforts to restore the Southern Uralian foreland fold and thrust belt based on surface geologic information [*Brown* et al., 1996, 1997, 1998; *Perez-Estaun* et al., 1997], here we present an attempt to restore on a crustal-scale a fairly detailed cross section of this part of the URSEIS profile west of the Zilair fault (Figure 3b). Interpretation of deep reflectors in the crust provides the geometrical constraints on the position and extent of the lithological/ structural boundaries.

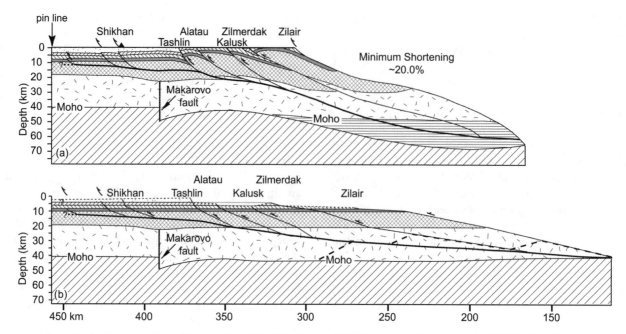

Figure 3. (a) Cross-section of the Southern Uralides foreland fold and thrust belt along the URSEIS profile, west of the Zilair fault. Dashed pattern below the Moho represents the eclogitized crustal root. (b) Crustal-scale restoration of the cross-section displayed in a. The master detachment is shown in bold line both in a and b. Triangle shows the position of the Shikhan well. Legend the same as in Figure 2.

A whole-crust balance requires knowledge of the thickness of the crust prior to the deformation [*Allmendinger* et al., 1990]. A fairly reliable constraint on the crustal thickness beneath the East European platform along the URSEIS profile is provided by the Makarovo fault underneath the East European platform toward the western end of the Uralian Foreland fold and thrust belt [*Diaconescu* et al., 1998]. The Makarovo fault was interpreted to be a relic Precambrian (1.6 Ma) high angle fault that disrupts the Moho, but not the overlying Late Proterozoic sediments. Consequently, from this cross-cutting relationship, the Moho for this part of the orogen was interpreted to be Late Proterozoic or older in age (Figures 2 and 3) [*Diaconescu* et al., 1998]. Based on its preserved reflective character and the lack of a thick pile of Paleozoic or younger sediments, we interpret that the Moho depth (42–45 km) has not changed significantly beneath the East European margin during or after the Uralian deformation.

The fold thrust geometry of the Southern Uralides indicates that the dominant deformation mechanism of the fold and thrust belt appears to be fault propagation folding [*Brown* et al., 1997]. The crustal-scale restoration presented here (Figure 3) was based on the assumption that the stratigraphic thicknesses were maintained con-

stant throughout the sedimentary section. Information on the thicknesses of the sedimentary layers was mainly provided by the Shikhan well that reached the Upper Riphean strata (Figure 3b) [*Skripiy and Yunusov*, 1989]. The Uralian and pre-Uralian deformation were not separated in the restoration. The sedimentary portion of the cross-section was bed-length balanced, whereas constant cross-sectional area balancing was used for the restoration of the crystalline basement while maintaining the slip on the faults constant. The pin line was placed at the western tip of the westernmost detected thrust (Figures 3a and b).

The position and geometry of the master detachment (Figures 3a and b) were interpreted on the basis of the seismic reflection character and agrees with some of the previous interpretations [*Berzin* et al., 1996]. The master detachment was located within the crystalline basement in the central-eastern side of the foreland fold and thrust belt and ramps up to ~12–16 km with the Tashlin thrust, approaching the sedimentary portion of the section (Figures 3a and b). West of the Zilmerdak thrust, the Upper and Middle Riphean rocks were involved in thrusting, suggesting that the basal detachment should be at least at the level of Middle Riphean in the section (~15 km depth). According to this interpretation

(Figure 3), the Southern Uralides are underlain by a root at ~53–70 km depth, which originated from continental material of the East European craton (~70%). This restoration of the Southern Uralides foreland fold and thrust belt west of the Zilar thrust predicts a shortening of ~20% during the Uralian orogeny [*Perez-Estaun et al.*, 1997]. However, since we only restored the Southern Uralide fold and thrust belt west of the Zilair fault due to the high complexity of the geology between this fault and the MUF, we interpret that this is an underestimated value.

4. GEOPHYSICAL AND GEOLOGICAL DATA

This study of the geodynamic evolution of the Southern Uralides draws heavily on a series of geophysical data and geological observations. Included in our study are (1) published near-vertical incidence/wide-angle URSEIS seismic profile, (2) crustal-scale balanced cross-sections (3) published gravity, (4) topography, (5) published fission track data, and (6) published thermal modeling. Recent reprocessing of the URSEIS wide-angle data [*Carbonell et al.*, 1998, 2000] suggested that the seismically defined Moho, corresponding to an increase in the P-wave velocity from ~7.2 km/s to more than 8.0 km/s, occurs along a subhorizontal boundary at ~53 km depth across the central portion of the orogen (50–300 km in Figures 2a and b). This boundary, picked on the basis of first arrivals of PmP waves on stacked versions of the wide-angle data, corresponds well with the downward disappearance of the well-defined Moho reflection on the eastern and western ends of the near-vertical incidence URSEIS profile (~50 and 300 km in Figure 2a).

The gravity signature along the URSEIS profile (Figure 2c) indicates a subdued (−50 mgal) long wavelength regional Bouguer gravity minimum across the axis of the Southern Uralides [*Kruise and McNutt*, 1988; *Döring et al.*, 1997; *Döring and Götze*, 1999]. Accounting for previous structural interpretations of the URSEIS profile [*Echtler and Hetzel*, 1997] as well as velocity information from the wide-angle data, *Döring and Götze* [1999] performed a gravity modeling. Although not uniquely constrained, this model indicates high density material within the orogenic root to account for isostatic balance, with densities varying gradationally from 3.25 to 3.45 g/cm^3 (Figure 2c). This model is in agreement with previous studies including a finite-difference flexural modeling approach for a simplified lithospheric model [*Kruse and McNutt*, 1988] which suggested that the lack of a significant negative Bouguer gravity anomaly above the Southern Uralides could be best explained by a sub-

stantial subsurface load. The short wavelength Bouguer gravity maximum (~10 mgal) observed between 150–300 km distance along URSEIS is consistent with the interpretation of high density material in the upper crust [*Döring et al.*, 1997].

The Southern Uralides show a subdued topographic relief (Figure 2a) for a non-extended orogen, with maximum elevations of ~1600 m [*Berzin et al.*, 1996; *Piwowar et al.*, 1996]. Moreover, most of the topographic relief occurs in the foreland fold and thrust belt, west of the Main Uralian fault (260–400 km in Figure 2a), and is shifted westwards from the orogenic axis. The asymmetry of the topographic relief with respect to the crustal root suggests that the present-day topography is unrelated to the crustal thickness, and most likely represents remnant relief from the Paleozoic Uralian deformation [*Piwowar et al.*, 1996]. Zircon fission-track ages for rocks exposed at the surface along the URSEIS profile [*Seward et al.*, 1997] (Figure 2b) group at about 250 Ma on both sides of the Main Uralian fault, suggesting that there has been little or no differential movement identifiable through fission-track analysis along this line since Triassic time. Similarly, apatite fission track ages range from ~180 to 210 Ma in the Southern Uralian foreland fold and thrust belt, suggesting that little differential movement within the footwall of the Main Uralian fault occurred since Jurassic time. Although slightly older in the central part of the Magnitogorsk volcanic arc (Figure 2b) the apatite fission track ages confirm that there has been very little tectonic activity along the URSEIS transect including significant erosion or uplift recorded in fission track-data since Triassic time. The fission track data implies that the present topographic relief of the Southern Uralides has not significantly changed since Triassic time.

The central part of the Uralian orogen displays very low heat flow with typical values below 30 mWm^{-2} [*Kukkonen et al.*, 1997]. Geotherms calculated from heat flow density measured in boreholes along the central axis of the Southern Uralides indicate temperatures of ~500–550°C at depths exceeding 50 km. Since the Southern Uralides seem to have maintained their lithospheric structure and composition throughout their post-orogenic evolution, we believe that the present-day crustal scale geotherms may have not significantly changed since Late Paleozoic time.

5. THE CASE FOR PHASE-CHANGE MOHO

The role of the phase-change Moho to higher density eclogite facies rocks has been increasingly emphasized in the past years in relation to the geodynamic evolution of orogenic systems [e.g., *Austrheim*, 1991; *Fountain* et al., 1994a; *Poli and Schmidt*, 1997]. The temperature (500–600°C) and pressure (>10 kbar) conditions required for eclogite facies occurrences worldwide indicate that they form in subduction or overthickened crustal zones at depths exceeding ~50 km [*Austrheim*, 1991; *Spear*, 1993; *Fountain* et al., 1994a; *Hynes and Snyder*, 1995; *Schreyer and Stöckhert*, 1997]. Although considered to be anhydrous garnet-clinopyroxene (± quartz/coesite ± rutile) assemblages [*Poli and Schmidt*, 1997], formation of eclogites may critically depend on the presence of fluids [*Austrheim*, 1991].

The bivergent geometry of the Southern Uralides from the URSEIS seismic profile (Figures 2a–c) and the slightly dipping Moho reflections toward the central part of the orogen imply that the crustal root was perhaps depressed to depths exceeding ~53 km [*Carbonell* et al., 1998]. This depth favors the high pressure conditions required by the metamorphic phase-change to eclogite facies rocks [*Austrheim*, 1991]. In addition, the low geotherms (~500–550°C) at the Moho as derived from modeling of the heat flow density [*Kukkonen* et al., 1997], if similar throughout the post-orogenic evolution, are favorable to eclogite formation within the Uralian root zone [*Spear*, 1993]. The subhorizontal wide-angle PmP Moho reflection beneath the main axis of the Southern Uralides corroborated by the lack of a clearly defined near-vertical incidence Moho reflection suggest that the Moho is a gradational boundary that was perhaps superimposed by a metamorphic phase-change developed across the structural fabrics produced during the Uralian orogeny. However, the diffuse (versus clearly defined, kilometers length coherent) zone of reflectivity within the Southern Uralides root (175–300 km distance in Figure 2a) may suggest a mixture of rocks in different metamorphic phases i.e. mafic granulites and eclogites [*Austrheim* et al., 1997]. This would imply a partial metamorphic phase-change to higher density eclogites. Such mixing of metamorphic facies in the lower continental crust at similar depths was proposed in the Bergen Arc of western Norway where granulite and eclogite facies assemblages from the lowermost Caledonian continental crust were exposed at the surface [*Boundy* et al., 1992; *Fountain* et al., 1994a].

Eclogite facies rocks are known from laboratory studies to have elastic properties similar to mantle peridotites (P-wave velocity of 7.8 to 8.5 km/s; density of 3.1 to 3.6 g/cm^3) [*Austrheim*, 1991; *Fountain* et al., 1994a]. Despite the fact that eclogites derive from rocks of crustal origin [*Kern and Richter*, 1981; *Austrheim*, 1991; *Mengel and Kern*, 1992] such similarities in velocity and density make them practically indiscernible from mantle lithologies by seismic techniques. Large increases in P-wave velocity (7.4 to 8.3 km/s) and density (3.0 to 3.6 g/cm^3), resulting in an increase of ~4–9% in acoustic impedance, were observed at the transition from granulite to eclogite facies rocks exposed in the Bergen Arcs of western Norway [*Fountain* et al., 1994b]. The lack of a reflective Moho boundary was interpreted to be a consequence of eclogite facies metamorphism in some continent-continent collision zones, including the Central Alps [*Laubscher*, 1990; *Austrheim*, 1991] and the Trans-Hudson orogen [*Baird* et al., 1995]. According to *Furlong and Fountain* [1986], the juxtaposition of eclogite facies rocks with peridotitic mantle material would produce very small reflection coefficients that are hardly observable on seismic data.

A significant constraint in support of our proposed model for eclogitization of the Southern Uralian crustal root is provided by balanced restoration of the Uralian crust. The crustal-scale restoration of the foreland fold and thrust belt along the URSEIS profile provides support for the crustal origin for the material at ~53–70 km depth beneath the central part of the orogen (horizontal line pattern in Figures 2b and 3a). We interpret this portion of the section to be the crustal root, despite its position below the inferred Moho from the wide-angle data, and thus making the Moho a phase-change boundary. From the crustal-scale restoration (Figure 3) there is an indication that the Southern Uralides root originated primarily from continental material belonging to the East European craton (~70%), and specifically lower crustal rocks of probable mafic granulitic composition. This interpretation is somewhat different from other recent studies [*Stadtlander* et al., 1999] that interpreted the higher density and velocity rocks of the Southern Uralian root as remnant oceanic crust or a mix of oceanic crust and mantle material. Compared to other orogens like the Alps where mass balance analysis suggested deficit of crustal material interpreted to have been recycled into the mantle [*Laubscher*, 1990], we suggest that the Southern Uralides have preserved their crustal root, but as higher density eclogite facies rocks contradicting some recent models which argue otherwise [*Leech*, 2001]. Therefore, the base of the root (Figures 2 and 3) is deeper (~70 km) than it was previously interpreted (~55–60 km) [*Carbonell* et al., 1996; *Knapp* et al., 1996; *Steer* et al., 1998] due to the higher velocity eclogitic material (7.6–8.2 km/s) [*Carbonell* et al., 1998]. The total shortening calculated for the Southern Uralides foreland fold and thrust belt is ~20%, slightly larger than previously estimated (~17%) from shallow crustal restoration [*Brown* et al., 1997; *Perez-Estaun* et al.,

1997]. This relatively reduced shortening was previously interpreted as one of the causes for the long-lived orogenic structure of the Southern Uralides [e.g., *Berzin* et al., 1996].

There have been several models of the subdued long-wavelength gravity signature over the Uralides [*Kruse and McNutt*, 1988; *Döring* et al., 1997; *Döring and Götze*, 1999]. The flexural model proposed by Kruse and McNutt [1988] argued for the presence of a subsurface crustal load to account for the subdued (−50 mgal; Figure 2c) negative Bouguer gravity anomaly above the central part of the orogen. More recently, *Döring and Götze* [1999] modeled the gravity field across the URSEIS profile (Figure 2c), and they suggested the presence of high density rocks (3.25–3.45 g/cm^3; Figure 2c) within the root. Therefore, this discrepancy of a lack of a significant negative gravity anomaly across a preserved, non-extended orogen, could be accounted for, if in fact, the original root has been transformed into a higher density eclogite consistent with the model proposed in this paper and the densities derived from the gravity modeling [*Döring and Götze*, 1999; Figure 2c]. The Magnitogorsk volcanic arc in the hanging wall of the Main Uralian fault, with high density (~3.0 g/cm^3) rocks, appears to account for the short wavelength local Bouguer gravity maximum (~10 mgal) across the axis of the orogen [*Döring and Götze*, 1999].

Maximum topographic elevations across the Uralides indicate relatively low (~1600 m) relief for a non-extended orogen, implying that the compensation mechanism is not related exclusively to the crustal thickness. We interpret the lack of significant topographic relief across the Southern Uralides as additional evidence for major post-orogenic changes within the root. The short wavelength topography in the Southern Uralian foreland fold and thrust belt appears to be mainly a result of the shallow geologic structure and lithology, with no evident correlation to the crustal root. Since the Southern Uralides still preserve the Paleozoic structure and escaped orogenic collapse, we interpret this "lack" of orogenic root as in fact a metamorphic phase-change to higher density eclogite facies rocks. Moreover, the asymmetry in the gravity about the topographic peak (Figures 2a and c) provides additional evidence that the mountain load is not supported exclusively by local thickening of the crust [*Kruse and McNutt*, 1988]. This may serve as a substantiation for additional load in the lower crust provided by higher density eclogites. Furthermore, the zircon and apatite fission-track data suggest minimum cooling ages for rocks exposed at the surface along the URSEIS section of Late Triassic-Early Jurassic (Figure

2b). This analysis corroborated by preservation of surficial geologic features at low metamorphic grade [*Echtler* et al., 1996; *Echtler and Hetzel*, 1997] in the footwall of the Main Uralian fault suggest that very little tectonic activity, including uplift and erosion, has been recorded in the post-tectonic development of the Southern Uralides. Thus, we put forth a model that the inferred metamorphic phase change to higher density eclogite facies rocks of the Southern Uralian crustal root perhaps served to stabilize the orogenic architecture, preventing it from orogenic collapse. This geo-dynamic setting is very different from other orogenic systems, where the eclogitization of the orogenic roots caused post-orogenic collapse and extension [*Austrheim*, 1991; *Laubscher*, 1990; *Baird* et al., 1995].

The timing we propose for the eclogitization of the Southern Uralides lower crust bears on the interpretation of zircon and apatite fission-track data and the position of the Moho relative to the Uralian structures. The fairly flat Moho at ~53 km depth from the URSEIS wide-angle data (Figure 2b) overprints the Uralian orogenic fabric [*Carbonell* et al., 1998], and consequently it must be younger than Uralian. The zircon and apatite fission-track data [*Seward* et al., 1997] indicate that the cooling ages for rocks exposed now at the surface cluster in the Late Triassic to Early Jurassic time (200–260 Ma), indicating that no significant erosion or uplift have occurred in the Southern Uralides since that time. Therefore, we propose that the eclogitization of the Uralian crustal root perhaps occurred at or after the end of the collisional process between Late Triassic and Early Jurassic time.

The presence of the high-grade metamorphic Maksyutov Complex in the footwall of the Main Uralian fault implies that we cannot rule out the occurrence of eclogite facies metamorphism in the lower crust of the Southern Uralides at earlier stages of Uralian orogenic deformation. There is independent geologic evidence of continental collision in the Late Precambrian, which may have resulted in eclogite formation [*Gee* et al., 1996; *Giese* et al., 1999]. In addition, there are eclogitic rocks exposed at the surface and preserved in the high-grade metamorphic Maksyutov Complex [*Beane* et al., 1995; *Lennykh* et al., 1995, 1997; *Leech and Stockli*, 2000]. These eclogites have been dated as Devonian in age (377–384 Ma) based on U-Pb decay ages of rutile within the mafic eclogite [*Beane* et al., 1995, 1997; *Beane and Conelly*, 2000]. From apatite fission track data, *Leech and Stockli* [2000] proposed that the Maksyutov Complex was exhumed in Early Permian time (~300 Ma), therefore it appears to be very little related to our proposed Late Triassic to Early Jurassic eclogite facies assemblages within the Southern Uralian lower crust.

The eclogite facies phase change appears to require fluids to trigger the reaction kinetics, in addition to suitable pressure and temperature conditions [*Austrheim*, 1987; *Fountain* et al., 1994b]. Study of eclogites from the Bergen Arc suggested that the amount of eclogite versus granulite in the lower crust is dependent primarily on fluid access and existing deformation rather than only pressure, temperature, and rock composition. Similarly, research on the Precambrian granulites of the Western Gneiss region of Norway indicated that granulites may remain metastable in the eclogite field if water is not available [*Austrheim* et al., 1997]. Although highly speculative, pathways for fluids in the Southern Uralides might have been provided by a later westward subduction to the east of the Main Uralian fault along either the Kartaly fault or the structures further to the east (Figure 2b) [*Echtler and Hetzel*, 1997]. The presence of early Permian Chebik granites at the surface within the Uralian hinterland (Figure 2b) [*Echtler* et al., 1996; *Steer* et al., 1998] may suggest that the underlying Kartaly fault may be younger than early Permian, and implicitly younger than the MUF.

6. GEODYNAMIC EVOLUTION OF THE SOUTHERN URALIDES IN RELATION TO OTHER OROGENS

Tectonic evolution of the Southern Uralides stands in apparent contrast to other orogens where either delamination of the lower crust and uppermost mantle or significant subsidence were interpreted to result from eclogitization of crustal roots (Figure 4) [*Austrheim*, 1991; *Laubscher*, 1990; Baird et al., 1995]. In the Early Tertiary Alps (Figure 4a), the proposed eclogitized European crustal root is thought to be depressed to depths in excess of 60 km based on deep seismic reflection data [*Laubscher*, 1990]. The metamorphic phase-change to higher density eclogites was suggested to occur concurrently with the collision between the European and African plates. The eclogitization of the lower crust, the indentation of the European crust by wedges of the African crust protruding northwards beneath the Alps as well as the ultramafic composition of the protolith were interpreted as triggering factors for the delamination of the European lower crust and lithospheric mantle [*Frei* et al., 1989; *Bousquet* et al., 1997]. Among the strongest evidence for the subduction of the European continental lithosphere and delamination of the Alpine crustal root was provided by material balance calculations [*Laubscher*, 1989] and tomographic studies

[*Spakman* et al., 1993]. Unlike the model proposed for the Southern Uralides, the estimated shortening in the Alps exceeds the length of the restored section of the Alpine foreland fold and thrust belt, and delamination of the Alpine crustal root was interpreted to account for this deficit.

Some of the best studied orogenic belts, particularly in relation to deep crustal processes, is the early Paleozoic Caledonian belt. The post-orogenic evolution of the Scandinavian Caledonides indicates a similar tectonic progression with the Alps, but the Caledonides are probably in a more advanced geodynamic setting having already experienced orogenic collapse [*Austrheim*, 1987; *Andersen* et al., 1991; *Austrheim*, 1991; *Boundy* et al., 1992; *Fountain* et al., 1994a; *Austrheim* et al., 1997]. In the Scandinavian Caledonides, Precambrian granulite facies rocks were interpreted to have undergone fluid-controlled eclogitization on a regional scale, which conceivably destabilized the isostatic equilibrium due to a much heavier root. As a result, the Caledonides perhaps dropped their root, which triggered subsequent collapse and extension.

Eclogitization of orogenic roots has also been proposed for Proterozoic age orogenic belts such as the Trans-Hudson orogen of North America (Figure 4c) [*Baird* et al., 1995, 1996]. Here, eclogitization of the Hudsonian crustal root was proposed as a much later event in the orogenic development, some ~1.2 Ga after the termination of the collisional process. From deep seismic reflection profiling, the eclogitization of the Trans-Hudson root was interpreted as a mechanism of triggering post-orogenic subsidence of the overlying upper crust, resulting in the formation of the Williston sedimentary basin [*Baird* et al., 1995, 1996].

Quite a different evolution of the lower crust characterizes the Tibetan Plateau where in fact retrogression of eclogite to granulite facies rocks was proposed as a mechanism to cause large-scale uplift without surface shortening [*Le Pichon* et al., 1997]. An alternative model for the high elevations of the Himalayas was proposed by *Henry* et al. [1997] who suggested that the eclogitization of the underthrust Indian lower crust at ~75 km, as opposed to ~55 km for the Alps, enabled the mountain belt to maintain its higher average altitude (5 km). This latter study proposed that the depth of the granulite to eclogite transition may play an important role in the geodynamic evolution of the orogens.

The model put forth in this paper certainly does not provide all of the answers with regard to the post-tectonic stabilization of the Southern Uralides. Yet, it is widely accepted that the Southern Uralides have preserved their collisional architecture for more than 250 Ma, and our

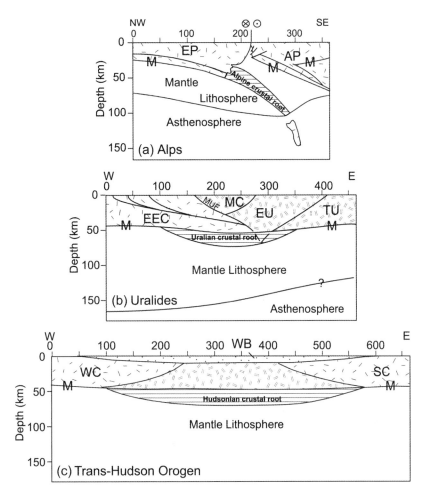

Figure 4. Contrasting geodynamic models resulting from the eclogitization of orogenic roots: (a) the Alps (delamination), after Marchant and Stampfli (1997). M — the Moho, EC — European Plate, AP — Adriatic plate. (b) the Uralides (stabilization). EEC — East European Craton, EU — East Uralian zone, MC — Maksyutov complex, TU — Trans Uralian zone, MUF — Main Uralian fault; and (c) the Trans-Hudson orogen (subsidence), after *Baird* et al. (1995). WB – Williston Basin, WC — Wyoming Craton, SC — Superior Craton. Horizontal line pattern shows the portion of the continental crust interpreted as eclogite-facies rocks.

model presents a possible scenario for this anomaly. The metamorphic phase-change of the Southern Uralian crustal root into higher density eclogite facies rocks possibly caused the stabilization and preservation of the Late Paleozoic Uralian orogenic architecture, and built an isostatically balanced system that restrained the eclogitic crustal root from sinking into the mantle. Based on this study, we further suggest that the timing of the eclogitization of crustal roots may play a significant role in the geodynamic evolution of the orogens. This interpretation adds a new possible explanation to previous attempts to decipher the causes for stabilization and preservation of the Uralian orogen, including (1) abundance of island arcs or/and (2) incomplete or

"arrested" collisional process [*Berzin* et al., 1996]. While this interpretation is in agreement with some of the previous models put forth for the geodynamic evolution of the Southern Uralides [*Artyushkov* et al., 2000], it contradicts others [*Leech*, 2001] that suggested that the Southern Uralides orogenic root has not undergone metamorphic phase-change to higher density eclogites.

There is still a question why the Uralides did not loose their heavy root as proposed for other orogenic belts [e.g., *Platt and England*, 1994; *Bousquet* et al., 1997; *Marotta* et al., 1998], or alternatively, why the eclogitic root did not retrogress to higher temperature granulites. A possible scenario is that the Southern Uralian root is made of lighter andesitic eclogites as opposed to heavier gabbroic eclogites favoring gravitational equilibrium in contrast to gravitational instability and delamination [*Bousquet* et al., 1997] although this model stays at odds with the lack of abundance reflectivity at the lower crustal levels beneath the axis of the orogen. However, the geotherms at ~70 km depth beneath the central part of the Southern Uralides are below 700°C [*Kukkonen* et al., 1997], indicating that the root is within the eclogite stability field, and too low to allow retrogression to lower grade granulites [*Henry* et al., 1997]. However, the timing of eclogitization, perhaps driven by the availability of fluid to flux the reaction kinetics, may be as important as the depth of the metamorphic phase-change.

7. CONCLUSIONS

Crustal-scale restoration of the Southern Uralide fold and thrust belt corroborated by wide-angle/near-vertical incidence URSEIS seismic profile, gravity, topography, fission track data, and thermal modeling provide the basis for a model involving metamorphic phase-change to higher density eclogite facies assemblages within the orogenic root. Our model predicts that the Southern Uralian lower crust should be eclogitized at a depth of ~53–70 km where the wide-angle PmP arrivals indicate an increase of the P-wave velocity to ~8.0 km/s and the near-vertical incidence seismic reflection Moho is lost due to presumably mantle-like density and velocity. From the crustal scale restoration, we predict that ~70% of the existent Southern Uralides root originates from continental crust of East European affinity, and only 30% derives from accreted terranes west of the Main Uralian fault. The loss of the Moho reflection character could be interpreted that massive eclogitization occurred within the Southern Uralian root to raise the velocity and density to mantle values. This would further imply that sufficient water was released in the crust to allow massive eclogitization, possibly from adjacent west-dipping subduction zone of the accreted terranes in the Uralian hinterland.

Earlier studies [*Döring and Götze*, 1999; *Carbonell* et al., 2000] suggested the presence of a high density body at the Southern Uralian crust/mantle boundary. Here, we do not only provide a more quantitative model in support of the eclogitization of a substantial crustal root, but we suggest that this process occurred between the Late Triassic and Early Jurassic time, and may have served to stabilize and preserve the collisional orogenic structure. The Southern Uralides represent a unique case for studying the long-term stability of orogenic systems, being the only Paleozoic mountain belt which escaped post-orogenic collapse. In addition to previous studies

[*Austrheim* et al., 1997; *Henry* et al., 1997] which argued that the depth of the granulite to eclogite transition may play a significant role in the orogenic evolution, we suggest that the timing of eclogitization may be an important factor in the geodynamic development of orogenic systems. The inferred post-collisional eclogitization of the Southern Uralides lower crust perhaps built an isostatically balanced orogenic system, with subdued topography, and served to stabilize the orogenic architecture.

Formation of eclogite facies rocks in overthickened orogenic roots and the resulting geodynamic processes are still under debate. It has been suggested that formation of eclogites at the crust-mantle boundary triggered delamination of the lower crust and upper mantle in the Alps [*Laubscher*, 1990] or delamination followed by post-orogenic collapse in the Caledonides [*Austrheim*, 1991], or only subsidence of the overlying upper crust in the Trans-Hudson orogen [*Baird* et al., 1995, 1996]. Unlike these other models of eclogitization of the orogenic roots (Figure 4), the Southern Uralides appear to be an intermediate case in which there is no evidence for either delamination or subsidence of the overlying upper crust. Conversely, examples from the Tibetan Plateau [*Henry* et al., 1997; *Le Pichon* et al., 1997], the Variscan granulites of the French Massif Central [*Pin and Vielzeuf*, 1983; *Mercier* et al., 1991], and eastern Australia [*Smith*, 1982] suggested that retrogression from eclogite to granulite facies rocks due to the gradual increase in temperature could be a mechanism of triggering epeirogenic events including regional scale uplift. The present-day low temperatures in the Southern Uralides at ~70 km (<700°C) [*Kukkonen* et al., 1997] are within the eclogite stability field, well below the stability temperatures for granulite facies rocks (~800°C), preventing the occurrence of retrogression to granulites. Since it is widely accepted that eclogites commonly have densities higher than the surrounding mantle peridotites [*Mengel and Kern*, 1992] there is still an open question why the Southern Uralides root has not recycled yet into the mantle [e.g., *Platt and England*, 1994; *Dewey*, 1998].

The Southern Uralides represent yet another example that orogenic loads could be supported by density heterogeneities in the lower crust or upper mantle rather than thickening of the crust or lateral density variations. However, in contrast with the compensation model proposed for the Sierra Nevada, where the high elevations may be supported by low density bodies in the upper mantle [e.g., *Ducea and Saleeby*, 1996; *Jones and Phinney*, 1998], we suggest that the low elevations of the Southern Uralides resulted from high density material in the upper mantle, specifically eclogite facies assemblages that perhaps served to stabilize the orogen.

Acknowledgments. This work benefited from funding provided by National Science Foundation grant EAR-9418251, and formed part of the Ph.D. thesis research of C. C. Diaconescu while at Cornell University. We thank all of the participants in the URSEIS project for their contributions to its success. L. D. Brown, R. W. Allmendinger, D. N. Steer, and E. Sandvol helped to shape our ideas through engaging discussions. J. McBride and D. Fountain provided useful insight on an earlier version of the manuscript. Special thanks are due to D. Brown, C. Hurich, and H. Austrheim for their helpful comments and thorough reviews of the manuscript.

REFERENCES

Ahrens, T. J. and G. Schubert, Gabbro-eclogite reaction rate and its geophysical significance, *Rev. of Geophys. Space Phys.*, 13, 383–400, 1975.

Allmendinger, R. W., D. Figueroa, D. Snyder, J. Beer, C. Mpodozis and B. L. Isacks, Foreland shortening and crustal balancing in the Andes at 30° S latitude, *Tectonics*, 9, 789–809, 1990.

Andersen, T. B., B. Jamtveit, J. F. Dewey and E. Swensson, Subduction and eduction of continental crust; major mechanisms during continent-continent collision and orogenic extensional collapse, a model based on the South Norwegian Caledonides, *Terra Nova*, 3, 303–310, 1991.

Artyushkov, E. V., M. A. Baer, P. A. Chekhovich and N. A. Morner, The Southern Urals. Decoupled evolution of the thrust belt and its foreland: a consequence of metamorphism and lithospheric weakening, *Tectonophysics*, 320, 271–310, 2000.

Austrheim, H., Eclogitization of lower crustal granulites by fluid migration through shear zones, *Earth Planet. Sci. Lett.*, 81, 221–232, 1987.

Austrheim, H., Eclogite formation and dynamics of crustal roots under continental collision zones, *Terra Nova*, 3, 492–499, 1991.

Austrheim, H., M. Erambert and A. K. Engvik, Processing of crust in the root of the Caledonian continental collision zone; the role of eclogitization, *Tectonophysics*, 273, 129–153, 1997.

Baird, D. J., J. H. Knapp, D. N. Steer, L. D. Brown and K. D. Nelson, Upper-mantle reflectivity beneath the Williston basin, phase change Moho, and the origin of the intracratonic basins, *Geology*, 23, 431–434, 1995.

Baird, D. J., K. D. Nelson, J. H. Knapp, J. J. Walters and L. D. Brown, Crustal structure and evolution of the Trans-Hudson orogen: Results from seismic reflection profiling, *Tectonics*, 15, 416–426, 1996.

Beane, R. J. and J. N. Connelly, ^{40}Ar/^{39}Ar, U-Pb and Sm-Nd constraints on the timing of metamorphic events in the Maksyutov Complex, Southern Ural Mountains, *J. Geol. Soc. London*, 157, 811–822, 2000.

Beane, R. J., Petrologic evolution and geochronologic constraints for high-pressure metamorphism in the Maksyutov Complex, South Ural mountains, *Ph.D. thesis*, Stanford University, Stanford, 1997.

Beane, R. J., J. G. Liou, R. G. Coleman and M. J. Leech, Petrology and retrograde P-T path for eclogites of the Maksyutov Complex, Southern Ural Mountains, Russia, *The Island Arc*, 4, 254–266, 1995.

Berzin, R., O. Oncken, J. H. Knapp, A. Pérez-Estaún, T. Hismatulin, N. Yunusov and A. Lipilin, Orogenic evolution of the Ural mountains: results from an integrated seismic experiment, *Science*, 274, 220–221, 1996.

Boundy, T. M., D. M. Fountain and H. Austrheim, Structural development and petrofabrics of eclogite facies shear zones, Bergen arcs, western Norway; implications for deep crustal deformational processes, *J. Metamorph. Geol.*, 10, 127–146, 1992.

Bousquet, R., B. Goffe, P. Henry, X. Le Pichon and C. Chopin, Kinematic, thermal and petrological model of the Central Alps; Lepontine metamorphism in the upper crust and eclogitization of the lower crust, *Tectonophysics*, 273, 105–127, 1997.

Brown, D., V. Puchkov, J. Alvarez-Marron and A. Perez-Estaun, The structural architecture of the footwall to the Main Uralian Fault, southern Urals, *Earth Sci. Rev.*, 40, 125–147, 1996.

Brown, D., V. Puchkov, J. Alvarez-Marron and A. Perez-Estaun, Preservation of a subcritical wedge in the south Urals foreland thrust and fold belt, *J. Geol. Soc., London*, 154, 593–596, 1997.

Brown, D., C. Juhlin, J. Alvarez Marron, A. Perez-Estaun and A. Oslianski, Crustal-scale structure and evolution of an arc-continent collision zone in the Southern Urals, Russia, *Tectonics*, 17, 158–171, 1998.

Carbonell R., A. Perez-Estaun, J. Gallart, J. Diaz, S. Kashubin, J. Mechie, R. Stadtlander, A. Schultze, J. H. Knapp and A. Morozov, Crustal root beneath the Urals: Wide-angle seismic evidence, *Science*, 274, 222–224, 1996.

Carbonell, R., D. Lecerf, M. Itzin, J. Gallart and D. Brown, Mapping the Moho beneath the Southern Urals with Wide-angle Reflections, *Geophys. Res. Lett.*, 25, 4229–4232, 1998.

Carbonell, R., J. Gallart, A. Perez-Estaun, J. Diaz, S. Kashubin, J. Mechie, F. Wenzel and J. H. Knapp, Seismic wide-angle constraints on the crust of the Southern Urals, *J. Geophys. Res.*, 105, 13,755–13,777, 2000.

Cook, F. A., D. S. Albaugh, L. D. Brown, S. Kaufman, J. E. Oliver and R. D. Hatcher, Jr., Thin-skinned tectonics in the crystalline southern Appalachians; COCORP seismic-reflection profiling of the Blue Ridge and Piedmont, *Geology*, 7, 563–567, 1979.

Dahlstrom, C. D. A., Structural geology in the eastern margin of the Canadian rocky mountains, *Bull. Canadian Petrol. Geol.*, 18, 332–406, 1970.

Dewey, J. F., Extensional collapse of orogens, *Tectonics*, 7, 1123–1139, 1988.

Dewey, J. F., P. D. Ryan and T. B. Andersen, Orogenic uplift and collapse, crustal thickness, fabrics and metamorphic phase changes; the role of eclogites in Magmatic processes and plate tectonics, edited by H. M. Prichard, T. Alabaster, N. B. W. Harris and C. R. Neary, *Geological Society Special Publications*, 76, 325–343, 1993.

Diaconescu, C. C., J. H. Knapp, L. D. Brown, D. N. Steer and M. Stiller, Precambrian Moho offset and tectonic stability of the East European platform from the URSEIS deep seismic profile, *Geology*, 26, 211–214, 1998.

Dobretsov, N. L., V. S. Shatsky, R. G. Coleman, V. I. Lennykh, P. M. Valizer, J. Liou, R. Zhang and R. J. Beane, Tectonic setting and petrology of ultrahigh-pressure metamorphic rocks in the Maksyutov Complex, Ural Mountains, Russia, *Int. Geol. Rev.*, 38, 136–160, 1996.

Döring, J., H. J. Götze and M. K. Kaban, Preliminary study of the gravity field of the Southern Urals along the URSEIS '95 seismic profile, *Tectonophysics*, 276, 49–62, 1997.

Döring, J. and H. J. Götze, The isostatic state of the southern Urals crust, *Geol. Rundsch*, 87, 500–510, 1999.

Ducea, M. N. and J. B. Saleeby, Buoyancy sources for a large, unrooted mountain range, the Sierra Nevada, California; evidence from xenolith thermobarometry, *J. Geophys. Res.*, 101, 8229–8244, 1996.

Druzhinin, V. S., S. N. Kashubin, S. V. Avtoneev and V. M. Rybalka, New data on deep structure of the Southern Urals according to results of investigations on the Troitsk DSS profile, *Soviet Geol. Geophys.*, 29, 79–82, 1988.

Druzhinin, V. S., S. V. Avtoneev, S. N. Kashubin and V. M. Rybalka, New data on the deep structure of the northern area of the Southern Urals in section of the Taratash DSS profile, *Soviet Geol. Geophys.*, 31, 113–116, 1990.

Echtler, H. P., M. Stiller, F. Steinhoff, A. Suleimanov, J. H. Knapp and J. Alvarez-Marron, URSEIS '95 — Crustal architecture of the southern Urals by vibroseis CMP-profiling, *Science*, 274, 224–226, 1996.

Echtler, H. P. and R. Hetzel, Main Uralian thrust and Main Uralian normal fault; nonextensional Paleozoic high-P rock exhumation, oblique collision, and normal faulting in the Southern Urals, *Terra Nova*, 9, 158–162, 1997.

Fountain, D. M., T. M. Boundy, H. Austrheim and P. Rey, Eclogite-facies shear zones — deep crustal reflectors, *Tectonophysics*, 232, 411–424, 1994a.

Fountain, D. M., P. Rey and H. Austrheim, Seismic velocities across the garnet granulite-eclogite facies transition, *Eos, Trans., Am. Geophys. Union*, 75 (44, Suppl.), 676 pp., 1994b.

Frei, W., P. Heitzmann, P. Lehner, S. Muller, R. Olivier, A. Pfiffner, A. Steck and P. Valasek, Geotraverses across the Swiss Alps, *Nature*, 340, 1989.

Furlong, K. P. and D. M. Fountain, Continental crustal underplating; thermal considerations and seismic-petrologic consequences, *J. Geophys. Res.*, 91, 8285–8294, 1986.

Gee, D. G., M. Beckholmen, M. Friberg, A. N. Glushkov and V. Sokolov, Baikalian-age structural control of Uralian tectonics-evidence from the Middle Urals: Scientific Program and Abstracts, *EUROPROBE Workshop Uralides Session*, Granada, Spain, 7 pp., 1996.

Giese U, U. Glasmacher, V. I. Kozlov, I. Matenaar, V. N. Puchkov, L. Stroink, W. Bauer, S. Ladage and R. Walter, Structural framework of the Bashkirian anticlinorium, SW Urals, *Geol. Rund.*, 87, 526–544, 1999.

Hamilton, W., The Uralides and the motion of the Russian and Siberian platforms, *Geol. Soc. of Amer. Bull.*, 81, 2553–2576, 1970.

Henry, P., X. Le Pichon and B. Goff, Kinematic, thermal and petrological model for the Himalayas: constraints related to metamorphism within the underthrust Indian crust and topographic elevation, *Tectonophysics*, 273, 31–56, 1997.

Henstock, T. J., A. Levander, C. M. Snelson, G. R. Keller, K. C. Miller, S. H. Harder, A. R. Gorman, R. M. Clowes, M. J. A. Burianyk and E. D. Humphreys, Probing the Archean and Proterozoic lithosphere of western North America, *GSA Today*, 8, 1–5, 1998.

Hetzel, R., Geology and geodynamic evolution of the high-P/low-T Maksyutov Complex, southern Urals, Russia, *Geol. Rundsch.*, 87, 577–588, 1999.

Hetzel, R., H. P. Echtler, W. Seifert, B. A. Schulte and K. S. Ivanov, Subduction- and exhumation-related fabrics in the Paleozoic high-pressure/low-temperature Maksyutov Complex, Antingan area, Southern Urals, Russia, *GSA Bulletin*, 110, 916–930, 1998.

Hynes, A. and D. B. Snyder, Deep-crustal mineral assemblages and potential for crustal rocks below the Moho in the Scottish Caledonides, *Geophys. J. Int.*, 123, 323–339, 1995.

Jones, C. H., H. Kanamori and S. W. Roecker, Missing roots and mantle "drips"; regional P (sub n) and teleseismic arrival times in the southern Sierra Nevada and vicinity, California, *J. Geophys. Res.*, 99, 4567–4601, 1994.

Jones, C. H. and R. A. Phinney, Seismic structure of the lithosphere from teleseismic converted arrivals observed at small arrays in the southern Sierra Nevada and vicinity, California, *J. Geophys. Res.*, 103, 10,065–10,090, 1998.

Juhlin, C., J. H. Knapp, S. Kashubin and M. Bliznetsov, Crustal evolution of the middle Urals based on seismic reflection and refraction data, *Tectonophysics*, 264, 21–34, 1996.

Kern, H. and A. Richter, Temperature derivatives of compressional and shear wave velocities in crustal and mantle rocks at 6 kbar confining pressure, *J. Geophys.*, 49, 47–56, 1981.

Knapp, J. H., C. C. Diaconescu, M. A. Bader, V. B. Sokolov, S. N. Kashubin and A. V. Rybalka, Seismic reflection fabrics of continental collision and post-orogenic extension in the Middle Urals, central Russia, *Tectonophysics*, 288, 115–126, 1998.

Knapp, J. H., D. N. Steer, L. D. Brown, R. Berzin, A. Suleimanov, M. Stiller, E. Lüschen, D. Brown, R. Bulgakov, S. N. Kashubin and A. V. Rybalka, Lithosphere-scale seismic image of the southern Urals from explosion-source reflection profiling, *Science*, 274, 226–227, 1996.

Kruse, S. E. and M. McNutt, Compensation of Paleozoic orogens; a comparison of the Urals to the Appalachians, *Tectonophysics*, 154, 1–17, 1988.

Kukkonen, I. T., I. V. Golovanova, Y. V. Khachay, V. S. Druzhinin, A. M. Kosarev and V. A. Schapov, Low geothermal heat flow of the Urals fold belt — implication of low heat production, fluid circulation or palaeoclimate?, *Tectonophysics*, 276, 63–85, 1997.

Laubscher, H., The problem of the Moho in the Alps, *Tectonophysics*, 182, 9–20, 1990.

Leech, M., Arrested orogenic development; eclogitization, delamination and tectonic collapse, *Earth Planet. Sci. Lett.*, 18, 149–159, 2001.

Leech, M. L. and F. Stockli, The late exhumation history of the ultrahigh-pressure Maksyutov Complex, South Ural mountains, from new apatite fission-track data, *Tectonics*, 19, 153–167, 2000.

Lennykh, V. I., P. M. Valizer, R. Beane, M. Leech and W. G. Ernst, Petrotectonic evolution of the Maksyutov Complex, Southern Urals, Russia: Implications for ultrahigh-pressure metamorphism, *Int. Geol. Rev.*, 37, 584–600, 1995.

Le Pichon, X., P. Henry and B. Goffe, Uplift of Tibet; from eclogites to granulites; implications for the Andean Plateau and the Variscan belt, *Tectonophysics*, 273, 57–76, 1997.

Meissner, R., T. Wever and R. Bittner, Results of DEKORP 2-S and other reflection profiles through the Variscides, *Geophys. J. Royal Astronom. Soc.*, 89, 319–324, 1987.

Marchant, R. H. and G. M. Stampfli, Crustal and lithospheric structure of the Western Alps; geodynamic significance, In *Results of NRP 20; Deep Structure of the Swiss Alps*, edited by O. A. Pfiffner, P. Lehner, P. Heitzman, S. Mueller and A. Steck, pp. 326–337, 1997.

Marotta, A. M., M. Fernandez and R. Sabadini, Mantle unrooting in collisional settings, *Tectonophysics*, 296, 31–46, 1998.

Matte, Ph., H. Maluski, R. Caby, A. Nicolas, P. Kepezhinskas and S. Sobolev, Geodynamic model and $^{39}Ar/^{40}Ar$ dating for the generation and emplacement of the high pressure (HP) metamorphic rocks in SW Urals, *C. R. Acad. Sci. Paris*, 317, 1667–1674, 1993.

Mengel, K. and H. Kern, Evolution of the petrological and seismic Moho; implications for the continental crust-mantle boundary, *Terra Nova*, 4, 109–116, 1992.

Mercier, L., J.-M. Lardeaux and P. Davy, On the tectonic significance of retrograde P-T-t paths in eclogites of the French Massif Central, *Tectonics*, 10, 131–140, 1991.

Nelson, K. D., J. H. McBride, J. A. Arnow, D. M. Wille, L. D. Brown, J. E. Oliver and S. Kaufman, Results of recent COCORP profiling in the Southeastern United States, *Geophys. J. Royal Astronom. Soc.*, 89, 141–146, 1987.

Perez Estaun, A., J. Alvarez Marron, D. Brown, V. N. Puchkov, Y. Gorozhanina and V. Baryshev, Along-strike structural variations in the foreland thrust and fold belt of the Southern Urals, *Tectonophysics*, 276, 1997.

Pin, C. and D. Vielzeuf, Granulites and related rocks in Variscan median Europe; a dualistic interpretation, *Tectonophysics*, 93, 47–74, 1983.

Piwowar, T. J., J. H. Knapp and G. Danukalova, Long wavelength Cenozoic flexural uplift across the Southern Urals and Central Eurasia, *Geol. Soc. Am.*, Denver, Colorado, Abst., A172, 1996.

Platt, J. P. and P. C. England, Convective removal of lithosphere beneath mountain belts; thermal and mechanical consequences, *Am. J. Sci.* 294, 307–336, 1994.

Poli, S. and M. W. Schmidt, The high-pressure stability of hydrous phases in orogenic belts: an experimental approach on eclogite-forming processes, *Tectonophysics*, 273, 169–184, 1997.

Puchkov, V. N., Geodynamic control of regional metamorphism in the Urals, *Geotectonics*, 30, 97–113, 1996.

Rodgers, J., Fold-and-thrust belts in sedimentary rocks; Part 1, Typical examples, *Am. J. Sci.*, 290, 321–359, 1990.

Sapin, M. and A. Hirn, Seismic structure and evidence for eclogitization during the Himalayan convergence, *Tectonophysics*, 273, 1–16, 1997.

Schreyer W. and B. Stöckhert, High-Pressure metamorphism in nature and experiment, *Lithos*, 41, 1997.

Sengör, A. M. C., B. A. Natal'in and B. S. Burtman, Evolution of the Altaid tectonic collage and Paleozoic crustal growth in Eurasia, *Nature*, 364, 299–307, 1993.

Seward, D., A. Perez-Estaun and V. Puchkov, Preliminary fission-track results from the southern Urals-Sterlitamak to Magnitogorsk, *Tectonophysics*, 276, 281–290, 1997.

Skripiy, A. A. and N. K. Yunusov, Tension and compression structures in the articulation zone of the Southern Urals and the East European Platform, *Geotectonics*, 23, 515–522, 1989.

Smith, A. G., Late Cenozoic uplift of stable continents in a reference frame fixed to South America, *Nature*, 296, 400–404, 1982.

Spakman, W., S. van der Lee and R. van der Hilst, Travel-time tomography of the European-Mediterranean mantle down to 1400 km, *Phys. Earth Planet. Interiors*, 79, 3–74, 1993.

Spear, F. S., Metamorphic phase equilibria and pressure-temperature-time paths, Monograph Series, 1, *Mineral. Soc. Am.*, Washington, D.C., 799 pp., 1993.

Stadtlander, R., J. Mechie and A. Schulze, Deep structure of the Southern Ural mountains as derived from wide-angle seismic data, *Geophys. J. Int.*, 137, 501–515, 1999.

Steer, D. N., J. H. Knapp, L. D. Brown, H. P. Echtler, D. Brown and R. Berzin, Deep structure of the continental lithosphere in an unextended orogen: An explosive-source seismic reflection profile in the Urals (Urals Seismic Experiment and Integrated Studies (URSEIS 1995)), *Tectonics*, 17, 143–157, 1998.

Thouvenot, F., S. N. Kashubin, G. Poupinet, V. V. Makovskiy, T. V. Kashubina, P. Matte and L. Jenatton, The root of the Urals; evidence from wide-angle reflection seismics, *Tectonophysics*, 250, 1–13, 1995.

Wernicke, B. P., R. W. Clayton, M. N. Ducea, C. H. Jones, S. K. Park, S. D. Ruppert, J. B. Saleeby, J. K. Snow, L. J. Squires, M. M. Fliedner, G. R. Jiracek, G. R. Keller, S. L. Klemperer, J. H. Luetgert, P. E. Malin, K. C. Miller, W. D. Mooney, H. W. Oliver and R. A. Phinney, Origin of high mountains in the continents; the southern Sierra Nevada, *Science*, 271, 190–193, 1996.

Zakharov, O. A. and V. N. Puchkov, On the tectonic nature of the Maksutovo complex of the Ural-Tau zone, 28 pp., report, Ufimian Sci. Cent., Russ. Acad. of Sci., Ufa, Russia, 1994 (in Russian).

Zonenshain, L. P., V. G. Korinevsky, V. G. Kazmin, D. M. Pechersky, V. V. Khain and V. V. Matveenkov, Plate tectonic model of the south Urals development, *Tectonophysics*, 109, 95–135, 1984.

Zonenshain, L. P., M. I. Kuzman and L. M. Natopov, Geology of the USSR: A plate-tectonic synthesis, Geodyn. Ser., 21, *Amer. Geophys. Un.*, Washington, D.C., 25–54, 1990.

Camelia C. Diaconescu and James H. Knapp, Department of Geological Sciences, University of South Carolina, Columbia, SC 29208

Tectonic Processes During Collisional Orogenesis From Comparison of the Southern Uralides with the Central Variscides

J. Alvarez-Marrón

Institute of Earth Sciences, Spain

A great deal of available multidisciplinary data that focused around URSEIS and DEKORP 2N-2S deep seismic profiles contribute to an advanced understanding of the respective Southern Uralide and Central Variscide transects. Both transects cross a continental margin and several terranes that accreted to it by subduction/collision processes during the Paleozoic. However, there are important differences in the present crustal architecture between the two orogens, mainly with respect to the geometry of the terrane boundaries, and in the structure at lower crustal and Moho depths. Major differences during the Devonian to Permian tectonic evolution recorded in each orogen are; the nature of the material accreted, the order of accretion, the orientation of the subduction zones, and the location/distribution of active deformation through time. The comparison of the respective orogenic evolutions highlights the different tectonic processes that are dominant during continental assemblage in each orogenic system. In the Uralides, significant crustal growth occurred by accretion of newly formed volcanic arc crust to the continental margin and subsequent accretion of outboard subduction related accretionary material during the imbrication of the continental margin. In the Variscides, existing crustal material progressively accreted towards the continental margin by the forward advance of a growing orogenic wedge. The soft or rigid behaviour of the major crustal boundaries seems to have played an important role in the different nature and geometry of the deformation between the Uralides and Variscides. The collisional architecture in the central Variscides has been overprinted by post-orogenic processes and subsequent tectonic events, while that of the southern Uralides has been preserved from any major subsequent tectonic process.

1. INTRODUCTION

Mountain Building in the Uralides: Pangea to the Present
Geophysical Monograph 132
Copyright 2002 by the American Geophysical Union
10.1029/132GM06

An important advance in the understanding of orogen evolution has been achieved by the worldwide acquisition of multidisciplinary transects across a number of orogenic belts. Most of these transects are focused around deep seismic reflection/refraction profiles

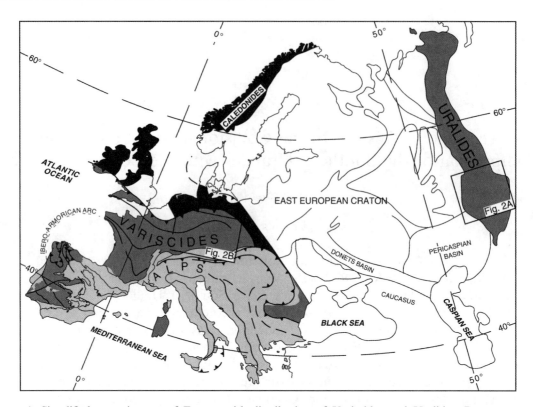

Figure 1. Simplified tectonic map of Europe with distribution of Variscides and Uralides. Boxes represent locations of the Central Variscides and Southern Uralides transects discussed in this paper and areas in Figure 2.

because they allow the architecture of the entire crust to be interpreted by establishing a link between surface and subsurface structures [e.g., *Meissner* et al., 1991; *Clowes* et al., 1996]. The integration of other geophysical, geochemical and geocronological data within this crustal-scale framework permits the development of models that address the tectonic evolution of an orogenic belt [e.g., *Ando* et al., 1984; *Clowes* et al., 1996; *Cook* et al., 1999]. Furthermore, the comparisons of multidisciplinary transects allow to distinguish different orogens in terms of the orogenic processes involved [*Cook and Varshek*, 1994]. In this paper, the comparison of multidisciplinary transects attempts to solve a fundamental question; is the current crustal architecture the consequence of the particular orogenic processes that operated during orogenic evolution, or were the inherited (pre-orogenic) crustal configurations and late orogenic processes more fundamental factors?

Two well studied transects across the Paleozoic southern Uralides and central Variscides orogens are especially well suited for this comparative study (Figure 1). The Uralides are a N-S trending, linear orogen at the geographic divide between Europe and Asia that seems to have been preserved from any subsequent tectonic

processes since the Early Triassic. In contrast, the Variscides in Western Europe exhibit a highly curved trend and appear dismembered by subsequent Cenozoic tectonics. Both orogens were fundamental during the process of continental assemblage that led to formation of the Pangea supercontinent in the Paleozoic. The southern Uralides have previously been compared with the Iberian transect of the Variscides [*Matte*, 1995].

The transects being compared are focused around available deep seismic reflection sections that have not sampled the full extent of either orogen. However, they both offer a complete view of one flank, crossing a deformed continental margin and the major tectonic boundaries, or sutures that record the collision of several magmatic and continental terranes against the margin. Fundamental structural differences between the two sections are drawn from the comparison of the crustal-scale architecture as deduced from available interpretations of deep seismic reflection profiles. The recorded tectonic history of continental assemblage from Devonian to Permian times is compared using models that illustrate the different processes that were active during various stages of evolution common to both orogenic systems.

The comparison between the southern Uralides and central Varisicides allows the role of dominant processes active during the tectonic evolution and the building of the final architecture of the orogenic edifices to be discussed, along with the importance of the differences in the kind of material involved in the orogeny and in the orientation of subduction zones. Other aspects such as inherited mechanical properties together with the soft or rigid behaviour of major crustal boundaries seem to have played an important role in producing the different nature and geometry of the deformation.

2. GEOLOGICAL FRAMEWORK

Both the Uralide and Variscide orogens are divided into several orogenic zones that represent a terrane collage resulting from subduction/collision processes (Figure 2). In the southern part of the Uralide orogen the western flank records the Early Devonian to Early Triassic history of accretion and final collision of an island arc complex (Magnitogorsk zone), and volcanic arcs mixed with possible continental fragments to the east (East Uralian and Trans Uralian zones) with the continental margin of the East European Craton. The northern flank of the central Variscides is subdivided into three zones that form more or less independent belts [*Kossmat*, 1927] separated by two major zonal boundaries [e.g., *Franke* et al., 2000]. The Saxothuringian and Moldanubian zones to the south assembled in pre-early Famennian times and subsequently accreted and finally collided during Early Carboniferous (Tournaisian) to Late Carboniferous times to the continental margin represented in the Rhenohercynian Zone [e.g., *Franke*, 2000; *Onken*, 1997].

The following is a review of the main geological features along both transects (summarized in Table 1). Emphasis is given to the lithological record, the structural style, and main tectonothermal events in each zone, together with the significance of the sutures. More detailed and comprehensive syntheses for the Uralides can be found in *Zonenshain* et al. [1984, 1990], *Puchkov* [1997], *Brown* et al. [1996], and for the Variscides in *Franke* [2000], *Matte* [1986, 1991], and *Franke* et al. [2000].

2.1. Southern Uralides

In the west, the southern Uralide transect extends across the foreland areas that are comprised of the Permian to Early Triassic foreland basin and the foreland thrust and fold belt that deforms the East European Craton margin in the footwall to the Main Uralian fault

(SU1 in Figure 2). The transect continues across three hinterland zones towards the east, the Magnitogorsk, East Uralian and Trans Uralian zones, that are juxtaposed along major tectonic boundaries, or sutures (SU2 and SU3 in Figure 2).

The foreland thrust and fold belt forms a west directed thrust stack divided into two superposed thrust systems that developed during different stages of the orogen's evolution [*Brown* et al., 1997a; *Alvarez-Marrón* et al., 2000]. The upper thrust system is composed of variably metamorphosed material from the outer edge of the continental margin, allochthonous fragments of ophiolites, and syntectonic Late Devonian volcanoclastic sediments forming the Late Devonian to Early Carboniferous arc-continent collision accretionary complex [*Alvarez-Marrón* et al., 2000; *Puchkov*, 2002]. The lower thrust system includes Precambrian basement rocks of the East European Craton that in the eastern edge contain a Cadomian-age terrane [*Puchkov*, 1997; *Giese* et al., 1999], Ordovician to Late? Carboniferous thin (1500–2000 m), predominantly carbonates of the continental margin sequences, and syntectonic sediments of Late Carboniferous to Permian age [*Puchkov*, 2002]. This lower thrust system caused less than 20 km shortening and consists of an imbricate fan developed by forward thrust propagation along a gently east dipping sole thrust during Late Carboniferous to Permian [*Brown* et al., 1997a]. Fission track dating yielded stratigraphic ages with only partial resetting, suggesting that the foreland thrust and fold belt has not experienced a large amount of erosion and exhumation [*Seward* et al., 1997]. The foreland basin consists of up to 3000 metres of westward-thinning and younging Upper Carboniferous through Permian-age siliciclastic and bioclastic sediments, marls, limestones, reefal limestones, and evaporites, locally topped by Lower Triassic conglomerates and sandstones [*Chuvashov and Diupina*, 1973; *Chuvashov* et al., 1993; *Einor* et al., 1984]. The easternmost margin of the foreland basin is deformed by the frontal structures of the foreland thrust and fold belt.

The Magnitogorsk island arc is sutured to the continental margin along the Main Uralian fault (SU1 in Figure 2), an up to 10 km wide serpentinite mélange that incorporates Late Devonian and Lower Carboniferous age fragments tectonically eroded from the Magnitogorsk volcanic arc, including a number of oceanic crust and mantle fragments [e.g., *Savelieva* et al., 1997]. A late undeformed phase of the Syrostan batholith (327 ± 2 Ma. [*Montero* et al., 2000]) intrudes the fault zone, suggesting that tectonic activity along it had ended by the end of the Early Carboniferous.

Table 1. Comparative diagram between Southern Uralides and Central Variscides. Data of the different crustal blocks and sutures integrated in the orogenic architecture are from references in text.

	SOUTHERN URALIDES	CENTRAL VARISCIDES

Continental margin +	*External Areas*:
	 Cover: Ord. to Dev. (1500 m)
	 Foreland basin: Late Carboniferous to Early Triassic
	   (3000 m)
	 Structure: Imbricate thrust system
	 20 km shortening
	 Late Carb to Permian
		Rhenohercynian Zone:
		 L.Dev. to E. Carb. (3–12 km)
		 Namurian to late Wesfalian
		   (>10000 m)
		 Imbricate thrust system
		 180-200 km shortening
		 Visean to Westfalian

Accretionary Complex (Allochthon)	*Zilair Nappe* Famenian Flysh greywackes + Sil. Offiolites + Sil. Pelites + Famenian HP belt	*Giessen-Harz Nappes* Frasn.-L. Carb. Flysh greywackes + Devonian MORB -like metabasalts Mid.Dev-Frasn. Pelites HP belt (Phyllite Zone) 325 Ma
SUTURE 1	**SU1** - High angle fault (*melange*, Not reworked) Frontal collision, closed by 330 Ma.	**SV1** - Low angle fault (reworked) Oblique collision
Magmatic Arc	*Magnitogorsk Zone*: OCEANIC Devonian volcanics (no metamorphism)	*Saxoturingian Zone*: CONTINENTAL low Carb. volcanics LP/HTmetamorphism
SUTURE 2	**SU2** - High angle fault, belt of serpentinites Oblique collision	**SV2** - Low angle fault Ductile shear zone Frontal collision, closed by 320 Ma
Crustal Block	*East Uralian Zone*: Precambrian-Paleozoic rocks Subduction-accretion complexes Greenschist-amphibolite facies metamorphism	*Moldanubian Zone*: Teplá-Barrandian Unit CONTINENTAL Cadomian basement NW directed nappes Barrowian metamorphism
SUTURE 3	**SU3** - Melange+ dextral strike-slip fault Closed by 295 Ma	Possible suture ? 340 Ma. batholith
Crustal Block	*Transuralian Zone*: Precambrian-Paleozoic rocks Subduction-accretion complexes Ofiolites, Hp rocks Devonian-Lower Carboniferous seds. east vergent folds	Moldanubian *sensu stricto* SE directed nappes Eclogites, HP granulites, mantel rocks

The Magnitogorsk zone consists of Early to Late Devonian age arc-tholeiite to calc-alkaline volcanic units [*Seravkin* et al., 1992] that form the basement, and up to 5000 m of fore arc sediments of Middle Devonian to Early Carboniferous age [*Maslov* et al., 1993; *Brown* et al., 2001]. The fore arc basin sediments are weakly deformed by open, volcanic basement-cored synsedimentary folds and minor west directed thrusting and the metamorphic grade is very low [*Brown* et al., 2001]. The Magnitogorsk arc was not active during most of the Carboniferous, although Lower Carboniferous magmatic rocks are found [*Fershtater* et al., 1997]. The

eastern contact with the East Uralian zone is a highly imbricated belt with large amounts of serpentinites (SU2 in Figure 2).

The East Uralian zone is composed of highly deformed and metamorphosed Precambrian and Paleozoic rocks that are tectonically juxtaposed, and show continental and island arc affinities [*Puchkov*, 1997]. Metamorphic grade ranges from greenschist to amphibolite facies, and is characterized by high temperatures and low pressures [*Puchkov*, 1997; *Echtler* et al., 1997]. These rocks have been extensively intruded by granitoids ranging from Lower Devonian to Permian in age [*Bea*

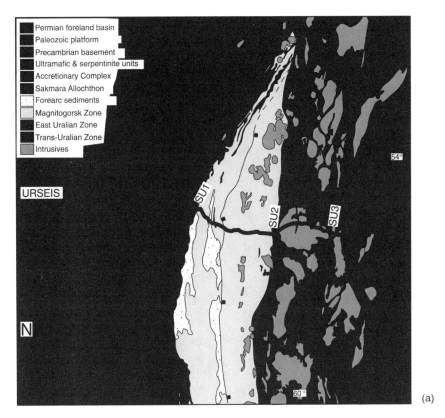

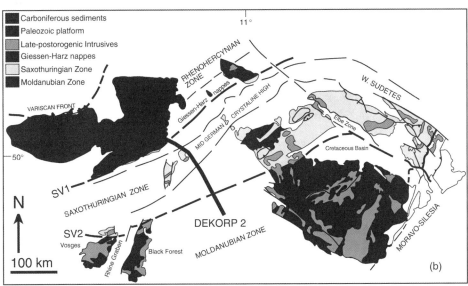

Figure 2. (a) Simplified geological map of the Southern Urals [*Brown* et al., 2000]. The main sutures are marked by SU1, SU2 and SU3. (b) Simplified geological map of the Central Variscides [*Franke* et al., 2000]. The main sutures are marked by SV1 and SV2. Thick lines correspond to seismic profiles shown in Figure 3.

et al., 1997, 2002]. The Upper Devonian to Lower Carboniferous granitoids are interpreted to have formed in an Andean type setting [*Bea* et al., 2002], whereas

Upper Carboniferous to Permian age granitoids are of continental origin [*Fershtater* et al., 1997; *Bea* et al., 1997; *Gerdes* et al., 2002]. The boundary between the

East Uralian and Trans Uralian zones corresponds to a several kilometer wide serpentinite melange that contains harzburgite locally (SU3 in Figure 2). A late, undeformed phase of the Dzhabyk granite that intrudes into the melange has been dated at 291 ± 4 [*Montero* et al., 2000]. The eastern margin of this melange corresponds to the dextral strike-slip Troisk fault.

The Transuralian zone, the easternmost zone of the Uralides, is poorly exposed and is not well known. It is composed of variably metamorphosed, tectonically imbricated Paleozoic volcanic arc fragments and Precambrian and Paleozoic rocks of possible continental origin [*Puchkov*, 2000]. Ophiolitic rocks and evidences of high-pressure metamorphism have been reported [*Ivanov* et al., 1975]. The best known units are Devonian and Carboniferous calc-alkaline volcano-plutonic complexes including volcaniclastics and lava flows that are overlain by Triassic terrigeneous red beds and evaporites [*Puchkov*, 1997]. Devonian and Lower Carboniferous rocks are deformed by open to tight folding and east vergent thrusting.

2.2. The Central Variscides

The transect across the central Variscides crosses in the north, the Rhenohercynian Belt that includes the Avalonian continental margin and a complex association of rocks of the Northern Phyllite zone, overlain by the Giessen-Harz allochthon, in the footwall to the SV1 zonal boundary (Figure 2). To the south, SV2 separates the northern (Saxothuringian zone) and southern (Moldanubian zone) members of the Armorican Terrane assemblage. The Moldanubian zone is in turn divided in the Teplá-Barrandian (Bohemia) and the Moldanubian *sensu stricto* [*Franke*, 2000].

The Rhenohercynian zone is composed of a north-directed foreland thrust and fold belt that deform the sedimentary cover and some basement rocks of the continental margin (Figure 2). The continental margin basement sequences include Cambro-Ordovician rocks with a weak Caledonian age metamorphism, Ordovician and Silurian shales and some Precambrian crystalline rocks [*Franke*, 2000], overlain by 3–12 km of mostly siliciclastic rocks that were deposited during Early Devonian intracontinental rifting [*Oncken* et al., 1999]. The Rhenohercynian foreland thrust and fold belt is composed of a set of imbricated thrusts branching from a basal detachment that dips gently southwards from 10 to 15 km, and it records 180–200 km of shortening [*Oncken* et al., 1999]. It developed between 350 and 300 Ma [*Ahrendt* et al., 1983]. Foreland basin evolution includes two major stages, the Late Viséan to Early Namurian Rheno-Hercynian Turbidite Basin evolution,

and the Namurian to Late Westphalian molasse in the sub-Variscan foreland basin [*Ricken* et al., 2000]. The southern part of the sub-Variscide foreland basin includes the deformation front.

The Rhenohercynian zone also includes exotic terranes of the Giessen-Harz Nappe that is interpreted to represent the remains of an accretionary complex [*Oncken*, 1997]. This nappe constitutes an allochthon that carries Devonian MORB-like metabasalts, Middle Devonian through Frasnian age condensed pelites, and Frasnian to Lower Carboniferous flysch greywackes [*Franke*, 1992]. The suture between the Rhenohercynian and the Saxothuringian zone to the south is a narrow belt (Phyllite Zone), including a complex association of rocks, characterised by the presence of phyllites with pressure-dominant metamorphism of Carboniferous age (about 325 Ma, [*Oncken* et al., 1995]). This belt is bound to the northwest by a late strike-slip fault that truncates its internal structure [*Franke*, 2000].

The Saxothuringian zone is composed of a complex rock assemblage whose early orogenic architecture has been overprinted by large-scale wrenching and basin formation during the Upper Carboniferous [*Oncken*, 1997]. A belt of magnetic anomalies marks the Mid German Crystalline Rise that trends oblique to the orogenic strike, separating the Saar Basin to the North andÂthe Saxothuringian Basin to the South. The Rise is interpreted to represent the deeply exposed basement of an active magmatic margin that developed above a south dipping subduction zone [*Franke*, 2000]. It includes a Lower Carboniferous low pressure-high temperature magmatic arc with Lower Paleozoic basement tectonically stacked on metamorphic rocks of inferred Rhenohercynian origin [*Oncken*, 1997]. Post-orogenic undeformed granitoids of late Viséan to late Carboniferous age crosscut all units and faults. The basement complex beneath the Permo-Carboniferous intramontane Saar Basin is interpreted to represent a fore arc relict [*Oncken*, 1997].

The Saxothuringian Basin includes Late Devonian to Early Carboniferous (Viséan) synorogenic clastic sediments suggesting that the southern suture (SV2, Figure 2) had been closed by Late Devonian (Famennian) times [*Schäfer* et al., 1997]. The basement to the synorogenic clastic sediments includes Cadomian crystalline rocks overlain by Cambro-Ordovician to Late Devonian (Famennian) sediments. The basin is deformed by south-directed high angle thrust faults [*Oncken*, 1997; *Schäfer* et al., 2000] that are superposed on the widespread northwest-directed nappes derived from the southern suture [*Franke*, 2000]. Tectonic klippen within the basin include mafic rocks metamorphosed to eclogite

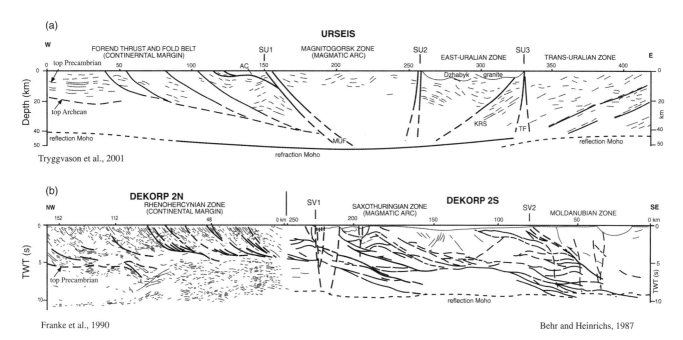

Figure 3. (a) Interpreted section of vibroseis URSEIS profile across the Southern Uralides [redrafted from *Trygvasson* et al., 2001]. (b) Interpreted section of profiles DEKORP 2N and 2S across the Central Variscides [redrafted from *Franke* et al., 1990, and *Behr and Heinrichs*, 1987, respectively]. The sections have been adapted for display at the same horizontal scale (note vertical scale in URSEIS profile is in km, and DEKORP sections in seconds). In URSEIS and DEKORP 2N thin lines represent the main reflections, and thick lines are interpreted structures. The DEKORP 2S profile includes only interpreted lines; reflections of the layered lower crust are omitted as in original figure by *Behr and Heinrichs* [1987]. SV1 and SV2 are added to represent the limit between orogenic zones at surface. For original images of reflection profiles and more detailed interpretations the reader is referred to the original papers. Location of profiles in Figure 2.

facies at around 395 Ma and a 380–365 Ma medium pressure metamorphic assemblage, which relate to the northwest-directed nappes.

The Moldanubian zone consists of piles of nappes comprised of highly strained and polymetamorphic rocks to the south of the SV2 zonal boundary [*Franke*, 2000]. Most nappes include lithological units that represent slices of continental crust with a complex Variscide deformation history and Barrowian-type metamorphism that varies from one area to another, and shows differences in kinematics, including northwest directed and south directed emplacements [e.g., *Schulmann* et al., 1991; *Hegner* et al., 2001; *Zulauf*, 2001]. In the Teplá-Barrandian unit, pre-Variscide basement rocks underwent older Cadomian deformation, and are overlain by Cambrian to Middle Devonian sedimentary and volcanic rocks [*Zulauf* et al., 1997]. Variscan deformation occurred at c. 380 Ma in the northwest margin, including exhumation of high pressure rocks, and around 340 Ma at the southeast margin with the Moldanubian *sensu*

stricto [*Franke*, 2000]. The Moldanubian *sensu stricto* is composed of a complex association of rocks (including eclogites, high pressure granulites and mantle rocks) dominated by large scale, southeast-directed thrusting and inversion of metamorphic facies. A large, c. 340 Ma, batholith intruded at the boundary, between the Teplá-Barrandian and Moldanubian *sensu stricto* [*Franke*, 2000], and younger post-collisional granitoids (325–305 Ma) intruded largely throughout the Moldanubian and Saxothuringian zones.

3. CRUSTAL ARCHITECTURE FROM SEISMIC PROFILES

In the Southern Urals of Russia, the Urals Reflection Seismic Experiment and Integrated Studies (URSEIS) recorded 465 km of coincident vibroseis and explosive-source near-vertical deep reflection seismic data, complemented with explosive source, wide-angle reflection and refraction seismic recording [*Berzin* et al., 1996;

Carbonell et al., 1996; *Echtler* et al., 1996; *Knapp* et al., 1996]. In Germany, the DEKORP Research Group recorded two vibroseis-source near-vertical deep reflection seismic profiles that amount to a 470 km long transect through the central and northern parts of the Central Variscides (DEKORP-2N [*Behr and Heinrichs*, 1987], and DEKORP-2S [*Franke* et al., 1990a]). Much of the interpretation of the URSEIS section is constrained by a direct link with surface outcropping geology [*Trygvasson* et al., 2001], although in the poorly exposed eastern part gravity and magnetic data provide important constraints [*Brown* et al., 2002; *Kimbell* et al., 2002]. Along part of its length, the DEKORP sections image areas covered by post-Variscan basins (see Figure 2) thus hampering a direct correlation between surface geology and structures at depth. In particular, the surface location of the zonal boundaries along these profiles relies on geophysical data such as gravity and magnetics [e.g., *Franke* et al., 1990b].

The transects of the Uralides and Variscides sampled by the URSEIS and DEKORP experiments provide two crustal-scale sections of approximately the same length (~420 km) that image one flank of each orogen (Figure 3). In both cases, the seismic section images a deformed continental margin and the crustal blocks that accreted to it during the orogeny. A large-scale comparison of the interpreted URSEIS and DEKORP sections shows important differences in the geometry and distribution of the reflectivity. While in the URSEIS vibroseis section reflectivity varies laterally, defining several crustal-scale panels, in the DEKORP section the variation of reflectivity defines a sub-horizontal division of the crust into upper and lower panels. These differences in reflectivity distribution suggest major differences in the crustal architectures of these two orogens. In particular, important differences are in the geometry of the main crustal boundaries, and in the structure at lower crustal and Moho depths.

In the URSEIS section the crustal architecture can be divided into four zones that extend from the surface down to the Moho. The boundaries between these zones coincide with major fault zones at the surface (SU1, SU2, SU3, Figure 3) and their interpretation at depth is based on the truncation of sets of reflections, with the exception of SU3 that is interpreted to be above the high amplitude reflections of the Kartaly reflection sequence [*Trygvason* et al., 2001]. The boundaries between the zones have been interpreted as crustal-scale high angle fault zones that are traced from the surface down to the Moho [*Trygvason* et al., 2001; *Brown* et al., 2002]. The difference in dip direction of the interpreted sutures, together with the internal structure

of the zones at both sides of the section defines a bivergent crustal architecture [*Echtler* et al., 1996]. In the footwall to the east dipping SU1 suture (Main Uralian fault), the foreland thrust and fold belt is imaged by upper crustal reflectivity interpreted to consist of west directed imbricate units that merge into an eastward dipping sole thrust. In the footwall to the west dipping SU3 suture, the Trans-Uralian zone consists of west dipping highly reflective bands merging into a highly reflective Moho defining east directed imbricate units at the scale of the whole crust. The crustal thickness across the URSEIS section increases from more than 40 km on both sides to 53 km in the central part [*Tryggvason* et al., 2001; *Steer* et al., 1998; *Carbonell* et al., 1998]. The reflection Moho in the west (poorly imaged in the vibroseis section in Figure 3) is interpreted as the inherited pre-orogenic Moho of the continental margin [*Echtler* et al., 1996; *Tryggvason* et al., 2001]. The well-defined reflection Moho beneath the Trans-Uralian zone has been interpreted as an orogenic Moho that possibly acted as a crustal-scale decollement level [*Tryggvason* et al., 2001]. The Moho in the central part of the section has been interpreted to correspond to a collisional Moho that is undergoing processes of reequilibration [*Carbonell* et al., 1998].

In the DEKORP sections the crustal architecture is interpreted to consist of three zones juxtaposed along two gently southeast dipping crustal-scale boundaries [*Behr and Heinrichs*, 1987; *Franke* et al., 1990b]. Although these boundaries are not directly linked to surface structures because of covering sediments they have been correlated with the zonal boundaries SV1 and SV2 at the surface [*Franke* et al., 1990b]. The crustal-scale boundaries are interpreted at depth as low angle reverse fault zones (ductile shear zones) that coincide with south-dipping bands of high reflectivity beneath large-scale upper crustal antiforms [*Behr and Heinrichs*, 1987]. The fault zones are interpreted by these authors to crosscut the whole crust, merging into the flat-lying lower crustal reflectivity and penetrating down to Moho depths as wide anastomosing shear zones. The upper crust of the Saxothuringian zone is dominated by south-dipping, fault-related reflections and some important north-dipping backthrusts that are interpreted to merge into a sub-horizontal decollement level at 15–20 km [*Behr and Heinrichs*, 1987; *Franke* et al., 1990b; *Schäfer* et al., 2000]. Along DEKORP 2N, the upper panel includes sub-horizontal and curved reflections truncated by dominant inclined reflections, interpreted as stratigraphic horizons cut by listric thrust faults [*Franke* et al., 1990a]. The inclined reflections merge into flat reflection bands at mid-crustal depths (between

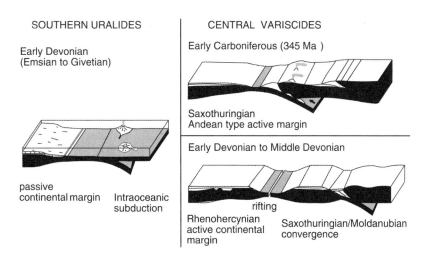

SOUTHERN URALIDES

Early Devonian
(Emsian to Givetian)

passive
continental margin   Intraoceanic
                          subduction

CENTRAL VARISCIDES

Early Carboniferous (345 Ma)

Saxothuringian
Andean type active margin

Early Devonian to Middle Devonian

rifting
Rhenohercynian   Saxothuringian/Moldanubian
active continental   convergence
margin

Figure 4. Initial stages of convergence. In the Uralides (left), intraoceanic subduction was responsible for the building of a new volcanic arc that finally constituted the Magnitogorsk block [*Brown and Spadea*, 1999]. In the Variscides (right), while Saxothuringian and Moldanubian (Teplá-Barrandian unit) blocks were approaching, rifting was occurring at the Rhenohercynian margin [*Franke and Oncken*, 1990].

10–15 km) that correspond with the decollement level to the Rhenohercynian foreland thrust and fold belt [*Oncken*, 1998; *Oncken* et al., 2000].

Along the DEKORP transect crustal thickness varies from 28 to 30 km [*Prodehl and Giese*, 1990]. The lower crust is very reflective with dominant subhorizontal reflections and the Moho is interpreted to be locating at its base. Only in the northernmost part does the lower crust appear to be relatively transparent (Figure 3). This transparent region is interpreted as pre-Paleozoic basement of the London/Brabant Massif [*Franke* et al., 1990a/b]. Several proposals have been made for the origin and age of formation of the lower crustal reflectivity along the rest of the section. It has been interpreted to be the result of crustal re-equilibration processes during the latest stages of orogeny (mostly beneath the Saxothuringian and Moldanubian zones) [e.g., *Meissner* et al., 1991]. Also, the lower crustal reflectivity beneath the Rhenohercynian foreland thrust and fold belt was interpreted as a pre-Variscan feature of the continental margin by *Oncken* et al. [1999] and *Franke* [2000]. Crustal scale balancing of the Rhenohercynian belt and analyses of material paths indicate that the lower crust to the Rhenohercynian continental margin should extend beneath a large part of the Saxothuringian zone [*Oncken*, 1998; *Oncken* et al., 2000], suggesting that lower crustal reflectivity there may be a combination of a pre-Variscan feature related to rifting processes and formation of the margin and post-orogenic reequilibration processes.

4. COMPARISON OF EVOLUTIONARY MODELS

In both the southern Uralides and central Variscides the orogenic system evolved by progressive accretion and final collision of several crustal blocks to a passive continental margin. Fundamental differences include the nature of the material accreted, the order of accretion, the orientation of the subduction zones, and the location/distribution of active deformation through time.

Considering a period of 110 Ma that spans from late Early Devonian to Permian times, the orogens evolved through three common tectonic episodes that are used for the comparison. These episodes correspond to the initial stages of convergence, the arc-continent collision evolution, and the development of the foreland thrust and fold belt. The main processes active in each orogen during these three episodes are compared using schematic models that represent snapshot views of a continuous history of tectonic evolution. The Variscides models are redrawn from *Oncken* [1997, 1998], *Franke and Oncken* [1990], and [*Franke*, 2000]. Uralides models are based on *Brown* et al. [1998], *Brown and Spadea* [1999], and *Alvarez-Marrón* et al. [2000].

4.1. Initiation of Convergence

In the Uralides, this stage corresponds with intra-oceanic subduction (Figure 4). The onset of subduction is marked by Early Devonian (Emsian) age boninite-bearing basalts believed to have erupted in a suprasubduction zone setting [*Spadea* et al., 1998]. East directed (present coordinates) intra-oceanic subduction lasted

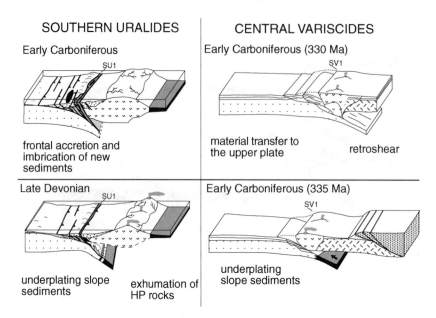

Figure 5. Arc-continent collision processes in the Uralides and Variscides. In the Uralides (left) accreted and underplated material accumulates to form the accretionary complex *Brown* et al. [1998], *Brown and Spadea* [1999], *Alvarez-Marrón* et al. [2000]. In the Variscides underplated material is transferred to the upper plate *Oncken* [1997, 1998].

until Givetian times [*Brown* et al., 2001]. For a period of ~10 Ma, a volcanic front was active, contributing to intra-oceanic island arc formation and building of the Magnitogorsk arc. During this period the East European Craton margin remained passive [*Puchkov*, 2002], and little is known about the processes occurring in the areas corresponding to the East Uralian and Trans-Uralian zones. Because they contain Precambrian and Paleozoic rocks amalgamated within possible subduction-accretion complexes [*Puchkov*, 1997, 2000], these hinterland blocks might have been evolving along outboard subduction zones or somehow laterally along older sections of the same subduction zone long before the Magnitogorsk arc developed [e.g., *Sengor and Natal'in*, 1996].

In the Variscides, the convergence between the northern part of Moldanubian (Teplá-Barrandian unit) and Saxothuringian blocks is interpreted to have occurred along a southward-directed subduction zone at the northern active margin of the former (present coordinates). This geometry is deduced from structural vergence and direction of migration of syntectonic sedimentation [*Franke*, 1992; *Schäfer* et al., 1997]. Variscan plate convergence between Moldanubian and Saxothuringian continental blocks is shown by the widespread 380 Ma age metamorphism found within the Bohemian Massif nappes, both within the Saxothuringian and Moldanubian zones [*Kreuzer* et al., 1989]. Subduction of oceanic crust in all basins was probably

accomplished during the Devonian, so that the tectonic development, at least from the Early Carboniferous onwards, can be taken to represent the collisional stage [*Franke*, 1992].

A remarkable occurrence in the Variscides, in contrast with the Uralides, is that by Middle Devonian times compressional deformation associated with convergence between the Saxothuringian and Moldanubian blocks was coeval with rifting and development of the Rhenohercynian continental margin in the north. This rifting resulted in the opening of the Rhenohercynian oceanic basin [*Franke*, 1995] (Figure 4).

4.2. Arc-Continent Collision

Here, the initiation of arc-continent collision is established to be at the moment that the continental margin crust entered the subduction zone beneath the active volcanic edifice. In the Uralides, the East European Craton margin was underthrust beneath the Magnitogorsk island arc along an east directed subduction zone (Figure 5). Arc-continent collision in the Variscides was accomplished by southward subduction of the Rhenohercynian continental margin beneath the Saxothuringian active continental margin (Figure 5). At this stage in their evolution, an important difference between the Uralide and Variscide orogenic systems was the type of volcanic arc involved in the arc-continent collision process. In the Uralides it was an intra-oceanic volcanic edifice

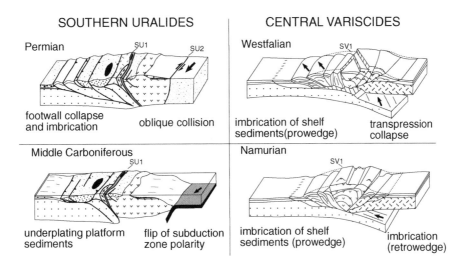

Figure 6. Evolution of the foreland thrust and fold belt. Variscides models are redrawn from *Oncken* [1997, 1998]. Uralides models are based on *Brown* et al. [1998], *Brown and Spadea* [1999].

(Magnitogorsk arc [*Brown* et al., 1998]), and in the Variscides it was an Andean type arc built on continental basement [e.g., *Oncken*, 1997].

At the initiation of arc-continent collision in the Uralides by the Middle Devonian [*Brown and Spadea*, 1999], the volcanic activity in the accreting arc shifted outboard of the collision zone to form a new volcanic front, leaving the extinct volcanic front in a forearc position [*Brown* et al., 2001]. Volcanic activity lasted during the Givetian to earliest Famennian (about 15 My) until it was shut off at the entry of the full thickness of the continental margin into the subduction zone [*Brown and Spadea*, 1999]. In the Variscides, initial stages of arc-continent collision between the Rhenohercynian continental margin and Saxothuringian active arc are represented by a development of a northwest-dipping retroshear to the south of the active arc, coeval with underplating and uplift of the active margin (about 340 My, [*Schäfer* et al., 2000]). Active magmatism lasted from 340 to 325 Ma, or about 15 My [*Franke*, 2000]. As within the Uralides, the magmatism is shut-off in the Saxothuringian block at the entrance of the full thickness of Rhenohercynian continental margin in the subduction zone [*Oncken* et al., 2000].

In both the Uralides and Variscides transects the arc-continent collision process is accompanied by formation of accretionary complexes that appear in both cases as allochthonous units onto the deformed continental margin wedge (Figure 2). Both allochthons, the Zilair accretionary prism in the Uralides [*Brown* et al., 1998] and the Giessen-Harz nappe in the Variscides [*Franke*, 2000] include similar rock type associations such as large

accumulations of syntectonic turbidites, deformed sediments offscraped from the underthrusted continental slope, and pieces of ophiolites. The age of syntectonic sediments indicates that the development of the accretionary complex in the Uralides and Variscides possibly occurred during Late Devonian to Early Carboniferous [*Brown* et al., 1998; Oncken et al., 2000]. In the Uralides, high pressure rocks derived from the continental margin were being subducted by the Middle Devonian and finally exhumed by Givetian-Frasnian times, during the early stages of arc-continent collision [*Hetzel*, 1999; *Brown* et al., 2000]. In the early collision stage between the Rhenohercynian and Saxothuringian blocks in the Variscides, units of the Phyllite zone were subducted and underwent pressure-dominated metamorphism [*Oncken* et al., 2000; *Franke*, 2000].

A fundamental difference between the Uralides and the Variscides orogenic segments during the arc-continent collision stage is the behaviour of the tectonic boundary between the colliding blocks. In the Uralides, the Main Uralian fault (SU1) acted as the backstop to the growing accretionary complex (Figure 5), and processes of underplating and frontal accretion occurred only within the growing accretionary complex [*Alvarez-Marrón* et al., 2000]. The SU1 was a fundamental rigid boundary at a crustal scale that stopped any material transfer from the lower plate into the Magnitogorsk block during the whole tectonic evolution. After arc-continent collision stopped the Main Uralian fault became inactive at about 330 Ma [*Montero* et al., 2000]. The present boundary between the Rhenohercynian and Saxothuringian blocks had a very different evolution.

Similar to the Uralides, SV1 at the surface is located behind a belt of high pressure rocks that may indicate, as is the case in the Uralides, that this suture acted as a backstop during the building of the accretionary complex. However, it acted as a soft plate boundary because addition of continental lower plate material to the upper plate (Saxothuringian block) occurred by tectonic underplating (Figure 5) [*Oncken*, 1997, 1998]. Also, it remained an active transcurrent tectonic boundary during continent-continent collision as the Saxothuringian block continued to overthrust the Rhenohercynian block and the foreland thrust belt developed (see below) [*Oncken*, 1997]. In the Uralides, the process of arc-continent collision came to a halt soon after the full thickness of the continental crust entered the subduction zone, and the arc edifice subsided (Figure 5) [*Brown* et al., 2001]. At this stage, the deposition of the Early to Late Carboniferous stable platform carbonate sediments on the arc and the continental margin suggests that there was very little, if any, uplift and erosion [*Brown and Spadea*, 1999].

4.3. Foreland Thrust and Fold Belt Evolution

The original architecture of the continental margins involved in the Uralide and Variscide collisions was significantly different. Along the URSEIS transect the sedimentary cover in the continental margin is less than 2000 m thick, and is dominated by shallow water platform carbonates accumulated during the passive margin stage [*Puchkov*, 2002]. Along the DEKORP transect, the Renohercynian sedimentary cover reaches 3 to 12 km in thickness and is composed mostly of clastic sediments accumulated during the synrift stage of evolution of the continental margin [*Oncken* et al., 1999].

The foreland thrust and fold belt along the URSEIS transect progressed by foreland directed imbrication of basement and cover rocks from Late Carboniferous to Permian times (Figure 6). The imbrication progressed under subcritical wedge conditions above a basal thrust that cuts continuously up section and is deeply rooted in the basement, resulting in small amounts of shortening (17 km) [*Brown* et al., 1997a,b]. Among the possible explanations to this unusual evolution of the foreland thrust and fold belt are the non-wedge shaped geometry of the Paleozoic sedimentary package, its small thickness, and the near-horizontal dip of the basement/ sedimentary cover contact, which may have resulted in the location and geometry of the basal thrust within very rigid basement [*Brown* et al., 1997b]. Deformation of the continental margin basement started with the development of a basement cored antiformal stack at the rear of the thrust belt, beneath the accretionary complex

allochthon (Figure 6). This initial stacking is interpreted to be connected to the final stages of advance of the accretionay complex above the continental margin [*Brown* et al., 1997a; *Alvarez-Marrón* et al., 2000]. The initiation of subsequent frontal accretion and imbrication is marked at the rear of the wedge by a thrust that has breached through the whole accretionary complex [*Brown* et al., 1997a]. The amount of work needed to uplift the dense allochthon in order to form the topography required for a critical taper may have exceeded the energy budget of the system and possibly was one of the main factors controlling the mechanical development of the wedge [*Brown* et al., 1997b].

The development of the Rhenohercynian foreland thrust and fold belt occurred immediately following the arc-continent collision stage, further advancing the Saxothuringian block above the Rhenohercynian continental margin [*Oncken* et al., 2000] (Figure 6). This episode of northwest propagation of the orogenic wedge continued until 300 Ma, some 25 My after arc magmatism had stopped [*Oncken*, 1998]. The continental margin was imbricated by foreland directed thrusting and progressive frontal accretion above a general decollemen level at 10–15 km depth, and resulted in 180–200 km of shortening [*Oncken* et al., 2000]. These authors propose that the decollement propagated along the brittle-plastic transition during large-scale flexural bending of the lower plate under the advancing Saxothuringian block. Also, during the latest stages of evolution, the localization of the decollement was achieved by thermal softening, and the dominant frontal accretion acted to decrease the wedge taper. Propagation of the decollement and formation of imbricate fans was controlled by the inherited structure of the previous rift basin [*Oncken* et al., 2000].

The evolution of the Rhenohercynian foreland thrust and fold belt as a foreland directed prowedge was coeval with the development of a retrowedge that propagated southwards into the Saxothuringian basin (about 310 Ma) [*Oncken*, 1998; *Schäfer* et al., 2000]. During this period, the arc and upper plate material underwent transpressional dismembering and the SV1 tectonic boundary was reactivated in a strike-slip mode [*Oncken*, 1998]. Higher sedimentation rates in the Sub-Variscan foreland basin during the Carboniferous (Westphalian) were associated with higher erosion rates of the uplifting orogenic wedge [*Ricken* et al., 2000]. By Permian times most of the hinterland zones were overprinted by processes that resulted in the formation of late-orogenic basins, large scale crustal melting and magmatism, and a reduction of crustal thickness [*Franke*, 2000]. Important strike-slip displacement along orogen-parallel shear zones caused severe disruptions at zonal boundaries [*Franke*, 2000].

In contrast, the southern Urals foreland thrust and fold belt may have evolved in a retrowedge position relative to a possible west directed subduction zone along which the eastern crustal blocks finally collided (Figure 6). The east-directed imbrication of Trans Uralian zone crustal-scale units suggests that they accreted along a westward subduction boundary, and the final closure of the west dipping SU3 by 295 Ma [*Montero et al.*, 2000] indicates that the accretion-collision processes in the east were coeval with imbrication of the continental margin in the west. At this stage, the suture between the arc and the continental margin (SU1) was inactive, and minor compression and uplift occurred in the eastern Magnitogorsk block. By Permian-Triassic times the whole southern Uralide edifice sampled in this transect was assembled when the latest stages of foreland basin sedimentation occurred in the west. Along this segment of the Uralides, the thin foreland basin sedimentation dominated by bioclastic, marly and evaporitic sediments contrasts, with the thick accumulation of siliciclastic sediments in the Sub-Variscide foreland basin.

5. DISCUSSION

The orogenic architecture of the Uralides along the URSEIS transect was built by initial accretion of an oceanic island arc against the continental margin along an eastward subduction zone and subsequent oblique accretion of outboard crustal blocks to the East during the west-directed imbrication of the foreland thrust and fold belt. One of the still open questions in the tectonic evolution of the Uralides is the mechanism by which the East Uralian and Trans Uralian zones collided with each other and with the Magnitogorsk arc. One possible mechanism is that of a west-directed subduction zone in the eastern part of the Uralide segment, also proposed by *Puchkov* [2000], although this remains an unconstrained interpretation. Widespread Devonian to Carboniferous granitoid intrusions within the East Uralian zone have been interpreted to have formed above an Andean type active margin [*Bea* et al., 2002]. In this possible scenario, the west dipping reflectivity imaged in the Trans Uralian zone may be the result of whole crust east-directed imbrication against that upper plate active margin. Subsequent Permian magmatism of continental origin within the East Uralian zone [*Fershtater* et al., 1997; *Bea* et al., 1997; *Gerdes* et al., 2002] occurred during final stages of collision, after final closure of the Uralian ocean. The foreland thrust and fold belt grew simultaneously with the Permian magmatism, during

large-scale strike-slip movement and final collision in the interior part of the orogen.

Deformation in the Uralides did not evolve by the advance of a deformation front as in the Variscides. Instead, the deformation jumps within the orogenic system, and the final orogenic architecture resulted from a process of accretion on both sides, and at the scale of the whole crust. In contrast with the Variscides, large-scale decollements seem to be lacking in the Uralides (see below). One of the key features of the Uralides is the large amount of crustal growth that occurred as newly formed volcanic arc crust and subduction related accretionary material was accreted to the continental margin, forming a Turkic Type orogen [*Sengör and Natal'in*, 1996]. However, large-scale tectonic superposition, the development of topographic relief, and large amounts of shortening within the external areas did not take place in the Uralides. Furthermore, in the western part of the orogen the small amount of metamorphic rocks at the surface and the thin foreland basin suggests a low orogenic load, minor erosion and exhumation and consequently small supply of sediments to the foreland basin [e.g., *Matte*, 1995; *Brown* et al., 1997]. In the west, metamorphic rocks within the accretionary complex were exhumed during arc-continent collision by a possible process of extrusion between the colliding blocks [*Chemenda* et al., 1997; *Brown* et al., 2000]. The Main Uralian fault in the southern Uralides represents the original rigid backstop to the arc-continent collision accretionary complex [*Brown* et al., 1998; *Ayarza* et al., 2000], and across which there was little or no material transfer. The evolution of the eastern zonal boundaries is not clear. For example, the SU2 appears to be a late strike-slip boundary, whereas SU3 may be marking an original subduction zone plate boundary.

In the Variscides, the progressive accretion and underthrusting of crustal material beneath the growing orogenic wedge occurred as the continental plates were imbricated along two southward dipping subduction zones that were progressively active towards the continental margin. This process resulted in a crustal architecture consisting of a northwest directed thrust stack in which the different zones represent major crustal slices juxtaposed (imbricated) along south-east dipping zonal boundaries [*Franke*, 2000]. This crustal-scale imbricate now appears above an orogen-scale decollement with ramp/flat geometry. From a flat at the Moho in the hinterland areas, the decollement ramps up through the lower crust to mid crustal levels and follows a large flat beneath the Rhenohercynian foreland thrust and fold belt. The final architecture, with a layered crust, is reflecting the importance of intracrustal decoupling as a deformation mechanism during the Variscide evolution [*Oncken* et al., 2000], and suggests that the Variscan crust responded to stress as a layered medium, similar to other orogens analysed by *Cook and Varshek* [1994]. The inherited, layered mechanical properties of the Rhenohercynian crust appear to have imposed an important control on the nature and geometry of the deformation [*Oncken* et al., 2000].

One important process that was active during the arc-continent collision in the Variscides was the thickening of the edifice by underplating and transfer of lower plate material into the upper plate [*Oncken*, 1998]. This process of material transfer had an important effect on the evolution of the arc-continent collision suture. The initial plate boundary became progressively an inactive suture that was uplifted to higher levels in the crust. The present SV1 zonal boundary observed as a crustal scale thrust in the DEKORP transect formed during the latest stages of collision [*Oncken*, 1997]. The southern SV2 boundary may have formed by a similar process [*Oncken*, 1998], being the basal thrust at the Munchberg and Erbendorf-Vohenstrauss klippen part of an uplifted original suture [*Franke*, 2000; *Vollbrecht* et al., 1989]

The combination of large scale decollements and material transfer through "soft" zonal boundaries in the Variscides acted to generate significant amounts of tectonic superposition and shortening, large amounts of crustal thickening, and topographic relief [e.g., *Oncken* et al., 2000; *Ricken* et al., 2000], that resulted in prevalent prograde metamorphism and a large sediment supply. Transpressive deformation during the final stages of collision facilitated the final exhumation of metamorphic rocks to the surface and provided the largest amounts of synorogenic sediments to the foreland basin [*Oncken*, 1998; *Ricken* et al., 2000]. The progressive re-equilibration of the crust to average continental thickness occurred during melt generation in the late orogenic transpressive stage [*Franke*, 2000].

A comparison between the Uralides and Variscides does not provide any answers to the reasons for the present differences in lower crustal and Moho architectures imaged by reflection seismic sections. While some authors interpreted the lower crustal structure of the Varsicides as the result of late orogenic processes [e.g., *Meissner* et al., 1991; *Henk*, 1999], others have suggested that part of it is an inherited continental margin feature beneath large overthrusting of the upper plate, with some overprinting due to crustal thickness reequilibration during late and post-orogenic stages [*Oncken* et al., 1999, 2000; *Franke*, 2000]. The Uralides lack both the large tectonic superposition and the

important late orogenic processes. In addition, it is important to mention here that the Variscides have been affected by subsequent Mesozoic and Cenozoic tectonic processes to various degrees, although most authors agree that the area considered here underwent only minor overprinting [e.g., *Henk*, 1999]. In contrast, the Uralides seem to have been preserved within a large, stable continental mass, away from any subsequent tectonic processes. Thus, the present crustal architecture of the Uralides is a direct result of its orogenic evolution, although the current topography may be the effect of recent large scale uplift [*Piwowar*, 1997]. How the Mesozoic and Cenozoic tectonic overprinting in the Varisicides may have been responsible for any of the observed differences with the crustal architecture of the Uralides remains an open question. However, the tectonic evolution and dominant processes that were active during orogeny in each transect may explain part, if not all of the resulting differences in their crustal architecture.

6. CONCLUSIONS

The comparison between the southern Uralides and central Variscides transects has shown that major differences in their final crustal architecture are the result of the modes of evolution of each orogenic system, and the dominant processes active during their evolution. In the southern Uralides transect, the final orogenic architecture resulted from a process of accretion on both sides of the orogen, and at the scale of the whole crust, whereas in the central Variscides the final orogenic architecture resulted from the progressive advance of a deformation front and growth of an orogenic wedge. Processes such as large-scale intracrustal decoupling and material transfer through zonal boundaries resulted in important amounts of shortening and large tectonic superposition between crustal slices in the central Variscides. In contrast, the accretion of juvenile arc material against rigid zonal boundaries dominated in the southern Uralides, resulting in crustal growth and smaller amounts of shortening and tectonic superposition. Although the central Variscides underwent important post-orogenic processes, the orogenic architecture remains a fundamental imprint. The southern Uralides, on the other hand, preserved its collisional architecture intact. The southern Uralides and central Variscides may be viewed as two end member examples in which the orogeny acted to provide different results. The Uralides may be seen as a factory for "making" new continental crust and the Variscides as a factory for "recycling" existing continental crust.

Acknowledgments. The author gratefully acknowledges the thorough review by W. Franke, which greatly improved the Variscides part of the manuscript, and V. Puchkov for the Uralides. Also, thanks to D. Blundell for his comments and suggestions. I would like to thank D. Brown for asking me to prepare this paper and for his continuous encouragement.

REFERENCES

Ahrendt, H., N. Clauer, J. C. Hunziker and K. Weber, Migration of folding and metamorphism in the Rheinisches Schiefergebirge deduced from K-Ar and Rb-Sr age determinations, in *Intracontinental Fold Belts. Case Studies in the Variscan Belt of Europe and the Damara Belt in Namibia*, edited by H. Martin and F. W. Eder, pp. 323–338. Springer, Berlin, Heidelberg, New York, 1983.

Alvarez-Marrón, J., D. Brown, A. Pérez-Estaún, V. Puchkov and Y. Gorozhanina, Accretionary complex structure and kinematics during Paleozoic arc-continent collision in the southern Urals, *Tectonophysics*, 325, 175–191, 2000.

Ando, C. J., B. L. Czuchra, S. L. Klemperer, L. D. Brown, M. J. Cheadle, F. A. Cook, J. E. Oliver, S. Kaufman, T. Walsh, J. B. Thompson, J. B. Lyons and J. L. Rosenfeld, Crustal profile of mountain belt: COCORP deep seismic reflection profiling in New England Appalachians and implications for architecture of convergent mountain chains. *AAPG Bull.*, 68, 819–837, 1984.

Ayarza, P., D. Brown, J. Alvarez-Marrón and C. Juhlin, Contrasting tectonic history of the arc-continent suture in the Southern and Middle Urals: Implications for the evolution of the orogen, *J. Geol. Soc. London*, 157, 1065–1076, 2000.

Bea, F., G. Fershtater, P. Montero, V. Smirnov and E. Zin'kova, Generation and evolution of subduction-related batholiths from the central Urals: constraints on the P-T history of the Uralian orogen, *Tectonophysics*, 276, 103–116, 1997.

Bea, F., G. Fershtater and P. Montero, Variscan granitoids of the Urals: Implications for the evolution of the orogen, this volume, 2002.

Behr, H. J. and T. Heinrichs, Geological interpretation of DEKORP 2S: A deep seismic reflection profile across the Saxothuringian and possible implications for Late Variscan structural evolution of Central Europe, *Tectonophysics*, 142, 173–202, 1987.

Berzin, R., O. Oncken, J. H. Knapp, A. Pérez-Estaún, T. Hismatulin, N. Yunusov and A. Lipilin, Orogenic evolution of the Ural Mountains: Results from an integrated seismic experiment, *Science*, 274, 220–221, 1996.

Brown, D., V. Puchkov, J. Alvarez-Marrón and A. Pérez-Estaún, The structural architecture of the footwall to the Main Uralian fault, southern Urals, *Earth-Sci. Rev.*, 40, 125–147, 1996.

Brown, D., J. Alvarez-Marrón, A. Pérez-Estaún, Y. Gorozhanina, V. Baryshev and V. Puchkov, Geometric and

kinematic evolution of the foreland thrust and fold belt in the southern Urals, *Tectonics*, 16, 551–562, 1997a.

Brown, D., J. Alvarez-Marrón and A. Pérez-Estaún, Preservation of a subcritical wedge in the south Urals foreland thrust and fold belt. *J. Geol. Soc. London*, 154, 593–596, 1997b.

Brown, D., C. Juhlin, J. Alvarez-Marrón, A. Pérez-Estaún and A. Oslianski, Crustal-scale structure and evolution of an arc-continent collision zone in the southern Urals, Russia, *Tectonics*, 17, 158–171, 1998.

Brown, D. and P. Spadea, Processes of forearc and accretionary complex formation during arc-continent collision in the southern Ural Mountains, *Geology*, 27, 649–652, 1999.

Brown, D., J. Alvarez-Marrón, A. Perez-Estaun, V. Puchkov, P. Ayarza and Y. Gorozhanina, Structure and evolution of the Magnitogorsk forearc basin: Identifying upper crustal processes during arc-continent collision in the southern Urals, *Tectonics*, 20, 3, 364–375, 2001.

Brown, D., C. Juhlin, A. Tryggvason, D. Steer, P. Ayarza, M. Beckholmen, A. Rybalka and M. Bliznetsov, The Crustal Architecture of the Southern and Middle Urals from the URSEIS, ESRU and Alapaev Reflection Seismic Profiles, this volume, 2002.

Carbonell, R., A. Pérez-Estaún, J. Gallart, J. Díaz, S. Kashubin, J. Mechie, R. Stadtlander, A. Schulze, J. H. Knapp and A. Morozov, A crustal root beneath the Urals: Wide-angle seismic evidence, *Science*, 274, 222–224, 1996.

Carbonell, R., D. Lecerf, M. Itzin, J. Gallart and D. Brown, Mapping the Moho beneath the Southern Urals, *Geophys. Res. Lett.*, 25, 4229–4233, 1998.

Chemenda, A., P. Matte and V. Sokolov, A model of Paleozoic obduction and exhumation of high-pressure/low-temperature rocks in the Southern Urals, *Tectonophysics*, 276, 217–228, 1997.

Chuvashov, B. I. and G. V. Diupina, The Upper Paleozoic terrigenous sediments of the Western slope of the Middle Urals (in Russian), Nauka, Moscow, 208 pp., 1973.

Chuvashov, B. I., V. A. Chermynkh, V. V. Chernykh, V. J. Kipin, V. A. Molin, V. P. Ozhgibesov and P. A. Sofronitsky, The Permian System: Guides to Geological Excursions in the Uralian Type Localities. Part 2 — Southern Urals, *Occasional Publications ESRI, New Series, No.* 10, 1993.

Clowes, R. M., A. J. Calvert, D. W. Eaton, Z. Hajnal, J. Hall and G. M. Ross, LITHOPROBE reflection studies of Archean and Proterozoic crust in Canada, *Tectonophysics*, 264, 65–88, 1996.

Cook, F. A., A. J. van der Velden and K. W. Hall, Frozen subduction in Canada's Northwest Territories: Lithoprobe deep lithospheric reflection profiling of the western Canadian Shield, *Tectonics*, 18, 1–24, 1999.

Cook, F. A. and J. L. Varshek, Orogen-scale decollements, *Rev. of Geophys.*, 32, 1, 37–60, 1994.

Echtler, H. P., M. Stiller, F. Steinhoff, C. M. Krawczyk, A. Suleimanov, V. Spiridonov, J. Knapp, Y. Menshikov,

J. Alvarez-Marrón and N. Yunusov, Preserved collisional crustal architecture of the southern Urals – Vibroseis CMP-profiling, *Science*, 274, 224–226, 1996.

Echtler, H. P., K. S. Ivanov, Y. L. Ronkin, L. A. Karsten, R. Hetzel and A. G. Noskov, The tectono-metamorphic evolution of gneiss complexes in the Middle Urals, Russia: A reappraisal, *Tectonophysics*, 276, 229–252, 1997.

Einor, O. L., I. I. Sinitsyn, N. M. Kochetkova, M. A. Kamaletdinov and V. M. Popov, *Guidebook for the South Urals, Excursion 047, "Upper Paleozoic of Southern Urals"*, 135 pp., 27th Geol. Congress, USSR, Moscow, Nauka, 1984.

Fershtater, G. B., P. Montero, N. S. Borodina, E. V. Pushkarev, V. N. Smirnov and F. Bea, Uralian magmatism: An overview, *Tectonophysics*, 276, 87–102, 1997.

Franke, W., Phanerozoic structures and events in Central Europe, in *A continent revealed, The European Geotraverse*, edited by D. Blundell, R. Freeman and S. Mueller, pp. 164–180. Cambridge University Press, 1992.

Franke, W., Rhenohercynian foldbelt: Autochthon and non-metamorphic nappe units – stratigraphy, in *Pre-Permian Geology of Central and Western Europe*, edited by D. Dallmeyer, W. Franke and K. Weber, Springer, Berlin, pp. 33–49, 1995.

Franke, W., The mid-European segment of the Variscides: tectonostratigraphic units, terrane boundaries and plate tectonic evolution, in *Orogenic Processes: Quantification and Modelling in the Variscan Belt*, edited by W. Franke, V. Haak, O. Oncken and D. Tanner, *Geol. Soc. Spec. Publ.* 179, 35–61, 2000.

Franke, W., R. K. Bortfeld, M. Brix, G. Drozdzewski, H. J. Dürbaum, P. Giese, W. Janoth, H. Jödicke, Chr. Reichert, A. Scherp, J. Schmoll, R. Thomas, M. Trünker, K. Weber, M. G. Wiesner and H. K. Wong, Crustal structure of the Rhenish Massif: results of deep seismic reflection lines DEKORP 2-N and 2-North-Q, *Geol. Rundsch.*, 79, 523–566, 1990a.

Franke, W., P. Giese, S. Grosse, V. Haak, H. Kern, K. Mengel and O. Oncken, Geophysical imagery of geological structures along the central segment of the EGT, in *The European Geotraverse: Integrative Studies*, edited by R. Freeman, P. Giese and St. Mueller, pp. 177–186, European Science Foundation, Strasbourg, France, 1990b.

Franke, W., V. Haak, O. Oncken and D. Tanner (Eds.), *Orogenic Processes: Quantification and Modelling in the Variscan Belt. Geol. Soc. Spec. Publ.* 179, 459 pp., 2000.

Gerdes, A., P. Montero, F. Bea, G. Fershtater, N. Borodina, T. Osipova and G. Shardakova, Peraluminous granites frequently with mantle-like isotope compositions: The continental-type Murzinka and Dzhabyk batholiths of the eastern Urals, *Int. J. Earth Sci.*, 91, 1–17, 2002.

Giese, U., U. Glasmacher, V. I. Kozlov, I. Matenaar, V. N. Puchkov, L. Stroink, W. Bauer, S. Ladage and R. Walter, Structural framework of the Bashkirian anticlinorium, SW Urals, *Geol. Rundsch.*, 87, 526–544, 1999.

Hegner, E., F. Chen and H. P. Hann, Chronology of basin closure and thrusting in the internal zone of the Variscan belt in the Schwarzwald, Germany: evidence from zircon ages, trace element geochemestry and Nd isotopic data, *Tectonophysics*, 332, 169–184, 2001.

Henk, A., Did the Variscides collapse or were they torn apart? A quantitative evaluation of the driving forces for postconvergent extension in central Europe, *Tectonics*, 18, 774–792, 1999.

Hetzel, R., Geology and geodynamic evolution of the high-P/low-T Maksyutov Complex, southern Urals, Russia, *Geol. Rundsch.*, 87, 577–588, 1999.

Ivanov, S. N., A. S. Perfiliev, A. A. Efimov, G. A. Smirnov, V. M. Necheukhin and G. B. Fershtater, Fundamental features in the structure and evolution of the Urals, *Am. J. Sci.*, 275, 107–130, 1975.

Kimbell, G., C. Ayala, A. Gerdes, M. Kaban, V. Shapiro and Y. Menshikov, Insights into the Architecture and Evolution of the Southern and Middle Urals from Gravity and Magnetic Data, this volume, 2002.

Knapp, J. H., D. N. Steer, L. D. Brown, R. Berzin, A. Suleimanov, M. Stiller, E. Lüschen, D. Brown, R. Bulgakov and A. V. Rybalka, A lithosphere-scale image of the Southern Urals from explosion-source seismic reflection profiling in URSEIS '95, *Science*, 274, 226–228, 1996.

Kreuzer, H., E. Seidel, U. Schüssler, M. Okrusch, K. L. Lenz and H. Raschka, K-Ar geochronology of different tectonic units at the northwestern margin of the Bohemian Massif, *Tectonophysics*, 157, 149–178, 1989.

Kossmat, F., Gliederung des varistischen Gebirgsbaues. *Abhandlungen des Sächsischen Geologischen Landesamtes*, 1, 1–39, 1927.

Maslov, V. A., V. L. Cherkasov, V. T. Tischchenko, A. I. Smirnova, O. V. Artyushkova and V. V. Pavlov, *On the Stratigraphy and Correlation of the Middle Paleozoic Complexes of the Main Copper-pyritic Areas of the Southern Urals* (in Russian), Ufimsky Nauchno Tsentr, Ufa, Russia, 1993.

Matte, Ph., Tectonics and plate tectonic models for the Variscan Belt of Europe, *Tectonophysics*, 126, 329–374, 1986.

Matte, Ph., Accretionary history and crustal evolution of the Variscan belt in Western Europe, *Tectonophysics*, 196, 309–337, 1991.

Matte, Ph., Southern Uralides and Variscides: Comparison of their anatomies and evolutions, *Geologie en Mijnbouw*, 1–16, 1995.

Meissner, R. and DEKORP Research Group, The DEKORP surveys: Major results in tectonic and reflective styles, in *Continental Lithosphere: Deep Seismic Reflections*, edited by R. Meissner, L. Brown, H. J. Dürbaum, W. Franke, K. Fuchs and F. Seifert, pp. 69–76, *AGU Geodynamic Series* Vol. 22, Washington, D.C., 1991.

Meissner, R., L. Brown, H. J. Dürbaum, W. Franke, K. Fuchs and F. Seifert (Eds.), *Continental Lithosphere: Deep Seismic Reflections*, AGU Geodynamic Series Vol. 22, 450 pp., Washington, D.C., 1991.

Montero, P., F. Bea, A. Gerdes, G. Fershtater, E. Zin'kova, N. Borodina, T. Osipova and V. Smirnov, Single zircon evaporation ages and Rb-Sr dating of four major Variscan batholiths of the Urals: A perspective on the timing of deformation and granite generation, *Tectonophysics*, 317, 93–108, 2000.

Oncken, O., Transformation of a magmatic arc and an orogenic root during oblique collision and its consequences for the evolution of the European Variscides (Mid-German Crystalline Rise), *Geol. Rundsch.*, 86, 2–20, 1997.

Oncken, O., Orogenic mass transfer and reflection seismic patterns — evidence from DEKORP sections across the European Variscides (central Germany), *Tectonophysics*, 286, 47–61, 1998.

Oncken, O., H. J. Franzke, U. Dittmar and T. Klügel, Rhenohercynian foldbelt: metamorphic units, in *Pre-Permian Geology of Central and Western Europe*, edited by D. Dallmeyer, W. Franke and K. Weber, pp. 108–117, Springer, Berlin, 1995.

Oncken, O., C. von Winterfeld and U. Dittmar, Accretion of a rifted passive margin: The Late Paleozoic Rhenohercynian fold and thrust belt (Middle European Variscides), *Tectonics*, 18, 75–91, 1999.

Oncken, O., A. Plesch, J. Weber, W. Ricken and S. Schrader, Passive margin detachment during arc-continent collision (Central European Variscides), in *Orogenic Processes: Quantification and Modelling in the Variscan Belt*, edited by W. Franke, V. Haak, O. Oncken and D. Tanner, *Geol. Soc. Spec. Publ.* 179, 199–216, 2000.

Piwowar, T. J., Long-wavelength Neogene flexural uplift of the Southern Urals and central Eurasia, unpublished M.Sc., Cornell University, U.S.A., 74 pp., 1997.

Prodehl, C. and P. Giese, Seismic investigations around the EGT in Central Europe, in *The European Geotraverse: Integrative Studies*, edited by R. Freeman, P. Giese and St. Mueller, pp. 77–97, European Science Foundation, Strasbourg, France, 1990.

Puchkov, V. N., Structure and geodynamics of the Uralian orogen, in *Orogeny Through Time*, edited by J. P. Burg and M. Ford, *Geol. Soc. Spec. Publ.*, 121, 201–236, 1997.

Puchkov, V. N., *Paleogeodynamics of the Central and Southern Urals* (in Russian), 145 pp., Ufa Dauria, 2000.

Puchkov, V. N., Paleozoic evolution of the East European continental margin involved in the Uralide orogeny, this volume, 2002.

Ricken, W., S. Schrader, O. Oncken and A. Plesch, Turbidite basin and mass dynamics related to orogenic wedge growth; the Rheno-Hercynian case, in *Orogenic Processes: Quantification and Modelling in the Variscan Belt*, edited by W. Franke, V. Haak, O. Oncken and D. Tanner, *Geol. Soc. Spec. Publ.* 179, 257–280, 2000.

Savelieva, G. N., A. Y. Sharaskin, A. A. Saveliev, P. Spadea and L. Gaggero, Ophiolites of the Southern Uralides adjacent to the East European Continental Margin, *Tectonophysics*, 276, 117–138, 1997.

Sengör, A. M. C and B. A. Natal'in, Turkic-Type orogeny and its role in the making of the continental crust, *Annu. Rev. Earth Planet. Sci.*, 24, 263–337, 1996.

Seravkin, I., A. M. Kosarev and D. N. Salikhov, *Volcanism of the Southern Urals* (in Russian), 195 pp., Nauka, Moscow, 1992.

Schäfer J., H. Neuroth, H. Ahrendt, W. Dörr and W. Franke, Accretion and exhumation at a Variscan active margin, recorded in the Saxothuringian flysch, *Geol. Rundsch.*, 86, 599–611, 1997.

Schäfer, F., O. Oncken, H. Kemnitz and R. Romer, Upper-plate deformation during collisional orogeny: A case study from the German Variscides (Saxo-Thuringian Zone, in *Orogenic Processes: Quantification and Modelling in the Variscan Belt*, edited by W. Franke, V. Haak, O. Oncken and D. Tanner, *Geol. Soc. Spec. Publ.*, 179, 281–302, 2000.

Schulmann, K., P. Ledru, A. Autran, R. Melka, J. M. Lardeaux M. Urban and M. Lobkowicz, Evolution of nappes in the eastern margin of the Bohemian Massif: A kinematic interpretation, *Geol. Rundsch.*, 80, 73–92, 1991.

Seward, D., A. Perez-Estaun and V. Puchkov, Preliminary fission-track results from the Southern Urals — Sterlitamak to Magnitogorsk, *Tectonophysics*, 276, 281–290, 1997.

Spadea, P., L. Y. Kabanova and J. H. Scarrow, Petrology, geochemistry and geodynamic significance of Mid-Devonian boninitic rocks from the Baimak–Buribai area (Magnitogorsk Zone, southern Urals), *Ofioliti*, 23, 17–36, 1998.

Steer, D. N., J. H. Knapp, L. D. Brown, H. P. Echtler, D. L. Brown and R. Berzin, Deep structure of the continental lithosphere in an unextended orogen: An explosive-source seismic relection profile in the Urals (Urals Seismic Experiment and Integrated Studies (URSEIS 1995), *Tectonics*, 17, 143–157, 1998.

Tryggvason, A., D. Brown and A. Pérez-Estaún, Crustal architecture of the southern Uralides from true amplitude processing of the URSEIS vibroseis profile, *Tectonics*, 20, 1040–1052, 2001.

Zonenshain, L. P., V. G. Korinevsky, V. G. Kazmin, D. M. Pechersky, V. V. Khain and V. V. Mateveenkov, Plate tectonic model of the south Urals development, *Tectonophysics*, 109, 95–135, 1984.

Zonenshain, L. P., M. I. Kuzmin and L. M. Natapov, Uralian Foldbelt, in *Geology of the USSR: A Plate-Tectonic Synthesis*, edited by B. M. Page, pp. 27–54, *AGU Geodynamics Series*, Volume 21, Washington, D.C., 1990.

Zulauf, G., Structural style, deformation mechanisms and paleodifferential stress along an exposed crustal section: Constraints on the rheology of queartzofeldespathic rocks at supra- and infrastructureal levels (Bohemian Massif), *Tectonophysics*, 332, 211–237, 2001.

Zulauf, G., W. Dörr, J. Fiala and Z. Vejnar, Late Cadomian crustal tilting and Cambrian transtension in the Teplá-Barrandian unit (Bohemian Massif, Central European Variscides), *Geol. Rundsch.*, 86, 571–584, 1997.

J. Alvarez-Marrón, Instituto de Ciencias de la Tierra "Jaume Almera", c/ Lluis Sole i Sabaris s/n, 08028 Barcelona, Spain (jalvarez@ija.csic.es)

Arc-Continent Collision in the Southern Urals: Petrogenetic Aspects of the Forearc-Arc Complex

P. Spadea[1], M. D'Antonio[2], A. Kosarev[3], Y. Gorozhanina[3] and D. Brown[4]

In the Southern Urals a well-preserved, rapidly created, arc-continent collisional orogen developed during the Devonian to Early Carboniferous. The Magnitogorsk arc includes a forearc represented by lavas, dikes and shallow intrusive rocks (boninites, arc tholeiites and calc-alkaline extrusive rocks), and a fully developed intraoceanic arc that developed during Early Devonian time (Emsian to Eifelian). The suture zone between the arc and the continent, the Main Uralian fault, is a mélange zone containing ophiolite fragments, volcanic rocks derived from the arc, and sediments from the forearc basin. A petrogenetic characterization of the igneous suites from the Magnitogorsk zone and the Main Uralian fault has been based on rare earth element distributions, abundance of immobile, incompatible elements (Nb, Ta and Th), and Sr-Nd isotope systematics. The Magnitogorsk magmatic suites were generated from depleted mantle sources variably enriched with fluids or melts derived from a subducting slab. A Nb enrichment and high Th/Yb ratios with moderate high-field strength element depletion may reflect addition of an enriched ocean-island basalt-type component to the mantle wedge. The boninitic magmatism was related to an exceptionally high thermal anomaly responsible for high degree of melting and high rate of magma production, similarly to the Eocene magmatic event occurred in the Izu-Bonin-Mariana system in western Pacific Ocean. Arc-related igneous rocks similar to those of the Magnitogorsk zone occur in the Main Uralian fault mélange. Here, ocean-island basalt-type intrusive rocks and late dikes indicate distinct magmatic sources with a typical enriched signature, suggesting a deeper origin

[1]Dipartimento di Georisorse e Territorio, University of Udine, Italy
[2]Dipartimento di Scienze della Terra, University of Napoli Federico II, Italy
[3]Institute of Geology, Russian Academy of Sciences, Ufa, Russia
[4]Instituzto de Ciencias de la Tierra "Jaume Almera" CSIC, Barcelona, Spain

Mountain Building in the Uralides: Pangea to the Present
Geophysical Monograph 132
Copyright 2002 by the American Geophysical Union
10.1029/132GM07

with respect to most arc lavas, unrelated to subduction. This enriched component was also involved in arc magma generation. It is inferred that the Devonian igneous rocks from the Main Uralian fault were intruded into already obducted lherzolitic ophiolites, and that the arc has been built on, or near, an older lithosphere including oceanic crust and undepleted mantle.

1. INTRODUCTION

The Uralides formed in the Late Paleozoic by collision of the East European Craton platform with oceanic and island arc crust generated in the Uralian ocean basin. The north-south trending Main Uralian fault is the fossil suture between the East European Craton and the island arcs. To the east, and adjacent to the Main Uralian fault, island arc assemblages record voluminous magmatism during Silurian and Devonian times [*Savelieva and Nesbitt*, 1996]. Intra-oceanic, east-directed (opposite to the East European Craton) subduction of the closing Uralian ocean basin has been proposed to explain formation of the island arc assemblages [*Zonenshain* et al., 1984, 1990; *Brown* et al., 1998; *Brown and Spadea*, 1999].

The Magnitogorsk arc of the Southern Urals has been compared [*Brown* et al., 1998; *Brown and Spadea*, 1999] with sites of the western Pacific characterized by initiation of subduction in a non-accretionary forearc setting along the Izu-Bonin-Mariana belt [*Hawkins* et al., 1984; *Crawford* et al., 1981; *Stern* et al., 1991; *Bloomer* et al., 1995], and oceanward subduction at continental margins, e.g., Papua New Guinea, Timor, and Taiwan [*Abbott* et al., 1994; *Huang* et al., 1997], has produced complex magmatism of boninitic, tholeiitic and calc-alkaline type, and sometimes of enriched ocean island basalt type [*Bloomer* et al., 1987; *Beccaluva* et al., 1988; *Crawford* et al., 1989; *Hickey-Vargas*, 1989; *Falloon and Crawford*, 1991; *Murton* et al., 1992; *Pearce* et al., 1992; *Sobolev and Danyushevsky*, 1994; *Danyushevsky* et al., 1995; *Taylor and Nesbitt*, 1995]. In the western Pacific, ongoing collisional orogeny has accreted these magmatic assemblages to the continent [*Hawkins* et al., 1984; *Hall*, 1996], providing a modern analogue for the Uralides orogen [*Brown and Spadea*, 1999].

The petrographic and chemical variability of the Uralides island arc assemblages requires a large spectrum of magmatic sources (i.e., variously depleted and metasomatized lithospheric mantle, asthenospheric mantle) all of them rapidly evolving through time, and therefore sampled in any possible transient stage during eruption of single volcanoes. From the western Pacific model [*Stern* et al., 1988; *Lin* et al., 1990; *Bloomer* et al., 1995], we can argue that the arc assemblages were formed by distinct, short lived chains of island. Erosion and tectonic disruption of the sequences of the Uralides island arc rocks make it difficult, for instance, to investigate even on fractionation processes, which are the background for identification of parental magmas and a starting point for magma genesis. Furthermore, most Uralides arc rocks are affected by seafloor hydrothermal metamorphism and, to a lesser extent, by low-grade orogenic metamorphism which have largely obliterated the rock primary textures, minerals, and chemical compositions.

Our approach in investigating the Uralides island arc assemblages has been to concentrate on the best exposed sector of the Devonian Magnitogorsk arc, and to address the rock petrogenesis by using major, trace elements and radiogenic isotopes (Sr and Nd) compositions of rocks classified petrographically on the basis of relict protolith features. The large set (to account for the rock variety) of both available and new analyzed samples includes:

(1) extrusive rocks from arc sequences of known stratigraphic position,
(2) arc related extrusive and intrusive rocks which are present as disrupted sequences or as fragments in the Main Uralian fault mélange, and
(3) dikes and shallow plutonic bodies intruding the Main Uralian fault ophiolites at different stratigraphic levels and representing syn- and post-arc magmatism [*Gaggero* et al., 1997; *Pertsev* et al., 1997].

Using geochemical data, we aim at reconstructing the geodynamic setting of the magmatism, its evolution through time and possible sources.

An interpretation of primary processes is largely inhibited by the geochemical overprint caused by pervasive hydrothermal alteration. In particular, the problem of identifying and evaluating the role of fluids is essential to explain the petrogenesis of arc and back arc assemblages [*Pearce* et al., 1992, 1995; *Pearce and Parkinson*, 1993; *Thirlwall* et al., 1994]. A fluid phase, consisting of volatiles and/or melts deriving from subducted oceanic lithosphere, is an important component of subduction related magmas. The subducted slab-derived volatiles can be mostly considered as recycled seawater and therefore in principle not different from the hydrothermal fluids which circulate and alter the oceanic and arc crust. However, their capability of selective transport of elements to the magmatic sources may sometimes provide a key to evaluate their role and amounts [*Stern* et al., 1991; *Elliott* et al., 1997; *Taylor and Nesbitt*, 1998]. The subducted slab-derived melts may have a specific signature, which is indicative of melted sediments of different provenance (volcanogenic versus continental), or a somewhat more ambiguous signature, which is indicative of partially melted oceanic crust. On the other hand, post magmatic hydrothermal alteration is also fluid controlled and able to deeply change even the whole composition of the rocks [*Alt* et al., 1986]. Therefore, there may not be large differences between modern and ancient arc rocks, even if affected by orogenic metamorphism, which is often chemically conservative, because in most cases the largest changes occur prior to tectonic events, during the oceanic stage, by water-rock interaction in the seafloor.

Keeping in mind the aforementioned difficulties, we shall address questions about the sources and involvement of oceanic crust in magma genesis.

2. GEOLOGICAL SETTING

The Southern Uralides provide one of the best examples of well preserved, Late Paleozoic arc-continent collision zone developed by subduction and collision between Baltica (i.e., East European Craton [*Matte*, 1985]) to the west and the Magnitogorsk island arc to the east [*Brown* et al., 1997, 1998]. The whole Uralide orogen, from its Polar Urals segment at 66°–70° N to the Southern Urals segment at 52°–48° N, displays diachronous intraoceanic subduction, arc development and subsequent arc-continent collision. The southern Uralides (Figure 1) contain the best exposed sector of the arc-continent collisional orogen. This includes a forearc complex represented by an upper section of lavas, dikes and shallow intrusive rocks [*Spadea* et al., 1998; *Spadea and Scarrow*, 2000]; a fully developed intraoceanic arc [*Savelieva and Nesbitt*, 1996; *Seravkin*, 1997], an accretionary wedge that includes high-pressure rocks of continental provenance [*Lennykh* et al., 1995; *Hetzel* et al., 1998] which was emplaced over the subducting slab [*Puchkov*, 1997a, 1997b, 2000; *Brown* et al., 1998], and the suture zone between the arc and the continental margin marked by a mélange containing ophiolite fragments [*Savelieva* et al., 1997]. The syn- to post-collision history is recorded by syn-tectonic deposits (the Late Devonian to Early Carboniferous Zilair Formation [*Brown* et al., 1998, 2001; *Alvarez-Marròn* et al., 2000], and alkaline volcanic [*Frolova and Burikova*, 1977;

Salikhov and Yarkova, 1992] and plutonic rocks [*Yazeva and Bochkarev*, 1998].

Remnants of the Uralian oceanic crust may be represented by Ordovician tholeiitic basalts of mid ocean ridge type (MORB), considered to represent the 2B oceanic layer of the Main Uralian fault dismembered ophiolites [*Savelieva and Nesbitt*, 1996; *Savelieva* et al., 1997]. However, the absence of a coherent ophiolite stratigraphy makes the relations of the Ordovician MORBs with the mantle peridotites uncertain.

3. STRATIGRAPHY OF THE MAGNITOGORSK ZONE

A general review of the stratigraphy and structure of the western Magnitogorsk arc is given in *Brown* et al. [2001], and the reader is referred there for details. In this section we present detail description of those areas where rocks were sampled for this study.

3.1. Khvorostyanka Region

In the Khvorostyanka region (Figures 2 and 3) the Emsian [*Maslov* et al., 1993] Baimak–Buribai Formation is best exposed along the Tanalyk River. The lower part of the Baimak-Buribai Formation in this area varies from 500 to 1400 m in thickness, and consists of boninitic and basaltic to andesitic massive and pillow lavas, dikes and sills, volcanic breccias and hyaloclastites. This is overlain by 250 to 670 m of pillowed basalts. The uppermost part of the formation consists of boninitic and high-Mg basaltic and andesitic lavas and volcanoclastic rocks, and rhyodacitic to dacitic extrusive rocks. The latter is composed of 10–30 m thick flows interbedded with volcanic breccias and sandstones. Massive Cu sulfide ore deposits of hydrothermal origin are associated mostly with the acidic volcanic rocks.

The Tanalyk Formation (Emsian) stratigraphically overlies the Baimak–Buribai Formation and includes a lower unit of basaltic to andesitic lavas and breccias and an upper unit composed of acidic lavas, breccias and volcanoclastic rocks intruded by rhyodacitic dikes and sills. The total thickness is about 1200 m. Volcanic massive sulfide deposits are associated with andesites and rhyodacites of the upper part of the formation.

Locally, the Early to Middle Devonian Turat Formation (exposed near Samarskoe) also overlies the Baimak–Buribai Formation. It consists of a condensed sequence of siltstone, sandstone and black shale accumulated in distal position from volcanic belt and is the distal equivalent of the lower part of the Aktau Formation [*Maslov* et al., 1993].

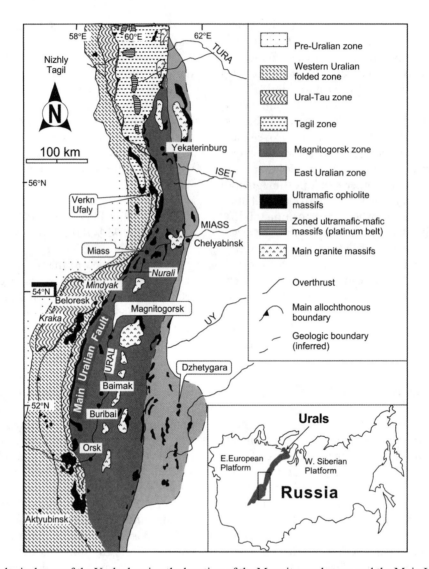

Figure 1. Geological map of the Urals showing the location of the Magnitogorsk zone and the Main Uralian fault.

In the Khvorostyanka region the Emsian to Eifelian Irendyk Formation [*Maslov* et al., 1993] consists of three members. The lower member is represented by 0–300 m of volcano sedimentary rocks, overlain by 300 m of volcanic breccias and volcanoclastic sediments and lavas of andesitic, dacitic and basaltic andesitic composition. The upper member whose thickness ranges from 100–450 m to 1000 m [*Kosarev*, 1986] consists of intermediate to acidic lavas (andesitic basalt, andesite, dacite and rhyodacite) interbedded with volcanic breccias and volcanoclastic sediments.

The Irendyk Formation is overlain by 15–25 m of the Eifelian Yarlykapovo red jasper unit [*Maslov* et al., 1993], which is in turn overlain by 1500–1800 m of the Givetian volcano-sedimentary Ulutau Formation. The Ulutau Formation is composed of turbiditic sandstone and siliceous shale with volcanoclastic interbeds with andesite and rhyodacite clasts.

The Frasnian Mukas chert [*Maslov* et al., 1993] is widespread and overlies the Ulutau Formation. It is composed of about 100 m of black chert with interlayers of polygenic sandstone, siltstone and cherty shale.

The Famennian Zilair Formation [*Maslov* et al., 1993] consists of polygenic graded bedded sandstone, siltstone, siliceous shale of fore arc origin [*Brown* et al., 2001]. The clasts of arenite are dominantly volcanic (basalt, diabase, diorite, rhyodacite), and in lesser amount plutonic (granodiorite), metamorphic (phyllite, quartzite), and sedimentary (siliceous shale and chert).

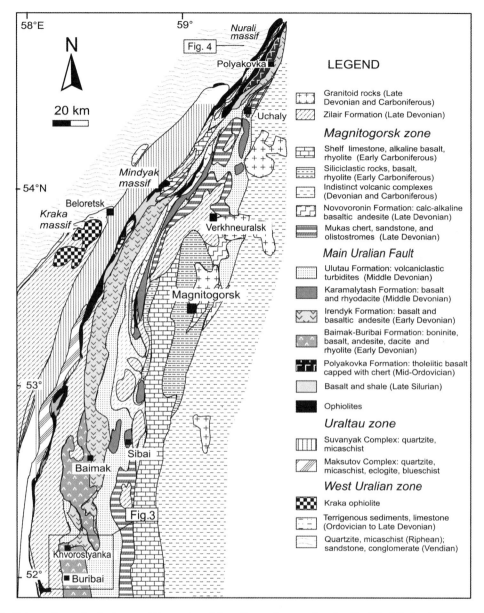

Figure 2. Schematic geological map of the western Magnitogorsk zone showing the Sibai, Baimak, Khvorostyanka, and Buribai sample localities and the Main Uralian fault Nurali, Mindyak, and Kraka ophiolitic massifs.

3.2. Sibai Region

In the Sibai region (Figure 2) the lowermost volcanic unit is the Irendyk Formation. It is composed of up to 3000 m of epiclastic turbidites [*Gorozhanina*, 1993], with coarse volcanic breccias interbedded with andesitic-basaltic and basaltic lavas intruded by dikes of the same composition in the middle part of the formation. To the north, east of the Main Uralian fault zone, the Irendyk Formation has the same characteristics. In the uppermost part of the Irendyk Formation an olistostrome unit

consisting of coarse breccia with mixed volcanic material and reef limestone blocks is present. Its age is Eifelian [*Maslov* et al., 1993].

The Eifelian Karamalytash Formation [*Maslov* et al., 1993] crops out from the Sibai region to the northeastern part of the Magnitogorsk zone near Uchaly (Figure 2). The lower part of the Karamalytash Formation consists of more than 1000 m diabase and pillow basalt with hyaloclastite and chert interbeds. This is overlain by about 1000 m of bimodal basaltic to

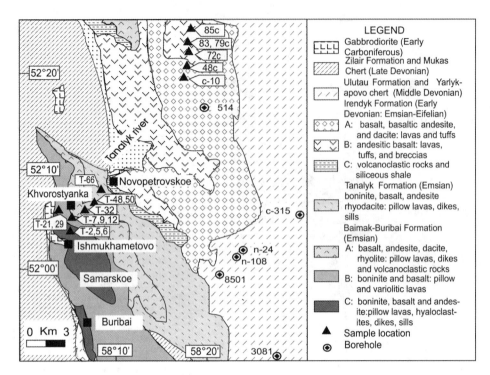

Figure 3. Detail geological map of the Baimak, Khvorostyanka, and Buribai sample localities.

rhyodacitic volcanic rocks [*Seravkin* et al., 1992], which in turn is overlain by about 500 m of volcanoclastic and volcanic rocks (basalt, andesite, dacite and rhyolite). Polymetallic volcanic massive sulfide deposits from Sibai and Uchaly are associated with dacitic and rhyolitic volcanic rocks from the middle and upper Karamalytash Formation. Up to 100 m thick red jasper bed (Bugulygyr jasper) rests on the top of the Karamalytash Formation.

In the Sibai region, the Ulutau Formation overlies the Bugulygyr jasper or, in areas where the Karamalytash Formation is absent, it directly overlies the Irendyk Formation. In the northern part of the Sibai area and far to the north, the Yarlykapovo red jasper horizon, of the same age as the Karamalytash Formation [*Maslov* et al., 1993], is interposed between the Ulutau and Irendyk units. The Ulutau Formation in the Sibai area is Givetian in age [*Maslov* et al., 1993] and has a total thickness of 1800–2000 m. It consists of epiclastic coarse-grained to fine-grained volcanic sandstone with chert interbeds, and conglomerate representing debris flow sediments, and turbidites. The volcanic clasts consist mostly of andesite and in minor amount of basalt and rhyolite and are thought to derive from volcanoes developed in the northeastern part of the Magnitogorsk Zone [*Frolova and Burikova*, 1977].

The Famennian Zilair Formation in the Sibai area overlies the Late Frasnian Mukas chert [*Maslov* et al., 1993] which is interposed between the Ulutau and Zilair formations. At its base, the Zilair Formation contains a thick (up to 100 m) olistostrome bed (Biyagoda olistostrome) with big blocks of limestone, basalt, rhyolite, chert and sandstone. This olistostrome is correlated with the Koltubanian volcano sedimentary formation, which crops out in the western part of Magnitogorsk zone. The middle and upper parts of the Zilair Formation consist mostly of graywacke turbidites with carbonate component increasing up the section.

4. IGNEOUS ROCKS FROM THE MAIN URALIAN FAULT

Diverse types of rocks from dismembered arc and oceanic crust are present in the Main Uralian fault within the mélange, or intruding mantle peridotites and the crust-mantle transition zone of ophiolite slabs.

One group includes arc-derived rocks which compose blocks, fragments and chips of polygenic tectonic breccias and small slabs (clast nomenclature according to *Hsü* [1969]) from the Nurali and Voshnezenka massifs (Figure 3) and the Mindyak massif located south of

LEGEND

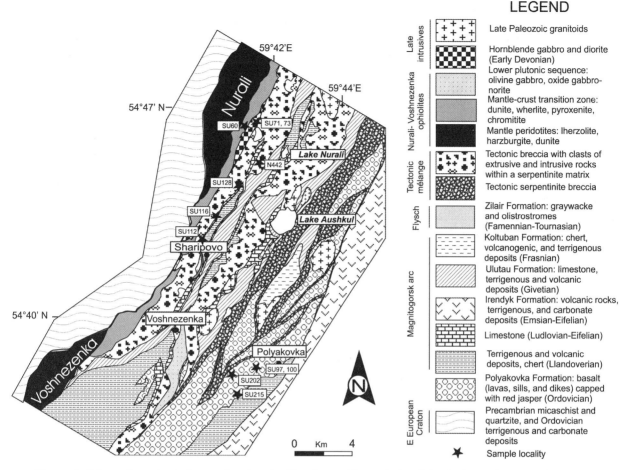

Figure 4. Schematic geological map showing the Nurali massif sample localities (redrawn from *Saveliev* et al. [1998]).

Nurali (Figure 1). This group is represented by lavas, dikes and tuffs of basaltic andesitic and andesitic to dacitic composition. The lavas and tuffs are often inter-layered or capped by chert, shale and carbonate rocks. Near Polyakovka (Figure 4) the mélange is in contact with the Irendyk Formation from which most of these fragments are derived [*Saveliev* et al., 1998] and with which they share similar petrographical and chemical characteristics [*Gaggero* et al., 1997].

A large spectrum of plutonic and hypabyssal rocks intrude at various levels the Nurali-Voshnezenka and Mindyak peridotites and the crust-mantle transition zone. They can be attributed to distinct generations.

A first generation is composed of hornblende gabbros and diorites which intrude the crust-mantle transition zone in the Nurali massif [*Gaggero* et al., 1997]. These gabbros and diorites contain xenoliths of transition zone pyroxenites [*Pertsev* et al., 1997]. They have been dated

radiometrically at about 400 Ma [*Smirnov*, 1995]. Similar gabbros and diorites are present as dispersed fragments within the Nurali mélange [*Gaggero* et al., 1997]. Blocks of quartz diorite and trondhjemite described by *Gaggero* et al. [1997] and identified as plutonic arc crust are also found in the Nurali massif. Also, blocks of ultramafic rocks (highly phyric ankaramite) from the Nurali massif may represent hypabyssal equivalent of intrusive ultramafic magmas intruding the Nurali mantle-crust transition zone [*Pertsev* et al., 1997; *Pertsev*, personal communication].

Two late dike generations are distinguished in the Voshnezhenka and Polyakovka areas. One, near Vosh-nezenka, is comprised of mafic dolerite dikes which intrude ophiolitic fragments, consisting of upper oxide gabbro, and are in turn intruded by andesite [*Saveliev* et al., 1998]. Another generation of undeformed, mafic dikes and sills near Polyakovka with variable texture from microgabbro to microlitic basalt, rich in amphibole, intrude the deformed and low metamorphic grade Ordovician Polyakovka Formation basalts (Figure 3). Their age is unknown; absence of deformation indicate that they are younger than the metamorphism affecting the basalts they intrude.

Dikes similar in petrography and chemistry to those intruding the Ordovician Polyakovka basalts, consisting of amphibole bearing microgabbro to dolerite are also present in the Mindyak massif. Some of them intrude the crust-mantle transition zone in northern Mindyak massif, others are dispersed as blocks in thrust zones.

In the Kraka massif located beyond the Main Uralian fault (Figure 1), the mélange flanking to the east the peridotite bodies includes blocks of extrusive and intrusive rocks similar to those from the Main Uralian fault [*Gaggero* et al., 1997]. These rocks have been included in this geochemical study.

5. ANALYTICAL TECHNIQUES

Bulk rock major element analyses were performed by X-ray fluorescence (XRF) at Udine and Padua Universities on fused beads prepared with lithium tetraborate flux to sample ratio of 1:10 to reduce matrix effects. Loss on ignition was determined by gravimetry. Rare Earth Elements (REE) and other 23 trace elements of most samples were analyzed by Inductively Coupled Plasma Mass Spectrometry (ICP-MS) at Granada University and measured in triplicate. For a set of samples, trace elements Sc, V, Ni, Cu, Zn, Rb, Y, Sr, Zr, Nb and Ba were analyzed by XRF on powder pellets, using the Compton scattering technique to correct matrix absorption; on the same sample set 10 REE were determined with a sequential ICP spectrometer at Udine University. Details on analytical techniques and precision are reported in *Spadea* et al. [1998] and *Spadea and Scarrow* [2000]. ICP-MS data on three Reference Rock Standards analyzed at Granada University are reported in Appendix 1. A comparison of trace element analyses done by XRF and ICP at Udine University with those done by ICP-MS at Granada University is presented in Appendix 2.

$^{87}Sr/^{86}Sr$ and $^{143}Nd/^{144}Nd$ analyses were done on whole rock samples after isolation of Sr and Nd by conventional ion exchange resin separation techniques. Measurements were made by Thermal Ionization Mass Spectrometry at the University of Granada on a Finnigan MAT 262 RPQ spectrometer. Instrument errors for determinations of $^{87}Sr/^{86}Sr$ and $^{143}Nd/^{144}Nd$ were $\pm$ 0.002–0.004 rel.% ($\equiv$ 14–28 ppm) and $\pm$ 0.0013–0.0030 rel.% ($\equiv$ 7–15 ppm), respectively. External precision (2σ) for $^{87}Sr/^{86}Sr$ from successive replicate measurements of

standards was $\pm$ 0.0028 rel.% ($\equiv$ 20 ppm) for the standard WS-E [*Govindaraju* et al., 1994], and was better than 0.0007 rel.% for the NBS-987 International Reference Standard of Sr. For $^{143}Nd/^{144}Nd$ the external precision (2σ) for WS-E was $\pm$ 0.0016 rel.% ($\equiv$ 8 ppm) and that for La Jolla was better than 0.0014 rel.% ($\equiv$ 10 ppm). The normalization value for fractionation of $^{143}Nd/^{144}Nd$ was $^{146}Nd/^{144}Nd = 0.7219$. The values of $^{87}Rb/^{86}Sr$ and $^{147}Sm/^{144}Nd$ were calculated from direct ICP-MS measurements (technique described above) of $^{85}Rb/^{88}Sr$ and $^{147}Sm/^{145}Nd$. The spectrometer was calibrated from pure dissolution of Rb and Sr ($^{87}Sr/^{86}S = 0.707265$). The precision of the method calculated on 10 replicates of standard WS-E, analyzed over 2 months, was better than 1.2% (2σ). For Sm and Nd the spectrometer was also calibrated from pure dissolution ($^{143}Nd/^{144}Nd = 0.512240$). For this ratio the precision of the $^{147}Sm/^{144}Nd$ measurements, also calculated on 10 replicates of standard WS-E analyzed over 2 months, was better than 0.9% (2σ). In order to evaluate the extent of $^{87}Sr/^{86}Sr$ changes related to alteration, Sr isotope determinations on leached samples were made at the Centro Inter-dipartimentale di Servizio per Analisi Geomineralogiche (University Federico II of Naples). 0.3 mg of powder were leached with warm 6N HCl for 30 minutes, then rinsed thoroughly in pure sub-boiling doubly distilled water, and finally dissolved with high purity HF-HNO_3-HCl mixtures. Sr was extracted by conventional ion exchange chromatographic techniques. Measurements were made using a VG 354 double collector thermal ionization mass spectrometer running in jumping mode, by normalizing to $^{86}Sr/^{88}Sr = 0.1194$ for mass fractionation effects. The internal precision (2σ) for $^{87}Sr/^{86}Sr$ on the analyzed samples ranges between ± 0.0014 rel.% ($\equiv$ 10 ppm) and ± 0.0018 rel.% ($\equiv$ 13 ppm). Replicate measurements of the NBS-987 International Reference Standard gave a mean value of $^{87}Sr/^{86}Sr = 0.71024$ (N = 50), with an external precision (2σ) of ± 0.0018 rel.% ($\equiv$ 13 ppm). The total blank was on the order of 6 ng during the period of measurements, thus negligible for the measured samples. $^{87}Sr/^{86}Sr$ ratios of unleached samples measured at Granada and $^{87}Sr/^{86}Sr$ ratios of leached samples measured at Naples are listed in Table 3. The comparison shows that leached samples have $^{87}Sr/^{86}Sr$ always lower than that in unleached samples, suggesting that leaching was effective in removing some secondary strontium. With the exception of sample SU 202, the difference between unleached and leached $^{87}Sr/^{86}Sr$ ratios is much larger than the external precision, being mostly between -100 and -1200 ppm, making negligible any possible bias between the two laboratories.

6. RESULTS

6.1. Major and Trace Elements

Major and trace element data for the extrusive rocks from the western Magnitogorsk zone and the igneous rocks (volcanic, hypabyssal and plutonic) from the Main Uralian fault zone at Nurali, Mindyak and Kraka ophiolite massifs are listed in Tables 1 and 2. Further analytical data used for this study are published in *Spadea and Scarrow* [2000]. Some major oxides analyses of the Main Uralian fault igneous rocks have been previously published by *Gaggero* et al. [1997] and reported in Tables 1 and 2. The new chemical data include new analyses of volcanic rocks from the Tanalyk and Karamalytash Formations, and new determinations of trace elements by ICP-MS for samples studied by *Gaggero* et al. [1997] and for reference Paleozoic MORB's, represented by the Middle Ordovician Polyakovka Formation basalts (Table 1).

Sample locations, petrographical classification, and modal or relict phenocryst mineralogy of volcanic rocks are reported in Tables 1 and 2. As hydrothermal alteration affects the majority of the arc samples, most having low grade prehnite-pumpellyite and pumpellyite-actinolite assemblages, few relics of magmatic minerals (spinel, clinopyroxene) are found, while olivine, orthopyroxene, and plagioclase are preserved only as textural relics. Also the original glass is identified only texturally. Detail descriptions of the metamorphic mineral assemblages, primary textures, and chemical composition of primary minerals of the analyzed rocks are reported in *Spadea* et al. [1998] for the Magnitogorsk arc rocks, and *Gaggero* et al. [1997] for the Main Uralian fault igneous rocks.

The rock classification is based mostly on petrography. Chemical criteria were adopted for volcanic high-Mg, strongly altered, and originally glassy rocks following the IUGS reclassification by *Le Bas* [2000].

Petrologic groups were distinguished chemically on the basis of the immobile, or less mobile, elements during hydrothermal alteration processes. Firstly we used major oxides (SiO_2, TiO_2, MgO and Mg#) which, although mobile during alteration, can provide criteria for comparisons among rocks affected by similar alteration. Secondly we used REE, as well as immobile very highly and highly incompatible high-field strength (HFS) elements abundance.

In Tables 1 and 2 the petrologic groups are classified into the boninitic, tholeiitic, calc-alkaline, and transitional (to alkaline) series, with further subdivisions according to the relative abundance of incompatible elements. Major oxides do not provide discrimination among Magnitogorsk arc rocks. This is shown by plots of TiO_2 versus Mg# (Figure 5a), although these plots evidence the overall low-Ti characteristics of the Magnitogorsk arc rocks. As expected for altered rocks, alkalis and in particular K_2O, though having non-erratic variations among groups, cannot be used to distinguish further the character of the arc rocks.

A variety of element plots to illustrate the chemical variations and main serial characteristics of the rocks from the western Magnitogorsk zone and the Main Uralian fault zone units are presented in Figures 6 to 9 and 11. For each rock group, MORB-normalized incompatible element abundance, and chondrite-normalized REE patterns are shown. In the following section the relative importance and significance of the selected chemical data will be discussed and evaluated.

6.2. Sr and Nd Isotopes

Sr and Nd isotopes were determined on representative samples belonging to most petrologic groups from the western Magnitogorsk arc and the Main Uralian fault. In addition, reference Ordovician MORB's have been also analyzed.

For the samples from the Magnitogorsk arc and Main Uralian fault the isotopic data were calculated back to 400 Ma, which is the average age resulting from stratigraphic data, with reference to the revised time scale for the Devonian time by *Tucker* et al. [1998], and radiometric dating of Main Uralian fault intrusive rocks (Sm-Nd [*Edwards and Wasserburg*, 1985] and U/Pb [*Smirnov*, 1995] ages). For the Ordovician MORB's, the ratios have been calculated back to 470 Ma, corresponding to the Lanverian (Middle Ordovician) age of the Polyakovka basalts [*Ryazantsev* et al., 1999]. Results of Sr and Nd isotope ratio determinations are listed in Table 3, along with initial isotope ratios, both present day and initial ε_{Nd} values (according to the CHUR model by *De Paolo* [1988]), as well as Rb, Sr, Sm and Nd contents (ICP-MS data).

$(^{87}Sr/^{86}Sr)_{400}$ for the whole set ranges from 0.7034 to 0.7066, and $(^{143}Nd/^{144}Nd)_{400}$ ranges from 0.51223 to 0.51276. Data are plotted on the conventional $^{87}Sr/^{86}Sr$ vs $^{143}Nd/^{144}Nd$ correlation diagram (Figure 12a, b), where they are compared to Atlantic, Pacific and Indian MORB data from literature, to Uralian MORB [*Edwards and Wasserburg*, 1985] and to the Bulk Earth, all calculated at 400 Ma ago ($^{87}Sr/^{86}Sr = 0.70403$; $^{143}Nd/^{144}Nd = 0.51212$; [*De Paolo*, 1988]).

It is noteworthy that much of the observed scatter in $^{87}Sr/^{86}Sr$ values has to be attributed to seawater alteration of the analyzed rocks, as also suggested for Uralian MORB from the Kempersay massif by *Edwards and Wasserburg* [1985]. During Devonian time the isotopic

Table 1. XRF major oxides and ICP-MS (except[a]) trace elements data for the Magnitogorsk zone extrusive rocks (wt% major oxides, ppm trace elements).

Unit Series	Polyakovka Formation (Mid-Ordovician) tholeiitic					Baimak-Buribai and Tanalyk Formations boninitic to island arc tholeiitic					
Sample Rock type	SU97 aphyr basalt	SU99 aphyr basalt	N429[a] aphyr basalt	SU33[a] aphyr basalt	SU35[a] aphyr basalt	LK148-4 aphyr basalt	AKT6V glassy boninite	LK29-2 glassy boninite	AKT2V pl px ph basalt	AKT66 pl px ph andesite	AK8501A pl px ph basalt
SiO_2	43.65	50.55	41.53	52.32	51.29	51.78	51.96	57.70	55.42	55.22	55.96
TiO_2	1.83	1.08	1.94	1.52	0.83	0.26	0.51	0.46	0.52	0.49	0.55
Al_2O_3	13.71	14.10	16.31	15.25	15.65	9.34	13.37	12.17	17.01	16.11	15.62
Fe_2O_3	16.56	11.07	15.91	11.08	8.76	9.20	10.44	7.95	7.42	9.11	7.99
MnO	0.26	0.20	0.23	0.18	0.20	0.18	0.17	0.17	0.11	0.13	0.31
MgO	7.78	8.86	7.80	5.14	5.89	18.65	10.74	9.63	8.46	7.48	6.78
CaO	13.91	9.15	14.04	7.75	10.70	8.88	9.00	6.56	3.90	4.78	6.47
Na_2O	1.01	2.35	0.72	6.79	5.40	1.63	3.28	5.50	6.08	5.91	5.49
K_2O	0.66	2.02	0.90	0.07	1.09	0.04	0.10	0.10	0.21	0.13	0.57
P_2O_5	0.17	0.12	0.18	0.21	0.18	0.08	0.06	0.09	0.07	0.10	0.08
Total	99.54	99.50	99.56	100.31	99.99	100.04	99.63	100.33	99.20	99.46	99.82
LOI	3.46	3.67	4.27	5.19	7.71	4.43	2.50	1.89	3.35	4.60	7.39
Mg#[b]	50.35	63.34	51.41	50.03	59.21	81.40	68.95	72.33	71.11	63.93	64.68
Li	29.5	13.7				7.55	3.97	2.11	2.28	27.4	11.9
Rb	8.10	45.7	12	2	10	0.59	2.39	0.71	2.97	0.76	7.02
Cs	0.45	0.73				0.23	0.08	0.04	0.07	0.03	0.19
Be	0.81	0.38				0.57	0.30	0.44	0.65	0.55	0.53
Sr	83.1	99.9	123	69	254	27.7	143	58.7	157	21.3	127
Ba	180	79.6	318	4	174	3.13	22.5	15.4	59.2	27.8	278
Sc	57.7	41.1	65	40	41	32.4	38.5	34.1	28.2	28.8	27.9
V	443	272	476	266	240	143	239	223	179	240	150
Cr	120	332	137	81	188	1186	422	129	71.5	13.0	16.0
Co	60.0	51.9				60.8	41.6	366	29.4	22.6	28.4
Ni	66.6	88.4	2100	39	45	428.7	92.8	82.8	65.2	12.6	25.2
Cu	67.8	74.4	50	52	47	8.0	14.7	8.8	10.1	162	79.6
Zn	127	98.4	118	88	68	46.6	79.9	50.9	112	62.2	49.8
Ga	14.7	13.3				7.92	10.4	8.17	14.2	12.4	11.5
Y	39.1	21.5	37	47	29	6.4	13.6	12.1	13.2	11.9	7.7
Nb	3.63	1.67	4	4	3	1.17	0.57	1.89	0.86	0.76	1.97
Ta	0.55	0.14				0.10	0.07	1.20	0.09	0.07	0.18
Zr	107	64	115	157	91	23	26	48	44	35	31
Hf	2.72	1.53				0.53	0.94	1.17	1.39	1.18	0.97
Mo	0.82	0.17				0.00	0.22	0.54	0.18	0.41	0.45
Sn	1.55	1.22				0.00	0.58	0.03	1.14	0.19	1.15
Tl	0.037	0.103				0.018	0.003	0.076	0.007	0.004	0.023
U	0.132	0.039				0.145	0.130	0.163	0.217	0.375	0.169
Th	0.280	0.136				0.180	0.253	0.297	0.404	0.331	0.318
La	4.28	2.16	4.28	7.30	4.85	0.94	1.39	1.59	2.14	1.39	1.50
Ce	13.24	6.83	12.90	19.70	12.56	1.29	4.04	4.22	5.70	4.04	4.01
Pr	2.21	1.09				0.39	0.67	0.70	0.94	0.70	0.70
Nd	11.49	5.98	11.60	16.30	9.60	1.55	3.57	3.45	4.68	3.72	3.44
Sm	4.02	2.11	4.26	5.40	3.05	0.56	1.31	1.12	1.52	1.34	1.07
Eu	1.31	0.80	1.42	1.67	0.92	0.20	0.50	0.39	0.65	0.48	0.27
Gd	5.91	3.03	5.65	7.03	3.97	0.78	1.81	1.61	2.15	1.79	1.29
Tb	1.02	0.56				0.14	0.33	0.30	0.38	0.30	0.22
Dy	7.04	3.77	6.84	7.96	4.54	1.00	2.20	1.96	2.42	1.97	1.37
Ho	1.56	0.81				0.22	0.50	0.44	0.52	0.44	0.29
Er	4.31	2.31	3.92	4.88	2.74	0.64	1.49	1.26	1.38	1.23	0.90
Tm	0.65	0.36				0.10	0.23	0.20	0.20	0.19	0.14
Yb	4.14	2.24	3.83	4.66	2.73	0.64	1.44	1.28	1.28	1.24	0.88
Lu	0.64	0.35	0.60	0.72	0.44	0.11	0.23	0.20	0.20	0.19	0.14

Table 1. (continued)

Sample Rock type	Calc-alkaline				Irendyk Formation tholeiitic Island arc tholeiitic				
	AKT7A px phyr basalt	AK3039 pl phyr basalt	AK514A px pl phr basalt	AKT-50 cpx phyr basalt	AKYO10 cpx pl ph basalt	AKn108 aphyr basalt	AK308A px pl ph basalt	AKYO7 px pl ph basalt	AK48C aph bas. andesite
SiO_2	57.00	55.40	51.54	54.24	47.62	52.69	48.80	52.14	56.92
TiO_2	0.50	0.99	0.54	0.45	1.90	0.51	0.52	0.46	0.36
Al_2O_3	16.76	17.56	17.13	14.02	16.76	16.87	17.53	15.26	17.07
Fe_2O_3	9.52	12.61	9.37	8.46	14.17	9.10	11.96	9.49	6.74
MnO	0.14	0.18	0.12	0.12	0.19	0.15	0.15	0.16	0.14
MgO	3.93	10.73	7.31	9.55	5.02	4.90	7.87	7.02	3.62
CaO	9.95	0.31	11.31	8.94	9.85	12.56	9.93	11.34	10.92
Na_2O	1.55	0.10	1.79	2.87	3.82	2.27	2.66	3.49	3.95
K_2O	0.03	1.32	0.20	0.63	0.08	0.03	0.07	0.30	0.13
P_2O_5	0.10	0.12	0.05	0.10	0.25	0.14	0.12	0.07	0.11
Total	99.48	99.32	99.36	99.38	99.66	99.22	99.61	99.73	99.96
LOI	4.99	6.58	4.43	4.00	4.73	7.48	5.68	3.26	3.43
Mg#[b]	47.12	64.75	62.74	70.90	43.33	53.75	58.68	61.49	53.69
Li	2.85	3.59	46.5	12.9	8.66	30.02	30.97	5.57	3.32
Rb	5.52	16.82	2.38	5.51	1.22	2.46	3.10	2.36	5.02
Cs	0.09	0.19	0.38	0.13	0.43	0.13	0.41	0.34	0.15
Be	0.50	0.71	0.64	0.61	0.60	0.45	0.70	0.79	0.69
Sr	130	20.4	98.0	114	42.7	37.0	46.3	36.7	41.0
Ba	39.7	471	19.1	46.1	21.5	56.9	18.5	47.9	22.8
Sc	35.3	32.2	37.8	39.7	35.2	45.3	50.7	46.2	33.1
V	227	343	147	217	345	319	347	256	282
Cr	569	4.02	431	192	0	112	207	200	46.0
Co	39.6	25.5	34.4	29.7	42.3	27.3	41.8	29.6	18.1
Ni	170	5.7	150	65.2	17.7	29.9	48.3	26.1	22.3
Cu	72.3	11.2	68.9	5.4	66.2	509	303	67.2	110
Zn	240	61.5	77.8	85.2	95.7	201	102	50.7	54.3
Ga	11.2	30.4	12.7	13.3	24.7	11.6	13.1	14.3	15.9
Y	12.5	10.8	12.1	10.5	38.5	7.89	9.01	12.5	13.6
Nb	0.58	1.92	1.75	1.20	2.31	0.27	0.37	1.71	0.30
Ta	0.06	0.20	0.04	0.10	0.11	0.03	0.04	0.03	0.03
Zr	25	56	77	47	97	6	9	20	12
Hf	0.89	1.69	1.32	1.32	2.92	0.37	0.47	0.79	0.70
Mo	1.14	0.64	0.00	0.51	0.07	3.06	3.54	0.04	0.46
Sn	0.97	1.32	0.00	0.91	0.00	0.71	0.68	0.00	1.20
Tl	0.022	0.040	0.020	0.007	0.017	0.060	0.053	0.019	0.016
U	0.270	0.561	0.192	0.189	0.100	0.612	0.330	0.252	0.223
Th	0.379	0.598	0.467	0.569	0.114	0.200	0.267	0.237	0.267
La	1.83	5.41	6.87	4.23	3.76	1.66	1.66	2.05	1.90
Ce	4.74	11.57	15.22	9.66	11.66	3.84	3.93	4.98	4.72
Pr	0.73	1.67	1.94	1.37	2.07	0.59	0.59	0.78	0.79
Nd	3.64	7.36	7.49	5.95	11.14	2.88	2.94	3.81	3.84
Sm	1.24	2.04	1.84	1.51	4.08	0.99	1.00	1.28	1.27
Eu	0.51	0.22	0.65	0.62	1.53	0.46	0.47	0.48	0.40
Gd	1.79	2.53	1.98	1.69	5.57	1.19	1.32	1.76	1.72
Tb	0.32	0.38	0.34	0.28	0.98	0.21	0.24	0.32	0.31
Dy	2.09	2.22	2.04	1.70	6.69	1.39	1.55	2.00	2.10
Ho	0.46	0.49	0.45	0.38	1.46	0.30	0.34	0.46	0.47
Er	1.36	1.39	1.29	1.09	3.94	0.84	0.99	1.32	1.42
Tm	0.21	0.21	0.20	0.16	0.63	0.13	0.16	0.22	0.22
Yb	1.32	1.32	1.23	1.09	3.95	0.81	1.01	1.31	1.42
Lu	0.22	0.21	0.20	0.16	0.58	0.13	0.15	0.21	0.23

Table 1. (continued)

Sample Rock type	IAT-CA		Karamalytash Formation island arc tholeiitic					calc-alkaline	
	AKn24 aphyr basalt	AK83C aph bas. andesite	Kn315 pl ph basalt	YO27 aphyr basalt	97-16 andesite	97-18 andesite	97-17 rhyolite	Kn35 pl ph basalt	97-11 bas. andes.
SiO_2	58.77	73.19	52.62	53.10	53.43	53.51	84.57	52.75	55.58
TiO_2	0.39	0.41	0.84	0.81	0.49	0.43	0.21	0.72	0.63
Al_2O_3	16.08	12.69	18.86	16.47	15.64	15.97	8.81	16.45	18.21
Fe_2O_3	5.80	3.63	13.72	12.57	13.01	10.66	2.43	12.86	8.05
MnO	0.16	0.04	0.24	0.16	0.16	0.15	0.05	0.21	0.14
MgO	1.47	1.30	4.97	5.41	6.40	7.00	0.40	3.70	4.20
CaO	10.30	2.39	1.69	5.89	9.57	9.40	0.61	8.81	10.58
Na_2O	5.10	3.87	5.58	4.11	1.43	1.42	3.83	1.75	2.42
K_2O	1.27	1.49	0.70	0.83	0.13	1.61	0.11	1.83	0.62
P_2O_5	0.12	0.19	0.19	0.08	0.03	0.06	0.04	0.20	0.17
Total	99.46	99.20	99.41	99.43	100.29	100.21	101.06	99.28	100.60
LOI[b]	7.54	1.61	3.89	4.00	5.23	1.60	0.69	5.64	2.01
Mg#[b]	35.36	43.60	43.88	48.16	51.50	58.63	26.22	38.31	52.97
Li	9.90	4.50	16.3	16.8	27.1	20.1	10.3	12.3	6.17
Rb	26.7	30.9	14.2	10.9	2.07	24.1	1.14	37.3	5.46
Cs	0.86	0.49	0.59	1.23	0.49	0.55	0.11	0.78	0.25
Be	1.16	0.38	0.49	0.56	0.43	0.41	0.63	1.07	0.81
Sr	242	137	144	341	138	395	47.7	93.0	450
Ba	326	137	395	185	21.6	106	20.0	453	181
Sc	25.7	19.3	30.4	46.2	56.4	46.5	10.8	24.2	31.4
V	147	67.6	131	317	409	276	2.65	227	248
Cr	38.0	9.52	312	66.9	58.6	126	137	797	128
Co	10.5	3.7	38.1	32.3	43.4	38.2	2.6	24.1	21.2
Ni	12.4	4.4	48.6	17.2	37.9	54.2	60.5	79.5	52.9
Cu	20.1	8.8	443	128	142	99.7	12.3	122	98.2
Zn	56.0	10.1	418	211	75.4	68.9	62.3	230	70.5
Ga	11.2	8.86	14.6	16.2	14.1	11.5	9.05	16.4	19.0
Y	21.1	34.6	24.9	19.1	9.66	12.0	17.0	25.6	17.4
Nb	0.88	0.59	3.59	1.33	0.18	0.20	0.63	2.10	2.28
Ta	0.08	0.05	1.21	0.02	0.02	0.02	0.03	0.17	0.15
Zr	57	37	50	41	20	28	65	57	66
Hf	1.73	1.27	1.60	1.37	0.36	0.63	2.18	1.87	1.32
Mo	0.48	0.52	4.95	0.51	1.15	1.03	3.31	1.81	1.08
Sn	1.09	0.97	0.00	0.00	2.55	1.10	2.28	1.72	1.70
Tl	0.738	0.071	0.156	0.089	0.003	0.053	0.001	0.188	0.052
U	0.432	0.288	0.361	0.173	0.047	0.212	0.185	0.782	0.543
Th	0.881	0.394	0.686	0.145	0.125	0.207	0.417	1.658	1.122
La	5.48	3.17	2.85	1.69	0.41	1.15	2.16	6.05	6.81
Ce	12.51	8.06	7.73	4.89	1.29	3.27	6.24	14.22	16.24
Pr	1.97	1.46	1.26	0.89	0.23	0.51	1.00	2.12	2.27
Nd	8.88	7.64	6.60	4.69	1.44	2.77	5.53	9.88	9.98
Sm	2.36	2.53	2.51	1.78	0.58	0.99	1.73	2.99	2.56
Eu	0.45	0.71	0.55	0.67	0.28	0.44	0.61	0.68	0.91
Gd	2.82	3.67	3.31	2.48	1.01	1.57	2.56	3.77	2.84
Tb	0.52	0.67	0.62	0.48	0.22	0.28	0.46	0.63	0.48
Dy	3.35	4.58	4.20	3.07	1.54	1.96	3.13	4.08	3.14
Ho	0.79	1.09	0.96	0.71	0.36	0.45	0.66	0.90	0.65
Er	2.41	3.21	2.72	2.03	1.08	1.29	1.88	2.68	1.83
Tm	0.37	0.49	0.42	0.33	0.18	0.21	0.29	0.39	0.28
Yb	2.45	3.36	2.63	2.07	1.16	1.28	1.80	2.44	1.81
Lu	0.40	0.53	0.40	0.34	0.18	0.20	0.28	0.38	0.27

[a]Trace elements analyzed by XRF and ICP-AES (REE), [b]mol $MgO/(MgO + FeO)$; $Fe_2O_3/FeO = 0.10$).

Table 2. XRF major oxides and ICP-MS (except[a]) trace elements data for the Main Uralian fault extrusive and intrusive rocks (wt% major oxides, ppm trace elements).

Locality Group	Nurali ophiolite MORB intrusives		Nurali massif depleted IAT		Nurali and Mindyak mélange polygenic breccias island arc tholeiitic extrusives							
Sample Rock type	SU295[a] micro-gabbro	SU306[a] basaltic andesite	N442 ankar-amite	N442-1[a] ankar-amite	SU205 aphyr andesite	SU206/1 cpx ph andesite	SU116 cpx ph andesite	SU240[a] qtz andesite	SU128 aphyr andesite	SU215 phyr andesite	SU235[a] pl phyr basalt	SU371 aphyr basalt
SiO_2	47.52	52.15	49.05	48.03	61.85	62.51	53.67	60.41	54.34	51.12	52.05	51.15
TiO_2	0.63	0.59	0.16	0.17	1.28	1.29	0.46	0.68	0.57	0.51	0.64	1.07
Al_2O_3	15.92	15.65	4.83	5.16	13.06	12.92	15.25	15.33	16.04	16.60	18.79	16.01
Fe_2O_3	11.10	11.05	6.83	7.48	8.16	8.81	8.37	10.71	9.30	10.19	8.97	9.42
MnO	0.17	0.18	0.14	0.17	0.14	0.14	0.14	0.36	0.14	0.17	0.23	0.16
MgO	9.81	7.16	21.37	22.83	4.79	4.00	8.25	3.91	6.64	6.49	5.68	7.10
CaO	12.76	9.96	17.52	16.05	4.88	4.39	10.94	2.26	9.51	10.84	8.43	12.41
Na_2O	1.93	3.37	0.00	0.05	4.36	4.62	2.43	5.43	2.65	3.40	3.53	2.41
K_2O	0.07	0.60	0,01	0.01	0.56	0.68	0.20	0.76	0.47	0.44	1.48	0.08
P_2O_5	0.05	0.09	0.04	0.05	0.18	0.17	0.07	0.09	0.09	0.09	0.11	0.13
Total	99.96	100.80	99.94	100.00	99.26	99.53	99.78	99.94	99.75	99.85	99.91	99.94
LOI	1.45	2.36	3.79	4.83	2.98	2.13	3.72	3.56	3.80	2.43	2.50	3.80
Mg#[b]	65.61	58.31	87.10	86.82	55.89	49.50	68.03	44.07	60.65	57.89	57.75	61.93
Li			25.9		7.98	10.1	10.2		9.26	10.58		10.2
Rb	2	8	0.73	2	10.50	8.31	2.19	18	5.93	5.61	20	1.48
Cs			0.20		0.48	0.27	0.81		0.67	0.38		0.30
Be			0.24		0.47	0.58	0.31		0.38	0.33		0.41
Sr	135	410	15.0	4	215	162	365	142	314	307	387	61.6
Ba	28	141	0	5	85.3	58.4	48.2	207	158	87.5	5417	0
Sc	75	50	41.8	45	36.8	39.6	45.6	47	36.4	50.5	41	35.4
V	318	241	125	127	247	241	260	244	245	301	320	252
Cr	121	71	1269	2180	230	191	205	64	71.4	77.6	60	200
Co			54.7		29.3	48.5	34.4		61.1	30.5		33.1
Ni	71	41	306	371	35.8	51.1	60.3	26	42.5	34.9	16	67.3
Cu	43	585	32.3	40	36.2	10.1	92.7	13	89.5	91.0	21	56.2
Zn	58	82	49.7	45	53.8	87.3	54.1	99	62.9	63.6	87	63.8
Ga			5.16		15.4	13.3	12.7		13.0	16.8		16.7
Y	22	18	4.5	5	19.5	15.9	11.3	14	16.7	14.3	16	24.0
Nb	2	2	1.03	2	1.64	0.95	0.61	3	1.07	1.10	3	4.24
Ta			0.04		0.06	0.00	0.04		0.09	0.04		0.15
Zr	19	42	13	13	38	36	28	40	37	39	33	75
Hf			0.24		1.20	1.09	0.69		1.07	0.78		1.96
Mo			0.00		0.12	0.00	0.00		0.50	0.00		0.18
Sn			1.52		0.00	0.00	1.12		0.37	2.26		1.83
Tl			0.014		0.225	0.093	0.017		0.029	0.034		0.013
U			0.099		0.193	0.066	0.108		0.215	0.225		0.150
Th			0.136		0.000	0.000	0.207		0.438	0.298		0.351
La	1.21	1.46	1.27	1.16	3.14	1.84	1.28	2.75	1.99	1.89	3.27	4.11
Ce	4.02	3.14	2.11	1.94	8.86	5.12	3.39	7.16	5.48	4.96	7.03	10.85
Pr			0.25		1.44	0.95	0.51		0.86	0.82		1.65
Nd	4.26	3.30	1.34	1.44	7.49	4.88	2.49	4.86	4.62	4.05	5.02	8.15
Sm	1.76	1.22	0.44	0.48	2.39	1.71	0.98	1.40	1.54	1.44	1.43	2.70
Eu	0.69	0.68	0.20	0.20	1.03	0.91	0.42	0.51	0.52	0.54	0.66	1.00
Gd	2.88	2.00	0.53	0.63	3.14	2.21	1.49	1.76	2.17	1.91	1.99	3.55
Tb			0.10		0.54	0.40	0.28		0.39	0.34		0.60
Dy	3.71	2.46	0.69	0.70	3.43	2.62	1.98	2.05	2.71	2.28	2.22	4.01
Ho			0.16		0.74	0.58	0.45		0.61	0.51		0.89
Er	2.32	1.66	0.46	0.54	2.04	1.65	1.30	1.29	1.77	1.46	1.50	2.44
Tm			0.07		0.31	0.25	0.21		0.28	0.24		0.38
Yb	2.20	1.84	0.46	0.51	1.83	1.55	1.33	1.22	1.78	1.54	1.64	2.40
Lu	0.31	0.29	0.07	0.08	0.26	0.24	0.22	0.19	0.28	0.24	0.16	0.37

Table 2. (continued)

	Kraka	Nurali mélange polygenic breccia calc-alkaline extrusive rocks								OIB intrusive rocks		
Sample Rock type	K-170 qtz dolerite	SU117 Phyr andesite	SU129 phyr andesite	N-429A aphyr dacite	SU141 cpx phyr basalt	SU145 ol px ph basalt	N-437 pl phyr basalt	SU211[a] dolerite	SU362[a] andesite	SU71 hbl gabbro	SU73 hbl gabbro	SU89 dolerite
SiO_2	52.42	60.35	53.71	66.93	46.27	44.45	44.63	51.12	60.95	48.07	43.15	56.32
TiO_2	0.59	0.86	0.82	0.84	0.61	0.69	0.62	1.89	0.54	0.83	1.08	0.82
Al_2O_3	15.90	16.15	16.85	13.03	13.04	13.84	14.70	14.29	16.15	21.02	21.15	14.32
Fe_2O_3	9.78	7.10	9.31	7.99	11.10	12.34	9.27	11.33	7.47	10.80	14.13	10.53
MnO	0.18	0.16	0.14	0.27	0.19	0.23	0.22	0.24	0.15	0.22	0.23	0.20
MgO	6.98	3.46	6.22	2.85	8.11	7.98	10.71	7.55	4.01	4.96	5.52	4.59
CaO	11.50	5.61	9.01	2.94	20.13	19.73	19.93	8.24	5.89	9.44	10.60	6.94
Na_2O	2.00	4.90	3.11	4.39	0.16	0.15	0.06	3.03	3.27	3.14	1.54	5.76
K_2O	0.24	0.74	0.55	0.53	0.02	0.04	0.00	1.86	1.38	0.79	1.79	0.23
P_2O_5	0.09	0.15	0.14	0.19	0.21	0.23	0.10	0.19	0.18	0.24	0.33	0.12
Total	99.68	99.48	99.86	99.96	99.84	99.68	100.24	99.74	99.99	99.51	99.52	99.83
LOI	0.95	2.78	4.06	2.16	3.68	3.23	4.39	2.44	3.15	1.91	2.18	0.86
Mg#[b]	60.64	51.26	59.05	43.50	61.20	58.26	71.38	58.99	53.68	49.78	45.75	48.48
Li	0.65	5.80	10.00	12.5	8.65	10.2	17.3			2.93	3.44	4.67
Rb	4.55	7.90	13.8	7.29	0.00	2.82	0.64	50	19	4.68	9.65	2.18
Cs	0.06	0.67	0.93	0.28	0.06	0.07	0.67			0.09	0.12	0.14
Be	0.39	0.76	0.67	1.39	0.63	0.42	0.46			0.63	0.58	0.45
Sr	139	189	357	233	49.1	58.7	130	131	309	476	436	77.8
Ba	57.4	371	125	152	37.6	9.36	13.3	275	393	91.4	132	86.2
Sc	42.8	39.5	36.7	21.1	35.3	35.8	43.1	46	29	28.4	34.4	35.1
V	286	250	289	140	274	302	275	301	170	292	297	307
Cr	88.5	125	193	24.3	175	241	318	260	65	20.3	15.7	32.7
Co	39.1	26.6	27.6	11.3	49.2	37.8	35.3			40.1	37.9	27.8
Ni	48.6	38.0	75.8	16.7	50.3	89.3	67.6	62	12	16.9	23.1	20.1
Cu	101	45.9	68.2	58.0	91.4	160	83.3	104	92	42.5	15.7	88.2
Zn	71.9	95.0	92.9	90.2	61.9	72.5	68.5	77	70	89.7	77.2	57.6
Ga	14.6	12.7	18.7	18.2	9.37	10.5	11.5			20.9	22.1	14.8
Y	16.4	22.0	23.3	28.0	13.3	15.1	16.1	38	20	22.3	30.2	19.6
Nb	0.64	6.24	6.05	6.42	1.86	1.09	3.49	9	5	4.04	3.67	2.64
Ta	0.00	0.40	0.38	0.40	0.58	0.02	0.20			0.30	0.70	0.13
Zr	38	76	73	135	33	30	51	132	100	63	35	51
Hf	1.03	1.80	1.70	3.93	0.83	0.99	1.33			0.72	0.75	1.41
Mo	0.25	1.80	1.24	0.03	0.05	2.60	0.30			0.15	0.27	0.22
Sn	0.82	0.93	0.00	2.64	0.43	0.00	1.22			0.81	0.44	0.85
Tl	0.024	0.046	0.046	0.05	0.006	0.015	0.004			0.005	0.027	0.014
U	0.154	0.623	0.425	0.96	0.164	0.214	0.532			0.107	0.120	0.414
Th	0.384	1.279	1.147	3.85	0.000	0.000	1.327			0.233	0.249	0.964
La	2.00	7.66	10.20	17.42	4.31	4.92	5.86	7.15	15.33	7.76	6.67	5.79
Ce	5.12	17.73	18.57	42.78	9.78	11.60	12.74	19.75	29.45	19.10	18.31	12.95
Pr	0.79	2.34	2.53	5.73	1.47	1.67	1.65			2.77	2.93	1.79
Nd	4.02	10.36	10.83	24.24	7.27	7.67	7.43	14.65	14.82	13.00	14.93	7.94
Sm	1.48	2.90	2.98	6.08	1.94	2.16	2.09	4.42	3.07	3.27	4.25	2.55
Eu	0.61	0.91	1.01	1.34	0.70	0.76	0.66	1.48	0.87	1.10	1.36	0.87
Gd	2.29	3.52	3.31	5.63	2.30	2.50	2.39	5.67	3.05	3.59	5.07	3.03
Tb	0.39	0.60	0.55	0.85	0.40	0.44	0.40			0.61	0.84	0.50
Dy	2.73	3.88	3.51	5.14	2.45	2.82	2.70	6.14	2.81	3.82	5.38	3.39
Ho	0.65	0.84	0.78	1.08	0.51	0.60	0.60			0.85	1.16	0.75
Er	1.76	2.47	2.27	2.93	1.37	1.70	1.68	3.29	1.76	2.50	3.35	2.16
Tm	0.27	0.36	0.36	0.46	0.21	0.26	0.26			0.38	0.49	0.34
Yb	1.77	2.31	2.17	2.87	1.29	1.58	1.76	2.97	1.78	2.22	3.00	2.05
Lu	0.30	0.36	0.34	0.43	0.19	0.23	0.28	0.40	0.28	0.36	0.45	0.32

Table 2. (continued)

Sample Rock type	Nurali, Mindyak, Kraka massifs OIB intrusives			Polyakovka late dikes intruding Polyakovka MORB basalts					
	SU112 qtz diorite	SU313 micro-gabbro	K153 qtz diorite	SU100 micro-gabbro	N430[a] dolerite	SU101[a] micro-gabbro	SU202 dolerite	SU212 micro-diorite	SU250[a] hbl dolerite
SiO_2	55.26	52.51	55.30	44.90	45.17	48.13	48.86	54.61	47.89
TiO_2	0.71	0.57	0.58	1.09	1.29	1.03	2.48	1.12	1.01
Al_2O_3	18.75	16.95	16.83	10.68	11.14	13.44	15.28	12.42	14.75
Fe_2O_3	8.91	9.54	9.78	11.42	11.37	9.98	13.92	7.90	10.24
MnO	0.19	0.17	0.18	0.17	0.18	0.15	0.21	0.22	0.17
MgO	3.98	6.66	4.39	18.40	17.96	14.14	8.43	9.45	11.08
CaO	7.52	9.79	8.52	10.71	10.57	8.68	5.90	7.25	10.33
Na_2O	3.08	3.68	3.17	0.99	0.78	0.94	3.59	1.99	0.73
K_2O	0.91	0.88	0.74	0.80	1.11	2.80	1.20	4.64	3.24
P_2O_5	0.23	0.18	0.23	0.35	0.41	0.43	0.28	0.49	0.46
Total	99.54	100.93	99.72	99.51	99.98	99.72	100.15	100.09	99.90
LOI	2.54	2.57	1.53	3.44	3.37	4.38	3.63	1.84	2.76
Mg#[b]	49.09	60.11	49.21	77.67	77.32	75.36	56.66	72.08	70.02
Li	3.33	5.24	1.71	23.36			18.6		
Rb	11.1	12.0	6.52	14.11	19	63	14.7	70	95
Cs	0.20	0.14	0.13	0.27			0.21		
Be	0.75	0.68	0.97	1.26			1.05		
Sr	466	426	586	320	313	201	147	370	911
Ba	302	128	176	184	285	452	436		1006
Sc	23.0	24.0	32.3	29.9	33	34	33.1	40	36
V	222	229	277	256	275	201	318	234	244
Cr	6.61	32.1	25.6	1229	1122	1164	60.7	181	782
Co	46.6	44.5	23.5	74.6			65.1		
Ni	2.7	36.4	9.1	602	536	626	52.8	76	252
Cu	45.4	87.2	103	98.7	145	89	110.3	35	65
Zn	82.3	64.5	72.4	76.9	83	89	113.3	62	92
Ga	17.7	13.0	17.6	15.4			19.65		
Y	20.8	15.2	18.4	13.8	15	19	27.9	20	22
Nb	6.59	1.68	3.89	7.02	7	7	17.73	8	8
Ta	1.16	0.48	0.23	0.46			1.20		
Zr	70	60	58	103	114	145	165	176	145
Hf	0.91	1.34	1.46	2.70			4.07		
Mo	0.30	0.92	0.00	0.00			0.08		
Sn	3.58	0.00	0.51	1.85			3.64		
Tl	0.050	0.046	0.034	0.065			0.093		
U	0.765	0.620	2.602	0.998			0.430		
Th	2.866	1.009	6.225	3.837			1.469		
La	12.23	7.08	13.88	21.66	23.62	23.63	13.98	20.58	33.19
Ce	24.91	16.12	27.35	45.84	53.00	48.80	34.97	46.04	64.27
Pr	3.23	2.17	3.58	5.78			4.79		
Nd	13.19	9.66	15.24	22.90	25.10	25.10	21.02	24.46	32.88
Sm	3.18	2.45	3.60	4.80	5.14	5.10	5.49	4.76	5.75
Eu	0.97	0.79	0.97	1.39	1.45	1.42	1.81	1.39	1.75
Gd	3.42	2.52	3.19	3.68	4.30	4.36	5.96	4.37	4.89
Tb	0.55	0.42	0.50	0.50			0.94		
Dy	3.37	2.71	3.06	2.70	3.01	3.16	5.48	3.24	3.37
Ho	0.73	0.60	0.68	0.51			1.09		
Er	2.08	1.60	1.99	1.28	1.39	1.52	2.72	1.62	1.63
Tm	0.34	0.25	0.31	0.19			0.41		
Yb	2.23	1.55	1.91	1.18	1.23	1.30	2.35	1.52	1.53
Lu	0.34	0.23	0.28	0.17	0.17	0.20	0.36	0.22	0.22

[a]Trace elements analyzed by XRF and ICP-AES (REE), [b]mol MgO/(MgO+FeO); $Fe_2O_3/FeO = 0.10$).

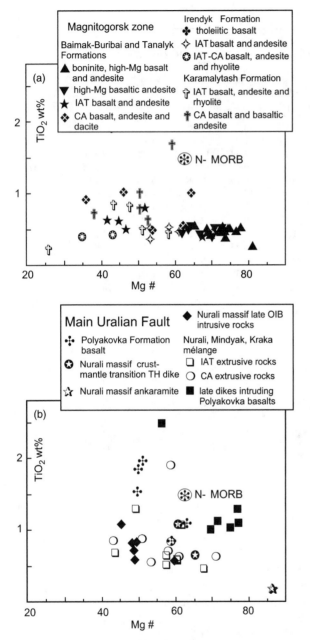

Figure 5. TiO$_2$ versus Mg# for Magnitogorsk zone Devonian extrusive rocks (a), and Ordovician Polyakovka Formation MORB and Main Uralian fault Devonian intrusive and extrusive rocks (b). N-MORB values are from *Sun and McDonough* [1989].

composition of seawater Sr was about 0.7087 [*Burke et al.*, 1982], and that of Nd was about 0.51175 [*Shaw and Wasserburg*, 1985]. Interaction with seawater derived hydrothermal fluids, accompanied by low grade metamorphism, is known to strongly affect the Sr isotope composition of deep seated igneous rocks, resulting

in higher than expected ^{87}Sr/^{86}Sr values compared to fresh rocks. This effect can be only slightly reduced by using acid leaching procedures on powdered samples (see analytical techniques), especially when low grade metamorphism, leaching resistant mineral phases such as epidote and pumpellyite, are present. However, no significant effect has been detected for Nd isotopes, essentially because of the low Nd content of seawater (on the order of 10^{-6} ppm [*Faure*, 1986]), and of the general low mobility of REE [*Faure*, 1986, and references therein].

Seawater contamination most probably is the main cause of the shift of most samples to the right of the MORB field (Figure 12a, b). Thus, on the basis of the above considerations, only the Nd isotope ratios can be confidently used to infer source characteristics.

Taking into account the discriminations and grouping of different samples made on the basis of petrographic and chemical features, inter group differences based on Nd isotopic compositions are detected, and will be discussed below.

6.3. Evidence and Origin of Serial Characteristics

Background information for a generalized genetic grouping of the studied rocks is provided by the petrographic features, either preserved or inferred. In particular for extrusive rocks, abundance and size of phenocrysts, their composition and textural relations have been considered to infer the crystallization order and estimate the crystal/melt proportion of the erupted magmas. The analyzed plutonic rocks are non-cumulate shallow level magmatic bodies, mostly dikes, and therefore may be considered representative of melt composition, while their late magmatic and subsolidus minerals may provide indications on the presence and role of fluids during crystallization. The petrographic criteria have been selected as minimum requirements to distinguish homogeneous petrogenetic groups.

Chondrite-normalized REE and MORB-normalized incompatible element diagrams have provided useful criteria to compare the trace element abundance in the studied igneous rocks, evidence intra- and inter-group variations and, finally, investigate on rock petrogenesis by a preliminary qualitative approach.

6.3.1. Extrusive Rocks from the Baimak–Buribai and Tanalyk Formations.
The extrusive rocks from the Baimak–Buribai and Tanalyk formations include boninites, and high-Mg basalts and andesites (boninite-andesite transitional rocks) characterized by originally glassy, aphyric or sparsely phyric textures with small phenocrysts, and generalized order of crystallization

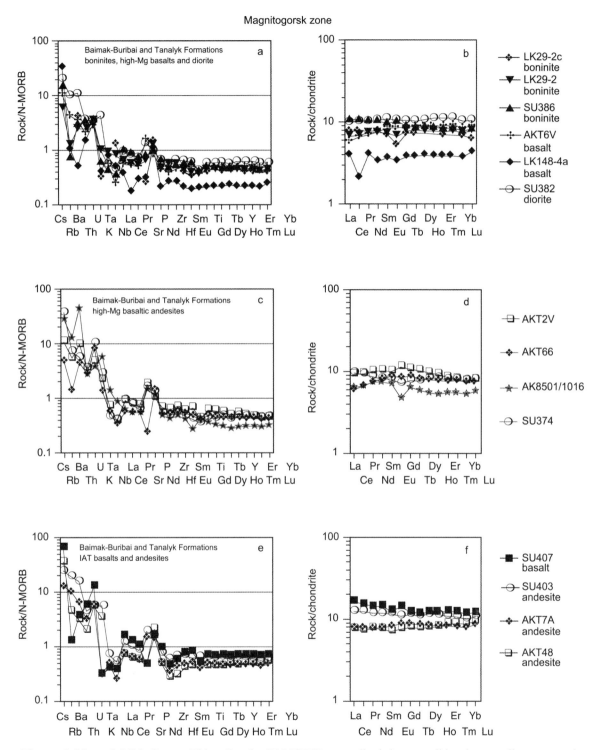

Figure 6. Normal Mid Ocean Ridge Basalt (N-MORB)-normalized incompatible element diagrams and chondrite-normalized rare earth elements diagrams for representative lower Baimak-Buribai and Tanalyk formation boninites and high-Mg extrusive rocks and diorite (a, b), high-Mg extrusive rocks (c, d), and island arc tholeiitic rocks (e, f). In N-MORB normalized diagrams the large ion lithophile elements (LILE) and high-field strength elements (HFSE) are in order of increasing incompatibility according to *Pearce and Parkinson* [1993]. Normalization factors are those of *Sun and McDonough* [1989] for N-MORB and of *Nakamura* [1974] for REE.

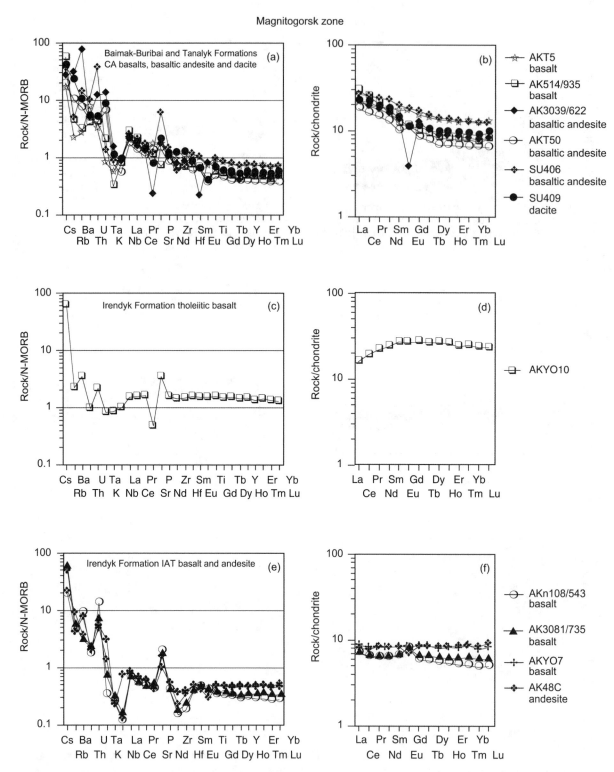

Figure 7. Normal Mid Ocean Ridge Basalt-normalized incompatible element diagrams and chondrite-normalized rare earth element diagrams for representative Baimak–Buribai and Tanalyk formations calc-alkaline rocks (a, b), and Irendyk Formation MORB-type basalt (c, d), and island arc tholeiitic rocks (e, f). Normalizations as for Figure 6.

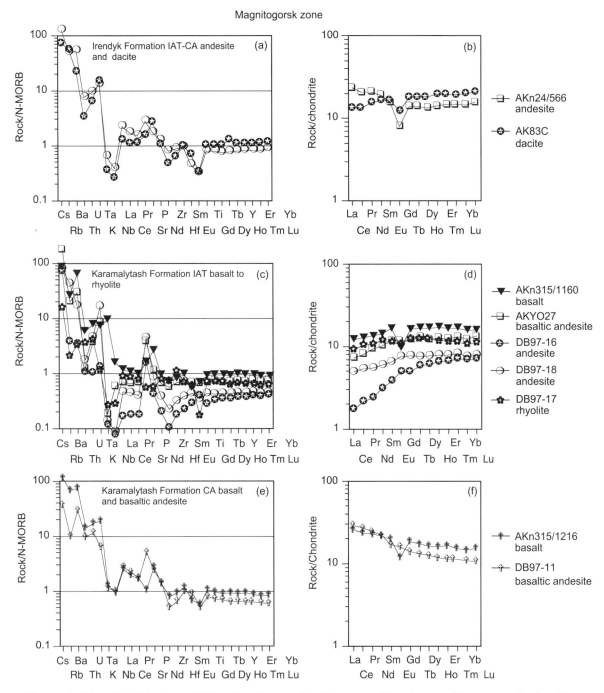

Figure 8. Normal Mid Ocean Ridge Basalt-normalized incompatible element diagrams and chondrite-normalized rare earth element diagrams for representative Irendyk Formation calc-alkaline rocks (a, b), and Karamalytash Formation island arc tholeiitic (c, d) and calc-alkaline rocks (e, f). Normalizations as for Figure 6.

chromite-olivine-?orthopyroxene-clino-pyroxene without (boninites) and with plagioclase (high-Mg andesites [*Spadea* et al., 1998]). They often show textural evidence of disequilibrium between phenocrysts and groundmass [*Spadea* et al., 1998]. Four major petrochemical groups

are distinguished, mostly showing chemical intra-group variations unrelated by simple fractional crystallization relationships.

The first group, boninites, are petrographically distinct from the other arc rocks. They are mostly glassy,

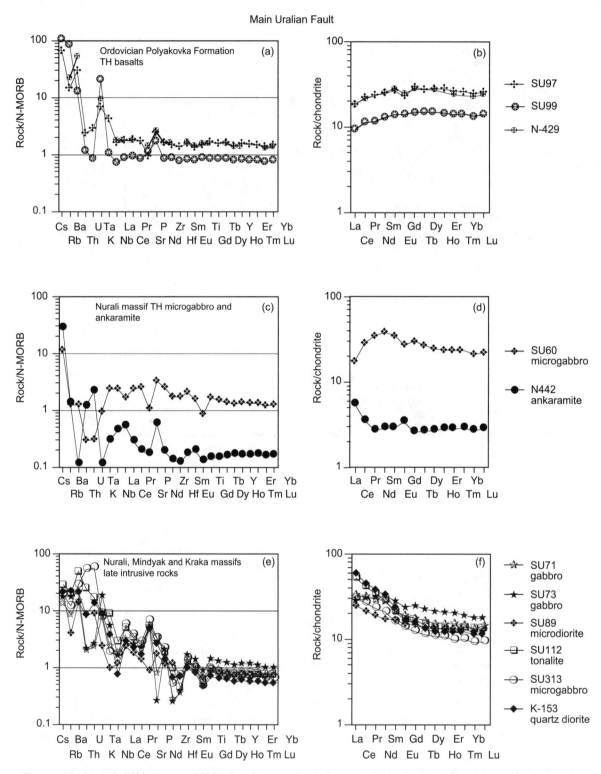

Figure 9. Normal Mid Ocean Ridge Basalt-normalized incompatible element diagrams and chondrite-normalized rare earth element diagrams for representative Ordovician Polyakovka Formation basalts (a, b), gabbro and ankaramite intruding the Main Uralian fault Nurali ophiolite (c, d), Main Uralian fault Devonian OIB gabbros and diorites (e, f). Normalizations as for Figure 6.

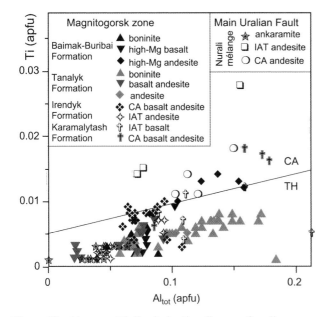

Figure 10. Al versus Ti discriminative diagram for clinopyroxene [*Leterrier* et al., 1982] for representative extrusive rocks from the Magnitogorsk arc and the Main Uralian fault. The boundary between pyroxene from tholeiitic (TH) and calc-alkaline (CA) rock series is shown. Source of data: unpublished microprobe analyses by *P. Spadea*, and by *Gaggero* et al. [1997].

aphyric and sparsely olivine and pyroxene phyric rocks, containing chromite, and without plagioclase. They are chemically primitive (Mg# 81–68) and very low in TiO_2 (Figure 5a), strongly depleted in high field-strength elements (HFSE) with respect to N-MORB (Figure 6a), with flat REE patterns (Figure 6b). As pointed out by *Spadea* et al. [1998] and *Spadea and Scarrow* [2000] the boninites are similar to high-Ca boninites from the western Pacific Izu-Bonin and Tonga forearc. We have included in the boninite group one sample of high-Mg diorite (SU 382) which may represent the shallow plutonic equivalent of the extrusive boninites [*Spadea* et al., 1998].

The second group, high-Mg basaltic basalts and andesites, are aphyric, or porphyritic rocks, the latter characterized by the presence of plagioclase phenocrysts. They are lower in MgO (Mg# 73–62) and slightly higher in TiO_2 than the boninites (Figure 5a), HFSE-depleted (Figure 6c) with flat REE to slightly light (L) REE-depleted patterns, and, in one sample, with a marked negative Eu anomaly (Figure 6d).

The third group, island arc tholeiite (IAT) type basaltic andesites and andesites, are aphyric rocks, or porphyritic rocks with pyroxene, or pyroxene plus plagioclase phenocrysts. They have Mg# in the 68–42

range, low TiO_2 (0.4–0.8 wt%), low and variable HFSE MORB-normalized values, and flat or slightly LREE-enriched patterns (Figures 6e, 6f).

The fourth group, calc-alkaline (CA) basalts, basaltic andesites and andesites, are mostly porphyritic rocks with variable amounts of pyroxene, plagioclase and olivine phenocrysts. They have Mg# in the 71–36 range, and are characterized by higher TiO_2 contents than the previous groups (0.5–1 wt%), depleted to slightly enriched N-MORB normalized patterns (Figure 7a) and relatively high REE contents with markedly LREE-enriched patterns (Figure 7b).

6.3.2. Extrusive Rocks from the Irendyk Formation. The Irendyk Formation lavas are mostly basalt and basaltic andesite with minor andesite and dacite characterized by phenocryst assemblages of pyroxene and plagioclase with pyroxene preceding plagioclase in the order of crystallization. The basalts are diverse in chemical characteristics. Three petrochemical groups are distinguished.

The first group consists of tholeiitic basalt with MORB-type characteristics. One representative sample is a pyroxene-plagioclase phyric lava high in Fe_2O_3tot, with Mg# 43, N-MORB type HFSE pattern (Figure 7c), about 20 times chondritic, LREE depleted, REE abundance (Figure 7d).

The second group consists of IAT basalts and basaltic andesites. These rocks are moderately fractionated (Mg# 61–54), typically low in TiO_2, incompatible HFSE depleted (except P) with respect to MORB (Figure 7e), with less than 10 times chondritic flat REE patterns (Figure 7f).

The third group is composed of IAT tholeiitic andesite and dacite. This group includes low Mg# (44–35), low TiO_2 (around 0.4 wt%) lavas, with MORB-type incompatible HFSE patterns and marked Ti depletion, more than 10 times chondritic REE contents with a marked Eu negative anomaly (Figures 8a, 8b). These rocks are similar in chemical characteristics to the second group. However, they are included in a separate group because they do not display simple relations by fractional crystallization.

6.3.3. Extrusive Rocks from the Karamalytash Formation. The Karamalytash Formation lavas include aphyric, and plagioclase or clinopyroxene phyric basalt, basaltic andesite, and quartz andesite and rhyolite. The rocks are classified in two groups.

The first group is IAT basaltic andesites with Mg# between 58 and 44. TiO_2 is in the 0.4–0.8 wt% range, and other HFSE are slightly to strongly depleted with respect

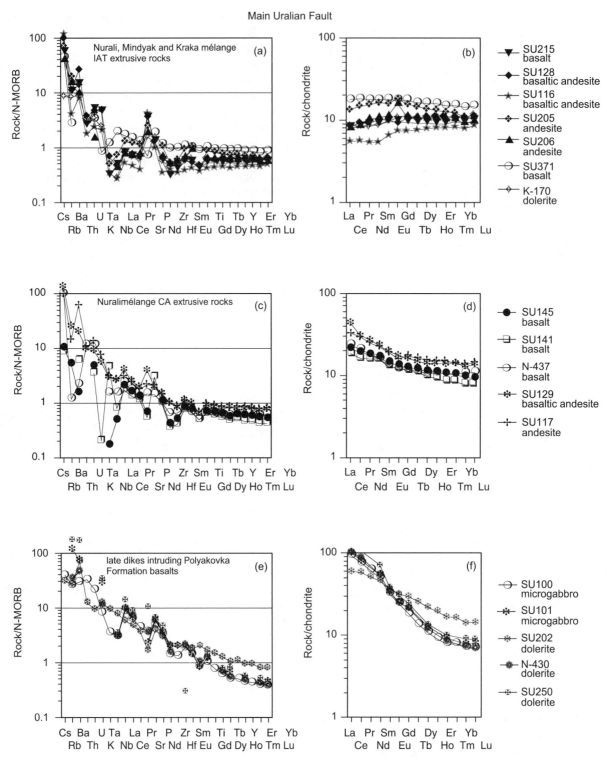

Figure 11. Normal Mid Ocean Ridge Basalt-normalized incompatible element diagrams and chondrite-normalized rare earth element diagrams for representative Main Uralian fault mélange arc-related tholeiitic (a, b) and calc-alkaline (c, d) extrusive rocks, and for late calc-alkaline dikes intruding the Ordovician Polyakovka Formation basalts and the mantle and crust-mantle transition zone rocks at Mindyak massif (e, f). Normalizations as for Figure 6.

Table 3. Whole-rock Sr and Nd isotope data. Trace element data from Tables 1 and 2 and *Spadea and Scarrow* [2000].

Sample	ICP-MS data (ppm)			Sm	Measured		^{143}Nd ^{144}Nd	Recalculated to 400 Ma and 470 Ma			
	Rb	Sr	Nd		$^{87}Sr/^{86}Sr$ unleached[a]	$^{87}Sr/^{86}Sr$ leached[b]		$^{87}Sr/^{86}Sr$ unleached	$^{87}Sr/^{86}Sr$ leached	^{143}Nd ^{144}Nd	εNd 400Ma
Ordovician Polyakovka Formation MORB											
SU97	8.10	83.1	11.49	4.02	0.70661		0.51313	0.70472°		0.51248°	8.69°
Magnitogorsk zone											
LK-148-4a	0.59	27.7	1.55	0.56	0.70694	0.70577	0.51287	0.70659	0.70542	0.51230	3.44
LK-29-2	0.71	58.7	3.45	1.12	0.70576	0.70556	0.51295	0.70556	0.70536	0.51244	6.29
SU-382	5.73	121	4.91	1.69	0.70534	0.70518	0.51300	0.70457	0.70443	0.51245	6.45
SU-407	0.69	40.9	6.65	1.93	0.70595	0.70532	0.51282	0.70567	0.70504	0.51237	4.74
AKT-48	2.47	40.7	3.47	1.07	0.70585	0.70463	0.51295	0.70485	0.70363	0.51247	6.74
AK-514A	2.38	98.0	7.49	1.84	0.70487	0.70446	0.51283	0.70447	0.70406	0.51245	6.33
AKT-5	1.21	150	8.57	2.62	0.70483	0.70474	0.51291	0.70470	0.70461	0.51243	5.98
AKY-010	1.22	42.7	11.14	4.08	0.70547	0.70481	0.51312	0.70499	0.70434	0.51254	8.23
AKY-07	2.36	36.7	3.81	1.28	0.70611	0.70579	0.51284	0.70505	0.70473	0.51231	3.69
AK-315/1160	14.18	144	6.60	2.51	0.70705	0.70666	0.51296	0.70543	0.70503	0.51236	4.57
AKY-027	10.94	341	4.69	1.78	0.70472	0.70465	0.51307	0.70420	0.70412	0.51247	6.78
DB97-16	2.07	138	1.44	0.58		0.70560	0.51339		0.70535	0.51276	12.36
DB97-18	24.09	395	2.77	0.99		0.70613	0.51313		0.70512	0.51257	8.66
DB97-17	1.14	47.7	5.53	1.73		0.70594	0.51317		0.70555	0.51268	10.81
DB97-11	5.46	450	9.98	2.56		0.70467	0.51305		0.70447	0.51265	10.22
Main Uralian fault											
SU60	0.67	92.6	17.67	5.25		0.70432	0.51284		0.70420	0.51237	4.79
N-442	0.73	15.0	1.34	0.44	0.70537		0.51299	0.70457		0.51248	6.95
SU116	2.19	365	2.49	0.98		0.70484	0.51302		0.70474	0.51240	5.38
SU128	5.93	314	4.62	1.54		0.70491	0.51300		0.70460	0.51247	6.77
SU-215	5.61	307	4.05	1.44	0.70471		0.51303	0.70441		0.51247	6.86
SU-371	1.48	61.6	8.15	2.70	0.70608		0.51303	0.70568		0.51251	7.53
K-170	4.55	139	4.02	1.48	0.70430		0.51307	0.70376		0.51249	7.11
SU117	7.90	189	10.4	2.90		0.70537	0.51277		0.70468	0.51233	3.95
SU129	13.77	357	10.8	2.98		0.70502	0.51278		0.70438	0.51234	4.33
N-429A	7.29	233	24.2	6.08		0.70553	0.51274		0.70501	0.51234	4.32
N-437	0.64	130	7.43	2.09	0.70519		0.51285	0.70511		0.51241	5.52
SU71	4.68	476	13.0	3.27		0.70411	0.51280		0.70395	0.51241	5.53
SU73	9.65	436	14.9	4.25		0.70471	0.51272		0.70434	0.51227	2.80
SU89	2.18	77.8	7.94	2.55	0.70604		0.51305	0.70558		0.51254	8.11
SU112	11.09	466	13.2	3.18		0.70450	0.51273		0.70411	0.51235	4.42
K-153	6.52	586	15.2	3.60		0.70394	0.51272		0.70376	0.51235	4.39
SU100	14.11	320	22.9	4.80		0.70416	0.51267		0.70344	0.51234	4.28
SU-202	14.66	147	21.0	5.49	0.70671	0.70669	0.51278	0.70507	0.70505	0.51237	4.78

[a]Analysis made at Granada University on unleached powder, [b]analysis made at Naples University on leached powder.

to MORB (Figure 8c). Zr and Ti negative anomalies are shown by two samples. The REE patterns are flat to variably LREE depleted (Figure 8d). One rhyolite sample with Mg# 26 is very similar in incompatible trace element abundance to some basaltic andesites.

The second group is calc-alkaline basalt and basaltic andesite with Mg# between 52 and 38. It is low in TiO_2 (0.63–0.72 wt%), characterized by low N-MORB-normalized HSFE patterns, and LREE enrichment (Figures 8e, f).

6.3.4. Mid-Ordovician MORB Lavas from the Main Uralian Fault. The Polyakovka Formation lavas are sparsely phyric olivine and plagioclase basalts whose petrography and chemistry has been previously described by *Gaggero* et al. [1997]. REE data on the Polyakovka

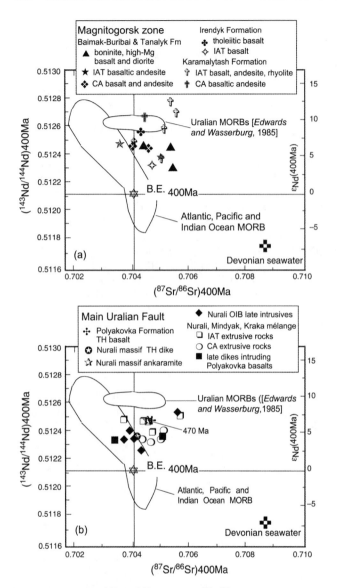

Figure 12. Initial $^{143}Nd/^{144}Nd$ versus $^{87}S/^{86}Sr$ (back calculated to 400 Ma) for representative Magnitogorsk zone Devonian extrusive rocks (a), and Main Uralian fault Devonian intrusive and extrusive rocks (b). In Figure 5b, the initial $^{143}Nd/^{144}Nd$ and $^{87}Sr/^{86}Sr$ values for a representative sample of Ordovician Polyakovka Formation MORB lava, back calculated to 470 Ma, are plotted. The fields of Atlantic, Pacific and Indian MORBs back calculated to 400 Ma, and Uralian MORBs (*Edwards and Wasserburg* [1985]) are shown. The isotopic composition of Devonian seawater is from *Burke* et al. [1982] for Sr, and from *Shaw and Wasserburg* [1985] for Nd.

basalts have also been reported previously by *Savelieva* et al. [1997]. New chemical analyses and new trace element data are included in Table 2 and plotted in the N-MORB normalized and chondrite normalized diagrams of Figures 9a, b. The new data confirm the

similarity of the Polyakovka basalt with depleted oceanic basalts from modern mid-ocean ridges, in particular, the TiO_2 content (1–2 wt%), the Ti/Zr ratio (around 100), and the La/Sm (0.65–0.69) and La/Yb ratios (0.69–0.89). In Table 2 and Figures 9a, b we have also included two samples (SU33, SU35) from the Middle Ordovician tholeiitic basalts sampled near Varna along the URSEIS 95 profile [*Berzin* et al., 1996].

6.3.5. MORB-Type Intrusive Rocks Cutting Peridotites from the Main Uralian Fault. A sample of mafic dike with medium grained textures from the Nurali massif has shown MORB-type chemical features (Figures 9c, d). Chemically similar rocks (hornblende microgabbros) have been found as dikes cutting peridotite in northern Mindyak massif (Table 2).

6.3.6. Depleted IAT Hypabyssal Rock (Ankaramite) from the Main Uralian Fault. Unusual rocks from the Nurali mélange are petrographically characterized by highly phyric textures with 30–35% of perfectly euhedral olivine and clinopyroxene phenocrysts. The groundmass has a relict hyalopilitic texture. The olivine phenocrysts are completely altered, and include small crystals of fresh chromite with average Cr# (Cr/(Cr + Al)) 0.83 and Mg# 0.64. The clinopyroxene phenocrysts have Mg# around 90 and average $Wo_{45}En_{49}Fs_6$ composition. Al-Ti relationship of clinopyroxene is in the tholeiitic field according to *Leterrier* et al. [1982] discriminative diagram, similarly to the pyroxene from the arc rocks (Figure 10). Consistent with the petrography, these rocks have high Mg# (87). One analyzed sample has shown, in N-MORB normalized diagrams, highly depleted HFSE contents with a marked positive anomaly of P, and LREE-enriched, concave upward, REE patterns with a positive Eu anomaly (Figures 9c, d). Given the high phenocryst content, possibly resulting from cumulus process, the whole rock chemistry is far from representative for that of a liquid and cannot be directly compared with that of the most depleted Magnitogorsk arc rocks (boninites and high-Mg basalts and andesites). The composition of the pyroxene, which appears in equilibrium with the groundmass, is similar to that of the boninites and high-Mg andesites, which are Ca-rich and have Mg# between 0.78 and 0.93 [*Spadea* et al., 1998] and poor in Ti (Figure 10), while that of chromite is significantly higher in Cr# (0.83 versus 0.57–0.78 values of the boninites and high-Mg andesites). Whole rock chemistry of the ankaramite is similar to that of high-Mg rocks from the Magnitogorsk arc in HFSE distribution, but has a distinct, concave upward, REE pattern which is typical of boninites [*Crawford* et al., 1989; *Shimizu* et al., 1992].

6.3.7. Enriched (Ocean Island Basalt-Type) Intrusive rocks from the Main Uralian fault.

Hornblende gabbro, diorite, and microgabbro intruding the Nurali massif crust-mantle transition zone have been studied previously by *Gaggero* et al. [1997] and *Pertsev* et al. [1997]. Similar gabbroic rocks and more evolved tonalites are present as blocks within the Nurali and Kraka massif mélanges [*Gaggero* et al., 1997]. The dikes intruding the northern Mindyak massif transition zone consist of hornblende diorite and microdiorite; from them a representative sample of microdiorite has been selected (SU313: Table 2). These intrusive rocks are moderately fractionated with Mg# values 60–49, with TiO_2 in the 0.6–1.1 wt% range, and are characterized by relatively high P_2O_5 contents (0.1–0.3 wt%). Their N-MORB normalized incompatible trace elements patterns show a marked Ti depletion, and variable enrichments in REE. The REE patterns are enriched in LREE (Figures 9e, f).

6.3.8. Island-Arc Tholeiitic Extrusive Rocks from the Main Uralian Fault Polygenic Breccias.

Representative samples of tectonic blocks and chips which make up the mélange flanking the Nurali, Mindyak and Kraka peridotite massifs to the east are aphyric and phyric basalt, basaltic andesite and andesite (Table 2). The porphyritic rocks from Nurali have plagioclase or clinopyroxene phenocrysts described and analyzed by *Gaggero* et al. [1997]; those from Mindyak have, in addition, hornblende among the phenocrysts. Chemically these rocks are characterized by variable TiO_2 contents (0.6–1.3 wt%), Mg# in the 44–68 range, slightly to moderately HFSE depleted N-MORB normalized patterns, and flat to slightly LREE depleted REE patterns (Figures 11a, b). Positive or weak negative Eu anomalies are displayed by some samples.

6.3.9. Calc-Alkaline Extrusive Rocks from the Main Uralian Fault Polygenic Breccias.

The representative, selected samples are basalt, andesite and dacite from the Nurali and Mindyak mélange. Those from Nurali include aphyric basalt and dolerite, porphyritic basalt and andesite, with orthopyroxene and/or, clinopyroxene and plagioclase phenocrysts analyzed by *Gaggero* et al. [1997], and porphyritic basalt with olivine and clinopyroxene phenocrysts. One sample from Mindyak is aphyric andesite. The basalts and andesites are variably fractionated (Mg# 51–71), with TiO_2 contents in the 0.6–1.9

wt% range. In terms of incompatible elements they are distinct from the previous IAT extrusive rocks mostly in higher REE abundance and marked enrichment in LREE (Figures 11c, d). The dacite is similar in incompatible elements abundance and distribution to the andesitic and basaltic rocks.

6.3.10. Late Dikes from the Main Uralian Fault.

The dikes intruding the Polyakovka Formation Ordovician basalts are mostly fine-grained gabbro and diorite characterized by pyroxene followed by hornblende as dominant mafic phases, the latter partly of subsolidus origin. In chemistry these rocks are quite distinct from the previous group. They are less fractionated (Mg# 78–57), higher in TiO_2 (1.1–2.5 wt%), and P_2O_5 (0.3–0.5 wt%). Their N-MORB normalized incompatible elements patterns show typical, strong enrichments in Zr, and LREE (Figure 11e). The LREE contents are up to 200 times chondritic, while those of heavy (H) REE are less than 10 times chondritic (Figure 11f).

6.4. Intergroup Comparison and Geochemical Characterization of the Magmatic Sources

Chemical criteria based on normalized abundance of HFSE, in particular REE have been used for preliminary grouping and serial characterization of the igneous rocks from the Magnitogorsk and Main Uralian fault zones.

We have based intergroup discriminations on abundance of immobile, very highly incompatible elements Th, Ta and Nb. The extrusive rocks with boninitic, island arc tholeiitic, and calc-alkaline characteristics from the Magnitogorsk arc are mostly and variably depleted in Ta and Nb, as shown in N-MORB normalized plots presented in Figures 6 to 8. The extrusive rocks from the Karamalytash Formation display the highest depletion in Ta and Nb among the arc rocks in the island arc tholeiitic group, as well as variable N-MORB normalized values in the calc-alkaline group (Figures 8c, 8e). A marked enrichment in Th is a common characteristic of all Magnitogorsk arc rocks, and is considered indicative of a generalized enrichment by a fluid phase which is also suggested, but not confidently valuable, by the mobile elements.

The intrusive and extrusive rocks from the Main Uralian fault are very diverse in Ta, Nb and Th abundance. The extrusive rocks with island arc tholeiite affinity from the Main Uralian fault mélange are similar to the Irendyk Formation lavas, displaying similar serial affinity for low Ta and Nb and relatively high Th contents (Figure 11a). On the contrary, the extrusive rocks with calc-alkaline affinity have higher Ta and Nb abundance with respect to the Magnitogorsk arc volcanics (Figure 11c). In normalized diagrams they also display up to 3 times N-MORB Ta and Nb values, and up to 10 times N-MORB Th values. It is noteworthy that the depleted ankaramitic rocks are different from the depleted arc rocks for relatively higher Ta and Nb proportions with respect to other HFSE (Figure 9c).

The intrusive rocks with ocean island basalt affinity and the late dikes from the Main Uralian fault all display high Ta and Nb abundance, and are distinct in Th, which is significantly higher in the late dikes than in the transitional ocean island basalt intrusive rocks.

In the discriminant diagram Th/Yb versus Ta/Yb by *Pearce* [1993] the analyzed extrusive rocks from the Magnitogorsk zone show evidence of a generalized Th enrichment with respect to Ta of most boninitic, IAT, and CA rocks, typical of the intraoceanic arcs (Figure 13a). This plot also confirms the depleted mantle origin of the Irendyk MORB-type basalt. On the contrary the Karamalytash calc-alkaline lavas have remarkable intra-group differences. They show either Th enrichment, with the highest Th/Yb values of all the arc rocks, or Th-Ta relations typical or close to those of the mantle array (Figure 13a).

The Main Uralian fault extrusive rocks with island arc tholeiite and calc-alkaline affinities and the ankaramite dike plot in the field of intra oceanic arcs (Figure 13b), similarly to most extrusive rocks from the Magnitogorsk arc. The Main Uralian fault intrusive rocks and late dikes are well in the mantle array, in particular the transitional Main Uralian fault intrusive rocks define a linear trend from depleted MORB to enriched ocean island basalt sources. The Ordovician MORB's plotted for reference in the same diagram evidence depleted mantle sources. These sources appear not to be as depleted as the Irendyk MORB basalt inferred source.

6.5. Intergroup Comparison by Combined Immobile Trace Element and Isotopic Characteristics

As discussed in a previous section, the isotopic characterization of the studied rocks has been mostly based on Nd isotopic ratios, whereas Sr isotopic ratios cannot be considered reliable because suspiciously affected by seawater contamination and/or low grade metamorphism. This is especially true for the whole set of Magnitogorsk arc samples (and related arc rocks from the Main Uralian fault) which define a distinct, broad field of higher initial $^{87}Sr/^{86}Sr$ values for the

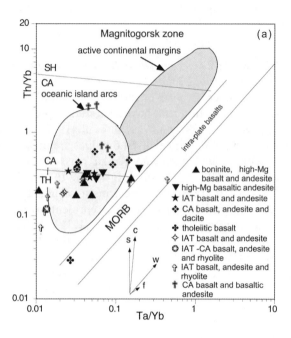

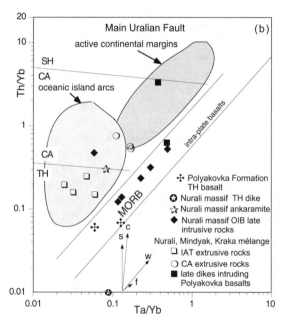

Figure 13. Th/Yb versus Ta/Yb discriminative diagram [*Pearce*, 1983] for representative Magnitogorsk zone Devonian extrusive rocks (a), Ordovician Polyakovka Formation MORB and Main Uralian fault Devonian intrusive and extrusive rocks (b). Dashed lines indicate the boundaries of the tholeiitic (TH), calc-alkaline (CA), and shoshonitic (SH) fields, vectors the influence of subduction components (s), within-plate enrichment (w), crustal contamination (c) and fractional crystallization (f).

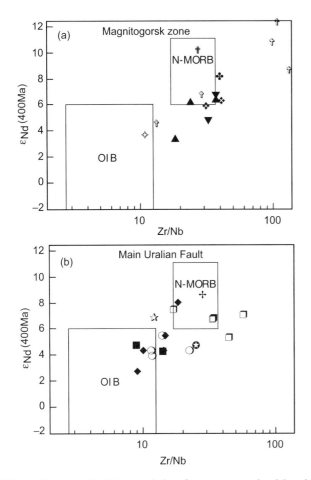

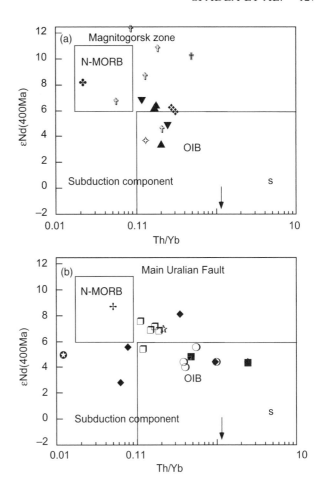

Figure 14. ε_{Nd} – Zr/Nb covariation for representative Magnitogorsk zone Devonian extrusive rocks (a), and Ordovician Polyakovka Formation MORB and Main Uralian fault Devonian intrusive and extrusive rocks (b). See text for sources of data. Symbols are those of Figure 12.

Figure 15. ε_{Nd} – Th/Yb covariation for representative Magnitogorsk zone Devonian extrusive rocks (a), Ordovician Polyakovka Formation MORB and Main Uralian fault Devonian intrusive and extrusive rocks (b). See text for sources of data. Symbols are those of Figure 12.

same initial ^{143}Nd/^{144}Nd values, outside the MORB field (Figure 12a). On the other hand, the transitional and OIB intrusive rocks from the Main Uralian fault mostly plot in the MORB field with a linear trend toward Bulk Earth (Figure 12b). Given that these rocks are mostly unaffected by hydrothermal metamorphism, their trend in the Nd-Sr diagram could reflect magmatic source characteristics, in particular variably enriched OIB sources.

In order to better highlight the mantle source characteristics of rocks from the Magnitogorsk zone and the Main Uralian fault, plots have been designed combining initial ε_{Nd} values with some trace element ratios discriminative of basalts derived from mantle sources depleted or enriched in incompatible trace elements (Figures 14 and 15). In these plots the southern Uralides igneous

rocks are compared with modern N-MORB and OIB taken from the literature [*Faure*, 1986, and references therein; *Le Roex*, 1987; *Hofmann*, 1988; *Pearce*, 1993; *Rollinson*, 1993, and references therein; *Weaver*, 1991], for which ε_{Nd} at 400 Ma has been calculated using suitable Sm/Nd ratios relative to depleted and enriched sources [*Faure*, 1986; *De Paolo*, 1988]. The diagrams of Figure 14a, b illustrate the relationship between initial ε_{Nd} and Zr/Nb ratio, the latter commonly thought to reflect mantle characteristics not modified by addition of slab derived components [e.g., *Tatsumi* et al., 1986; *Ellam and Hawkesworth*, 1988; *McCulloch and Gamble*, 1991; *Thirlwall* et al., 1994; *Davidson*, 1996]. It is clear that both Magnitogorsk arc and Main Uralian fault rocks are derived from heterogeneous mantle sources, variable from very depleted, in some cases even more depleted than modern N-MORB, to enriched. In detail, the MORB sample from the Magnitogorsk arc (Figure 14a) is slightly shifted to the right of the N-MORB field, whereas all samples with boninite, island arc tholeiite and calc-alkaline affinities are either clustered around the N-MORB field, or shifted towards the OIB field. Thus, the ε_{Nd} – Zr/Nb relationships confirm that these rocks are derived from a source with an enriched component, although only in a small amount according to this diagram.

It is noteworthy that the IAT basalt to andesite rocks from Main Uralian fault Nurali, Mindyak and Kraka mélange (Figure 14b) fall within, or cluster around, the N-MORB field, together with the Ordovician MORB lava and one transitional OIB gabbro cutting ophiolites. All other rocks are displaced towards, or fall well within, the OIB field, showing an enriched component in their source region. Thus, one peculiarity of arc derived rocks from the Main Uralian fault is that they display evidence for having been generated in both depleted and enriched source regions.

In the ε_{Nd} versus Th/Yb plot (Figure 15), it is also possible to highlight the occurrence of enriched components characterized by high Th/Yb ratios and relatively low ε_{Nd} values. These are certainly attributable to an OIB-like source, but could also testify to subduction components, namely slab derived fluids/melts enriched in Th relative to Yb, in accordance to *Pearce* [1983]. In this diagram, samples from both the Main Uralian fault and Magnitogorsk zone are much more displaced towards the enriched component field. However, given that the ε_{Nd} values are not very low, it is not possible to ascribe this enrichment to subduction components.

In summary, the evidence from the ε_{Nd} vs Zr/Nb and Th/Yb relationships strongly suggests that the mantle source regions of magmas represented by the diverse lithotypes from the Main Uralian fault and Magnitogorsk zone were variable from very depleted, N-MORB-like, to enriched, OIB-like and/or slab derived, components.

7. DISCUSSION

7.1. Magmatic Source Regions

7.1.1. Magnitogorsk Arc. The association of boninites, tholeiitic and calc-alkaline island arc lavas, and MORB basalts indicates that the magmatic sources and components of the arc complex were various. The short history of the Magnitogorsk arc development (about 22 Ma) also requires distinct sources to account for the large geochemical variations (e.g., of the Nd isotopic ratios). However, inferences on the primary magmas and fractionation processes to account for the highly variable covariations of major and trace elements cannot be provided with the low resolution of available sampling.

The assemblage of MORB basalts and tholeiitic island arc basalts from the Baimak–Buribai, Irendyk and Karamalytash formations indicates that a suboceanic mantle wedge was the likely source from which the arc crust was derived. Partial melting of a depleted mantle similar to the source of mid-ocean ridge basalts, together with the addition of an aqueous component from a subducting slab may account for the petrochemical characteristics of the magmas. This mantle source was similar to that of the Ordovician Polyakovka Formation MORB lavas. These Polyakovka lavas do not display a subduction component, and can be regarded as normal oceanic crust derived from a depleted source, similar to that of modern slow and medium spreading oceanic ridges.

The Baimak–Buribai boninite series rocks, like boninites elsewhere, require peculiar conditions for magma genesis, that are high heat flow, low pressure melting, a depleted source, and high hydrous flux flow to the source. The area and stratigraphic distribution of the boninites in the Magnitogorsk arc, and the fact that no boninites were sampled in the adjacent Main Uralian fault mélange suggest that the conditions necessary for boninite magmatism were restricted to the onset of widespread subduction, and only a small sector of the subduction zone had a high enough thermal anomaly. An anomalous high degree of partial melting is, however, inferred not only for the boninites, but also for most high-Mg tholeiitic lavas from the Magnitogorsk arc, as well as for high-Mg arc and arc related tholeiites from the Main Uralian fault. The Nd isotopes together with Th-Ta systematics suggest a subduction component, identified as a dominant hydrous fluid phase by other geochemical indicators, and an enriched component, identified as ocean island basalt-type melt, for the tholeiitic and transitional suites. Moreover, the rocks with calc-alkaline series affinity are not sharply distinct in terms of these geochemical indicators from the boninitic and tholeiitic rocks. So, we infer that the boninitic magmatism in the Southern Urals was related to an exceptionally high thermal anomaly in a subduction zone where a high degree of melting and high rate of magma production took place. In general, widespread subduction produced fully developed tholeiitic to calc-alkaline arc magmatism, which is, however, still characterized by a high degree of partial melting of a depleted mantle source. The consequence is a high crustal production rate, which is evidenced, in the southern Uralide orogen, by the extent of the rapidly built Devonian arc. Envisaging the origin of the thermal anomaly responsible for generalized high degrees of partial melting, and in particular boninite generation, is impossible for a fossil system. Recently, *Macpherson and Hall* [2001] suggested that a thermal anomaly or mantle plume was responsible for tectonic and magmatic processes in the active Izu-Bonin-Mariana system from Eocene to Recent. The size of the Southern Urals magmatic province, the rate of crustal accretion, which imply high magma budget, suggest that this model may be applicable to the Magnitogorsk arc.

A key to the interpretation of the magmatic processes in the Magnitogorsk arc system is boninite generation. In the Izu-Bonin-Mariana system boninites make a proto arc [*Pearce* et al., 1999] built at the onset of subduction and now preserved in forearc zones [*Bloomer* et al., 1995]. Well constrained interpretations based on modeling of geochemical and isotopic data by *Pearce* et al. [1992, 1999] include generation of boninites by interaction of depleted mantle with subduction related fluids, and the addition of fused mafic rocks. According to *Pearce* et al. [1999], this mafic component is geochemically distinct from an ocean island basalt component, which has been often considered to explain the peculiar thermal and geochemical features of the boninites. The subsequent Izu-Bonin-Mariana arc magmatism had sources related to a different system for magma generation, including a depleted mantle source and fluids derived from subducted pelagic and volcanogenic sediments.

For the Magnitogorsk arc, the magmatic components for boninite generation are identified by geochemistry as depleted mantle, fluids, and an additional component, which the available data suggest to be ocean island basalt-like. The same component is detected through the whole overlying arc sequence. This enriched component also characterizes the late stage magmatism affecting the already obducted Main Uralian fault ophiolites.

7.1.2. Main Uralian Fault Igneous Rocks. The
Polyakovka Formation Ordovician MORB lavas provide the evidence of the oldest normal oceanic crust in the Southern Urals. Further evidence of depleted MORB source is shown by the analyzed gabbros intruded in the crust-mantle transition zone in the Nurali massif. Apart from these gabbros, which could be older than Devonian, most records of igneous rocks in the Main Uralian fault zone point to a younger, Devonian age magmatism, and to a source distinct from that of the Ordovician MORB. The volcanic blocks from the mélange are clearly related to the Magnitogorsk arc, either as fragments of Irendyk and Karamalytash formation lavas, or derived from magmatic rocks currently unknown in the Magnitogorsk arc, represented by the ankaramites.

The ocean island basalt-type intrusive rocks and late dikes indicate spatially, or temporally, distinct magmatic sources with a typical enriched mantle signature. Chemical indicators (Mg#, high Ti, REE patterns, low Nd isotopes) suggest deeper magmatic sources with respect to most arc lavas, without addition of subduction related fluids. The occurrence of this enriched mantle component supports the inference that the additional enriched component detected in the Devonian arc rocks was widespread in the Uralides system. Thus, it can be envisaged that an enriched mantle source was involved first in magma generation during the whole Magnitogorsk arc development, and then it continued to feed late magmatism intruding the Main Uralian fault, at times when the subduction related components were fading, likely during the Carboniferous.

7.2. Implication for Main Uralian Fault Ophiolites Setting

In this paper we have studied various magmatic rocks present in the Main Uralian fault with the aim of classifying and interpreting their petrogenesis. To this purpose we have analyzed various tectonic slices in the Main Uralian fault and igneous fragments dispersed in the mélange. These provide important stratigraphic records for dismembered ophiolites like Nurali and Mindyak. Our data have shown that the plutonic rocks intruding the crust-mantle transition zone (in the Nurali massif) are MORB-type and OIB-type gabbros, and that late dikes intruding mantle or crustal rocks are OIB-type rocks. We have also shown that most igneous fragments in the mélange (lavas and hypabyssal rocks) are arc related, and mostly similar to the younger arc units (Irendyk and Karamalytash formations). The geochemical data have also indicated that the magmatic sources for the Main Uralian fault arc rocks were similar to those of the Magnitogorsk arc complex, as shown for instance by Nd isotope systematics.

For the study area, some inferences can be drawn from these data about the origin of the dominantly lherzolitic ophiolites. First, it is doubtful that the Ordovician MORB lavas are the upper crustal unit of a lherzolite ophiolite. In fact, records of ophiolitic material in the Main Uralian fault mélange do not include the expected sequences of sheeted dike, lower and upper gabbros of a normal mid-ocean ridge sequence, while OIB gabbros, dated to the Early Devonian, and arc extrusive rocks, related to the Devonian arc complex, are the most common igneous rocks in the Main Uralian fault mélange. Secondly, it is unlikely that the Main Uralian fault igneous rocks represent extracted melts leaving a lherzolite refractory residue, given the undepleted character of the Main Uralian fault lherzolites, with shown little evidence of metasomatism [*Savelieva* et al., 1997]. A depleted or highly depleted mantle could be genetically connected with the arc rocks, in particular the boninites. In fact depleted mantle ophiolites are present in other sectors of the Main Uralian fault, in the Polar Urals [*Savelieva and Nesbitt*, 1996] and in the Kempersay massif of the southernmost Urals [*Melcher* et al., 1999]. These ophiolites are also referred to as Devonian on the basis of radiometric dating [*Edwards and Wasserburg*, 1985]. Therefore we conclude that the Main Uralian fault igneous rocks of this study were intruded into already obducted ophiolites, and that the arc has been built on, or near, an older lithosphere including oceanic crust and undepleted mantle. The Early Devonian magmatic event, marked by boninites generation and ocean island basalt-type plutonism in the Main Uralian fault, represents therefore the onset of a new magmatic system in the Southern Urals. In this respect, the Ordovician MORB and, if coeval, the lherzolites, should represent trapped fragments of oceanic, or transitional, lithosphere.

8. CONCLUSIONS

The Magnitogorsk arc complex of the Southern Urals is the fossil record of a thermal pulse which initiated in Early Devonian time (Emsian), and caused an high rate of growth of oceanic crust until the Middle Devonian (Eifelian), that is between 410 and 388 Ma ago according to the Devonian time scale by *Tucker* et al. [1998]. With reference to recent models for the western Pacific Izu-Bonin-Mariana system, boninite generation requires a thermal pulse and mantle melting with addition of subduction related fluids and melts, the latter identified as shallow fused mafic rocks in the mantle, or ocean island basalt components added from a deep source. Boninites are early magmatic products present in a restricted zone of the arc; significantly they are missing in the Main Uralian fault arc-continent suture zone, where arc related magmatism is widespread, and where remnants of oceanic lithosphere are older and genetically unrelated to these arc rocks. Close spatial and temporal relationships, and some critical geochemical indicators suggest that boninitic, tholeiitic and calc-alkaline arc rocks are genetically similar, and record a unique event of oceanic crustal growth in forearc settings, which reached a mature stage before collision with a continental margin. Geochemical indicators suggest that an enriched ocean island basalt-type mantle component was involved throughout the Magnitogorsk arc development and fed late magmas that intruded the Main Uralian fault mélange likely until Carboniferous times.

Appendix 1. ICP-MS determinations (ppm element) at Granada University of 39 trace elements on three International Reference Rock Standards.

	Li	Rb	Cs	Be	Sr	Ba	Sc	V	Cr	Co
WS-E	13.654	25.03	0.457	1.179	413.4	337.27	28.412	338.66	94.218	44.536
PM-S	7.711	1.021	0.34	0.478	281.04	148.53	33.74	194.88	312.33	49.566
UB-N	27.255	1.79	10.041	0.169	9.274	27.42	13.66	67.589	2300.8	99.749

	Ni	Cu	Zn	Ga	Y	Nb	Ta	Zr	Hf	Mo
WS-E	52.26	67.416	119.98	22.783	30.707	17.937	1.153	196.68	5.305	3.728
PM-S	119.45	58.435	60.294	15.741	10.758	2.411	0.174	37.96	1.107	1.791
UB-N	1998.4	26.441	79.873	2.763	2.634	0.192	0.014	4.126	0.139	0.567

	Sn	Tl	Pb	U	Th	La	Ce	Pr	Nd	Sm
WS-E	17.731	0.176	13.239	0.634	3.169	26.886	60.214	7.786	32.99	8.864
PM-S	3.809	0.038	2.375	0.016	0.06	2.708	7.062	1.085	5.548	1.816
UB-N	1.335	0.042	13.17	0.064	0.076	0.344	0.854	0.123	0.647	0.225

	Eu	Gd	Tb	Dy	Ho	Er	Tm	Yb	Lu
WS-E	2.238	7.223	1.092	5.961	1.194	2.894	0.415	2.457	0.388
PM-S	1.089	2.063	0.345	2.005	0.441	1.137	0.161	0.993	0.151
UB-N	0.077	0.297	0.058	0.403	0.096	0.282	0.046	0.283	0.046

Appendix 2. Comparison of trace element data (ppm element) determined by ICP-MS at Granada University (UG) and by XRF and ICP (for 10 REE) at Udine University on selected samples analyzed in this study.

Sample	SU97		SU99		SU116		K170		N437		SU89	
Lab	UG	UU	UG	UU	UG	UU	UG	UU	UG	UU	UG	UU
Sr	83	81	100	108	359	342	139	133	130	124	77	79
Ba	180	193	79.6	102	49.0	62	57.4	70	13.3	19	82.6	93
Sc	57.7	63	41.1	56	44.4	47	42.8	50	43.1	45	35.7	45
V	443	435	272	278	263	235	286	263	275	271	306	272
Cr	120	119	332	349	206	193	88.5	85	318	364	41.6	47
Ni	66.6	67	88.4	110	58.7	57	48.6	47	67.6	71	32.8	36
Cu	67.8	69	74.4	89	98.4	105	101	108	83.3	79	91.6	104
Zn	127	117	98.4	89	55.3	59	71.9	68	68.5	71	59.3	61
Y	39.1	40	21.5	24	11.5	11	16.4	17	16.1	16	19.3	19
Nb	3.633	4	1.670	3	0.607	3	0.638	2	3.488	5	2.120	4
Zr	78.6	107	50.9	64	21.2	28	32.2	38	46.7	51	45.6	51
La	4.283	4.13	2.161	2.48	1.302	1.33	1.996	2.16	5.864	5.48	5.698	5.38
Ce	13.24	12.35	6.829	7.12	3.453	3.20	5.117	4.97	12.74	12.53	13.18	12.58
Nd	11.49	11.44	5.984	6.42	2.922	2.93	4.018	4.07	7.431	7.39	8.202	8.13
Sm	4.017	4.01	2.107	2.30	1.084	1.03	1.483	1.46	2.09	2.03	2.253	2.24
Eu	1.311	1.25	0.804	0.88	0.433	0.41	0.614	0.58	0.655	0.68	0.841	0.84
Gd	5.905	5.78	3.025	3.28	1.536	1.54	2.290	2.17	2.392	2.39	2.784	2.72
Dy	7.035	6.91	3.767	3.95	1.974	1.96	2.733	2.65	2.698	2.67	3.021	3.06
Er	4.308	4.05	2.309	2.21	1.329	1.30	1.758	1.72	1.676	1.72	1.991	1.98
Yb	4.142	3.90	2.240	1.98	1.366	1.35	1.773	1.79	1.757	1.76	1.950	2.05

Acknowledgments. We are grateful to F. Bea for trace elements and Sr and Nd isotope determinations, to L. Kabanova, A. Pertsev, V. Puchkov, G. Savelieva, A. Saveliev, and J. Scarrow for help with fieldwork, exchange of ideas on the project, suggestions, and useful discussions. We thank R.W. Nesbitt and R. Armstrong for their engaged and encouraging reviews which have substantially improved the manuscript. P. Belviso and A. Carandente are thanked for their support in isotope determinations at Naples. We acknowledge financial support from the European Commission (TMR-URO Program), the European Science Foundation (EUROPROBE Program), and the Italian Ministry of University and Scientific Research (Progetto Nazionale MURST-Cofin98). The Italian National Research Council is acknowledged for making possible the microprobe work in the Department of Mineralogy and Petrology of the University of Padua.

REFERENCES

Abbott, L. D., E. A. Silver and J. Galewsky, Structural evolution of a modern arc-continent collision in Papua New Guinea, *Tectonics*, 13, 1007–1034, 1994.

Alt, J. C., J. Honnorez, C. Laverne and R. Emmermann, Hydrothermal alteration at 1 km section through the upper ocean crust, Deep Sea Drilling Project Hole 504B: mineralogy, geochemistry and evolution of seawater-basalt interactions, *J. Geophys. Res.*, 91, 10309–10335, 1986.

Alvarez-Marròn J., D. Brown, A. Perez-Estaùn and Y. Gorozhanina, Accretionary complex structure and kinematics during Paleozoic arc-continent collision in the Southern Urals, *Tectonophysics*, 325, 175–191, 2000.

Beccaluva, L. and G. Serri, Boninitic and low-Ti subduction-related lavas from intraoceanic arc-backarc systems and low-Ti ophiolites: a reappraisal of their petrogenesis and original tectonic setting, *Tectonophysics*, 146, 291–315, 1988.

Berzin R., O. Oncken, J. H. Knapp, A. Perez-Estaùn, T. Hismatilin, N. Yunusov and A. Lipilin, Orogenic evolution of the Urals mountains: results from an integrated seismic experiment, *Science*, 274, 220–221, 1996.

Bloomer, S. H. and J. W. Hawkins, Petrology and geochemistry of boninite series volcanic rocks from the Mariana trench, *Contr. Mineral. Petrol.*, 89, 256–262, 1987.

Bloomer, S. H., B. Taylor, C. J. MacLeod, R. J. Stern, P. Fryer, J. W. Hawkins and L. Johnson, Early arc volcanism and the ophiolite problem: a perspective from drilling in the western Pacific, in *Active Margins and Marginal Basins of the Western Pacific*, edited by B. Taylor and J. Natland, 1–30, AGU Geophys. Monogr. 88, Washington, D.C., 1995.

Brown, D., J. Alvarez-Marròn, A. Perez-Estaùn, Y. Gorozhanina, V. Baryshev and V. Puchkov, Geometric and kinematic evolution of the foreland thrust and fold belt in the southern Urals, *Tectonics*, 16, 551–562, 1997.

Brown, D., J. Alvarez-Marròn, A. Perez-Estaùn, V. Puchkov, Y. Gorozhanina and P. Ayarza, Structure and evolution of the Magnitogorsk forearc basin: identifying upper crustal processes during arc-continent collision in the southern Urals, *Tectonics*, 20, 364–375, 2001.

Brown, D., C. Juhlin, J. Alvarez-Marron, A. Perez-Estaùn and A. Oslianski, Crustal-scale structure and evolution of an arc-continent collision zone in the southern Urals, Russia, *Tectonics*, 17, 158–171, 1998.

Brown, D. and P. Spadea, Processes of forearc and accretionary complex formation during arc-continent collision in the southern Urals, *Geology*, 27, 649–652, 1999.

Burke, W. H., R. E. Denison, E. A. Hetherington, R. B. Koepnick, H. F. Nelson and J. B. Otto, Variation of seawater $^{87}Sr/^{86}Sr$ through Phanerozoic time, *Geology*, 10, 516–519, 1982.

Crawford, A. J., L. Beccaluva and G. Serri, Tectono-magmatic evolution of the West Philippine-Mariana region and the origin of boninites, *Earth Planet. Sci. Lett.*, 54, 346–356, 1981.

Crawford, A. J., T. J. Falloon and D. H. Green, Classification, petrogenesis and tectonic setting of boninites, in *Boninites and Related Rocks*, edited by A. J. Crawford, 1–49, Unwin Hyman, London, 1989.

Danyushevsky, L. V., A. V. Sobolev and T. J. Falloon, North Tongan high-Ca boninite petrogenesis: the role of Samoan plume and subduction zone transform fault transition, *J. Geodynamics*, 20, 219–241, 1995.

Davidson, J. P., Deciphering mantle and crustal signatures in subduction zone magmatism, in *Subduction Top to Bottom*, edited by G. E. Bebout, D. W. Scholl, S. H. Kirby and J. P. Platt, AGU Geophys. Monogr. 96, Washington, D.C., pp. 251–262, 1996.

De Paolo, D. J., *Neodymium Isotope Geochemistry: An Introduction*, 186 pp., Springer Verlag, New York, 1988.

Edwards, R. L. and G. J. Wasserburg, The age and emplacement of obducted oceanic crust in the Urals from Sm-Nd and Rb-Sr systematics, *Earth Planet. Sci. Lett.*, 72, 389–404, 1985.

Elliott, T., T. Plank, A. Zindler, W. M. White and B. Bourdon, Element transport from subducted slab to volcanic front at the Mariana arc, *J. Geophys. Res.*, 102, 14991–15019, 1997.

Ellam, R. M. and C. J. Hawkesworth, Elemental and isotopic variations in subduction related basalts: evidence for a three component model, *Contr. Mineral. Petrol.*, 98, 72–80, 1988.

Falloon, T. J. and A. J. Crawford, The petrogenesis of high-calcium boninite lavas dredged from the northern Tonga ridge, *Earth Planet. Sci. Lett.*, 102, 375–394, 1991.

Faure, G., *Principle of Isotope Geology*, 2nd edition, 589 pp., Wiley, New York, 1986.

Frolova, T. I. and I. A. Burikova, *Geosyncline Volcanism (by the Example of the Eastern Slope of the Southern Urals)* (in Russian), 279 p., MSU, Moscow, 1997.

Gaggero, L., P. Spadea, L. Cortesogno, G. N. Savelieva and A. N. Pertsev, Geochemical investigations of the igneous rocks from the Nurali Ophiolite melange Zone, Southern Urals, *Tectonophysics*, 276, 139–161, 1997.

Gorozhanina, Y. N., Genetic types of volcanoclastic rocks of Irendyk volcanic arc Complex on the Southern Urals (in Russian), *Lithology and Resources*, 2, 99–112, 1993.

Govindaraju, K., P. J. Potts, P. C. Webb and J. S. Waston, Report on Win Sill Dolerite WS-E from England and Pitscurrie Microgabbro PM-S from Scotland: assessment by one hundred and four international laboratories, *Geost. Newslett.*, 18, 211–300, 1994.

Hall, R., Reconstructing Cenozoic SE Asia, in *Tectonic Evolution of Southeast Asia*, edited by R. Hall and D. J. Blundell, Geol. Soc. Spec. Pub. 106, pp. 153–184, 1996.

Hawkins, J. W., S. H. Bloomer, C. A. Evans and J. T. Melchior, Evolution of intra-oceanic, arc-trench systems, *Tectonophysics*, 102, 175–205, 1984.

Hetzel, R., H. P. Echtler, W. Seifert, B. A. Schulte and K. S. Ivanov, Subduction- and exhumation-related fabrics in the Paleozoic high-pressure/low-temperature Maksyutov Complex, Antingan area, southern Urals, Russia, *Geol. Soc. Am. Bull.*, 110, 916–930, 1998.

Hickey-Vargas, R., Boninites and tholeiites from DSPD Site 458, Mariana forearc, in *Boninites and Related Rocks*, edited by A. J. Crawford, pp. 288–313, Unwin Hyman, London, 1989.

Hofmann, A. W., Chemical differentiation of the Earth: the relationship between mantle, continental crust and oceanic crust, *Earth Planet. Sci. Lett.*, 90, 297–324, 1988.

Hsü, K. J., *Preliminary Report and Geologic Guide in Franciscan Mélange of the Morro Bay-San Simeon Area, California*, California Div. Mines and Geol. Spec. Pub. 35, 46 p., 1969.

Huang, C.-Y., W. Y. Wu, C. P. Chang, S. Tsao, P. B. Yuan, C. Lin and X. Kuan-Yuan, Tectonic evolution of an accretionary prism in the arc-continent collision terrane of Taiwan, *Tectonophysics*, 281, 31–51, 1997.

Kosarev, A. M., Petrochemical and geochemical features of the basalts on the Southern Urals and their significance for metallogenic reconstructions, in *Metallogeny of the South Urals* (in Russian), pp. 47–62, BFAN SSSR, Ufa, 1986.

Le Bas, M. J., IUGS reclassification of the high-Mg and picritic volcanic rocks, *J. Petrol.*, 41, 1467–1470, 2000.

Lennykh, V. I., P. M. Valiser, R. Beane, M. Leech and W. G. Ernst, Petrotectonic evolution of the Makysutov complex, southern Urals Mountains, Russia: implications for ultrahigh-pressure metamorphism, *Int. Geol. Rev.*, 37, 584–600, 1995.

Le Roex, A. P., Source regions of mid-ocean ridge basalts: evidence for enrichment processes, in *Mantle Metasomatism*, edited by M. A. Menzies and C. J. Hawkesworth, pp. 389–422, Academic Press, London, 1987.

Leterrier J., R. C. Maury, P. Thonon, D. Girard and M. Marchal, Clinopyroxene composition as a method of identification of the magmatic affinities of paleo-volcanic series, *Earth Planet. Sci. Lett.*, 59, 139–154, 1982.

Lin, P.-N., R. J. Stern, J. Morris and S. H. Bloomer, Nd- and Sr-isotopic compositions of lavas from the northern

Mariana and southern Volcano arcs: implications for the origin of island arc melts, *Contr. Miner. Petrol.*, 105, 381–392, 1990.

Macpherson, C. G. and R. Hall, Tectonic setting of Eocene boninite magmatism in the Izu-Mariana forearc, *Earth Planet. Sci. Lett.*, 186, 215–230, 2001.

Maslov, V. A., V. L. Cherkasov, V. T. Tischchenko, I. A. Smirnova, O. V. Artyushkova and V. V. Pavlov, On the stratigraphy and correlation of the Middle Paleozoic complexes of the main copper-pyritic areas of the Southern Urals (in Russian), *Ufimsky Nauch. Tsentr.*, Ufa, 1993.

Matte, Ph., Southern Uralides and Variscides: comparison of their anatomies and evolution, *Geologie en Mijnbow*, 74, 153–168, 1995.

McCulloch, M. T. and J. A. Gamble, Geochemical and geodynamical constraints on subduction zone magmatism, *Earth Planet. Sci. Lett.*, 102, 358–374, 1991.

Melcher, F., W. Grum, T. V. Thalhammer and O. A. R. Thalhammer, The giant chromite deposits of Kempirsai, Urals: constraints from trace elements (PGE, REE) and isotope data, *Mineralium Deposita*, 34, 250–272, 1999.

Murton, B. J., D. W. Peate, R. J. Arculus, J. A. Pearce and S. van der Laan, Trace-element geochemistry of volcanic rocks from Site 786: the Izu-Bonin forearc: *Proc. Ocean Drilling Program, Sci. Res.*, 125, 211–235, 1992.

Nakamura, N., Determination of REE, Ba, Fe, Mg, Na and K in carbonaceous and ordinary chondrites, *Geochim. Cosmochim. Acta*, 38, 757–775, 1974.

Pearce, J. A., The role of sub-continental lithosphere in magma genesis at destructive plate margins, in *Continental Basalts and Mantle Xenoliths*, edited by C. J. Hawkesworth and M. J. Norry, pp. 230–249, Shiva, Nantwich, 1993.

Pearce, J. A., P. E. Baker, P. K. Harvwey and I. W. Luft, Geochemical evidence for subduction fluxes, mantle melting and fractional crystallisation beneath the South Sandwich island arc, *J. Petrol.*, 36, 1073–1109, 1995.

Pearce, J. A., P. D. Kempton, G. M. Nowell and R. Noble, Hf-Nd element and isotope perspective on the nature and provenance of mantle and subduction components in western Pacific arc-basin systems, *J. Petrol.*, 40, 1579–1611, 1999.

Pearce, J. A. and I. J. Parkinson, Trace element models for mantle melting: application to volcanic arc petrogenesis, in *Magmatic Processes and Plate Tectonics*, edited by H. M. Prichard, T. Alabaster, N. B. Harris and C. R. Neary, *Geol. Soc. London, Spec. Pub.* 76, 373–403, 1993.

Pearce, J. A., S. R. van der Laan, R. J. Arculus, B. J. Murton, T. Ishii, D. W. Peate and I. J. Parkinson, Boninite and harzburgite from Leg 125 (Bonin-Marian forearc): a case study of magma genesis during the initial stage of subduction, *Proc. Ocean Drilling Program, Sci. Res.*, 125, 623–659, 1992.

Pertsev, A. N., P. Spadea, G. N. Savelieva and L. Gaggero, Nature of the transition zone in the Nurali ophiolite, southern Urals, *Tectonophysics*, 276, 163–180, 1997.

Puchkov, V. N., Structure and geodynamics of the Uralian orogen, in *Orogeny Through Time*, edited by J.-P. Burg and

M. Ford, pp. 201–236, Geol. Soc. London Spec. Pub. 121, 1997a.

Puchkov, V. N., Tectonics of the Urals: modern concepts, *Geotectonics*, 31, 294–312, 1997b.

Puchkov, V. N., *Paleogeodynamics of the South and Central Urals* (in Russian), edited by E. V. Chibrikova, Ufa, 144 pp., 2000.

Rollinson, H., *Using Geochemical Data: Evaluation, Presentation, Interpretation*, 352 pp., Prentice Hall, Harlow, England, 1993.

Ryazantsev, A. V., S. V. Dubinina and L. A. Kurkovskaya, Ordovician chert-basalt complex and ophiolites of the South Urals, in *General and Regional Problems of Geology*, GEOS, Moscow, pp. 5–23, 1999.

Salikhov, D. N. and A. V. Yarkova, *The Lower Carboniferous Volcanism of the Magnitogorsk Megasynclinorium* (in Russian), 137 pp., BSC Uralian Branch RAS, Ufa, 1992.

Saveliev, A. A., O. V. Astrakhantsev, A. L. Knipper, A. Ya. Sharaskin and G. N. Savelieva, Structure and deformation phases of the Northern Terminus of the Magnitogorsk Zone, Urals, *Geotectonika*, 3, 38–50, 1998.

Savelieva, G. N. and R. W. Nesbitt, A synthesis of the stratigraphy and tectonic setting of the Uralian ophiolites, *J. Geol. Soc. London*, 153, 525–537, 1996.

Savelieva, G. N., A. Ya. Sharaskin, A. A. Saveliev, P. Spadea and L. Gaggero, Ophiolites of the southern Uralides adjacent to the East European continental margin, *Tectonophysics*, 276, 117–138, 1997.

Seravkin, I. B., Southen Urals: tectono-magmatic zoning and position among orogenic systems of the Uralo-Mongolian Belt (in Russian), *Geotektonika*, 1, 32–47, 1997.

Seravkin, I. B., A. M. Kosarev and D. N. Salikhov, *Volcanism of the South Urals* (in Russian), 197 pp., Nauka, Moscow, 1992.

Shaw, H. F. and G. J. Wasserburg, Sm-Nd in marine carbonates and phosphates: implications for Nd isotopes in seawater and crustal ages, *Geochim. Cosmochim. Acta*, 1985.

Shimizu, H., H. Sawatari, Y. Kawata, YP. N. Dunkley and A. Masuda, Ce and Nd isotope geochemistry of island arc volcanic rocks with negative Ce anomaly: existence of sources with concave REE patterns in the mantle beneath the Solomon and Bonin island arcs, *Contr. Mineral. Petrol.*, 110, 242–252, 1992.

Smirnov, S. V., *Petrology of Wehrlite-Clinopyroxenite-Gabbro Association of Nurali Ultrabasic Massif and Related Platinum Ore* (in Russian), Ph.D. thesis, Ekaterinburg, 18 pp., 1995.

Sobolev, A. V. and L. V. Danyushevsky, Petrology and geochemistry of boninites from the north termination of the Tonga trench: constraints on the generation conditions of primary high-Ca boninite magmas, *J. Petrol.*, 35, 1183–1211, 1994.

Spadea, P., L. Ya. Kabanova and J. H. Scarrow, Petrology, geochemistry and geodynamic significance of Mid-Devonian boninitic rocks from the Baimak–Buribai area

(Magnitogorsk Zone, Southern Urals), *Ofioliti*, 23, 17–36, 1998.

Spadea, P. and J. H. Scarrow, Early Devonian boninites from the Magnitogorsk Arc, Southern Urals (Russia): Implications for early development of a collisional orogen, in *Ophiolites and Oceanic Crust: New Insights from Field Studies and Ocean Drilling Program*, edited by Y. Dilek, E. Moores, D. Elthon and A. Nicolas, *Geol. Soc. Am. Mem.* 349, 461–472, 2000.

Stern, R. J., S. H. Bloomer, P.-N. Lin, E. Ito and J. Morris, Shoshonitic magmas in nascent arcs, new evidence from submarine volcanoes in northern Marianas, *Geology*, 16, 426–430, 1988.

Stern, R. J., J. Morris, S. H. Bloomer and J. W. Hawkins, The source of the subduction component in convergent margin magmas: trace element and radiogenic isotope evidence from Eocene boninites, Mariana forearc, *Geochim. Cosmochim. Acta*, 55, 1467–1481, 1991.

Sun, S.-S. and W. F. McDonough, Chemical and isotopic systematics of oceanic basalts: implications for mantle composition and processes, in *Magmatism in the Ocean Basins*, edited by A. D. Saunders and M. J. Norry, pp. 313–345, Geol. Soc. London Spec. Pub. 42, 1989.

Tatsumi, Y., D. L. Hamilton and R. W. Nesbitt, Chemical characteristics of fluid phase released from a subducted lithosphere and origin of arc magmas: evidence from high-pressure experiments and natural rocks, *J. Volcanol. Geotherm. Res.*, 29, 293–309, 1986.

Taylor, R. N. and R. W. Nesbitt, Arc volcanism in an extensional regime at the initiation of subduction: a geochemical study of Hawajima, Bonin Islands, Japan, in *Volcanism Associated with Extension at Consuming Plate Margins*, edited by J. L. Smellie, 115–134, Geol. Soc. London Spec. Pub. 81, 1995.

Taylor, R. N. and R. W. Nesbitt, Isotopic characteristics of subduction fluids in an intra-oceanic setting, Izu-Bonin Arc, *Earth Planet. Sci. Lett.*, 164, 79–98, 1998.

Taylor, R. N., R. W. Nesbitt, P. Vidal, R. S. Harmon, B. Auvray and I. W. Crowdace, Mineralogy, chemistry and genesis of the boninite series volcanics, Chichijima, Bonin Islands, Japan, *J. Petrol.*, 35, 577–617, 1994.

Thirlwall, M. F., T. E. Smith, A. M. Graham, N. Theodorou, P. Hollings, J. P. Davidson and R. J. Arculus, High field strength element anomalies in arc lavas: source or process?, *J. Petrol.*, 35, 819–838, 1994.

Tucker, R. D., D. C. Bradley, C. Sa. Ver Straeten, A. G. Harris, J. R. Ebert and S. R. McCutcheon, New U-Pb zircon ages and the duration of Devonian time, *Earth Planet. Sci. Lett.*, 158, 175–186, 1998.

Weaver, B. L., The origin of ocean island basalt end-member compositions: trace element and isotopic constraints, *Earth Planet. Sci. Lett.*, 104, 381–397, 1991.

Yazeva, R. G. and V. V. Bochkarev, *Geology and Geodynamics of the Southern Urals*, 203 pp., Uralian Branch RAS, Ekaterinburg, 1998 (in Russian).

Zonenshain, L. P., M. L. Kazmin and L. M. Natapov, *Geology of the USSR: a Plate-tectonic Synthesis*, American Geophysical Union Geodynamics Ser., 21, Washington, D.C., 242 pp., 1990.

Zonenshain, L. P., V. G. Korinevsky, V. G. Kazmin, D. M. Pechersky, V. V. Khain and V. V. Matveenkov, Plate tectonic model of the South Urals development, *Tectonophysics*, 109, 95–135, 1984.

P. Spadea: Dipartimento di Georisorse e Territorio, University of Udine, Via Cotonificio 114, I-33100 Udine, Italy, E-mail spadea@uniud.it

M. D'Antonio: Dipartimento di Scienze della Terra, University of Napoli Federico II, Largo S. Marcellino 10, I-80138 Napoli, Italy, E-mail masdanto@unina.it

A. Kosarev and Y. Gorozhanina: Institute of Geology, Russian Academy of Sciences, ul. K. Max 16/2, 450000 Ufa, Russia, E-mail gorozhanin@anrb.ru

D. Brown: Instituto de Ciencias de la Tierra "Jaume Almera" CSIC, Lluìs Solé i Sabaris s/n, 08028 Barcelona, Spain, E-mail dbrown@ija.csic.es

Ophiolites and Zoned Mafic–Ultramafic Massifs of the Urals: A Comparative Analysis and Some Tectonic Implications

G. N. Savelieva[1], A. Ya. Sharaskin[1], A. A. Saveliev[1], P. Spadea[2], A. N. Pertsev[1]
and I. I. Babarina[1]

[1]Geological Institute, Russian Academy of Sciences, Moscow, Russia
[2]Dept. Georisorse Territorio, University of Udine, Udine, Italy

New data on structural geology, geochronology and geochemistry of ophiolites and zoned mafic-ultramafic massifs of the Urals have been analyzed. As is shown, the Southern Uralian ophiolites seem to be the relics of oceanic lithosphere that originated during Ordovician sea-floor spreading and associated rifting events in the adjacent passive margin of the East European Craton. Oceanic crust of the Ordovician ocean basin underwent incipient deformation about 420 My ago, and this triggered formation of Silurian–Devonian island arcs, relics of which are widespread in the Uralides. During the mid-Silurian time, plutons of the Platinum Belt intruded the basement of an old island arc, or that peripheral segment of the craton which was already transformed into the active continental margin. The Silurian–Devonian ophiolites of the Polar Urals may re-present fragments of the oceanic crust and underlying mantle of an inter-arc basin that split the pre-existing island arc.

1. INTRODUCTION

Investigation of gabbro-ultramafic massifs, which are widespread in the Uralides, was initiated long ago [e.g., *Levinson-Lessing*, 1900; *Vysotskii*, 1913; *Zavaritskii*, 1932], and then it was stimulated by the development of plate tectonic ideas that enabled the recognition of ophiolites in this foldbelt, which have been interpreted as relics of former oceanic crust [*Peyve*, 1969, 1977; *Ivanov* et al., 1975]. These ideas, and subsequent regional and inter-regional paleotectonic reconstructions [*Zonenshain* et al.,

1990; *Torsvik* et al., 1992; *Puchkov*, 1993; *Sengor* et al., 1993] gave rise to the understanding of the geological structure and history of the Uralides.

The structural position of Uralide ophiolites have been recently summarized by *Savelieva and Nesbitt* [1996]. However, some problems concerning the formation ages of ophiolite and zoned mafic-ultramafic massifs, the history of their emplacement into the general structure of the Uralides, and the reconstruction of their original settings remain enigmatic. For instance, the lherzolite massifs of the South Urals have been interpreted as rock associations originated either in rift basins of the East European continental margin [*Savelieva* et al., 1997], or in mid-ocean spreading centers [*Kamaletdinov and Kazantseva*, 1983]. Moreover, *Garuti* et al. [1997] suggested on the basis of geochemical data that these massifs represent an association of "orogenic lherzolites" and "wehrlites and pyroxenites" typical of the Alaskan-type platinum-bearing massifs. Problems of classification

Mountain Building in the Uralides: Pangea to the Present
Geophysical Monograph 132
Published in 2002 by the American Geophysical Union
10.1029/132GM08

and origin of mafic-ultramafic massifs in the Uralides have been considered in a series of publications [e.g., *Yefimov*, 1984]. The Platinum Belt massifs have been considered as peculiar analogues of ophiolites by *Yefimov* [1984, 1993] who attributed their ultramafic rocks to mantle restites that protrude in the crust as solid tectonic blocks. Other researchers [*Ivanov and Shmelev*, 1996; *Ivanov* et al., 1999] suggest that they represent intrusive bodies originated by island arc development in a supra-subduction environment. According to *Fershtater* et al. [1999], these were tubular differentiated intrusions now exposed at different erosion levels. In contrast, *Savelieva* et al. [1999] argued for an allochthonous position of the same massifs, and that they originally represented multi-chamber lenticular igneous bodies.

All these controversial interpretations require a comprehensive comparison of structural and petrological data characterizing ophiolites, on the one hand, and Platinum Belt massifs, on the other. Now that EURO-PROBE's multinational Uralides project has yielded new data on geophysics, structural geology, geochronology and geochemistry of the Uralides, such a comparison appears to be especially vital, as it can provide a deeper insight into the Paleozoic history of ophiolite and mafic-ultramafic massifs considered in this work.

2. GEOLOGICAL SETTING

The general structure of the Uralide thrust and fold belt was formed during the Paleozoic. Within the belt that is traditionally considered as a system of subme-ridional structural and lithological zones, the allochtho-nous ophiolite massifs are confined to tectonic sutures of different ranks (Figure 1), and the peculiar Platinum Belt of zoned mafic-ultramafic massifs is spatially restricted to the Tagil zone. The geological setting of the massifs are different in the Southern, Middle, Northern, and Polar Urals.

2.1. Southern Urals

In the Southern Urals, large and well studied ophiolite allochthons occur in the westernmost Sakmara zone and along the Main Uralian fault (Figure 2). The Sakmara zone includes a series of folded tectonic slices which are composed of the following Early to Middle Paleozoic rock complexes:

(1) Molasse sandstones and conglomerates that are
  . intercalated with trachybasalts and liparite tuffs that yield brachiopods, trilobites, and conodonts of Cambrian (?), Tremadocian, and Arenigian age. These rocks are considered as the rifting stage facies

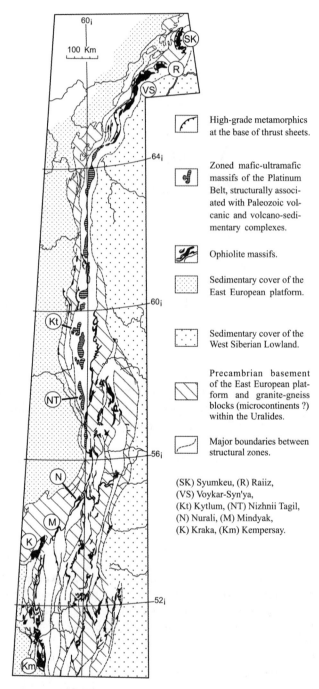

Figure 1. Simplified structural scheme of the Uralides showing the distribution of ophiolite and zoned mafic-ultramafic massifs. Massifs by names: (SK) Syumkeu, (R) Raiiz, (VS) Voykar-Syn'ya, (Kt) Kytlym, (NT) Nizhnii Tagil, (N) Nurali, (M) Mindyak, (K) Kraka, (Km) Kempersay.

[*Ivanov and Puchkov*, 1984; *Seravkin and Rodicheva*, 1990; *Puchkov*, 2000].

(2) Basaltic pillow lavas with jasper intercalations and phthanites of Middle Ordovician (Arenigian to

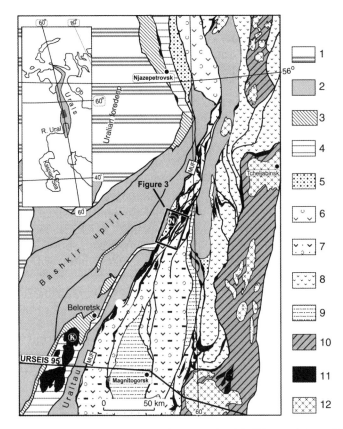

Figure 2. Tectonic sketch map of the contact zone between the Uralides and East European Craton [after *Saveliev* et al., 1998]: 1. sediments of the Uralian foredeep; 2. salients of Precambrian rocks; 3. Ordovician to Devonian shelf sediments of the Bel'sk-Elets zones; 4. Ordovician to Devonian pelagic sediments, volcanic rocks and molasses of the Sakmara-Burtym zone; 5. Upper Cambrian to Ordovician rock complexes of the rifting stage; 6. Ordovician to Devonian island arc and back arc complexes of the Tagil zone; 7. Ordovician to Devonian oceanic and island arc complexes of the Magnitogorsk zone; 8. Silurian to Devonian volcano sedimentary complexes of the Aramil zone; 9. Upper Devonian to Lower Carboniferous graywacke and olistostromes; 10. blocks of sialic crust; 11. ophiolite massifs and serpentinite mélange (abbreviations of massif names as in Figure 1); (MUF) Main Uralian fault.

Caradocian) age [*Korinevskii*, 1989; *Ivanov* et al., 1990], which overlie diabase dikes and gabbro of the Kempersay ophiolite massif and are assumed to be indicators of the spreading stage [*Saveliev and Savelieva*, 1991].

(3) Silurian to Devonian volcano sedimentary sequences, presumably of island arc origin, which include graywackes, cherts, shales, shallow water carbonates, and volcanics of tholeiitic, boninitic, and calc-alkaline series [*Rumyantseva* et al., 1989; *Seravkin and Rodicheva*, 1990; *Spadea* et al., 1998, 2000].

Tectonic sheets composed of these complexes are everywhere structurally intercalated with serpentinite slices that represent the ultramafic material mobilized at the time of Late Paleozoic collision events. The ophiolite

massifs *sensu stricto* occur as individual allochthons (Kempersay massif), as constituents of tectonic nappes (Kraka massifs), or as wedge shaped bodies marking the root zone of allochthons (Nurali and others massifs confined to the Main Uralian fault).

The Kempersay massif in the south of the Sakmara zone (Figure 1) occurs at the base of a package of tectonic slices thrust westward over the Ebeta microcontinent that is concealed under Paleozoic volcano sedimentary cover [*Saveliev and Savelieva*, 1991]. The basal plane of the massif is intricately deformed and dips gently or steeply westward along its northeastern and southeastern borders, respectively. A Rb-Sr and Sm-Nd isochron age of 397 ± 5 Ma is determined for the amphibolite sole of the massif [*Edwards and Wasserburg*, 1985]. With due regard for metamorphic fabrics (see next section), this date may indicate the time when the oceanic crust fragments experienced initial imbrication at the commencement of the basin closure. According to unpublished data of T.V. Thallammer and her colleagues, Sm-Nd isochron ages of two lherzolite samples from the massif are 487 ± 54 Ma, and a U-Pb zircon age of 420 ± 10 Ma characterizes dikes of pyroxenite gabbro composition crosscutting ultramafic rocks. In general, these radiometric dates are compatible with the Middle Ordovician time range inferred on the basis of paleontological data for basaltic pillow lavas with phthanite lentils (Sugrala Fm.), which overlie the sheeted dike complex of the Kempersay ophiolites.

In the northern part of the Sakmara zone, ophiolites are incorporated into the Kraka allochthon, a complex structure that is thrust onto the Zilair Formation (Late Devonian to Early Carboniferous) and Middle Ordovician to Late Devonian sedimentary cover of the East European Craton (Figure 2). Within the Kraka allochthon, ophiolites and associated serpentinite mélange over ride the lower tectonic slices of quartzites, shales, and cherts bearing Ordovician graptolites and chitinozoans, as well as Silurian to Devonian shales [*Puchkov*, 2000]. The base of the largest ultramafic bodies (North and South Kraka massifs) has an almost horizontal (dip angle of 5 to 20° eastward). In contrast, in the base of the Central and Uzyan Kraka massifs dips steeply (75–85°) eastward. Within these two massifs, the ophiolite sequence includes not only residual ultramafics, but also pyroxenites, wehrlites, and isotropic gabbro. Other members of the ophiolite sequence, such as basaltic pillow lavas and cherts, are detected here only among blocks in serpentinite mélange that includes also blocks of Silurian limestones [*Puchkov*, 2000]. Paleontological and radiometric data are not available for the Kraka massifs, but we can assume their disintegration after the Silurian period, during the accumulation time of the Late Devonian to Early Carboniferous Zilair Formation, which contains ophiolite detritus. It is generally accepted that this was a time of intense thrusting that was responsible for the westward transport of the Kraka and other ophiolite massifs.

To the east of the Sakmara zone and northeastward of the Kraka massif, a series of smaller, lenticular or wedge shaped ophiolite massifs (Nurali, Mindyak, Sakmara, and others) is confined to the Main Uralian fault. These east dipping (70–80°) bodies tectonically overlap either the crystalline schists of the East European Craton or its Paleozoic sedimentary cover. Toward the east, they are bound by serpentinite mélange of the western flank of the Magnitogorsk zone (Figure 3). The age of these massifs can be inferred from radiometric data obtained for garnet

pyroxenites (metagabbro) of the Mindyak massif. Zoned zircon crystals from these rocks were used to determine ages of their primary magmatic cores and metamorphic rims. The U-Pb age of zircon cores was 467 ± Ma for the zircon core [*Scarrow* et al., 1999] and 410 ± 5 Ma for the rims [*Saveliev* et al., 2001a], and the last date is concordant to the value 414 ± 4 obtained using the Sm-Nd isochron method [*Scarrow* et al., 1999]. Thus, plutonic rocks of the Mindyak massif originated most probably in the Middle Ordovician, and their metamorphic transformation took place close to the Silurian/Devonian boundary.

Serpentinite mélange of the Magnitogorsk zone includes blocks characterizing all rock complexes of the ophiolite association: residual peridotites, dunites, wehrlites, pyroxenites, gabbro, diabases, basalts and cherts [*Saveliev* et al., 1998]. Some large blocks of tholeiitic pillow lavas with chert intercalations yield the Early Ordovician conodonts and trilobites [*Ryazantsev* et al., 1999]. Other rock types of ophiolite association present in the mélange bodies are ophicalcites, breccias composed of gabbro and ultramafic clasts, and conglomerates almost lacking matrix and pre-dominantly consisting of serpentinite pebbles. The last rock types indicate that the disintegration and erosion of Ordovician oceanic basement took place presumably at Early Devonian time [*A. I. Voznesenskii*, unpublished data]. The initial phase of closure of the Ordovician ocean is indicated by blocks of extrusive rocks with island arc affinity [*Gaggero* et al., 1997], tephra and carbonate rocks within the same mélanges. Initial closure by intraoceanic subduction is clearly shown in the western Magnitogorsk zone by Early Devonian boninites associated with island arc volcanic rocks [*Spadea* et al., 1998; *Spadea and Scarrow*, 2000].

Metaophiolites of the Maksyutov Complex confined to the footwall of Main Uralian fault are attributed either to the Precambrian [*Lennykh*, 1966; *Krasnobaev* et al., 1996; *Lennykh and Valizer*, 1999] or to the Paleozoic, because the high-pressure metamorphism (the stage of oceanic crust subduction) affected these ophiolites in the Early to Middle Devonian time, and their exhumation happened in the Late Devonian [*Matte* et al., 1993; *Lennykh* et al., 1995; *Leech and Stockli*, 2000; *Leech and Ernst*, 2000].

In the eastern part of the Southern Urals, highly deformed and dismembered rock complexes of ophiolite association are concentrated in suture zones that bound a series of elongate blocks of the East Uralian uplift zone, which is thought to be underlain by continental crust. Tholeiitic pillow lavas with cherty intercalations yielding Ordovician conodonts are known in the Denisivo zone [*Puchkov*, 2000]. Dike complexes cross cutting Cambrian

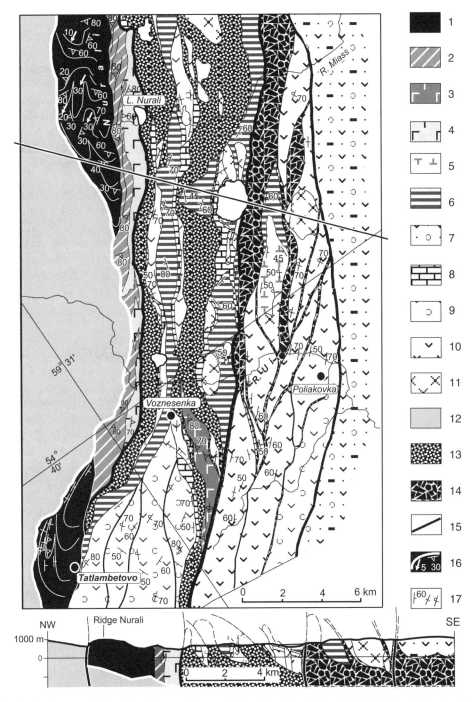

Figure 3. Geological map of the Nurali massif area: [after *Saveliev* et al., 1998]: 1. lherzolite, harzburgite, dunite; 2. dunite–wehrlite–pyroxenite series; 3. amphibole gabbro, gabbro-diorite; 4. layered gabbronorite; 5. Frasnian siliciclastic deposits and associated tuffs; 6. Givetian carbonate terrigenous deposits and volcanic rocks; 7. Emsian to Eifelian tuff, tuffite, and lava; 8. Ludlovian to Eifelian limestone; 9. Llandoverian terrigenous cherty deposits and basalt; 10. Ordovician pillow lava, chert, and diabase dikes; 11. Late Paleozoic granitoids; 12. Precambrian crystalline schists below Ordovician sediments; 13. polymictic serpentinite mélange; 14. monomictic serpentinite mélange; 15. fault; 16. lineation and banding in mantle rocks; 17. volcanic or sedimentary layering.

sedimentary sequences have been described by *Degtyarev* [1999] in the East Uralian zone. Paleozoic sequences in this zone are represented by rifting related arkoses, shales, and trachybasalts of Middle Ordovician age [*Korinevskii*, 1980; *Klyuzhina*, 1985], and by the Silurian to Devonian volcanic rocks, cherts, limestones, and terrigenous deposits [*Puchkov*, 2000].

2.2. Middle to Northern Urals

In the Tagil zone, strongly dismembered and metamorphosed ophiolite complexes occur as rare and small tectonic slices dipping eastward (Figure 1). Some of these ophiolite slices are tectonically overridden by zoned massifs (e.g., the Kytlym massif). Eastward, in the suture between the Tagil zone and the East Tagil uplift, ophiolite allochthons dip westwards. Tholeiitic basalts of the ophiolite sequence here include chert layers bearing Early Devonian conodonts [*Puchkov*, 2000]. The volcano sedimentary sequence of the Tagil zone is more than 5000 m thick and includes Late Ordovician bimodal basalt and rhyolite, lavas with black chert and carbonaceous silicilith interlayers (Kaban Formation). The overlying Silurian lavas and tuffs belong to basaltic andesite-dacite series. According to *Narkisova* et al. [1999], all volcanic rocks of the sequence, as well as picritic dikes crosscutting it, correspond in geochemical parameters to island arc volcanic rocks with a general geochemical trend that implies an increasing contribution of continental material to parental magmas. These features indicate that magmas could have been extruded through the thinned continental margin, or that the subducting East European Craton contributed material to the lavas. The abundance of tuff interlayers within lavas increases upsection; the Devonian, in fact, consists predominantly of trachyandesite lavas, pyroclastic rocks, and tuffite associated with diverse limestones [*Puchkov*, 2000]. The whole sequence is tectonically imbricated and deformed into west-verging folds.

The zoned massifs of the Platinum Belt occur within the lower tectonic sheets of the Tagil zone, which are thrust westward over the Paleozoic complexes of the former East European Craton continental margin (shelf and continental slope deposits). These zoned massifs are spatially associated with metabasalts, metarhyolites, greenschists, and metapelites of the Ordovician age Kaban Formation. According to *Narkisova* et al. [1999], the volcanic rocks are of calc-alkaline affinity. The multistage deformation recorded in the country rocks westward of the massifs includes: an early stage of south southwestward thrusting and subsequent asymmetric folding with axial planes that are conformable to tectonic contacts of allochthonous complexes (e.g., the Nizhnii

Tagil massif, Figures 4 and 5). Basal planes of the allochthons dip eastward according to field observations and geophysical data [*Sokolov*, 1992]. Within the inner contact zone, rocks of the massifs reveal features of intense ductile deformation, while in the peripheral zone they are also strongly deformed and metamorphosed under epidote amphibolite facies conditions. Thrust sheets above and eastward of the massifs are composed of Late Silurian to Middle Devonian island arc volcanic rocks and terrigenous deposits (andesites, trachybasalts, tuffs, and limestones).

Pb-Pb isotopic data for zircons from gabbronorite of the Kumba massif indicate that this pluton crystallized at 427 ± 7 ago [*Bosch* et al., 1997]. K-Ar dates for phlogopite clinopyroxenites of the Svetlyi Bor massif range from 437 to 415 [*Ivanov and Kaleganov*, 1992], and Sm-Nd isochron age for gabbro of the Chistop massif yielded an age of 419 ± 12 [*Ronkin* et al., 1997].

2.3. Polar Urals

In the Polar Urals, the Paleozoic sedimentary cover of the East European Craton is overridden by a system of allochthonous units shown in Figure 6. The ophiolite massifs Syumkeu, Raiiz, and Voykar-Syn'ya are sandwiched between two thrust sheets that are composed of complexes related in origin to the Tagil-Schych'ya (Late Ordovician-Early Silurian) and Voykar (Middle Silurian to Late Devonian) island arcs [*Saveliev* et al., 1999]. All these allochthonous units are thrust over the following rock complexes: (1) the Upper Cambrian to Tremadocian red sandstones and gravelstones of shallow water to terrestrial origin associated with basaltic to daciterhyolitic lavas; (2) Ordovician to Devonian black shales and cherts locally intercalated with pillow lavas of bimodal composition; (3) Silurian to Carboniferous terrigenous, carbonate, and siliceous deposits accumulated in deep shelf settings. All these deposits have been deformed into narrow linear asymmetric folds that are overturned toward the west and display an intense axial planar cleavage. Earlier folds have subhorizontal axial planes that dip the southeast, whereas later folds have subvertical, steeply east dipping axial planes. The extreme northeastern Syumkeu massif is overridden by tectonic units that dip east-southeast and are composed of Silurian pillow lavas with tuff and chert interlayers. Higher thrust slices consist of island arc volcanics and polymictic epiclastic rocks [*Saveliev and Samygin*, 1979; *Yazeva and Bochkarev*, 1984].

All ophiolite massifs of the Polar Urals have distinct metamorphic soles formed from volcano sedimentary country rocks and partially ophiolite gabbro and diabase protoliths. Metamorphism of these rocks is of eclogite

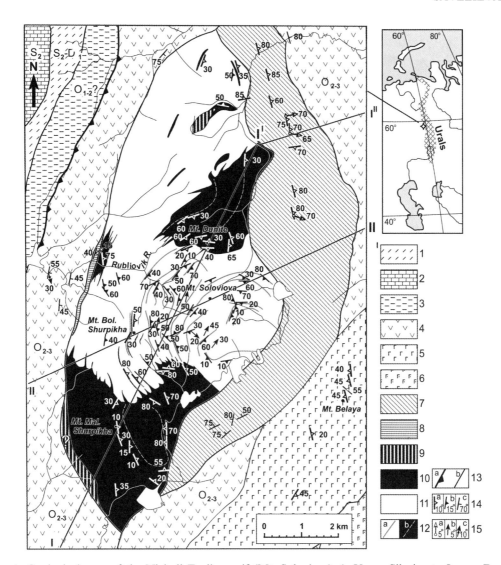

Figure 4. Geological map of the Nizhnii Tagil massif (Mt. Solov'eva): 1. Upper Silurian to Lower Devonian shales; 2. Upper Silurian limestones; 3. Lower to Middle Ordovician phyllites and quartz micaschists; 4. Middle to Upper Ordovician greenschist, basalt, and rhyolite; 5. gabbro and gabbronorite; 6. amphibolitized gabbro; 7. retrograde amphibolites after diabase and fine-grained gabbro; 8. blastomylonites after gabbro; 9. phlogopite-nepheline gabbro; 10. olivine clinopyroxenite and wehrlite; 11. dunite; 12. trajectories marked by foliation of Cr-spinel (a) and clinopyroxene (b); 13. tectonic boundary of the first (a) and second (b) orders; 14. dip of banding and foliation in pyroxenite, gabbro and metamorphic rocks (a), in dunite (b), and schistosity in volcano-sedimentary rocks (c); 15. lineation in wehrlite and gabbro (a), in dunite (b), and in greenschist (c).

(glaucophane), amphibolite, blueschist, and albite-lawsonite facies [*Savelieva*, 1987; *Dobretsov* et al., 1977]. The metamorphic mineral lineation in amphibolites suggests a sinistral shear sense and a south southwestern direction of early movement for the allochthons. The western vergence of the folds that deform the mineral lineation, and southeastern dip of the their axial planes in amphibolite and blueschists indicate a northwestern direction of transport during the late stages of deformation. Within and below the ultramafic bodies, there are jadeite-albite ($\pm$ muscovite or amphibole) metasomatic dikes. One of these ruby bearing veins near the base of the Raiiz massif has been dated at 358 ± 3 by Rb-Sr isochron (*J. Glodny*, unpublished data); obduction of the ophiolites is dated by the age of eclogite facies fluid-rock interaction at around 360 [*Glodny* et al., 2000].

The Sm-Nd isotopic data for a series of whole rock samples (gabbro, pyroxenite and diabase dikes) from the

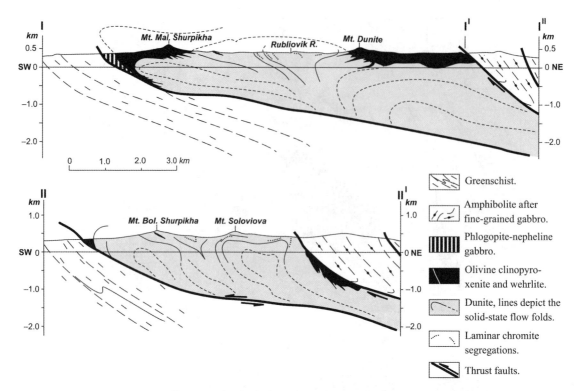

Figure 5. Geological cross sections to Figure 4.

northern part of the Voykar massif indicate that plutonic complexes of the massif could be 387 ± 34 [*Edwards and Wasserburg*, 1985; *Sharma* et al., 1995]. The Rb-Sr isochron age of an intrusive tonalite crosscutting the massif is 395 ± 5 [*Buyakaite* et al., 1983]. In addition, gabbro, ultramafic, and tonalite pebbles mixed in the Eifelian conglomerates [*Lupanova and Markin*, 1964; *Yazeva and Bochkarev*, 1984] also constrain the pre-Eifelian time of the tonalite emplacement.

West of the Paleozoic ophiolite allochthons, small outcrops of Riphean ophiolites [*Gee and Pease*, 1997] are known in the Enganope anticline (Figure 6). Ultramafic rocks, gabbro, and lavas, here metamorphosed to greenschist facies, were strongly deformed by the Late Vendian folding.

3. COMPARATIVE CHARACTERISTICS OF OPHIOLITE AND ZONED MASSIFS

The well known difference between ophiolites and zoned mafic-ultramafic massifs consists of the absence of residual mantle peridotites and sheeted dike complexes in the latter. In zoned massifs, rocks of harzburgite composition and of metasomatic origin [*Yefimov*, 1984] have been distinguished as the peculiar feature of the Kytlym massif only.

Allochthonous bodies of ophiolites and zoned plutons differ also in shape, as they are elongated in the first case and relatively isometric in the second. The successions of plutonic rocks are, however, similar in both cases. The base of ultramafic complexes consist of a dunite-wehrlite-pyroxenite series that is overlain by olivine gabbro, layered gabbronorite, and massif isotropic gabbro, the latter often bearing amphibole. This succession is frequently complicated by tectonic juxtaposition or by multistage intrusions of magma.

Residual mantle ultramafics, lherzolite, harzburgite and dunite compose a major part of the Uralide ophiolites. In various massifs, their total thickness ranges from 2 km to 9 km. Spinel and plagioclase lherzolite prevail in the Kraka, Nurali, Mindyak and other small massifs. The Kempersay massif consists instead of spinel lherzolite associated with harzburgites. In the Middle, Northern and Polar Urals, harzburgites are dominant, preserving only small lherzolite relics. In the Polar Urals (e.g., in the Voykar massif) original mantle harzburgites are locally impregnated with diopside (sometimes together with plagioclase), suggesting they are secondary lherzolites. Dunites are abundant within harzburgite, and rare within the lherzolite massifs. The dunites, however, are always present near the contact of residual mantle rocks and an overlying wehrlite-dunite-pyroxenite unit. Detailed field

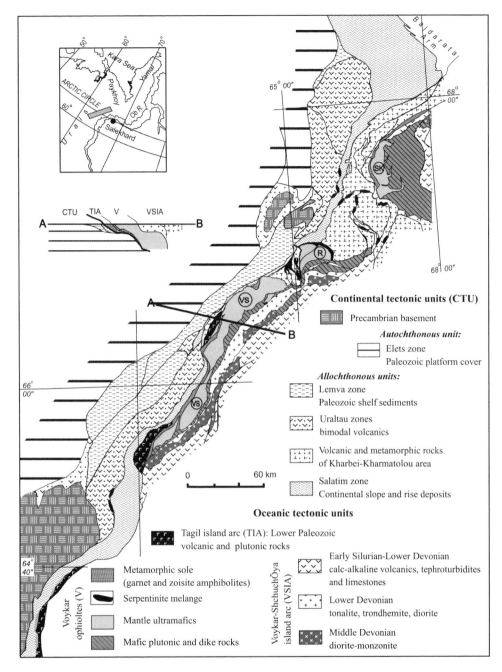

Figure 6. Structural position of the Voykar-Syn'ya (VS), Raiiz (R), and Syumkeu (SK) ophiolite massifs in the Polar Urals.

observations show that the lherzolite and harzburgite mantle sections have different internal structures. In lherzolites (e.g., Kraka and Nurali massifs), the high temperature lineation is parallel to hinges of small scale ductile flow folds, and persistently dips gently to the south southwest. The foliation is usually well shown and marked by the scattered blastomylonitic zones. A sharp

change of ductile flow patterns in the lherzolite massifs is confined to the boundary zone with the dunites that occur between the residual mantle and plutonic crustal complexes. The boundaries of this zone strike north northeast and are subvertical. A lineation defined by Cr-spinel is also subvertical. In harzburgites (e.g., Voykar massif), both banding and a lineation are well developed, but there

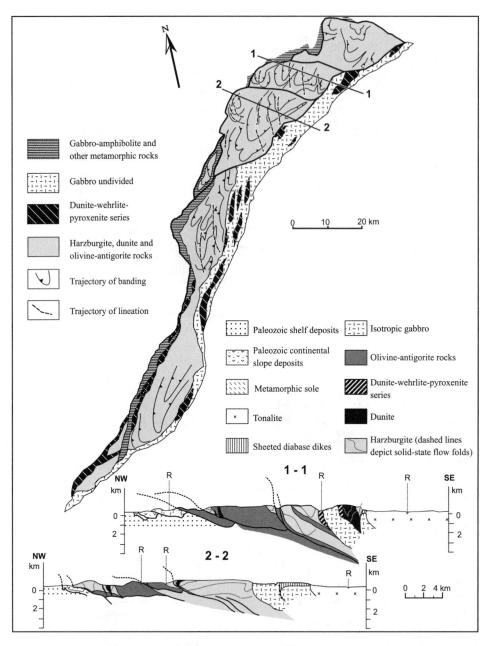

Figure 7. Schematic map and cross sections of the Voykar-Syn'ya massif.

is only a weak foliation. Large scale flow folds are associated with a subvertical system of dunite, pyroxenite and rare gabbro dikes. These dikes are not present in lherzolites. The common pattern of ductile deformation is traceable throughout the residual harzburgite sequence, including its most depleted dunite zone (Figure 7). In addition, lherzolites display widespread porphyroclastic and local mylonitic textures, whereas coarse-grained protogranular textures are dominant in harzburgites.

Residual ultramafics of the Polar Urals differ in composition from mantle rocks of the Southern Urals ophiolites. The higher degree of depletion of the Voykar mantle peridotites, as compared to their counterparts in the Southern Urals, is evident from the following data: (1) peridotites of the Voykar massif have lower Ca and Al content, (2) rare diopside is generally in neoblasts, (3) dunite bodies are more abundant and larger, and (4) the Al_2O_3 content of protogranular and porphyroclastic

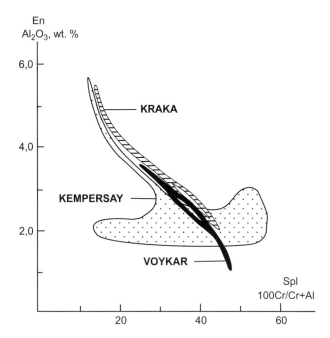

Figure 8. Compositional variations of enstatite and coexisting Cr-spinel in mantle peridotites of three ophiolite massifs (see text for explanation). Data from *Savelieva* [1987]; *Savelieva and Pertsev* [1995].

enstatite is much lower. In addition, the chromium number Cr# of accessory spinel varies from 25.2 to 48.7 in the Voykar peridotites and from 15.6 to 43.5 in the Southern Urals massifs (Figure 8). The Cr#-Mg# relations characterize the whole mantle sequences, thus implying not only different degree of depletion of mantle residues, but also different composition of the mantle involved in partial melting. Geochemical and isotopic data [*Sharma* et al., 1995] also show the extremely depleted nature of harzburgites and a very complex history of depletion and mantle metasomatism recorded in the Voykar ultramafics.

Plutonic complexes of the ophiolite massifs include dunite–wehrlite–clinopyroxenite series and layered to isotropic gabbro units. In lherzolite massifs, the former rock series, which is predominantly composed of olivine and clinopyroxene, is characterized by coarse to fine banding and layering. Magmatic petrofabrics of pyroxene dunite, wehrlite and isotropic gabbro are preserved in this case due to weak ductile deformation. The total thickness of the sequence is small (tens to a few hundreds of meters). Much thicker (up to 1.5 km) plutonic sequences from harzburgite massifs contain roughly banded, dunite-wehrlite-clinopyroxenite series and, higher in the section, layered cumulative gabbronorite, olivine gabbro, and subordinate troctolite. The lower part of the

sequence shows intense ductile and semibrittle deformation partly synchronous to melt intrusions, which have produced pyroxenite and gabbro dikes crosscutting banding of the rocks [*Savelieva*, 1987].

Gabbronorite and olivine gabbro are abundant in harzburgite massifs, but are absent in the lherzolite massifs. In the Voykar-Syn'ya massif, for instance, the layered gabbronorite unit includes dunite blocks impregnated with plagioclase and diopside. Xenoliths of residual metamorphosed ultramafics and wehrlites occur here within hornblende gabbro and in the sheeted dike complex (Figure 7, cross section 1-1). The early complex of gabbro and gabbronorite locally shows intrusive contacts with mantle rocks and signs of active interaction with the hot mantle residues of the Voykar massif, suggesting a complex history of Moho transposition that is also characteristic of other ophiolites (e.g., of those in Oman [*Boudier* et al., 1996]). The late complex of isotropic gabbro and related sheeted dikes with chilled margins, which have a total thickness of about 500 m, intruded the ultramafic and early mafic rocks after, and partially during the brittle-ductile deformation and hydration of mantle rocks [*Saveliev* et al., 1999]. Sheeted dikes are oriented parallel to the general strike of the boundary between mantle and crustal complexes.

In the Kempersay massif, the plutonic sequence is about 700 m thick. The magmatic history, however, has been complicated here by late intrusion of branching, amphibole- and phlogopite-bearing differentiated gabbro dikes and small intrusive bodies crosscutting ultramafics and earlier gabbro and diabase.

Plutonic rocks of the zoned mafic-ultramafic massifs of the Platinum Belt have been studied in the Kytlym massif (the most representative) by *Yefimov* [1977, 1984]. This author has described widespread high temperature deformation. According to recent data [*Savelieva* et al., 1999], this deformation is developed only locally and does not completely overprint the magmatic fabrics of the rocks. The lower, predominantly ultramafic rocks of the massif show strong layering. The layered dunite-wehrlite-pyroxenite sequence underlies a thick body of porphyritic melanocratic gabbro with olivine and clinopyroxene interlayers (Figure 9). The lower portion of the complex has been affected by intense high temperature ductile deformation and by brittle fracturing. Fine-grained blastomylonitic, blastoporphyritic, and porphyroblastic textures, and petrofabrics of the rocks suggest the temperature range of 850–1100°C for these deformation events. At the same time, melanocratic gabbros with zoned Cpx megacrysts within fine-grained clinopyroxene-olivine-plagioclase groundmass shows only local development of ductile deformation. Banding

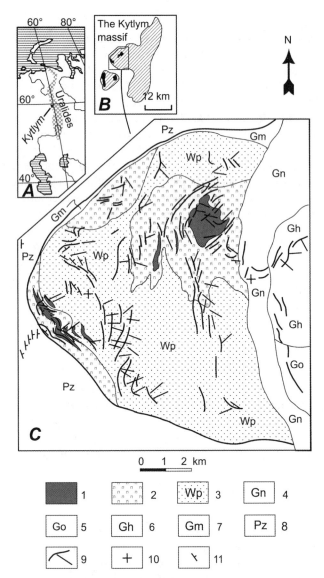

Figure 9. Structural map of the northwestern part of the Kytlym massif: 1. dunite; 2. olivine pyroxenite; 3. plagioclase wehrlite and melanocratic olivine gabbro; 4. gabbronorite; 5. olivine gabbro; 6. hornblende gabbro; 7. gabbro-migmatite, fine-grained pyroxene-hornblende gabbro; 8. Paleozoic greenschists; 9. dip of lineation (arrows) and trace of banding (solid lines); 10. horizontal banding; 11. schistosity of Paleozoic rocks. Data from *Savelieva* et al. [1999].

concluded that penetrative deformation developed immediately after the first stage of crystal fractionation in the magma chamber, which moved westward together with the country rocks.

Dunites of the lower sequence locally protrude into the overlying rocks and show evidence of syntectonic recrystallization in the presence of volatiles. Petrofabric patterns of these dunites probably reflect the lateral sliding of the pyroxenite roof that covered the dunite dome.

Olivine from the basal dunite-pyroxenite sequence reveal the dominant intracrystalline gliding system to be {0kl}[100], and very poor syntectonic recrystallization [*Savelieva* et al., 1999]. On the contrary, olivine from the dunite protrusion shows syntectonic recrystallization and poorly developed intracrystalline gliding. Stockworks of gabbro and wehrlite dikes are also typical of the lower sequence. They probably represented a feeding system of a later intrusive phase [*Pertsev* et al., 2000].

The upper part of the magmatic sequence mostly consists of layered gabbro and gabbronorite, and is distinct in origin from the lower one. The rocks are amphibolized in local zones that are affected by ductile deformation. The transition zone between ultramafic rocks and gabbro and gabbronorite sequence contains numerous dikes of fine-grained gabbronorites, hornblende-olivine gabbros.

Plutonic rocks of the Kytlym massif, though similar in composition, display diverse magmatic or metamorphic textures. For example, porphyritic melanocratic gabbros with fine-grained matrix are closely associated with high metamorphic grade blastoporphyritic or blastomylonitic metagabbros of similar composition. These features suggest: (i) a rather high level of crystallization, (ii) solid state flow deformation immediately after crystallization, and (iii) high gradient of deformation rate. The relatively shallow, hypabyssal crystallization of the pluton is also indicated by an abundance of dike stockworks, and by fine-grained porphyritic gabbronorites and hornblende gabbros localized at the periphery of the massif, where they include xenoliths of country rocks.

The basal dunite body is much thicker (about of 2 km) in the Nizhnii Tagil massif of the Platinum Belt (Figure 4). Here, dunites and overlying olivine pyroxenites are intruded by small bodies of phlogopite and nepheline-bearing olivine gabbro. The olivine-clinopyroxene rocks and clinopyroxenites represent fragments of a formerly continuous (?) higher horizon, about 0.2–0.5 km thick, that is preserved along the periphery of the massif. Irregular, solid state flow folds are outlined by the banding, mineral foliation and lineation observable in

and foliation both dip gently (10–30°) to the east and northeast in the western part of the massif near its base, or steeply (80–85°) in the east near the contact with gabbronorites. Petrofabric analysis by *Savelieva* et al. [1999] showed that sinistral ductile flow in the ultramafics near the base may be attributed to the development of an anticlinal fold in the country rocks. These authors

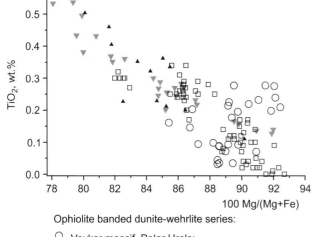

Ophiolite banded dunite-wehrlite series:

○   Voykar massif, Polar Urals;
□   Nurali massif, South Urals.

Uralian Platinum Belt plutons:

▲   Kytlym massif, Middle Urals;
▼   Nizhnii Tagil massif, Middle Urals.

Figure 10. TiO$_2$ versus Mg index diagram for clinopyroxenes (Voykar massif: Pertsev and Savelieva, unpublished data, Nurali massif: *Pertsev* et al. [1997], Kytlym massif: *Pertsev* et al. [2000], Nizhnii Tagil massif: *Saveliev* et al. [2001b]; *Chaplygina and Savelieva* [2001]).

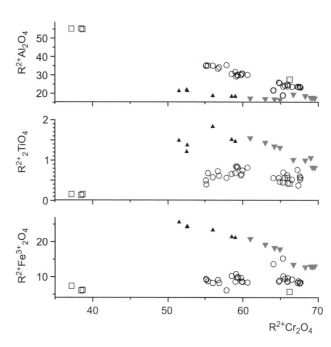

Figure 11. Compositional variations of spinels from dunites ($R^{2+} = Fe^{2+} + Mg + Mn$) (symbols and data sources as in Figure 10).

dunites and olivine-clinopyroxene rocks [*Saveliev* et al., 2001b]. The assumed asymmetric funnel-like structure or a synform between two flow folds in the central part of the massif (Figure 5) probably originated by plastic squeezing of early cumulative dunites up into the magma chamber. Recrystallized pegmatite and phlogopite-bearing dunites and associated chromite lenses are abundant in this part of the massif and probably mark the migration path of residual fluids. We infer that the solid state plastic deformation affected the rocks immediately after the pyroxenite crystallization above the dunites. Clinopyroxenite dikes are discordant relative to the main boundary separating dunites from pyroxenites, but concordant to flow planes in the early cumulates (Figure 5).

Compositional differences between plutonic rocks of the ophiolitic and the zoned massifs can be illustrated by chemical diagrams for clinopyroxene, spinel, and whole rock samples. The TiO$_2$ content of clinopyroxene from olivine-clinopyroxene ultramafic rocks of both plutonic sequences shows a general negative correlation with Mg# (Figure 10). This trend, corresponding to fractional crystallization, is well pronounced in the Tagil and Kytlym massifs, where compositional variations of pyroxene and spinel (Figure 11) are noticeably wider than in the ophiolite massifs. Both diagrams also indicate that

parental magmas of platinum-bearing massifs were comparatively rich in TiO$_2$ than the ophiolite massifs, and that they underwent crystallization under a higher oxygen fugacity. The positive Eu anomaly of gabbro-norites from the Voykar massif can also be related with plagioclase crystallization under reducing conditions [*Irving*, 1978]. In addition, plutonic rocks of the Voykar massif are depleted in incompatible elements as compared to their counterparts in the Kytlym massif (Figure 12).

Ultramafic and gabbroic rocks of the Kytlym massif show gradual enrichment in incompatible elements with negative anomalies of HFSE (Nb, Hf, Zr) and positive Ba anomaly, pointing to their comagmatic origin (Figure 12). Accordingly, their parental magma could have been generated in a suprasubduction setting [e.g., *Fershtater* et al., 1999]. The same environment of magma generation was inferred by *Saveliev* et al. [1999] for plutonic rocks of the Voykar massif. This is in contrast to those of the Kempersay and other southern ophiolite massifs, which represent oceanic crust of mid-ocean spreading or marginal rift zones.

4. ASSOCIATED METAMORPHIC ROCKS

In the Kempersay ophiolite, garnet and zoisite + garnet amphibolites of the metamorphic sole of the massif were derived from layered olivine gabbro and

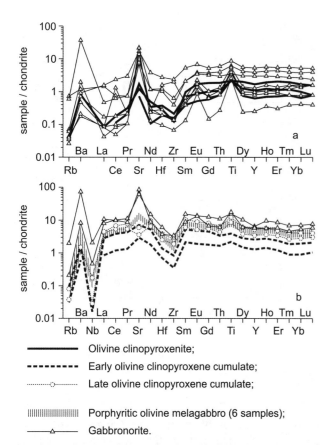

Figure 12. Relative abundances of REE, LILE, and HFSE in plutonic rocks of the ophiolite Voykar-Syn'ya (a) and platinum bearing Kytlym massif (b); chondrite normalizing values are those given by *Sun and McDonough* [1989] (Voykar-Syn'ya massif: *Saveliev* et al. [1999], Kytlym massif: *Nesbitt and Pertsev*, unpublished ICP-MS data).

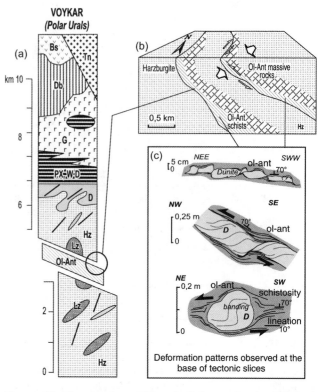

Figure 13. Structural position of olivine-antigorite (Ol-Ant) rocks in the Voykar-Syn'ya massif and their deformation patterns: (Hz) harzburgite; (Lz) lherzolite; (D) dunite; (W) wehrlite; (Px) pyroxenite; (G) gabbro; (Db) diabase; (Bs) basalt; (Tn) tonalite.

gabbronorite. The metamorphic grade decreases downward in the sole from amphibolite to greenschist facies [*Pertsev and Saveliev*, 1994]. In the Kraka allochthon, fragmented garnet amphibolites are found within the sheared wehrlite-pyroxenite-gabbro series and in a serpentinite mélange. Strongly deformed, porphyroclastic and mylonitic peridotites and metamorphosed volcano sedimentary rocks typical of a metamorphic sole of ophiolites [e.g., *Boudier* et al., 1982; *Moores*, 1982; *Michard* et al., 1991] are absent in both massifs. On the contrary, in the Polar Urals ophiolitic massifs the metamorphic sole is derived from volcano sedimentary rocks and, partially, from ophiolitic gabbro and diabase dikes. Metamorphism here ranges from blueschist, eclogite, amphibolite, to albite-lawsonite facies [*Dobretsov* et al., 1977; *Savelieva*, 1987]. Harzburgites above the sole are extremely foliated, have blastomylonitic textures, and contain pargasite.

Peculiar metamorphic rocks occur inside the ultramafic bodies of the Voykar, Raiiz and Syumkeu massifs. They are present in shear zones up to 400 m thick that separate tectonic slices of protogranular mantle peridotites (Figure 13a). Massive olivine-antigorite (Ol-Ant) rocks dominate in the upper part of these zones. The high temperature (850–450°C) and high pressure (around of 250 Mpa) hydration and metamorphism of ultramafic rocks began along the network of fractures (Figure 13b), where $Ol + En$ was replaced sequentially by $Ol + Trem$, $Ol + Tlc$ and $Ol + Ant$ parageneses. These rocks were transformed into Ol-Ant schists at the base of tectonic slices, where relic dunite boudins are partly recrystallized (Figure 13c). Intense ductile and semibrittle deformation led to the appearance of cleavage in olivine and porphyroclastic and mylonitic textures. Identical Ol-Ant rocks occur in ophiolites of the Mariana, Tonga, New Guinea and other island arcs [*Skornyakova and Lipkina*, 1975; *Zlobin and Puzhin*, 1989].

The established ranges of ^{18}O and D in antigorite from the Voykar massif are 4.6 to 6.8 and –40 to –78 ‰, respectively [*Buyakaite* et al., 1983], and indicate that metamorphic fluids might correspond in composition to sea water bearing some components derived from sediments.

5. DISCUSSION

Radiometric age dating and paleontological data constrain the Early to Middle Ordovician age of the Southern Urals ophiolites. These ophiolite massifs, which are especially widespread in this segment of the Uralides, differ from each other in composition and structure of their mantle and crustal complexes. Consequently, it seems logical to suggest that they originated in different tectonic settings of the Ordovician Uralian ocean. The complete (e.g., the Kempersay massif) or dismembered ophiolite sequences of the Magnitogorsk, East Uralian, and Denisovo zones include a succession of plutonic rocks, sheeted dikes, and tholeiitic lavas above the residual mantle peridotites of dominantly harzburgite composition. We infer that these rock associations were generated in a mid-ocean ridge system. The mantle peridotites show evidence of multistage high temperature deformation and variable signs of interaction with basaltic melts. The plutonic complexes, composed of banded wehrlite-pyroxenite series, lower layered and upper isotropic gabbro, are almost undisturbed. The diabase sheeted dike complexes strike conformably to steep foliation planes and lineation trajectories in residual peridotites, suggesting, as in ophiolite complexes elsewhere [Nicolas, 1989; Nicolas et al., 1994; Nicolas and Boudier, 1995; Boudier et al., 1996], the presence of a ridge axis, and characterizing a complete mid-ocean ridge segment. Massifs of this type also record recurrent events of brittle deformation of the crust away from the spreading axis and the synchronous emplacement of mafic dikes.

Other lherzolite massifs (e.g., Kraka, Nurali) reveal a more simple evolutionary history. Their fertile ultramafic rocks are characterized by the high gradient depletion profile that ends with a distinctive dunite zone at the contact with plutonic complexes. The latter are much smaller in volume than in the harzburgite massifs, and are composed of dunite-wehrlite-pyroxenite series and isotropic amphibole gabbro. Neither layered gabbro, nor sheeted dikes and lavas have been detected in these massifs. Similarly, pyroxenite-gabbro stockworks within mantle ultramafics have not been found. All these features may imply only minor development of partial melting in the mantle. In this case, magmatic melts presumably penetrated through the narrow and well heated dunite-harzburgite zone, where the deformation patterns are drastically different when compared to those

of lherzolites. As suggested Savelieva et al. [1997], massifs of this kind may have originated by the quick ascent of a mantle diapir such as those responsible for rifting events in a continental margin. If these speculations are correct, the mid-ocean ridge system of the Ordovician oceanic basin was somehow interrelated with the system of rift structures in the East European Craton margin.

The development of Silurian island arcs, whose rock complexes are traceable all along the Uralides, either left no remnants of obducted ophiolites, or they have not been recognized. Younger ophiolites that presumably originated close to the Silurian-Devonian boundary are known, however, in the Northern and Polar Urals, to the east of the Tagil zone (Figure 1).

In the Polar Urals, the Late Silurian to Early Devonian formation of ophiolites is constrained by available radiometric dates (see above) and by their structural association with a system of thrust sheets consisting of rock complexes of the Late Ordovician to Early Silurian Tagil-Shchuch'ya and Middle Silurian to Late Devonian Voykar island arcs [Saveliev et al., 1999]. All ophiolite massifs display complicated structural and magmatic evolution. Their residual peridotites are highly depleted, similar to mantle ultramafic rocks from marginal oceanic settings (e.g., beneath the Mariana and Tonga arcs [Peyve, Ed., 1980]). Such features as the structural association of flow folds and brittle to semibrittle deformation healed by dunite, pyroxenite and gabbro intrusions, and diopside (+ plagioclase) impregnations in mantle residues, indicate a deep interaction of the latter with percolating basaltic melts. Interaction with melts is typical of many harzburgite-type massifs, as is the appearance of localized deformation and focused melt flow in the upper mantle. For example, these features have been described in the Josephine ophiolite [Keleman et al., 1995].

The depleted character of mantle rocks is correlated with a considerable thickness of the crust. The crustal sequence consists of polyphase plutonic rock complexes indicating an intense magma generation in the mantle. The high temperature ductile deformation at the crust-mantle boundary reflects the unstable tectonic environments of crystallization of the earliest melt fractions derived from the mantle. These features are characteristic of ophiolite massifs of Newfoundland [Karson, 1989; Suhr, 1992] and Vourinos [Rassios et al., 1983], which are considered as rock associations of fast spreading centers on the basis of structural patterns [Nicolas et al., 1988; Nicolas, 1989; Nicolas et al., 1994; Nicolas and Boudier, 1995], or on the basis of geochemical data, as suprasubduction ophiolites [Batanova and Sobolev, 2000].

In addition, the magmatic events responsible for development of thick plutonic complexes were separated by metamorphic events that affected harzburgites and wehrlites, which are present as xenoliths in the upper gabbro. According to their structural position and composition, metamorphic olivine-antigorite rocks originated at the stage of tectonic delamination of oceanic lithosphere by the local high temperature hydration of harzburgites along shear zones similarly to the Josephine ophiolite [*Alexander and Harper*, 1992].

In general, the structural features of the Voykar massif, as well as the geochemistry of plutonic rocks and sheeted dikes, characterize a geodynamic environment that involved spreading above a subduction zone where the ascent of a mantle diapir was accompanied by intermittent intrusions of basaltic magma. The amagmatic period between pulses of mafic magma injections was marked by shearing and high-temperature hydration of oceanic lithosphere. We conclude that ophiolites of the Polar Urals were related in origin to the former marginal basin that split an old island arc structure for a short period during the Late Silurian to Early Devonian time.

Emplacement of platinum-bearing massifs in the mid-Silurian time predated the main Late Silurian to Early Devonian stage of island arc formation that was manifested all along the Uralides. Marginal rocks of the Kytlym and Uktus [*Pushkarev*, 2000] massifs include xenoliths of garnet-plagioclase-amphibole blastomylonites, which are interpreted as indicators of detachment faulting and shearing either in the thinned crust of continental margin [e.g., *Savelieva* et al., 1999], or in the island arc basement [e.g., *Ivanov and Shmelev*, 1996].

On the other hand, we suspect that the Tagil zone as a whole may represent a stack of thrust sheets that are composed of various rock complexes derived from the active continental margin of the East European Craton. This margin could have begun to develop after the Middle Ordovician. With this hypothesis we can easily explain why the Platinum Belt massifs are confined to this zone only, why the eastern part of the zone hosts a chain of granite-gneiss massifs (Californian type domes?), and why the volcanic series of the zone are so diverse in geochemical characteristics.

In any case, all rocks of the zoned massifs are, in our opinion, co-magmatic differentiates. However, it is not inconceivable that some fragments of residual mantle dunite could have been captured and transported by melt to higher crustal levels. Consequently, the high stress and high temperature deformation of the dunite is local, concentrated mostly in basal parts of the massifs where the deformed rocks are closely associated with rocks that preserve typical magmatic textures.

As mentioned above, olivine clinopyroxenite bodies, which follow the ductile flow planes inside the dunite, are discordant relative to the main dunite-pyroxenite boundary in the Nizhnii Tagil massif. We interpret this relationship to be the result of melt injection from a magmatic chamber into the early cumulates that had undergone deformation. Subsequent transposition to the crust dismembered the massifs into thin tectonic slices now occurring as separate dunite-pyroxenite (Nizhnii Tagil) or dunite (Svetlyi Bor, Veresovyi Bor) bodies. These massifs are allochthonous units and not the "roots" of eroded discordant plutons of tubular shape, as interpreted by *Fershtater* et al. [1999].

In summary, we interpret the ophiolite massifs of the Southern Urals to be relics of oceanic lithosphere that originated during Ordovician rifting and sea floor spreading events. The metamorphic soles of all ophiolite allochthons formed about 397 My ago, prior to their obduction onto the passive margin of the East European Craton in the Late Devonian to Early Carboniferous. Oceanic crust flooring the Ordovician ocean basin underwent incipient deformation about 420 My ago, and this triggered formation of Silurian and Devonian system of island arcs, relics of which are widespread in the Uralides. Plutons of the Platinum Belt could have been intruded during the mid-Silurian time into the basement of an island arc, or in the peripheral segment of the craton already transformed into an active continental margin.

Finally, we interpret the Silurian and Devonian ophiolites of the Polar Urals as fragments of the oceanic crust and underlying mantle of an intra-arc basin that originated by splitting of a former island arc. Their metamorphic soles were formed 360–380 My ago during obduction onto the continental active margin.

Acknowledgments.   We are grateful to F. Boudier and an anonymous reviewer for their helpful remarks and comments to our manuscript. This work was supported by the Russian Foundation for Basic Research, Project 00-05-64224, INTAS Project 94-1857, EUROPROBE Uralides Project, TMR-EC URO Project, and Italian MURST grant.

REFERENCES

Alexander, R. J. and G. D. Harper, The Josephine ophiolite: An ancient analogue for the slow to intermediate spreading oceanic ridges, in *Ophiolites and Their Modern Oceanic Analogues*, edited by L. M. Parson, B. J. Marton and P. Browing, *Geol. Soc. Spec. Publ.*, 60, 3–38, 1992.

Batanova, V. G. and A. V. Sobolev, Compositional heterogeneity in subduction-related mantle peridotite, Troodos massif, Cyprus, *Geology*, 28, 55–58, 2000.

Bosch, D., A. A. Krasnobayev, A. A. Yefimov, G. N. Savelieva and F. Boudier, Early Silurian ages for the gabbroic section of the mafic-ultramafic zone from the Urals Platinum Belt, *J. Conf. Abstr.*, 4, 122, 1997.

Boudier, F., A. Nicolas and J. L. Bouchez, Kinematics of oceanic thrusting and subduction from basal sections of ophiolites, *Nature*, 296, 825–828, 1982.

Boudier, F., A. Nicolas and B. Ildefonse, Magma chamber in the Oman ophiolites: Fed from the top and the bottom, *Earth Planet. Sci. Lett.*, 144, 239–250, 1996.

Brown, D., C. Juhlin, J. Alvarez-Marron, A. Perez-Estaun and A. Oslianski, Crustal-scale structure and evolution of an arc-continent collision zone in the southern Urals, Russia, *Tectonics*, 17, 158–171, 1998.

Brown, D. and P. Spadea, Processes of forearc and accretionary complex formation during arc-continent collision in the southern Urals, *Geology*, 27, 649–652, 1999.

Buyakaite, M. I., V. I. Vinogradov, V. V. Kuleshov, B. G. Pokrovskii, A. A. Saveliev and G. N. Savelieva, *Isotope Geochemistry in Ophiolite of the Polar Urals* (in Russian), Nauka, Moscow, 179 pp., 1983.

Chaplygina, N. L. and G. N. Savelieva, Magmatic nature of ultramafic rocks of the Mt. Solov'eva (the Nizhnii Tagil massif, Central Urals) (in Russian), in *General and Regional Problems of Geology*, edited by G. N. Savelieva and V. G. Nikolaev, GEOS, Moscow, 67–80, 2001.

Degtyarev, K. E., Tectonic evolution of Early Paleozoic continental margin of Kazakhstan (in Russian), Nauka, Moscow, 123 pp., 1999.

Dobretsov, N. L., Yu. E. Moldavantsev, A. P. Kazak, L. G. Pono-mareva, G. N. Savelieva and A. A. Saveliev, Petrology and metamorphism of ancient ophiolites (Polar Urals and West Sayan as examples) (in Russian), Nauka, Novosibirsk, 219 pp., 1977.

Edwards, L. R. and G. J. Wasserburg, The age and the emplacement of obducted oceanic crust in the Urals from Sm-Nd and Rb-Sr systematics, *Earth Planet. Sci. Lett.*, 72, 389–404, 1985.

Fershtater, G. B., F. Bea, E. V. Pushkarev, G. Garuti, P. Montero and F. Zaccarini, 1999, New data on geochemistry or the Uralian Platinum Belt: A contribution to insight into petrogenesis (in Russian), *Geochimiya*, 4, 352–370, 1999.

Gaggero, L., P. Spadea, L. Cortesogno, G. N. Savelieva and A. N. Pertsev, Geochemical investigation of the igneous rocks from Nurali ophiolite mélange zone, Southern Urals, *Tectonophysics*, 276, 139–161, 1997.

Garuti, G., G. Fershtater, F. Bea, P. Montero, E. V. Pushkarev and F. Zaccarini, Platinum-group elements as petrological indicators in mafic-ultramafic complexes of the central and southern Urals: preliminary results. *Tectonophysics*, 276, 181–194, 1997.

Glodny, J., Pease, P. Montero, H. Austrheim, A. Rusin and V. Lennykh, The Marun-Keu metamorphic complex, Polar Urals: protolith evolution and its geodynamics significance, *Abstracts INTAS EUROPROBE TIMPEBAR-Uralides Workshop, Oct. 19–22, 2000*, St. Petersburg, 13–14, 2000.

Gee, D. and V. Pease, Europrobe TIMPEBAR–Polar Urals Transect, Excursion 12–24 July, 1997, *EUROPROBE News*, 1, 4–7, 1997.

Ivanov, K. S., E. Anikina, A. A. Yefimov, E. V. Pushkarev, G. B. Fershtater and V. R. Shmelev, Uralian Platinum Belt, *Guidebook Int. Conf. "Paleosubduction Zones"*, *Yekaterinburg*, 95 pp., 1999.

Ivanov, K. S. and V. N. Puchkov, Geology of the Uralian Sakmara zone (new data) (in Russian), *Sverdlovsk*, 86 pp., 1984.

Ivanov, K. S., V. N. Puchkov and V. A. Babenko, Conodonts and Graptolites discovered in metamorphic complexes of the South Urals (in Russian), *Dokl. AN SSSR*, 310, 376–379, 1990.

Ivanov, K. S. and V. R. Shmelev, Platinum Belt of the Urals–magmatic trace of Early Paleozoic subduction (in Russian), *Dokl. RAN*, 347, 649–652, 1996.

Ivanov, O. K. and B. A. Kaleganov, K-Ar age of phlogopite pyroxenite of Svetlyi Bor massif (in Russian), *Ann. Rep. IGG UB RAS*, Yekaterinburg, 61–62, 1992.

Ivanov, S. N., A. S. Perfiliev, G. A. Smirnov, V. M. Necheukhin and G. B. Fershtater, Fundamental features in the structure and evolution of the Urals, *Am. J. Sci.*, 275, 107–130, 1975.

Irving, A. J., A review of experimental studies of crystal/liquid trace element partitioning, *Geochim. Cosmochim. Acta*, 42, 743–770, 1978.

Kamaletdinov, M. A. and T. T. Kazantseva, Allochthonous ophiolites of the Urals (in Russian), Nauka, Moscow, 168 pp., 1983.

Karson, J. A., Variations in structure and petrology in the Coastal Complex, Newfoundland: Anatomy of an oceanic fracture zone, in *Ophiolite and Oceanic lithosphere*, edited by I. G. Gass, S. J. Lippard and A. W. Shelton, 131–144, G. B. Alden Press, Oxford, 1989.

Kelemen, P. B. N. Shimizu and V. J. M. Salters, Extraction of mid-oceanic ridge basalt from the upwelling mantle by focused flow of melt in dunite channels, *Nature*, 375, 747–753, 1995.

Klyuzhina, M. L., Ordovician paleogeography of the Urals (in Russian), Nauka, Moscow, 189 pp., 1985.

Korinevskii, G., New data on the Late Ordovician stratigraphy and volcanism of the South Urals (in Russian), in *Pre-Ordovician History of the Urals*, UNTs AN SSSR, Sverdlovsk, 54–59, 1980.

Korinevskii, G., Paleozoic ophiolites of the Urals (in Russian), *Geotektonika*, 2, 34–44, 1989.

Krasnobayev, A. A., V. A. Davydov, V. I. Lennykh, N. V. Cherednichenko and B. I. Kozlov, Zircon and rutile ages of the Maksyutov complex (preliminary data) (in Russian), *Ann. Rep., IGG UB RAS*, Yekaterinburg, 13–16, 1995.

Leech, M. L. and D. F. Stockli, The late exhumation history of the ultrahigh-pressure Maksyutov Complex, south Ural Mountains, from new apatite fission track data, *Tectonics*, 16, 153–167, 2000.

Leech, M. L. and W. G. Ernst, Petrotectonic evolution of the high- to ultrahigh-pressure Maksyutov Complex, Karayanova area, south Ural Mountains: Structural and oxygen isotope constraints, *Lithos*, 52, 235–252, 2000.

Lennykh, V. I., Zoning and stages of metamorphism in miogeo-syncline area of the South Urals, (in Russian), in *Absolute Dating of Tectono-magmatic Cycle and Ore-forming Stages*, edited by S. N. Ivanov, Nauka, Moscow, 169–173, 1966.

Lennykh, V. I. and P. M. Valizer, High-pressure metamorphic rocks of the Maksyutov complex (Southern Urals), *Forth Int. Field Symp. Guide, Novosibirsk*, 64 pp., 1999.

Lennykh, I., P. M. Valizer, R. Beane, M. L. Leech and W. E. Ernst, Petrotectonic evolution of the Maksyutov complex, Southern Urals, Russia: Implications for ultrahigh-pressure metamorphism, *Int. Geol. Rev.*, 37, 584-600, 1995.

Levinson-Lessing, F. Y., Geology of Ujno-Zaozersky Dacha and Denezhkin Kamen in the North Urals (in Russian), *Trans. St. Petersburg Nat. Hist. Soc.*, 30, 257 pp., 1900.

Lupanova, N. P. and V. V. Markin, Greenstone Sequences in the Sob-Voykar Synclinorium (Eastern slope of the Polar Urals) (in Russian), Nauka, Moscow, 175 pp., 1964.

Matte, P., H. Maluski, R. Gaby, A. Nicholas, P. Kepezhinkskas and S. Sobolev, Geodynamic model and $^{39}Ar/^{40}Ar$ dating for generation and emplacement of high pressure metemorphic rocks in SW Urals, *C. R. Acad. Sci. Paris*, 317, Series II, 1667–1674, 1993.

Michard, A., F. Boudier and V. Goffe, Obduction versus subduction and collision in the Oman case and other Tethyan settings, in *Ophiolite Genesis and Evolution of the Lithosphere*, edited by T. J. Peters, 447–467, 1991.

Moores, E. M., Origin and emplacement of ophiolites, *Rev. Geophys.* 20, 735–760, 1982.

Narkisova V. V., L. V. Sazonova and A. A. Nosova, Magmatic bodies within the sequence of the Uralian ultradeep well (in Russian), in *Paleosubduction Zones:Tectonics, Magmatism, Metamorphism, Sedimentology, UrB RAS*, Yekaterinburg, 104–106, 1999.

Nicolas, A., *Structure of Ophiolites and Dynamics of Oceanic Lithosphere*, Kluwer Acad. Pub., 367 pp., 1989.

Nicolas, A. and F. Boudier, Mapping oceanic ridge segments in Oman ophiolites, *J. Geophys. Res.*, 100, 6179–6197, 1995.

Nicolas, A., G. Ceuleneer, F. Boudier and M. Misseri, Structural mapping in the Oman ophiolites: mantle diapirism along the oceanic ridge, *Tectonophysics*, 151, 27–56, 1988.

Nicolas, A., F. Boudier and B. Ildefonce, Evidence from the Oman ophiolite for active mantle upwelling beneath a fast-spreading ridge, *Nature*, 370, 51–53, 1994.

Pertsev, A. N. and A. A. Saveliev, Gabbro-amphibolites at the sole of Kempersay ophiolite massif, South Urals: petrological and tectonic aspects of their origin (in Russian), *Geotektonika*, 3, 21–35, 1994.

Pertsev, A. N., P. Spadea, G. N. Savelieva and L. Gaggero, Nature of the transition zone in Nurali ophiolite, soutern Urals, *Tectonophysics*, 276, 163–180, 1997.

Pertsev, A. N., G. N. Savelieva and O. V. Astrakhantsev, Magmatic origin of the ultramafic-mafic association of the Kytlym massif, Platinum Belt of the Urals, *Petrology*, 8, 370–393, 2000.

Peyve, A. V., Oceanic crust of past geological time (in Russian), *Geotektonika*, 4, 5–13, 1969.

Peyve, A. V. (Ed.), *Tectonics of the Urals (Explanatory Notes to Tectonic Map of the Urals, 1:1 000 000)* (in Russian), Nauka, Moscow, 120 pp., 1977.

Peyve, A. V. (Ed.), *Geology of the Philippine Sea Floor* (in Russian), 259 pp., Nauka, Moscow, 1980.

Puchkov, V. N., Paleooceanic structures of the Urals (in Russian), *Geotektonika*, 3, 18–33, 1993.

Puchkov, V. N., *Paleogeodynamics of the South and Central Urals* (in Russian), Ufa, 144 pp., 2000.

Pushkarev, E. V. and G. B. Fershtater, *Petrology of the Uktus Dunite-Clinopyroxenite-Gabbro Massif (The Middle Urals)* (in Russian), 296 pp., IGG UB RAS, 2000.

Rassios, A., E. M. Moores and H. W. Green, Magmatic structure and stratigraphy of the Vourinos ophiolite cumulate zone, Northern Greece, *Ofioliti*, 8, 377–410, 1983.

Ronkin, Yu. L., K. S. Ivanov, V. R. Shmelev and O. P. Lepekhina, On the problem of isotopic dating of the Uralian Platinum Belt: first Sm-Nd data, Geology and mineral resources of the Western Urals (in Russian), PGU, Perm, 66–68, 1997.

Rumiantseva N. A., G. A. Ushkova, K. L. Shmeleva and A. A. Kukui, Silurian boninitic unit of the Urals, (in Russian) *Dokl. Akad. Nauk SSSR, 304*, 4, 947–951, 1989.

Ryazantsev, A. V., S. V. Dubinina and L. A. Kurkovskaya, Ordovician chert-basalt complex and ophiolites of the South Urals (in Russian), *General and Regional Problems of Geology*, GEOS, Moscow, 5–23, 1999.

Saveliev, A. A., O. V. Astrakhantsev, A. L. Knipper, A. Ya. Sharaskin and G. N. Savelieva, Structure and deformation phases of the Northern Terminus of the Magnitogorsk Zone, Urals, (in Russian), *Geotektonika*, 3, 38–50, 1998.

Saveliev, A. A. and S. G. Samygin, Ophiolite allochthons of the Subarctic and Polar Urals (in Russian), in *Tectonic Evolution of the Earth Crust and Faults*, edited by Yu. M. Pushcharovsky and A. L. Janshin, Nauka, Moscow, 9–30, 1979.

Saveliev, A. A. and G. N. Savelieva, The Kempersay ophiolite massif: Main features of the structure and composition (in Russion), *Geotektonika*, 6, 57–65, 1991.

Saveliev, A. A., A. Ya. Sharaskin and M. D'Orazio. Plutonic to volcanic rocks of the Voykar ophiolite massif (Polar Urals): Structural and geochemical constraints on their origin. *Ofioliti*, 24, 21–30, 1999.

Saveliev, A. A., E. V. Bibikova, G. N. Savelieva, P. Spadea, A. N. Pertsev, J. Scarrow and T. I. Kirnozova, Garnet pyroxenites of the Mindyak massif, South Urals: Age and environment of their formation (in Russian), *Bull. MOIP, Ser. Geol.*, 76, 22–29, 2001a.

Saveliev, A. A., G. N. Savelieva, I. I. Babarina and N. L. Chaplygina, Tectonic contribution to the layering in dunite-pyroxenite bodies of the Platinum Belt massifs: The Nizhnii Tagil massif of the Urals as an example (in Russian), *Geotektonika*, 6, 2001b (in press).

Savelieva G. N., *Gabbro-ultrabasite assemblages of Uralian ophiolites and their analogues in the recent oceanic crust* (in Russian), Nauka, Moscow, 242 pp., 1987.

Savelieva, G. N. and R. W. Nesbitt, A synthesis of the stratigraphic and tectonic setting of the Uralian ophiolites, *J. Geol. Soc., London*, 153, 525–537, 1996.

Savelieva, G. N., A. N. Pertsev, O. Astrakhantsev, E. A. Denisova, F. Boudier, D. Bosch and A. V. Puchkova, The Kytlym Pluton, North Urals: Structure and emplacement history, *Geotectonics*, 33, 119–141, 1999.

Savelieva, G. N., A. Ya. Sharaskin, A. A. Saveliev, P. Spadea and L. Gaggero, Ophiolites of the Southern Uralides adjacent to the East European continental margin, *Tectonophysics*, 276, 117–137, 1997.

Scarrow, J. H., G. N. Savelieva, J. Glodny, P. Montero, A. Pertsev, L. Cortesogno and L. Gaggero, The Mindyak Palaeozoic lherzolite ophiolite, Southern Urals: Geochemistry and geochronology, *Ofioliti*, 24, 239–246, 1999.

Sengor, A. M. C., B. A. Natal'in and V. S. Burtman, Evolution the Altaid tectonic collage and Palaeozoic crustal growth in Eurasia, *Nature*, 364 (6235), 299–307, 1993.

Seravkin, I. B. and Z. I. Rodicheva, Kraka–Mednogorsk paleovolcanic belt (in Russian), Geol. Inst., Ufa, 53 pp., 1990.

Sharma, M., G. J. Wasserburg, D. A. Papanastassiou, J. E. Quik, E. V. Sharkov and E. E. Las'ko, High $^{143}Nd/^{144}Nd$ in extremely depleted mantle rocks, *Earth Planet. Sci. Lett.*, 135, 101–114, 1995.

Skorniakova N. S. and M. I. Lipkina, Mafic and ultramafic rocks of the Mariana trench, (in Russian), *Okeanologia*, 15, 1063–1066, 1975.

Sokolov, V. B., Crustal structure of the Urals (in Russian), *Geotek-tonika*, 5, 3–19, 1992.

Suhr, G., Upper mantle peridotite in the Bay of Island Ophiolite, Newfoundland: Formation during final stages of a spreading center?, *Tectonophysics*, 206, 31–53, 1992.

Sun, S. and W. F. McDonough, 1989, Chemical and isotopic systematic of oceanic basalts: Implications for mantle composition and processes: edited by A. D. Saunders and M. J. Norry, *Geol. Soc. Spec. Publ.*, 313–345, 1989.

Spadea, P., L. Ya. Kabanova and J. H. Scarrow, Petrology, geochemistry and geodynamic significance of Mid-Devonian boninitic rocks from the Baimak-Buribai area (Magnitogorsk Zone, Soutern Urals), *Ofioliti*, 23, 17–36, 1998.

Spadea, P. and J. H. Scarrow, Early Devonian boninites from the Magnitogorsk Arc, Soutern Urals (Russia): Implication for early development of a collisional orogren, *Ophiolites and Oceanic Crust: New Insights From Field Studies and Ocean Drilling Priogramm*, edited by Y. Dilek, E. Moors, D. Elthon and A. Nicolas, *GSA Mem.*, 349, 2000.

Torsvik, T. H., M. A. Smethurst, R. Van der Voo, A. Trench, N. Abrahamsen, and E. Halvorsen, Baltica: A synopsis of Vendian-Permian paleomagnetic data and their paleotectonic implications, *Earth Sci. Rev.*, 33, 133–152, 1992.

Vysotskii, N. K., Platinum locations in the Isovskii and Nizhnii-Tagil area of the Urals (in Russian), *Trans. Geol. Committee*, 62, 1–694, 1913.

Yazeva R. G. and V. V. Bochkarev, *The Voykar Volcano-Plutonic Belt* (in Russian), Ural. Nauch. Tsentr USSR, Sverdlovsk, 156 pp., 1984.

Yefimov, A. A., High-temperature tectonics of the Urals ultramafics and gabbro (in Russian), *Geotektonika*, 1, 14–24, 1977.

Yefimov, A. A., *Gabbro-hyperbasite complexes of the Urals and ophiolite problem* (in Russian), Nauka, Moscow, 230 pp., 1984.

Yefimov, A. A., L. P. Yefimova and V. I. Maegov, Tectonics of the Uralian Platinum Belt: Mineral Assemblages and Structuring Mechanism (in Russian), *Geotektonika*, 3, 34–46, 1993.

Zavaritskii, A. N., Peridotite massif Raiiz in the Polar Urals (in Russian), ONTI, Moscow, 221 pp., 1932.

Zlobin, S. K. and I. K. Puzhin, Composition and structure of the ophiolites from the Tonga trench (in Russian), *Geochemistry of Magmatic Rocks*, Nauka, Moscow, 106–117, 1989.

Zonenshain, L. P., M. L. Kazmin and L. M. Natapov, Geology of the USSR: A plate-tectonic synthesis. *Am. Geophys. Union, Geodyn. Ser.*, 21, Washington, D.C., 242 pp., 1990.

G. N. Savelieva, A. Ya. Sharaskin, A. A. Saveliev, A. N. Pertsev, and I. I. Babarina: Geological Institute, Russian Academy of Sciences, Pyzhevskii per. 7, Moscow, 109017 Russia

P. Spadea, Dept. Georisorse Territorio, University of Udine, Via Cotonificio 114, I-33100 Udine, Italy

Massive Sulfide Deposits in the South Urals: Geological Setting Within the Framework of the Uralide Orogen

R. J. Herrington[1], R. N. Armstrong[1], V. V. Zaykov[2], V. V. Maslennikov[2], S. G. Tessalina[1,2], J.-J. Orgeval[3] and R. N. A. Taylor[4]

The south Urals is host to more than 80 Paleozoic volcanic-hosted massive sulfide (VMS) deposits developed in four distinct metallogenic zones. From west to east these are: the Sakmara zone, Main Uralian fault zone, and the east and west Magnitogorsk zones. In the Sakmara zone, the chemistry of host volcanic suites is consistent with development of the zone in a Silurian oceanic arc. The Main Uralian fault marks a line of paleosubduction and contains VMS deposits similar to those formed in modern mid-ocean ridge settings. The Magnitogorsk zones contain VMS deposits formed in a Devonian fore-arc, arc and inter-arc or proto-back arc setting. The earliest volcanics of the Magnitogorsk zone, the Baimak–Buribai formation, form a boninitic fore-arc sequence, evolving later to more calc-alkalic volcanics with evidence for a contribution from subducted slab to the volcanics. Later, and farther east of the subduction suture, a rifted, more mature arc setting formed where the Karamalytash formation volcanics developed in an inter-arc or proto-back arc setting. The Karamalytash formation shows little evidence of contribution from subducted sediment to the melt. Stratigraphically overlying the Baimak–Buribai formation, and partly time equivalent to the Karamalytash formation, is the Irendyk formation. The Irendyk formation is VMS-poor, but contains abundant epiclastic volcanosediments and epithermal-like gold-barite deposits, indicative of shallower sea conditions. The Irendyk formation appears to form a long linear geographic feature, perhaps marking the line of an emerging arc sequence behind which the Karamalytash formation

[1]The Natural History Museum, London, UK
[2]Institute of Mineralogy, Miass, Russia
[3]BRGM, Orleans, France
[4]Southampton Oceanography Centre, Southampton, UK

Mountain Building in the Uralides: Pangea to the Present
Geophysical Monograph 132
10.1029/132GM09

developed in a rift. Previous authors suggest that the west and eastern Magnitogorsk zones developed as separate arcs, but the arc-like volcanics in the east Magnitogorsk zone may simply indicate the migration of the volcanic arc eastwards as the East European craton approached the Main Uralian fault.

1. INTRODUCTION

Volcanic-hosted massive sulfide deposits (VMS deposits) occur in rocks of all ages since the Archaean, and 80% of known deposits occur in arc-related rocks whilst the bulk of the remainder are hosted in mid-ocean ridge spreading zones [*Franklin* et al., 1981]. These settings have been directly compared to active spreading zones in the East Pacific and mid-Atlantic [e.g., *Rona* et al., 1993] and to rifted arc settings [e.g., *Ohmoto and Skinner*, 1983].

VMS deposits worldwide have been classified on the basis of metal content [*Large*, 1992; *Franklin* et al., 1981], host rock lithology [*Sangster and Scott*, 1976] and tectonic setting [*Sawkins*, 1976]. All of these classifications show a large degree of overlap, since studies of modern systems show that the bulk of metals for the deposits are derived by the leaching of the footwall lithologies. In most VMS systems the footwall is dominated by volcanic sequences, the chemistry of which is generally a reflection of the broader tectonic setting. It follows, therefore, that a classification based on the dominant footwall host-rock sequences or metal contents will give similar classifications. This also means that the chemistry and nature of VMS deposits themselves may be used as evidence for the tectonic setting of their formation in complex terrains. Table 1 shows a simple composite table

of deposit classification incorporating the Urals classifications [largely after *Barrie and Hannington*, 1999 with data from *Ivanov and Prokin*, 1988, 1992; *Franklin*, 1993; *Prokin and Buslaev*, 1999; *Smirnov*, 1988; *Zaykov* et al., 2000].

This paper presents an outline of the geodynamic setting of the major volcanic-hosted massive sulfide (VMS) deposits of the south Urals as deduced from their host volcanic packages. The paper briefly reviews the current literature on the geological setting of key VMS deposits in the south Urals. On the basis of petrochemistry we attempt to define the volcanic setting of the diverse VMS deposit types in a transect from the allocthonous Sakmara zone in the west and across the Magnitogorsk arc sequence. We show that different deposit types are clearly related to their host volcanic packages and that on a regional basis the variation of volcanic chemistry can be linked to regional tectonic setting. Lead isotope data supports different lead sources in the footwall volcanic sequences which are contributing metals to the hydrothermal systems of contrasting orebody types. Models for the development of the south Urals have incorporated modern plate tectonic models since the work of *Borodaevskaya* et al. [1977]. This tectonic framework was embraced and developed in later papers [*Zonenshain* et al., 1990; *Zonenshain and Kuz'min*, 1993; *Yazeva and Bochkarev*, 1995]. In the Russian

Table 1. Classification of VMS deposits of the Urals compared to classification of *Barrie and Hannington* [1999]. Data on Urals deposits sourced from *Prokin and Buslaev* [1999]; *Prokin* et al. [1992].

Barrie and Hannington [1999] Classification	Key Metal Contents	Uralian Classifications (alternates in brackets)	Urals Deposits
	Cu (Ni, PGE)	Atlantic-type	Ishkinino, Dergamish, Ivanovka
Mafic	Cu-Zn(Au)	Cyprus-type	Buribai, Letnye, Djarly Asha
Bimodal-mafic	Cu-Zn(Au,Ag)	Urals-type (Abitibi-type)	Yaman Kasy, Makan, Sibay, Uchaly, Uzelga, Moldezhnoe, Gai, Podolskoe
Mafic-siliciclastic	Cu(Zn, Ag)	Besshi-type	Mauk, Letnye
Bimodal-felsic	Cu-Pb-Zn-Ag(Au)	Baimak-type (Kuroko-type)	Balta Tau, Alexandrinka, Terensay, Bakr Tau, Uvariag
Bimodal-siliciclastic	Cu-Zn-(Pb, Ag, Au)	(Bathurst or Iberian Pyrite Belt-type)	

literature, the linking of such plate tectonic models to metallogenesis was outlined by *Necheukin* et al. [1986] and *Koroteev* [1997]. Until recently, papers in English that tackle the geodynamic aspects of the metallogeny of the Urals were scarce [e.g., *Koroteev* et al., 1997; *Sazonov* et al., 2001]. However, since the early 1990's, papers in English linking development of the Palaeozoic ocean, arc volcanism and massive sulfide formation have been published [e.g., *Kontar and Libarova*, 1997; *Prokin* et al., 1998; *Prokin and Buslaev*, 1999; *Zaykov* et al., 2000; *Maslennikov* et al., 2000; *Gusev* et al., 2000]. We discuss the formation of the VMS deposits in terms of the broadly accepted model invoking eastward dipping subduction [e.g., *Brown* et al., 1998], considering the case for a double arc with both westward and eastward subduction with interarc spreading proposed by some [e.g., *Surin*, 1991]. Our analysis suggests a model of a single, but possibly eastward migrating arc sequence that may explain the features of the Magnitogorsk arc in line with other authors [*Seravkin* et al., 1994]. A geological map of VMS deposits of the south Urals (Figure 1) shows that there is no simple geographical pattern in terms of deposit type, although some authors have suggested that the general distribution of deposits may support particular models of arc architecture [e.g., *Zaykov*, 1995].

2. GENERAL FEATURES OF VMS DEPOSITS IN THE SOUTH URALS

The Palaeozoic volcanic sequences of the middle and south Urals have over 80 recorded VMS deposits with at least 30 significant deposits in the south Urals alone [*Ivanov* et al., 1960; *Prokin and Buslaev*, 1999]. These deposits show a diverse range of volcanic associations, deposit morphologies, alteration, and ore mineralogies. Russian authors have classified the deposits into five broad types; a) Cyprus type, b) Urals type, c) Baimak type, d) Besshi type, and e) Atlantic type [*Smirnov*, 1988; *Zaykov*, 1991; *Zaykov* et al., 2000].

Cyprus type deposits in the Urals, as described in the literature, are hosted within dominantly tholeiitic basalt volcanic sequences where the ores are dominated by pyrite and chalcopyrite with subsidiary sphalerite. The deposits show characteristic enhanced Co and rather low Au values. These Cyprus type deposits show fairly simple quartz-chlorite and quartz-sericite-chlorite alteration zones in the footwall, developed in largely basaltic sequences [*Prokin and Buslaev*, 1999]. Many of these deposits are linked to mafic dominated sequences, or "spreading basins" [*Zaykov* et al., 2000] interpreted as

being primitive arc-related rift settings [*Ivanov* et al., 1973], similar to those defined in ophiolite sequences [e.g., *Galley and Koski*, 1999]. Cyprus type deposits are described in the middle Urals Ordovician to Devonian sequences, but are most notably developed in the Devonian in the regions of Dombarovsk and West Mugodzhary in the south Urals (Figure 1). Deposits in the Buribai region in the south Urals, immediately east of the Main Uralian fault, are also described as Cyprus type [*Prokin and Buslaev*, 1999]. No recent work has been carried out on these massive sulfide deposits, as none of them is currently accessible for study. Previous work shows them to be entirely hosted in basaltic and gabbroic rocks of lowermost Baimak–Buribai formation and are probably some of the oldest sequences of the Magnitogorsk zone east of the Main Uralian fault [*Prokin*, 1977; *Prokin* et al., 1985].

The Urals type deposits are the most abundant in the volcanic sequences east of the Main Uralian fault and form by far the most important group of active mines such as Gai, Sibay, Uchaly, Uzelga, Molodezhnoye and the large undeveloped Podolskoye deposit (Figure 1). The host sequences to these deposits are bimodal sodic rhyolitic basalt assemblages with mineralization usually located at the evolution from generally basic to acid volcanism. Urals type deposits are themselves diverse, ranging from single sulfide lenses to complex stacked orebodies [*Prokin and Buslaev*, 1999]. Mineralogically, the orebodies are dominated by pyrite-chalcopyrite-sphalerite, although they often form complex, stacked lenses and are mineralogically zoned with pyrrhotite common in the lower lenses and sulfosalts present in upper lenses. Characteristic zonation from the base of the deposit to the top grades from coarse pyritic ore to coarse-grained chalcopyrite (porphyritic)-pyrite ore to sphalerite-chalcopyrite-pyrite ore at the top and on the flanks [*Prokin and Buslaev*, 1999]. The sphalerite-rich zones have significantly elevated Ag, Ge, Cd values as well as higher reported gold grades [*Maslennikov*, 1999]. The Urals deposits can form extremely large sulfide masses and there are at least five orebodies (Gai, Sibay, Uchaly, Yubilenoe, Podolskoye) larger than 100 million tonnes of minable sulfides [*Herrington* et al., 1999]. Yaman Kasy is an example of a simple single lens Urals type deposit developed in the Mednogorsk region of the Sakmara zone. The body forms a simple lens of massive sulfide developed over a package of altered felsic volcanics (Figure 2). The sulfide lens clearly formed on the seafloor, as is shown by the presence of a distinctive vent fauna [*Zaykov* et al., 1995; *Little* et al., 1997] and vent chimney debris [*Herrington* et al., 1998]. Footwall alteration is simple, dominated by quartz-chlorite

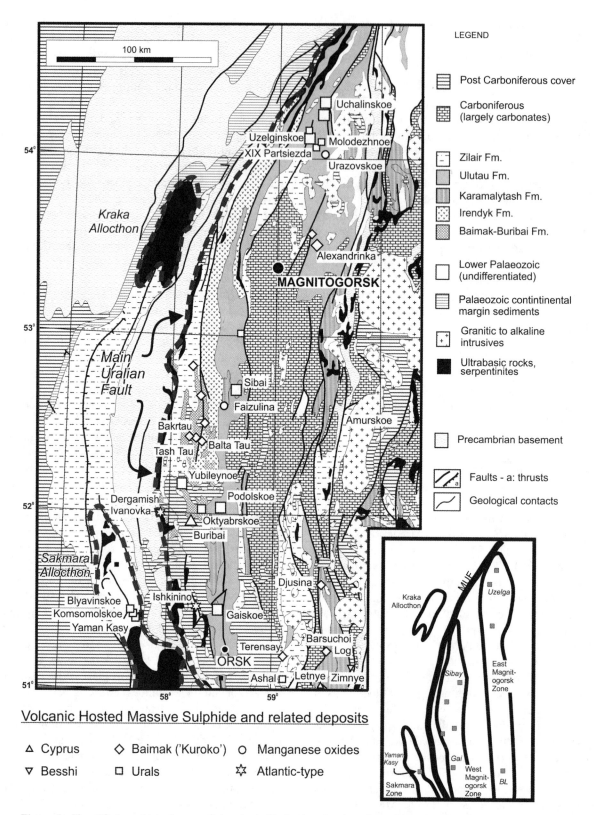

Volcanic Hosted Massive Sulphide and related deposits

△ Cyprus   ◇ Baimak ('Kuroko')   ○ Manganese oxides

▽ Besshi   □ Urals   ☆ Atlantic-type

Figure 1. Simplified geological map of the south Urals showing location of key VMS deposits (compiled from *Ivanov and Prokin* [1992] after digital basemap by *Shatov* et al. [2001]).

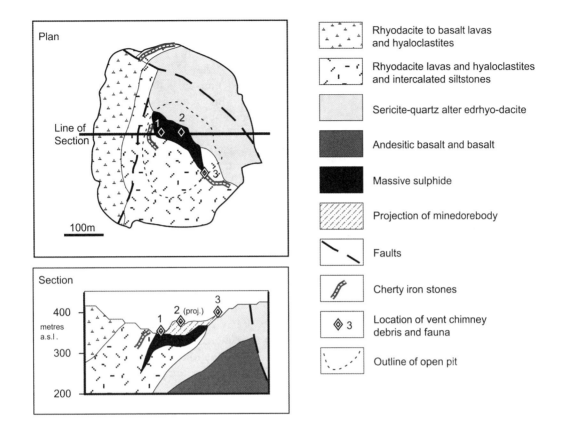

Figure 2. Simplified geology of the Silurian Yaman Kasy Urals type VMS deposit (after *Herrington* et al. [1998]).

and quartz-chlorite sericite. Uzelga lies south of Uchaly and is a typical large complex Urals type orebody developed in the Karamalytash formation beneath the basal section of the Ulutau formation (Figure 3). A general evolution of the volcanic pile from a basal basaltic pile (more than 1.5 km thick) into a sequence dominated by rhyolites and dacites is seen [*Petrov and Kazakova*, 1978], culminating with a series of fossiliferous limestones and cherts marking the cessation of volcanism. A strong structural control to the volcanism is evident in section (Figure 3) and there is good evidence of a semi-regional control to both formation of the volcanics and mineralized systems. Even the giant Urals type deposits show evidence of a seafloor origin. At Sibay, a community of vent fauna is testament to the deposit having formed at the seafloor [*Little* et al., 1998].

Baimak type deposits are typified by orebodies developed around the town of Baimak in the western part of the Magnitogorsk zone (Figure 1) [*Sopko*, 1973]. Baimak type deposits have also been referred to as

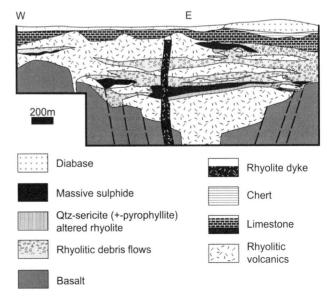

Figure 3. Simplified geological section of the Devonian Uzelga Urals type VMS deposit (after *Borodaevskaya* et al. [1977]).

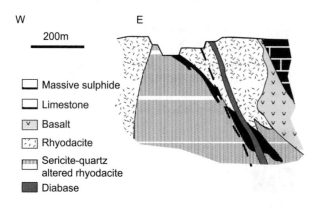

Figure 4. Simplified geological section of the Devonain Alexandrinka Baimak type VMS deposit (after *Tessalina* et al. [1998]).

"Kuroko type" by some authors [*Prokin and Buslaev,* 1999], due to some similarities with the deposits of the Hokuroko district of Japan. The Bakr Tau and Tash Tau deposits, which occur at the transition from footwall basaltic units to hanging wall rhyolites, are considered typical Baimak type deposits [*Prokin and Buslaev,* 1999; *Kuleyshov and Zaykov,* 1998]. The deposits are very variable in morphology and style, often characterised by extensive footwall stockwork zones. Zinc is much higher than copper in the deposits and the ore mineralogy is much more diverse, with abundant Ba, Pb, Ag, Au and Cd. The common association of what are interpreted as sub-volcanic diorite to granitic intrusions in the footwall is invoked [*Prokin and Buslaev,* 1999] leading to the assertion that they are "intermediate between massive sulfides and porphyry copper deposits". Marginal alteration patterns are more complex in the Baimak type deposits, with an abundance of carbonate alteration in many of the deposits. Baimak type deposits are also found in the eastern Magnitogorsk zone at Alexandrinka and Terensay (Figure 1). The Alexandrinka deposit occurs in the eastern part of the Magnitogorsk zone (Figure 4), along strike from the Uchaly–Verkneuralsk area. The host rocks are dominantly felsic and show the complex zoned alteration features, including extensive peripheral carbonate alteration [*Tessalina* et al., 1999]. The orebody sulfides are dominated by sphalerite and chalcopyrite, galena is common, and the ores are relatively gold rich. Some Baimak type deposits, such as Balta Tau, can have extremely gold-rich parts to the orebody [*Holland,* 2001].

Besshi type deposits are represented by the Zimnye deposit in the eastern Magnitogorsk zone (Figure 1) and the Mauk deposit (to the north of Figure 1) where the sulfides are hosted in fine clastic sediments (shales). Very little is published on this deposit type, but they show some general similarities with modern sulfide deposits of the Escanaba trough [*Zierenberg* et al., 1994] and Middle Valley [*Mottl* et al., 1994].

The unusual sulfide deposits in the Main Uralian fault zone remain enigmatic due to the ultramafic associations, the high nickel and PGE contents, and the degree of deformation [*Tessalina* et al., 2001]. *Zaykov* et al. [2000] termed these deposits "Atlantic type", comparing them to modern sulfides formed on serpentinites like those in the Mid-Atlantic active Logatchev field [*Wipfler* et al., 1999].

3. TECTONO-STRATIGRAPHIC FRAMEWORK TO THE SETTING OF VMS DEPOSITS OF THE SOUTH URALS

3.1. Metallogenic Zones

Early on, the Main Uralian fault in the south Urals was recognised as the probable site of a former Paleozoic subduction zone [*Ovchinnikov and Baranov,* 1978]. Therefore, spatial relationships relative to the Main Uralian fault are highly significant with regard to tectonic setting. The south Urals have been subdivided by Russian authors into three major sub-parallel tectono-stratigraphic metallogenic belts that follow the trend of the Main Uralian fault. From west to east these zones are: the Tagil-Sakmara zone, the Magnitogorsk-Mugodzhary zone, and the East Uralian metallogenic zone [*Maslennikov* et al., 2000]. The Tagil and Sakmara zones, although described together in the literature based on their comparable ages [*Prokin and Buslaev,* 1999], actually form two geographically separate zones. The Tagil zone, which occurs in the Middle and Northern Urals is not described here. The Sakmara zone lies in the south, to the west of the Main Uralian fault (see Figure 1) and it contains the important ophiolite-hosted chromite deposits of Kempirsai in Kazakstan [*Savlieva* et al., 1997]. Along the line of, and immediately east of the Main Uralian fault, lies a complex zone comprising slivers of ophiolitic material, melange and volcano-sedimentary sequences of Silurian to Devonian age that contain the small Atlantic type massive sulfide deposits [*Zaykov* et al., 2000]. The Magnitogorsk-Mugodzhary zone, to the east of the Main Uralian fault, contains the bulk of the massive sulfide deposits in the south Urals. The northern part of Magnitogorsk-Mugodzhary zone is termed the Magnitogorsk zone (Figure 1). VMS deposits in the Magnitogorsk zone form the focus of this paper. This broad zone can be subdivided into a western and eastern Magnitogorsk zone with a largely Carboniferous age sedimentary trough developed between and on top of them. Farther south of the area on Figure 1, this trough passes into the Mugodzhary zone, which has been interpreted as a "spreading zone" [*Zaykov* et al., 1996].

3.1.1. Sakmara Zone. The Sakmara zone is an allocthon containing the Mednogorsk ore district. From biostratigraphic dates [*Maslov* et al., 1993], confimed by *Little* [unpubl.] and *Buschmann* et al. [2001], the Mednogorsk ore district is hosted in an early Silurian volcano-sedimentary sequence. The mineralogical age of mineralisation is restricted to a single K-Ar date of 421 ± 3 Ma from Yaman Kasy [*Herrington* et al., 1998], and may be considered a minimum age for the deposits. The Kempirsai chromite deposits to the south are hosted in harzburgitic units of an Ordovician to Silurian ophiolite sequence [*Ivanov*, 1988].

3.1.2. The Main Uralian Fault. The Main Uralian fault is a 2 to 10 km wide, east-dipping fault system that is continuous for more than 2000 km along the Uralide orogen. On the basis of reflection seismic data it has been interpreted to extend into middle and perhaps even the lower crust [*Juhlin* et al., 1998; *Ayarza* et al., 2000; *Brown* et al., 1998; *Tryggvason* et al., 2001]. In the southern Urals, the Main Uralian fault zone is a serpentinitic melange containing predominately Lower to Middle Devonian volcanic fragments derived from the Magnitogorsk arc, together with fragments of oceanic crust and mantle [*Puchkov*, 1997; *Brown* et al., 1998; *Prokin and Poltavets*, 1996]. The Main Uralian fault is thought to be the boundary between the accretionary complex and back-stop (Magnitogorsk arc) during Devonian subduction [e.g., *Brown* et al., 1998; *Brown and Spadea*, 1999]. It currently represents the suture between the East European Craton and Magnitogorsk arc [*Brown* et al., 1998; *Prokin and Buslaev*, 1998; *Zonenshain* et al., 1984; *Chemenda* et al., 1997; *Ayarza* et al., 2000]. Related to the Main Uralian fault is a narrow belt of ophiolitic material, which runs north-wards from the Sakmara zone to the Tagil zone. These ophiolites are considered by a number of authors to be genetically linked relicts of sea floor material formed during rifting and ocean formation in the Silurian [e.g., *Zaykov* et al., 2000; *Maslennikov* et al., 2000]. The Main Uralian fault zone is host to three important VMS deposits; Ishkinino, Dergamish, and Ivanovka. Re-Os isotope analysis from the ores indicates an isochron age of 364 ± 10 Ma [*Tessalina* et al., 2001]. This age may record isotopic resetting during events related to the collision of the East European Craton with the Magnitogorsk arc in the Middle Devonian.

3.1.3. Magnitogorsk Zone. To the east of the Main Uralian fault lie the Devonian to Carboniferous complexes collectively referred to as the Magnitogorsk system or zone [*Gusev* et al., 2000]. Work on the major magmatic complexes related to the closure of the Urals palaeocean indicate a north-south linearity of magmatic activity that is zoned, in a broad sense, from west to east both in time and chemistry [*Ferstater* et al., 1997]. This has been used as key evidence for the east dipping polarity of the subducted slab [e.g., *Ferstater* et al., 1997]. Volcanic activity in the Magnitogorsk zone is expressed as a series of volcano-sedimentary complexes that, in broad terms, also show a similarly age zonation from west to east [*Seravkin* et al., 1994].

There are few published radiogenic ages from the Magnitogorsk zone. Late granitoids which cut the volcanic stratigraphy are dated at around 362 Ma [*Ferstater* et al., 1997]. Ages from the volcanic rocks are generally confined to K-Ar ages from sericite [*Prokin* et al., 1992]. Probable Baimak–Buribai or Irendyk volcanic rocks hosting the Podolskoe deposit have been dated at 360 ± 24 Ma [*Prokin and Buslaev*, 1999] and those at the Gai deposit at 376 ± 13 Ma [*Nesterenko*, 1978]. Both of these ages are close to, or slightly younger than, the timing of arc-continent collision (~375 Ma) calculated from the Maksutov Complex [*Brown* et al., 2000]. Another deposit, Molodyezhnoe, is dated at 390 ± 2 Ma, which fits well with the biostratigraphic dates from the host Karamalytash formation. Other published K-Ar dates are generally too young (<350 Ma) and are likely to represent minimum ages due to the problems of resetting. Zircon U-Pb dating is underway in the Institute of Geology, Ekaterinburg and the University of Freiberg, but no published results are yet available.

Biostratigraphic ages determined from conodont-bearing cherts intercalated with the volcanic sequences indicate that the Magnitogorsk arc developed from the Early to Late Devonian (Emsian to Fammenian) [*Maslov* et al., 1993; *Artyushkova and Maslov*, 1998]. The Late Devonian, volcaniclastic Zilair formation is thought to post date the collision of the East European continent with Magnitogorsk arc (Figure 5) [*Brown* et al., 1998; *Brown and Spadea*, 1999]. Conodont-bearing cherts put the Baimak–Buribai formation and its equivalents in the Gai region in the *serotinus* and *patulus* zones of the International Standard Conodont Scale, corresponding to uppermost Emsian (400–395 Ma). The Karamalytash formation and its eastern equivalents are in the *costatus*, *australis* and *kockelianus* zones, corresponding to uppermost Lower Eifelian to uppermost Eifelian (392–387.5 Ma). The Yarlikapovo chert, which overlies the Irendyk volcanic sequence, is in the *australis*

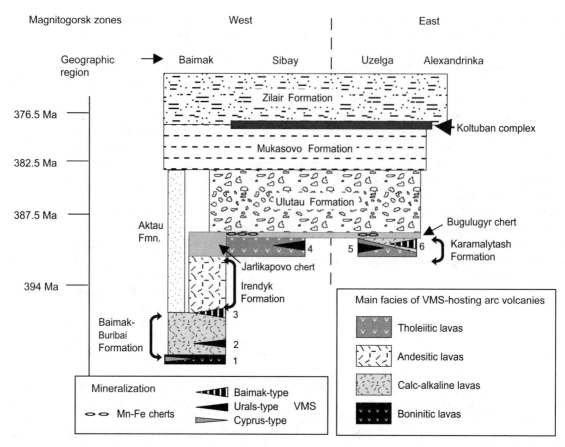

Figure 5. Schematic stratigraphic column of the Middle Devonian Magnitogorsk arc sequences. Numbered VMS deposit horizons 1: Buribai; 2: Makan; 3: Balta Tau; 4: Sibay; 5: Uzelga; 6: Alexandrinka (compiled from *Maslov* et al. [1993]).

and *kockelianus* zones [*Maslov* et al., 1993]. The Bugulygyr jasper, which overlies the Karamalytash formation, is correlated with the uppermost part of the Yarlikapovo chert. This chert unit forms the contact between the Karamalytash and Irendyk formations, and with the overlying Ulutau formation. In the area southwest of Sibay, at Faizulinskoe, manganese mineralisation is developed in a chert unit that overlies the Karamalytash formation in the north, and the Irendyk formation to the south, illustrative of this complex stratigraphic problem (Figure 1). Because of this, the exact age relationships between the Karamalytash and Irendyk formations still remains unclear and in places the Karamalytash formation is absent, with the time period represented only by cherts. The overlying Ulutau formation is dominated by flyschoid sediments and volcaniclastics linked to explosive volcanism that was occurring to the east. This explosive volcanism is believed to represent a shallowing of the ocean.

The Karamalytash formation in the western Magnitogorsk zone hosts the Sibay deposit and in the

east (in the Verkneuralsk region) the Uchaly, Uzelga and other deposits in a sequence of largely bimodal basalts and dacites with minor andesties [*Gusev* et al., 2000]. Almost without exception these Karamalytash-hosted deposits are of Urals type. However, farther south, in the eastern Magnitogorsk zone, the Alexandrinka deposit is interpreted as being hosted in the along-strike extension of the Karamalytash formation south east of Uchaly (Figure 1). Conodonts confirm that the Alexandrinka deposit is in rocks that are age equivalents to the Karamalytash formation [*Artyushkova and Maslov*, 1998]. However, the Alexandrinka deposit is mineralogically similar to the Baimak type VMS deposits developed in the Baimak–Buribai formation in the west, a deposit type to which it has been compared [*Tessalina* et al., 1999]. The host sequence at Alexandrinka contains a significant amount of andesitic volcanics [*Tessalina* et al., 1998] which may point to some differences in the nature of the Karamalytash formation in this region.

Clearly, further work on the nature of the igneous suites which host the diverse VMS deposits in the Uralide orogen would help unravel some of the uncertainties of tectono–stratigraphic setting.

4. IGNEOUS GEOCHEMISTRY OF THE HOST UNITS TO MASSIVE SULFIDE DEPOSITS

In order to better define the nature of the host volcanics in the Sakmara, Main Uralian fault, and Magnitogorsk zone, a west to east whole and trace element geochemical analysis transect was carried out. These analyses focussed on the host rocks to massive sulfide bodies of diverse type. Sampling is shown on the simplified stratigraphic columns indicated in Figure 6. Geographical locations relate to features on Figure 1. Analysis of major elements was carried out using ICP-AES and trace elements (including REE) by ICP-MS in the labs at the Natural History Museum, London.

4.1. Sakmara Zone

Three fresh samples from Silurian volcano-sedimentary rocks of the Mednogorsk ore district at Yaman Kasy were analysed. The samples were mafic extrusive pillow breccias and hyaloclastite, which infers a degree of sea water interaction. The rocks classify as basalt and show extremely low, flat chondrite normalised rare earth element (REE) patterns (Figure 7). One of the samples shows a distinctive negative Ce-anomaly, consistent with alteration by seawater. These data are in agreement with the published data of *Buschmann* et al. [2001] from the adjacent Blyava deposit. The geochemical data suggests that the basalt is similar to E-MORB-type basalt. Taken together with the data from felsic rocks in the Mednogorsk district [*Buschmann* et al., 2001], a bimodal nature is indicated for the host suite, as was previously suggested by *Seravkin* et al. [1992]. *Buschmann* et al. [2001] also note variability in the felsic units from tholeiitic to calc-alkaline, indicative of a degree of fractionation in sub-volcanic systems. These rocks are unlikely to be mid-ocean ridge generated and probably represent an arc assemblage.

4.2. East of the Main Uralian Fault

Twenty-five samples were collected from the western zone of the Magnitogorsk arc, in a traverse from the Baimak–Buribai formation to the Karamalytash formation. These results are shown in Tables 2, 3 and 4. Rocks from the Baimak–Buribai formation include samples from the Shankai river boninitic section previously reported in *Spadea* et al. [1998 and references therein]. Other samples come from massive sulfide deposits in the district and granitoids intruding the Baimak–Buribai formation at Bogachev. Rocks from the uppermost part

of the Baimak–Buribai formation immediately north of Baimak town were sampled together with pyroxene-bearing mafic volcanics from the Irendyk formation exposed northeast of Baimak. Three rocks were sampled at the Balta Tau VMS deposit, which may be at the contact between the Baimak–Buribai formation and overlying Irendyk formation [*Seravkin* et al., 1992]. Sections of the Karamalytash formation were sampled at Sibay where the unit is host to the giant VMS deposit of the same name, and across the Karamalytash formation type-section on Karamalytash mountain. A suite of volcanic rocks from the eastern zone of the Magnitogorsk arc were analysed from the Alexandrinka deposit, which is hosted in rocks that are equivalent in age to the uppermost Karamalytash formation.

4.2.1. Baimak–Buribai Formation. REE patterns from the Shankai river section show distinct primitive arc tholeiite and boninitic patterns (Figure 8). The next section, farther east associated with Baimak type massive sulfide deposits, is higher in the section of the Baimak–Buribai and here volcanics are bimodal, but are notable by the increased amount of pyroclastic and epiclastic rocks. REE patterns for the host rocks to these deposits shows dominant LREE enriched signatures typical of calc-alkaline arc rocks. Felsic extrusive and related sub-volcanic intrusives show similar patterns (Figure 9). The final sample, collected from what is interpreted as the uppermost section of Baimak–Buribai north of Baimak close to the massive sulfide deposit of Kul Yurt Tau, shows the most calc-alkaline signature of all.

4.2.2. Irendyk Formation. The Irendyk formation is best exposed in the region from east of Baimak town to the sequences of the Karamalytash formation exposed around the city of Sibay. In this section, the Irendyk formation comprises a fairly thick sequence of pyroxene phyric andesites and their hyaloclastic and epiclastic equivalents. The valley running north from the town of Baimak clearly follows a major structure and the contacts between the Baimak–Buribai and Irendyk formations may be largely tectonic in this region [e.g., *Brown* et al., 2001]. Pyroxene phyric basaltic andesite from immediately northeast of Baimak shows a very flat REE pattern, very similar to that of the lowermost Baimak–Buribai formation (Figure 10). REE patterns from basaltic andesites at Balta Tau reflect the presence of both calc-alkaline and flat normalised patterns. These patterns may confirm the location of Balta Tau at the contact between lower Baimak–Buribai and overlying Irendyk formations, in line with interpretations by previous authors [*Seravkin* et al., 1992]. The fresh basalts BT-18

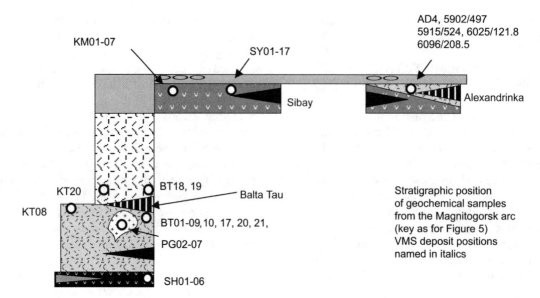

Figure 6. Stratigraphic location of samples for geochemical study in this paper. (Key similar to Figure 5).

and 19 are pyroxene-phyric, typical for the Irendyk formation seen in the north. In the hills above Balta Tau, the Yarlikapovo chert unit is exposed, which regionally forms the top of the Irendyk formation (Figure 5).

4.2.3. Karamalytash Formation. In the easternmost part of the west Magnitogorsk arc, the Karamalytash formation is host to the major Urals type massive sulfide

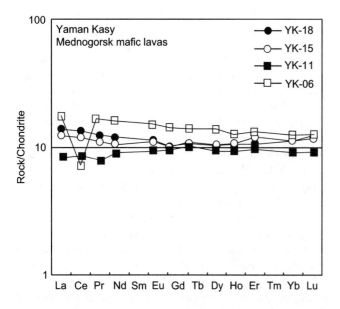

Figure 7. Chondrite normalised REE plot for Yaman Kasy volcanic rocks (data in Figures 7 to 14 normalised using chondrite values in *Anders and Grevesse* [1989]).

deposit of Sibay. The Sibay deposit was found in an exposed window of the Karamalytash formation beneath overlying Ulutau formation [*Prokin and Buslaev, 1999*]. Our data from the basalts and andesites of the Karamalytash formation type section (Figure 11) and sampling from the open pit of the Sibay deposit sequence itself (Figure 12) are very similar, and clearly distinct from volcanics of either the Baimak–Buribai or Irendyk formations. The patterns show depleted MORB-type REE patterns and whole-rock major patterns typical for volcanics developed in a rifted arc setting. Felsic rocks also show distinctive patterns (Figure 13) with Eu anomalies indicative of high-level fractionation.

4.3. Eastern Magnitogorsk Arc

Five samples from the host volcanics to the Alexandrinka deposit were analysed. Their REE patterns are shown in Figure 14. Although time equivalent to the uppermost Karamalytash formation of the Sibay and Uchaly region, and previously interpreted as the same volcanic suite [*Artyushkova and Maslov, 1998*], the REE patterns Alexandrinka samples contrast strongly with the rocks of the Karamalytash formation at Sibay. Light REE enrichment indicates a more arc-like signature, and a somewhat similar pattern to host rocks to the deposits in the Baimak region.

4.4. Summary

Geochemistry supports the notion that the mineralogy of the VMS deposits relates to the chemistry of the host volcanic suites, as is suggested in studies of VMS

Table 2. Major element wholerock geochemical data for volcanic host rocks from selected Urals VMS deposits.

Field No.	BT-09	BT-10	BT-17	BT-18	BT-19	BT-20	BT-21A	BT-21B	SH-01	SH-02	SH-03
Description	Epiclastic tuff	Epiclastic tuff	Silicified dacite	Green andesite	Purple andesite	Grey dacite	Dacite dyke	Dacite dyke	Dacite dyke	Gabbro	Gabbro
Formation	Upper BB	Upper BB	Upper BB	Irendyk	Irendyk	Upper BB	Upper BB	Upper BB	BB	BB	BB
Locality	Balta Tau	Balta Tau	Balta Tau	Balta Tau	Balta Tau	Balta Tau	Balta Tau	Balta Tau	Shankai	Shankai	Shankai
Weight%											
SiO_2	53.1	53.1	74.2	57.9	55.9	56.1	66.9	61.6	71.9	51.8	52.8
TiO_2	0.59	0.34	0.18	0.39	0.33	0.38	0.38	0.26	0.29	0.52	0.50
Al_2O_3	18.63	17.11	9.82	17.34	17.29	18.13	13.69	15.18	13.28	15.24	14.00
FeO	5.10	4.01	0.11	2.09	1.51	4.89	2.09	0.98	2.81	8.77	7.08
Fe_2O_3	1.99	2.94	6.05	3.05	4.87	1.24	1.55	5.28	1.21	3.11	0.84
MnO	0.14	0.11	0.01	0.14	0.16	0.13	0.21	0.20	0.10	0.23	0.26
MgO	5.95	5.52	1.90	4.07	2.97	4.27	4.70	2.64	1.61	8.20	9.11
CaO	3.65	8.24	0.05	3.05	4.27	1.39	1.68	9.36	0.45	1.91	4.65
Na_2O	1.32	1.99	0.29	1.79	1.91	4.46	1.96	0.39	5.97	2.98	4.03
K_2O	1.96	<	1.03	3.00	3.31	1.15	1.73	0.98	<	0.44	<
P_2O_5	0.17	0.12	0.14	0.14	0.12	0.15	0.16	0.06	0.12	0.11	0.09
LOI	6.96	5.89	6.97	6.33	5.87	7.49	4.78	3.06	1.79	6.07	5.61
Total	99.53	99.39	100.83	99.31	98.56	99.80	99.29	99.99	99.53	99.39	98.98
ppm											
Zr	44	26	21	22	27	30	32	30	74	26	28
Cr	94	<	106	<	<	<	<	<	<	97	249
Ni	<	<	<	<	<	<	<	<	<	<	<
Ba	1095	55	798	815	958	1124	363	202	20	222	20
Sr	269	289	18	147	123	110	167	1649	20	80	46

Table 2. (continued)

Field No.	SH-04	SH-05	SH-06	SY-01	SY-02	SY-03	SY-04	SY-05	SY-06	SY-07	SY-10
Description	Dacite	Qtz-rhyolite	Dolerite	Dacite	Dacite	Basalt	Basalt	Rhyodacite	Rhyodacite	Rhyolite	Dacite
Formation	BB	BB	BB	Kar	Kar	Kar	Kar	Kar	Kar	Kar	Kar
Locality	Shankai	Shankai	Shankai	Sibay o/pit	Sibay o/pit	Sibay o/pit	Sibay o/pit	Sibay o/pit	Sibay o/pit	Sibay o/pit	Sibay o/pit
Weight%											
SiO_2	66.8	73.5	52.6	68.6	82.4	45.8	48.3	71.1	76.9	75.4	77.0
TiO_2	0.64	0.24	0.59	0.33	0.17	0.60	0.63	0.20	0.20	0.18	0.21
Al_2O_3	13.02	12.97	15.52	13.40	7.08	16.82	17.99	11.62	10.75	9.88	10.71
FeO	4.46	1.16	5.38	3.60	1.92	8.03	7.52	2.54	1.98	1.83	0.96
Fe_2O_3	2.43	1.48	3.22	1.68	0.68	4.32	4.11	0.82	0.72	1.20	2.17
MnO	0.18	0.05	0.21	0.12	0.07	0.18	0.19	0.07	0.03	0.08	0.03
MgO	1.70	0.78	7.22	1.21	0.49	7.89	7.32	0.79	0.52	0.68	0.46
CaO	3.78	0.45	3.75	1.97	1.13	7.28	2.94	3.49	0.56	2.16	0.48
Na_2O	3.58	6.43	4.01	4.57	2.97	2.09	5.77	0.75	5.77	5.32	5.65
K_2O	v	0.05	v	0.58	0.38	v	v	3.18	v	v	v
P_2O_5	0.13	0.12	0.12	0.18	0.13	0.06	0.11	0.11	0.09	0.10	0.15
LOI	2.27	2.17	6.58	2.81	1.64	6.40	5.87	4.79	2.30	2.63	1.93
Total	99.02	99.40	99.19	99.04	99.11	99.54	100.72	99.49	99.84	99.44	99.76
ppm											
Zr	73	70	35	51	38	27	29	93	94	77	79
Cr	v	v	84	v	v	v	v	v	v	v	v
Ni	v	v	v	v	v	v	v	v	v	v	v
Ba	23	43	13	92	59	34	36	228	17	17	21
Sr	175	43	165	311	47	530	120	37	29	42	30

Table 2. (continued)

Field No.	SY-11	SY-12	SY-13	SY-14	SY-16	SY-17	KM-01	KM-02	KM-03	KM-04	KM-05
Description	Green dacite	H/W Green basalt	Basaltic Dyke	Basalt Hyaloclast.	Basalt Hyaloclast.	Basalt	Vesicular andesite	Andesite-basalt	Aphanitic Pillow Basalt	Aphanitic Pillow Basalt	Hyaloclast. andesite
Formation	Kar	Kar	Kar	Kar	Kar	Kar	Kar	Kar	Kar	Kar	Kar
Locality	Sibay O/pit	Sibay O/pit	Sibay O/pit	Sibay O/pit	Sibay O/pit	Sibay O/pit	Karamaly. mtn.	Karamaly. mtn.	Karamaly. mtn.	Karamaly. mtn.	Karamaly. mtn.
Weight%											
SiO_2	66.5	48.4	47.0	47.3	56.7	47.8	64.4	56.3	48.4	49.7	69.8
TiO_2	0.18	0.58	1.86	0.35	0.31	0.47	0.48	0.58	0.50	0.60	0.42
Al_2O_3	12.51	15.20	16.69	13.40	12.24	15.82	12.80	15.03	16.03	15.22	12.52
FeO	2.19	8.72	6.32	5.82	2.12	5.91	4.56	5.49	4.09	4.15	2.07
Fe_2O_3	1.19	2.51	2.98	2.89	5.52	4.21	2.40	3.44	6.75	7.65	3.46
MnO	0.13	0.17	0.17	0.13	0.17	0.43	0.12	0.13	0.15	0.14	0.10
MgO	0.79	7.07	8.51	4.66	2.95	8.31	2.70	5.12	4.66	4.55	0.82
CaO	5.99	6.07	9.21	11.71	5.06	4.20	1.69	3.23	9.97	9.19	3.13
Na_2O	0.34	4.23	3.11	1.40	2.38	3.41	3.63	3.99	3.70	3.67	5.08
K_2O	3.45	<	0.18	<	2.68	1.06	0.59	0.16	<	<	<
P_2O_5	0.12	0.10	0.26	0.10	0.11	0.08	0.19	0.13	0.13	0.05	0.25
H_2O+	7.19	6.34	3.09	10.96	10.01	8.62	6.12	5.51	4.82	4.17	2.12
Total	100.52	99.37	99.33	98.77	99.21	100.32	99.69	99.13	99.24	99.05	99.78
ppm											
Zr	77	23	189	6	9	20	48	24	20	24	30
Cr	<	76	272	<	<	224	<	<	<	<	<
Ni	<	<	138	<	<	<	<	<	<	<	<
Ba	193	27	69	60	278	240	123	62	22	17	30
Sr	38	76	413	42	72	85	73	100	52	43	35

Table 2. (continued)

	KM-06 Dacite Upper Kar Karamly. mtn.	KM-07 Rhyolite Upper Kar Karamly. mtn.	PG-02 Plagiogra-nite BB? Bogachev-sky massif	PG-03 Plagiogra-nite BB? Bogachev-sky massif	PG-04 Plagiogra-nite BB? Bogachev-sky massif	PG-07 Plagiogra-nite BB? Bogachev-sky massif	KT-20 Andesite Upper BB Kul Yurt Tau	KT-08 Andesite Upper BB Kul Yurt Tau
Field No. / Description / Formation / Locality								
Weight%								
SiO_2	77.3	84.2	72.4	71.0	73.5	71.9	53.8	64.2
TiO_2	0.30	0.21	0.27	0.26	0.26	0.26	0.32	0.38
Al_2O_3	10.32	8.59	13.31	13.58	13.12	13.64	15.52	15.96
FeO	0.78	0.74	1.14	1.04	0.86	1.06	3.29	3.65
Fe_2O_3	2.72	0.59	1.58	1.71	1.52	1.72	2.96	0.96
MnO	0.03	0.02	0.05	0.04	0.03	0.02	0.14	0.10
MgO	0.34	0.15	1.30	1.27	0.99	1.21	5.67	3.51
CaO	0.46	0.45	1.35	1.49	1.57	1.39	12.36	0.27
Na_2O	5.69	4.77	5.63	5.29	5.50	5.24	0.67	6.58
K_2O	<	<	0.64	0.23	0.13	0.23	<	<
P_2O_5	0.15	0.16	0.15	0.09	0.17	0.10	0.13	0.14
LOI	1.80	0.69	1.38	3.29	2.21	1.57	4.28	3.96
Total	99.92	100.55	99.21	99.30	99.87	98.38	99.17	99.68
ppm								
Zr	400	73	106	92	79	110	<	92
Cr	<	<	<	<	<	<	123	<
Ni	<	<	<	<	<	<	<	<
Ba	23	20	150	125	229	210	20	21
Sr	33	10	104	162	188	177	21	33

Table 3. Trace element geochemical data for volcanic host rocks from selected Urals VMS deposits.

ppm	BT-09	BT-10	BT-17	BT-18	BT-19	BT-20	BT-21A	BT-21B	SH-01	SH-02
Li	21.52	16.78	14.77	17.06	11.39	20.37	15.77	7.95	1.77	8.95
Be	0.42	0.09	nd	0.27	0.24	0.40	0.19	0.29	0.35	0.18
B	nd	nd	nd	1.18	nd	nd	nd	nd	nd	1.95
Ti	2854.64	1850.87	1016.40	2132.21	1882.44	2122.49	2103.10	1451.52	1622.24	2761.27
V	136.00	51.94	133.45	142.25	151.00	143.23	91.99	167.01	36.74	111.12
Mn	or	or	55.09	or	or	or	or	or	or	or
Co	21.38	22.89	20.70	21.18	18.50	22.35	19.23	11.32	4.31	44.05
Ni	40.45	19.30	31.01	21.21	18.74	21.48	45.90	48.56	1.01	41.80
Zn	713.64	77.67	262.60	135.15	131.63	5236.02	525.49	204.62	60.66	113.87
Ga	64.29	16.71	24.55	50.70	56.14	65.10	24.25	21.19	10.33	21.85
Ge	5.33	3.34	2.38	3.16	3.64	3.40	2.05	5.15	1.92	5.28
As	12.46	6.02	127.80	6.84	7.32	15.47	4.52	14.90	2.33	2.80
Se	nd	nd	nd	nd	nd	nd	nd	nd	nd	nd
Rb	25.93	or	9.79	32.19	35.21	15.64	27.34	15.58	or	4.79
Sr	or	or	17.37	136.98	121.06	110.18	129.59	or	23.50	82.22
Mo	nd	nd	5.65	nd	nd	nd	nd	nd	nd	0.04
Ru	nd	nd	nd	nd	nd	nd	nd	nd	nd	nd
Cd	0.10	0.08	3.10	nd	0.02	2.04	1.47	8.27	0.88	0.11
Sn	0.39	nd	nd	nd	0.28	0.39	nd	nd	nd	0.51
Sb	8.51	2.74	23.84	0.99	1.26	4.73	0.84	5.33	nd	nd
Te	nd	nd	0.13	nd	nd	nd	nd	0.19	nd	nd
Cs	0.46	nd	0.38	1.55	1.33	0.26	0.65	0.26	nd	nd
Ba	or	53.63	or	or	or	or	or	198.75	20.57	186.15
W	1.30	nd	1.37	nd	nd	0.74	nd	nd	nd	nd
Tl	0.99	nd	3.13	0.38	0.47	0.63	0.15	0.05	nd	nd
Bi	nd	nd	nd	nd	nd	nd	nd	0.27	nd	nd
Y	23.33	4.74	1.51	5.68	10.58	6.95	13.66	12.50	12.92	14.42
Zr	24.25	8.47	10.12	23.45	40.24	24.90	15.78	14.12	72.70	27.40
Nb	11.96	0.68	0.30	3.10	2.36	2.07	3.91	3.69	8.52	3.18
La	11.32	0.70	0.45	1.39	3.20	2.87	2.34	7.06	2.64	2.54
Ce	20.99	1.85	1.13	4.87	8.55	7.83	7.63	16.78	7.37	5.75
Pr	3.01	0.37	0.23	0.62	1.21	1.06	0.99	1.50	1.11	0.93
nd	13.51	1.66	0.95	2.81	5.52	4.72	4.48	6.12	4.95	4.49
Sm	3.56	0.63	0.32	0.94	1.57	1.40	1.48	1.61	1.56	1.51
Eu	1.23	0.27	0.11	0.42	0.56	0.67	0.52	0.64	0.38	0.74
Gd	4.15	0.76	0.30	1.09	1.78	1.52	1.86	2.24	1.90	2.04
Tb	0.74	0.20	0.11	0.25	0.37	0.31	0.42	0.40	0.41	0.44
Dy	4.35	0.99	0.40	1.31	2.06	1.59	2.55	2.26	2.48	2.60
Ho	0.93	0.25	0.13	0.31	0.48	0.35	0.59	0.50	0.59	0.61
Er	2.56	0.66	0.31	0.78	1.36	0.87	1.69	1.44	1.74	1.68
Tm	0.37	0.12	0.07	0.14	0.22	0.14	0.26	0.23	0.29	0.26
Yb	2.15	0.74	0.41	0.83	1.40	0.83	1.57	1.48	1.87	1.65
Lu	0.38	0.18	0.13	0.19	0.28	0.19	0.31	0.29	0.38	0.32
Hf	1.88	0.23	0.16	0.98	1.22	0.89	0.72	0.42	2.22	0.83
Ta	15.47	1.36	1.15	4.44	2.83	1.75	6.78	1.60	2.55	2.27
Pb	8.15	2.13	69.12	0.82	1.72	0.93	2.07	35.55	nd	0.30
Th	1.64	nd	nd	0.58	0.40	0.43	0.49	0.66	0.40	0.11
U	0.33	0.18	1.08	0.08	0.18	0.16	0.25	0.29	0.31	0.18

Table 3. (continued)

ppm	SH-03	SH-04	SH-05	SH-06	SY-01	SY-02	SY-03	SY-04	SY-05	SY-06	SY-07
Li	7.62	2.07	1.02	4.79	5.48	6.41	14.38	11.33	4.75	4.68	5.23
Be	0.31	0.23	0.50	0.15	0.14	0.07	nd	nd	0.39	0.22	0.16
B	nd	nd	nd	nd	nd	nd	nd	nd	nd	nd	nd
Ti	1888.38	3279.57	1353.45	3075.51	2662.94	948.00	3094.84	3302.99	1030.66	927.42	757.05
V	6.33	114.01	24.70	12.18	113.33	4.70	3.77	0.04	8.02	5.32	6.43
Mn	or	or	or	or	or	or	or	or	or	or	or
Co	1.50	11.37	3.52	28.90	28.97	0.55	44.17	42.80	0.73	0.46	0.78
Ni	1.10	2.49	1.94	35.20	59.82	0.58	25.50	22.44	1.13	0.69	0.78
Zn	112.88	168.60	43.26	645.04	133.44	75.60	107.47	126.55	126.20	84.77	87.61
Ga	17.70	12.54	13.20	15.11	13.47	8.37	16.87	13.37	30.54	12.37	11.51
Ge	4.57	3.94	1.74	4.39	3.65	2.06	5.42	5.05	2.88	2.02	2.21
As	8.89	1.62	1.10	1.94	2.04	4.20	24.87	19.15	4.40	1.98	2.35
Se	nd	nd	nd	nd	nd	nd	nd	nd	nd	nd	nd
Rb	7.99	or	or	or	or	5.32	or	or	42.60	or	or
Sr	or	or	or	137.32	or	47.46	or	105.74	38.92	25.25	44.08
Mo	nd	0.15	nd	nd	0.01	nd	0.01	0.06	0.22	nd	nd
Ru	nd	nd	nd	nd	nd	nd	nd	nd	nd	nd	nd
Cd	0.10	0.38	nd	0.77	0.04	nd	nd	nd	nd	0.03	0.13
Sn	0.07	0.07	nd	nd	nd	nd	nd	nd	0.38	0.03	nd
Sb	4.33	nd	nd	nd	nd	nd	2.68	0.97	2.33	nd	nd
Te	nd	nd	nd	nd	nd	nd	nd	nd	nd	nd	nd
Cs	nd	nd	nd	nd	nd	nd	nd	nd	1.01	nd	nd
Ba	or	23.93	44.20	13.24	or	53.15	44.06	38.72	219.08	8.92	10.23
W	nd	nd	nd	nd	nd	nd	nd	nd	nd	nd	nd
Tl	nd	nd	nd	nd	nd	nd	nd	nd	1.42	nd	nd
Bi	nd	nd	nd	nd	nd	nd	nd	nd	nd	nd	nd
Y	34.23	24.12	14.62	12.28	10.96	23.55	10.07	13.50	26.42	6.73	10.01
Zr	34.17	52.12	74.87	35.77	25.51	22.76	12.75	11.87	60.44	61.92	38.37
Nb	2.04	6.64	8.33	3.54	2.18	0.95	0.51	0.35	2.44	2.27	1.59
La	2.69	5.69	4.76	3.25	2.04	2.03	0.77	1.46	3.39	1.82	3.43
Ce	7.81	13.28	10.50	7.24	4.84	5.56	2.42	4.14	9.70	5.77	9.61
Pr	1.40	2.03	1.51	1.12	0.79	0.92	0.44	0.83	1.65	1.11	1.67
Nd	7.68	9.61	6.42	5.15	3.75	4.62	2.25	4.50	8.48	6.06	8.60
Sm	3.07	2.98	1.86	1.65	1.29	1.71	1.02	1.66	2.92	1.90	2.55
Eu	1.02	0.94	0.50	0.59	0.40	0.52	0.42	0.60	0.72	0.35	0.53
Gd	4.24	3.63	2.17	2.00	1.62	2.54	1.41	2.31	3.63	1.37	1.94
Tb	0.90	0.73	0.44	0.41	0.35	0.57	0.34	0.47	0.72	0.27	0.34
Dy	6.06	4.48	2.67	2.34	2.10	3.84	2.01	2.84	4.45	1.43	1.79
Ho	1.38	0.97	0.60	0.54	0.48	0.93	0.47	0.65	1.04	0.38	0.45
Er	4.12	2.83	1.77	1.51	1.36	2.95	1.32	1.83	3.25	1.26	1.40
Tm	0.63	0.43	0.30	0.24	0.22	0.47	0.22	0.27	0.54	0.24	0.26
Yb	4.05	2.68	1.95	1.50	1.36	3.02	1.37	1.64	3.66	1.83	1.85
Lu	0.69	0.46	0.38	0.30	0.28	0.55	0.28	0.32	0.70	0.42	0.40
Hf	1.42	1.68	2.25	1.06	0.74	0.77	0.38	0.38	2.18	2.18	1.30
Ta	1.60	2.16	2.29	1.94	1.45	1.21	1.59	1.37	1.55	1.40	1.42
Pb	3.37	0.16	nd	0.57	0.53	2.59	0.27	0.34	2.45	0.59	1.67
Th	0.26	0.56	0.86	0.20	0.07	0.04	nd	nd	0.37	0.17	0.23
U	0.27	0.34	0.44	0.19	0.12	0.14	0.06	0.06	0.38	0.25	0.28

Table 3. (continued)

ppm	SY-10	SY-11	SY-12	SY-13	SY-14	SY-16	SY-17	KM-01	KM-02	KM-03	KM-04
Li	3.30	3.56	13.14	18.17	13.38	14.73	46.19	10.17	16.52	12.36	9.08
Be	0.05	0.68	nd	1.11	nd	nd	nd	0.07	nd	0.07	0.08
B	nd	nd	0.05	nd	nd	nd	nd	nd	nd	nd	nd
Ti	964.42	903.85	3079.21	or	2001.91	1768.46	2440.03	2612.43	3137.59	2590.74	3139.15
V	16.97	5.10	nd	159.70	3.38	134.46	94.45	124.55	22.59	0.07	or
Mn	or	or	or	or	or	or	or	or	or	or	or
Co	2.15	0.63	42.66	41.66	38.80	23.07	55.10	15.30	37.70	34.51	35.00
Ni	2.43	1.56	39.53	146.95	22.46	14.67	37.84	5.86	16.23	22.99	18.06
Zn	46.49	86.99	107.74	85.02	81.48	79.42	129.45	122.35	102.20	90.89	88.29
Ga	9.54	25.03	14.42	17.76	14.43	20.64	21.49	18.80	13.07	16.94	15.85
Ge	1.62	3.16	5.25	4.53	3.88	3.34	4.03	2.92	3.60	5.17	5.44
As	2.65	4.84	28.77	13.26	27.00	3.80	3.84	3.84	3.71	6.77	10.20
Se	nd	nd	nd	nd	nd	nd	nd	nd	nd	nd	nd
Rb	or	34.62	or	3.08	or	48.23	20.01	5.40	5.78	3.21	2.88
Sr	20.27	33.47	73.42	or	48.83	75.32	83.03	63.93	94.39	55.45	43.94
Mo	nd	nd	0.05	0.39	nd	nd	nd	nd	nd	0.13	0.09
Ru	nd	nd	nd	nd	nd	nd	nd	nd	nd	nd	nd
Cd	nd	0.11	0.02	0.09	0.03	0.05	nd	nd	nd	0.11	nd
Sn	0.78	0.44	nd	0.92	nd	nd	nd	nd	nd	2.86	nd
Sb	0.26	0.41	2.76	nd	nd	nd	nd	nd	nd	nd	nd
Te	nd	nd	nd	nd	nd	nd	nd	nd	nd	nd	nd
Cs	nd	1.10	nd	0.15	nd	3.66	1.67	0.08	0.27	nd	nd
Ba	13.51	174.22	20.66	90.25	or	or	223.08	114.25	54.28	20.56	21.42
W	nd	nd	nd	nd	nd	nd	nd	nd	nd	nd	nd
Tl	nd	0.47	nd	nd	nd	nd	nd	nd	nd	nd	nd
Bi	nd	nd	nd	nd	nd	nd	nd	nd	nd	nd	nd
Y	3.14	18.84	9.53	19.57	6.58	4.65	5.20	19.61	11.29	11.35	11.94
Zr	26.69	51.32	8.47	145.20	4.60	5.46	7.40	24.87	14.42	12.76	14.40
Nb	1.66	1.73	0.36	11.48	nd	0.03	0.20	1.41	0.51	1.67	0.49
La	0.84	3.11	0.75	7.43	0.45	0.36	0.42	1.46	1.04	1.20	1.31
Ce	2.74	8.78	2.21	21.01	1.16	1.27	1.13	4.35	2.84	3.10	3.15
Pr	0.52	1.52	0.44	3.33	0.24	0.23	0.26	0.85	0.53	0.56	0.61
nd	2.64	7.68	2.27	15.43	1.10	1.09	1.24	4.67	2.67	2.77	3.19
Sm	0.85	2.57	1.00	4.11	0.55	0.55	0.63	1.90	1.11	1.14	1.27
Eu	0.18	0.70	0.42	1.47	0.28	0.17	0.20	1.43	0.32	0.42	0.44
Gd	0.70	3.04	1.42	4.44	0.81	0.73	0.86	2.69	1.56	1.54	1.77
Tb	0.17	0.59	0.34	0.78	0.22	0.20	0.24	0.60	0.36	0.36	0.40
Dy	0.81	3.44	2.02	4.54	1.25	0.99	1.32	3.85	2.21	2.17	2.41
Ho	0.23	0.78	0.50	0.94	0.33	0.25	0.34	0.89	0.53	0.51	0.56
Er	0.63	2.42	1.41	2.55	0.89	0.61	0.94	2.63	1.53	1.45	1.57
Tm	0.12	0.43	0.23	0.37	0.15	0.10	0.16	0.41	0.25	0.24	0.25
Yb	0.86	3.00	1.44	2.25	0.94	0.56	0.91	2.62	1.51	1.48	1.59
Lu	0.21	0.60	0.30	0.41	0.21	0.14	0.21	0.48	0.31	0.30	0.32
Hf	0.90	1.79	0.19	3.65	0.04	0.06	0.18	0.98	0.46	0.42	0.47
Ta	1.33	1.42	1.45	3.35	1.27	1.18	1.33	1.47	1.58	4.69	1.59
Pb	1.01	1.18	0.35	1.22	nd	nd	0.11	0.07	0.40	1.47	0.95
Th	nd	0.21	nd	0.14	nd	nd	nd	0.03	nd	nd	nd
U	0.11	0.27	0.08	0.18	0.16	0.09	0.12	0.34	0.54	0.08	0.07

Table 3. (continued)

ppm	KM-05	KM-06	KM-07	PG-02	PG-03	PG-04	PG-07	KT-08	KT-20	DL
Li	9.83	6.16	4.37	2.21	2.64	2.48	3.76	4.08	8.19	0.03
Be	0.24	0.24	0.14	0.56	0.74	0.71	0.75	0.32	0.23	0.05
B	nd	nd	nd	nd	nd	nd	nd	nd	1.26	0.10
Ti	2092.22	1513.16	1175.43	1463.83	1460.96	1454.39	1564.45	2029.97	1847.65	0.29
V	40.37	8.46	5.27	43.66	45.63	39.34	43.01	47.82	97.34	0.03
Mn	or	46.69	128.83	or	or	or	or	or	or	0.04
Co	6.14	1.47	0.72	4.23	4.52	3.22	4.54	6.76	26.47	0.03
Ni	2.53	1.49	1.48	1.89	1.36	1.20	2.17	5.28	35.80	0.17
Zn	77.80	71.90	37.90	34.09	25.81	29.28	50.89	76.24	78.82	0.56
Ga	14.80	8.88	8.18	17.70	17.00	22.13	21.20	15.73	13.96	0.03
Ge	3.10	1.96	1.65	1.93	1.98	2.21	2.20	2.13	3.77	0.13
As	1.84	1.96	2.28	1.36	1.46	0.92	1.68	2.06	3.22	0.09
Se	nd	nd	nd	nd	nd	nd	nd	nd	nd	0.97
Rb	or	or	or	4.79	2.55	2.65	2.37	or	or	0.02
Sr	34.68	30.16	11.69	88.10	131.12	or	144.12	26.67	28.30	0.04
Mo	0.10	0.04	0.40	nd	nd	nd	nd	nd	nd	0.07
Ru	nd	nd	nd	nd	nd	nd	nd	nd	nd	0.04
Cd	0.06	nd	nd	nd	nd	0.01	nd	nd	0.22	0.11
Sn	nd	nd	nd	nd	nd	nd	nd	nd	nd	0.17
Sb	nd	nd	nd	nd	nd	nd	nd	nd	nd	0.09
Te	nd	nd	nd	nd	nd	nd	nd	nd	nd	0.14
Cs	nd	nd	nd	nd	nd	nd	nd	nd	nd	0.03
Ba	27.10	15.44	12.59	141.68	117.86	219.61	190.51	29.78	26.38	0.04
W	nd	nd	nd	nd	nd	nd	nd	nd	nd	0.09
Tl	nd	nd	nd	nd	nd	nd	nd	nd	nd	0.04
Bi	nd	nd	nd	nd	nd	nd	nd	nd	0.52	0.04
Y	18.04	25.45	42.73	8.97	8.85	11.22	8.70	3.39	9.82	0.02
Zr	19.90	58.51	31.12	71.76	40.03	37.32	30.36	60.78	9.69	0.04
Nb	1.08	2.53	1.82	7.85	5.66	6.12	5.91	6.76	1.72	0.13
La	1.82	1.90	2.01	5.32	4.19	6.40	4.34	6.32	1.55	0.03
Ce	4.83	6.13	5.32	12.01	9.98	14.30	9.93	13.37	3.82	0.03
Pr	0.88	1.15	1.09	1.74	1.48	2.05	1.49	1.84	0.67	0.02
nd	4.68	6.13	5.93	6.91	6.28	8.64	6.38	6.98	3.32	0.07
Sm	1.84	2.43	2.44	1.67	1.66	2.10	1.66	1.24	1.14	0.06
Eu	1.06	0.64	0.68	0.46	0.48	0.53	0.48	0.30	0.36	0.03
Gd	2.53	3.31	3.80	1.64	1.70	2.19	1.79	1.05	1.45	0.06
Tb	0.54	0.74	0.91	0.33	0.34	0.41	0.35	0.19	0.31	0.02
Dy	3.36	4.87	6.58	1.78	1.81	2.22	1.85	0.88	1.75	0.05
Ho	0.77	1.12	1.61	0.42	0.41	0.50	0.42	0.23	0.42	0.02
Er	2.28	3.31	5.16	1.22	1.15	1.39	1.15	0.65	1.15	0.04
Tm	0.36	0.53	0.82	0.22	0.19	0.22	0.19	0.12	0.19	0.02
Yb	2.28	3.44	5.24	1.41	1.17	1.41	1.15	0.78	1.17	0.04
Lu	0.44	0.59	0.89	0.29	0.24	0.27	0.24	0.19	0.25	0.02
Hf	0.70	2.18	1.20	2.18	1.40	1.40	1.12	1.95	0.31	0.06
Ta	2.09	1.81	1.53	2.42	2.16	2.05	2.10	2.06	3.92	0.09
Pb	0.49	0.37	nd	0.14	nd	0.13	0.21	0.19	0.57	0.08
Th	nd	0.29	0.12	0.99	0.52	0.73	0.54	0.16	nd	0.05
U	0.13	0.23	0.14	0.44	0.24	0.41	0.20	0.16	0.16	0.05

Table 4. Rare earth data for Alexandrinka host volcanic rocks.

Field No. ppm	AD4 Gabbro dyke	5902/497 Footwall basalt	5913/163 Host Andesite	5915/524 Host Dacite	6025/121.8 Hangingwall basalt	6096/208.5 Footwall porphyry
La	12.66	2.41	7.94	15.36	8.80	9.73
Ce	32.67	6.61	16.41	33.99	23.75	24.1
Pr	5.08	1.08	2.31	4.676	3.81	3.45
Nd	24.51	5.74	9.84	19.99	20.02	15.21
Sm	6.69	1.81	2.32	4.63	5.91	3.63
Eu	2.32	0.76	0.81	1.12	2.17	1.04
Gd	7.53	2.27	2.42	4.22	6.96	3.40
Tb	1.10	0.35	0.62	0.60	0.99	0.46
Dy	7.53	2.60	2.47	4.37	6.68	3.17
Ho	1.30	0.47	0.45	0.83	1.14	0.54
Er	4.15	1.60	1.58	2.96	3.47	1.82
Tm	0.51	0.20	0.20	0.41	0.39	0.23
Yb	3.61	1.43	1.60	3.12	2.58	1.78
Lu	0.50	0.20	0.23	0.46	0.34	0.24

camps worldwide [*Barrie and Hannington*, 1999]. The "Cyprus type" deposits in the Buribai region are related to the most primitive fore-arc rocks, which contain lavas of boninitic affinity. The Baimak–Buribai formation evolves to a more arc-like calc-alkaline sequence that hosts the Baimak type deposits. Bimodal tholeiitic sequences of the Karamalytash formation host the Urals type deposits and the arc-like Karamalytash formation age equivalent volcanics host the Baimak type Alexandrinka deposit. The fairly homogeneous pyroxene-phyric basalt-dominated Irendyk formation is largely barren of

VMS, possibly due to the shallower submarine conditions suggested by the volcanic facies. VMS deposits are generally interpreted to have a deep submarine origin [*Franklin*, 1993].

Ce/Yb ratios have been previously used to discriminate subduction-related volcanic suites into two groups; those with high Ce/Yb, those with low Ce/Yb [*Hawkesworth* et al., 1991, 1994]. The high Ce/Yb group generally reflects a subducted sedimentary contribution to the melt, or that of an enriched mantle wedge. In our study, the majority of the Magnitogorsk volcanics have

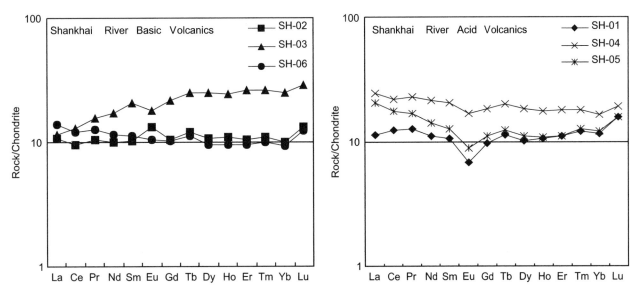

Figure 8. Chondrite normalised REE plot for Shankhai river lower Baimak–Buribai formation volcanics.

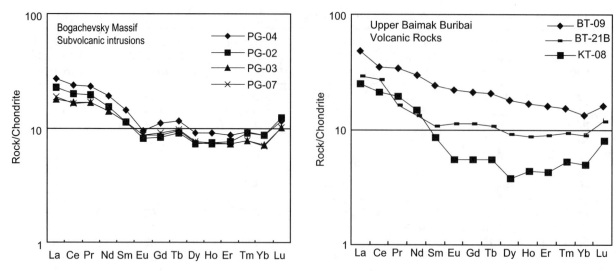

Figure 9. Chondrite normalised REE plot for upper Baimak–Buribai formation volcanics and sub-volcanic rocks.

very low (< 15) Ce/Yb ratios, suggesting that subducted sediments did not contribute significantly to the melts. Nevertheless, BT-09, BT-21B and related granitoids from the Bogachev region (PG 02, 03, 07, 08) have elevated Ce/Yb ratios, as does the sample from the uppermost Baimak–Buribai formation KT-08. A single sample from a late (Ulutau formation?) dyke at Sibay (SY-13) also shows higher Ce/Yb. These data may support a contribution to those rocks from an enriched source of either subducted sediment or mantle wedge.

5. LEAD ISOTOPES

5.1. Published Data

Published lead isotope data for the Urals massive sulfide deposits are rare [*Sunblad* et al., 1996; *Prokin* et al., 1999]. The general conclusions from this work indicated a degree of old crustal lead contribution to the sulfide systems, particularly in the case of the Baimak type deposits which contain relatively non-radiogenic lead [*Prokin* et al., 1999].

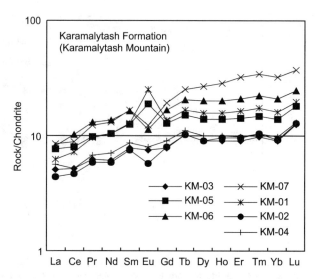

Figure 10. Chondrite normalised REE plot for Irendyk formation volcanic rocks.

Figure 11. Chondrite normalised REE plot for type section of Karamalytash volcanic rocks.

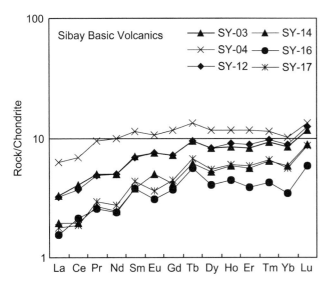

Figure 12. Chondrite normalised REE plot for Karamalytash volcanics hosting Sibay Urals type VMS deposit.

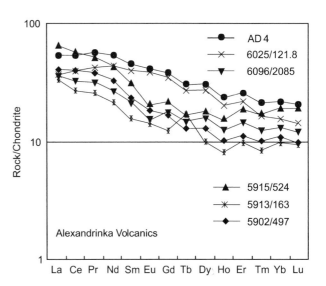

Figure 14. Chondrite normalised REE plot for volcanic rocks hosting Alexandrinka Baimak type VMS deposit.

5.2. Preliminary Analysis of New Data

Lead isotope data discussed here was collected at the laboratories of the BRGM, Orleans, France. The raw data were previously reported by *Orgeval* et al. [1999], and *Tessalina* et al. [2001]. In Figure 15 this data has been compiled and subdivided on the grounds of; a) deposit type, and b) current geographical grouping. The most obvious feature of the data is the broad spread between the mantle and upper crustal curves, with a range of data of 0.15 for $^{207}Pb/^{204}Pb$ and 1.3 for $^{206}Pb/^{204}Pb$, straddling the Stacey Kramers curve.

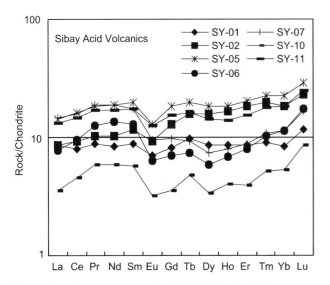

Figure 13. Chondrite normalised REE plot for felsic volcanic rocks hosting Sibay VMS deposit.

This spread is similar to typical values from arc-related VMS deposits in Japan [*Sasaki* et al., 1982], which contrasts with the closely clustered data from VMS deposits of similar age in the Iberian Pyrite Belt [*Marcoux*, 1998].

The most interesting initial observations from this data are when general separations of data on the grounds of deposit classification and on simple geography are made. The VMS deposits in the Main Uralian fault plot closest to the mantle curve and are distinctive in this regard. These deposits are compared to massive sulfides developed in mid-ocean ridge volcanic settings [*Zaykov* et al., 2000] and the Pb data appears consistent with a generally primitive mantle source. Plots of late vein galena clearly plot on the younger end of the Stacey Kramers curve, and are indicative of a possible younger age, post-dating the main massive sulfide formation. If the samples from the west versus the east Magnitogorsk arcs are compared, regardless of deposit type, there is around 80% overlap of data, with perhaps a drift toward more radiogenic lead in the eastern arc. However, when the data is plotted on the grounds of deposit type (i.e., Baimak versus Urals type) there is a clearer separation indicated. This indicates a difference in lead isotope values for the two deposit types, regardless of geographical location. This suggests that Urals and Baimak type deposits have different lead budgets, reflecting different relative contributions of lead from mixed mantle and crustal sources. The data from the Baimak deposits is less radiogenic, indicative of an "older" source for the lead, and a contribution from older subducted oceanic crust is

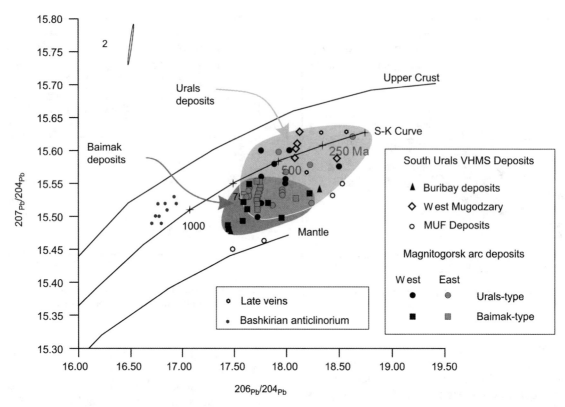

Figure 15. Plot of 207/204 Pb versus 206/204 Pb for galena from a range of VMS deposits from the south Urals (data from *Tessalina* et al. [2001]).

indicated. By contrast, the Urals type deposits show source lead ages that may be much closer to the age of the host volcanic rocks themselves, and hence may be sourced entirely from them. Clearly, more data is needed, particularly that from potential source rocks (volcanics and sediments), but the plot of data from deposits in the Bashkirian anticlinorium suggests other sources than these Precambrian rocks are implicated in having a contribution to the metal budget.

6. DISCUSSION

In the Sakmara zone, to the west of the Main Uralian fault, our data indicate that Silurian volcanics in the Mednogorsk ore district show strong affinities with volcanics formed in an oceanic setting, comparable to the work of previous authors [*Seravkin* et al., 1992; *Zaykov* et al., 2000; *Buschmann* et al., 2001]. *Savlieva* et al. [1997] interpreted the extensive ophiolites of the Kempirsai massif to indicate that the Sakmara zone developed in a rift which developed to a full oceanic basin. *Puchkov* [1997] also suggested the presence of island

arc magmatism in a marginal sea setting [*Zaykov* et al., 1995]. Interpretation of the Sakmara zone is complicated by its being a mélange that includes ophiolitic massifs such as those that host the chromite deposits of the Kempirisai massif [*Zonenshain* et al., 1984; *Seravkin* et al., 1992; *Zaykov* et al., 2000]. It has been suggested that the PGE-enriched chromites at Kempirsai formed in a supra-subduction zone setting of an oceanic arc at around 400–370 Ma [*Melcher* et al., 1999]. In the Mednogorsk VMS district, the volcanic sequences show a transition from tholeiite to calc-alkaline nature and a degree of alkaline enrichment which is distinctive for separating these from mid-oceanic ridge associated sulfides [*Buschmann* et al., 2001]. Very low REE and Zr/Y ratios of between 1 and 2 are indicative of a primitive arc setting, perhaps in the main arc or even forearc setting [*Taylor* et al., 1992; *Barrett and MacLean*, 1999].

The presence of small massive deposits in the Main Uralian fault zone is somewhat enigmatic as little is known about them [*Zaykov* et al., 2000]. The general associations have been compared to sulfides forming presently in the Logatchev field of the mid-Atlantic

[*Wipfler* et al., 1999] and a mid-oceanic ridge setting for the sulfide formation seems likely.

Current models that explain the irregular development of VMS deposits along a north to south lineament, and from west to east in the Magnitogorsk zone include arc segmentation [*Zaykov* et al., 1996] and the development of an opposing double arc [e.g., *Zaykov*, 1991; *Zaykov* et al., 1996; *Prokin and Poltavets*, 1996]. Geographically there are east and west branches to the Magnitogorsk arc (Figure 1) separated by the Carboniferous Magnitogorsk trough, which may be extrapolated south into the West Mugodzhar zone [*Prokin and Poltavets*, 1996]. The presence of oceanic rocks and a sheeted dyke complex in the West Mugodzhar zone has been forwarded as supporting evidence for the presence of two parallel arcs [*Ivanov* et al., 1973; *Prokin and Poltavets*, 1996]. The apparent symmetry of massive sulfide deposit types in the Devonian volcanic assemblages in the Magnitogorsk arc has also been cited as evidence to support this model [*Gusev* et al., 2000; *Prokin and Poltavets*, 1996].

The earliest of the Magnitogorsk arc sequences comprises the Baimak–Buribai formation exposed immediately east of the Main Uralian fault. In broad terms, this sequence shows a younging from near Buribai eastward to the contact with the Irendyk formation east of Baimak. Both the past and the current study shows that volcanics at the presumed base of the Baimak–Buribai mafic sequence are of boninitic affinity [*Spadea* et al., 1998] and by analogy with the Izu-Bonin arc, are likely to have formed in a fore arc setting. Tectonically the volcanics of the Buribai region which host the Buribai Cyprus type deposit occur in a distinctive structural zone [*Maslennikov and Zaykov*, 1998] that is separated from the Urals type deposits and the Baimak type deposits in the region to the east. Near Baimak, the two deposits of Tash Tau and Bakr Tau are good examples of the Baimak type deposit type [*Zaykov and Maslennikov*, 1987; *Prokin and Buslaev*, 1999].

Our geochemistry confirms that the evolution of the Baimak–Buribai formation towards a calc-alkaline affinity is reflected in the change in deposit type. The mafic, boninitic lower sequence is host to Cyprus type deposits, whereas the bimodal sequences host the Urals type deposits, and finally the more calc-alkaline sequences host the Baimak type deposits. *Prokin* et al. [1998] propose that the Baimak type deposits are associated with "the mature stage of ensimatic arc development". This is based, amongst other things, on the abundance of felsic volcanics in the sequence and the lead isotope data. Our data suggests that the Baimak

type deposits are in the more clearly calc-alkaline parts of the early arc. Of the seven samples that show more elevated Ce/Yb ratios, six are from what is interpreted as the uppermost Baimak–Buribai formation, which might support the model of a contribution to the melt source from subducted sediment or enriched mantle wedge.

The Irendyk formation has been considered by previous authors to have formed in a mature intra-oceanic arc setting [*Seravkin*, 1986; *Ferstater* et al., 1997]. Much of the Irendyk formation may have formed in relatively shallow water conditions (abundant epiclastic units, possible epithermal-like deposits) and is mostly barren of major VMS deposits.

Geochemistry shows the Karamalytash formation to be N-MORB like. Our data from the type section at Karamalytash mountain and from the Sibay open pit show low Zr/Ti (range 1–5 generally) and low Th, consistent with general tholeiitic affinity and typical for bimodal tholeiites in a rifted arc to back arc setting. There is no evidence of any calc-alkaline volcanics in the Karamalytash formation, except for a late cross-cutting dyke at Sibay, which appears to be related to the overlying Ulutau fromation. This dyke also shows a high Ce/Yb ratio, indicating a different composition to the melt source. In the eastern part of the Magnitogorsk zone all known massive sulfide deposits lie in what is interpreted to be the Karamalytash formation. The eastern zone stretches from the Verkneuralsk region (including Uchaly, Uzelga and Molodzheznoe) through to Alexandrinka to the south (Figure 1). The Alexandrinka deposit is the only example of a Baimak type deposit in the Karamalytash formation. Good stratigraphic control confirms the position of the deposit immediately below the Ulutau formation. However, results of this study and published data from rocks in the Alexandrinka region indicates the presence of both tholeiitic and calc-alkaline volcanics [*Surin*, 1992, 1993], unlike the Karamalytash formation elsewhere. *Bochkarev and Surin* [1996] also reported the presence of "boninite-like" high MgO basalts in the Alexandrinka district. REE data from this and other Karamalytash formation rocks from the Alexandrinka region are shown on Figure 16 (from a compilation by *Gusev* et al. [2000]). A clear saucer-like REE pattern typical for boninites is shown in sample 8. In fact, the patterns for these rocks are remarkably similar to the suite collected from the Shankai river and upper Baimak–Buribai formation, and contrast strongly with the Karamalytash rocks collected from the type section at Karamalytash mountain and from the Sibay open pit. In detail, the volcanic rocks from the Alexandrinska deposit show characteristic LREE enrichment, typical of more

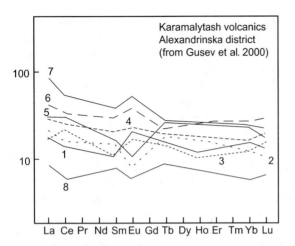

1-Basalt, 2-Basalticandesite, 3-Aphyricandesite, 4-Qtzandesite
5-Aphyricdacite, 6-Dacite, 7-Rhyolite, 8-'Boninite'

Figure 16. Chondrite normalised REE plot for volcanic rocks from Alexandrinka region of the south Urals (from *Gusev* et al. [2000]).

calc-alkaline volcanics [*Tessalina* et al., 1998]. Clearly the volcanics of the Alexandrinka region are complex and are worthy of more detailed study. South of Alexandrinka, the Karamalytash formation hosts no known massive sulfide bodies, a feature which has been linked to the original arc architecture [*Zaykov* et al., 1996].

7. CONCLUSIONS

In conclusion, the VMS deposits from the Sakmara zone, the Main Uralian fault zone, and the Baimak–

Buribai region in the west Magnitogorsk zone through to Sibay reflect the change from a primitive arc or fore arc setting (Sakmara allocthon) to an obducted ridge (Main Uralian fault) setting to fore arc (Buribai) setting to and arc (Baimak) setting, and finally to interarc rift or possibly a back arc (Sibay) setting (Figure 17).

In the Sakmara zone at Kempirsai, PGE enriched chromitite points to both a normal oceanic and a supra-subduction zone setting for the northwestern and south-eastern deposits, respectively [*Melcher* et al., 1999]. This implies the preservation of both sides of a Silurian arc at Kempirsai, with the upper plate to the east. This may correlate with the development of arc-related volcanics in the VMS-bearing Mednogorsk area farther north. This early arc would appear to predate development of the Devonian Magnitogorsk arc complex.

Deposits in the Main Uralian fault zone are enigmatic but they resemble mid-ocean ridge sulfides forming currently in the Atlantic, and may therefore point to the presence of obducted ridge rocks in the melange.

East of this, the Magnitogorsk arc formed in the Devonian. Rather than interpreting the VMS-hosting Baimak–Buribai formation as mature arc as has been suggested in the literature [*Prokin* et al., 1999], the Baimak–Buribai formation may represent a boninitic fore-arc sequence followed by early oceanic island arc sequence with typical light REE enrichment patterns. Later, and farther from the site of subduction, a mature arc rift appears to have developed, as is expressed by the Karamalytash formation. The N-MORB like affinity with characteristic REE patterns indicate that the Karamalytash formation was probably derived from an upper mantle source. Stratigraphically overlying the Baimak–Buribai formation and possibly partly

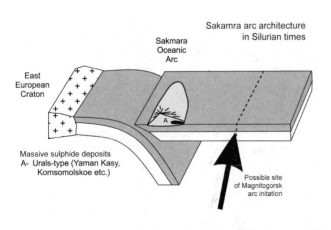

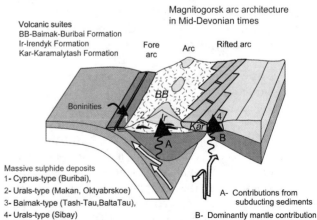

Figure 17. Cartoon representation of simplified arc architecture and setting of VMS deposits of the south Urals during Silurian and Middle Devonian times.

time-equivalent to the Karamalytash formation is the Irendyk formation. The Irendyk formation shows evidence of having been partly formed in shallow water conditions with epiclastic volcanosediments and epithermal-like gold-barite deposits found in the region east of Baimak. It may be that the Irendyk formation marked a long linear geographic feature marking the line of an emerging arc sequence behind which the back-arc Karamalytash formation developed. Other evidence for this is the restriction of a condensed forearc sequence to the west of the Irendyk ridge, which is time equivalent to the Karamalytash and Ulutau formations developed in the east [e.g., *Maslov* et al., 1993; *Brown* et al., 2001]. In the east, the distinctive chert units developed at the top of the Irendyk and Karamalytash formations (the Jarlikapovo and Bugulugyr cherts, respectively) are both immediately overlain by Ulutau formation. The Ulutau formation is a largely unproductive sedimentary to volcanic unit regionally covering the arc sequences in both the west and east.

The relationship between the proposed western and eastern "arcs" and the inter-arc basin is less clear. East of Magnitogorsk, the host rocks to the Alexandrinka deposit include calc-alkaline rocks quite unlike the tholeiitic rocks hosting the other deposits. The implications of this are that the arc may have migrated east at the end of inter-arc rifting represented by the Karamalytash formation, possibly in response to the approach of the East European Craton to the west [i.e., *Brown and Spadea*, 1999]. Such a model needs to be tested by further geochemistry, particularly systematic radiogenic isotope studies of the volcanic suites.

Acknowledgments. Part of the work presented in this publication has been funded by the European Community, Cordis-RTD projects, 5th Framework Programme INCO-2, project number ICA2-2000-10011. NATO-Royal Society support for S. Tessalina and ESF support for GEODE Urals workshops is acknowledged. The authors also acknowledge funding under Russian Grant RFFI 01-05-65329 and INTAS 96-1699. Jane Scarrow and Mike Meyer are thanked for their thorough reviews and unprecedented editorial patience is appreciated. This is an ESF GEODE Project publication.

REFERENCES

Anders, E. and N. Grevesse, Abundances of the elements, meteoritic and solar, *Geochim Cosmochim. Acta*, 53, 197–214, 1989.

Artyushkova, O. V. and V. A. Maslov, Palaeontological basis for the stratigraphic division of pre-Famennian volcanic complexes of the Verkne–Uralsk and Magnitogorsk Regions (in Russian), *Ufa, Ufinian Sci. Center, RAS*, 156, 1998.

Ayarza, P. D. Brown, J. Alvarez-Marron and C. Juhlin, Contrasting tectonic history of the arc-continent suture in the Southern and Middle Urals; implications for the evolution of the orogen, *J. Geol. Soc. London*, 157, 1065–1076, 2000.

Barrett, T. J. and W. H. MacLean, Volcanic sequences, lithogeochemistry and hydrothermal alteration in some bimodal volcanic-associated massive sulfide systems, *Rev. Econ. Geol.*, 8, 101–132, 1999.

Barrie, C. T. and M. D. Hannington, Classification of volcanic-associated massive sulfide deposits based on host-rock composition, *Rev. Econ. Geol.*, 8, 1–12, 1999.

Bochkarev V. V. and T. N. Surin, REE in Middle Devonian boninite-like basalts of the Urals (in Russian), *Yearbook-1995, Inst. Geol. Geochem., Yekaterinburg*, 1996.

Borodaevskaya, M. B., A. I. Krivtsov and E. P. Shirai, Provinces of massive sulfide deposits: principles of tectonic study (in Russian), Moscow, Nedra, 1977.

Borodaevskaya, M. B., A. G. Zlotnik-Chotkevich, P. L. Pirozhok and E. P. Shirai, Localization and formation conditions of massive sulphide ores on example of Uchaly deposit (in Russian), *Soviet. Geol.*, 3, 25–35, 1984.

Brown, D. and P. Spadea, Processes of forearc and accretionary complex formation during arc-continent collision in the southern Urals, *Geology*, 27, 649–652, 1999.

Brown, D., C. Juhlin, A. Alvarez–Marron, A. Perez-Estaun and A. Olianski, Crustal-scale structure and evolution of an arc-continent collision zone in the southern Urals, Russia, *Tectonics*, 17, 158–171, 1998.

Brown, D., R. Hetzel and J. H. Scarrow, Tracking arc-continent collision subduction zone processes from high pressure rocks in the southern Urals, *J. Geol. Soc. London*, 157, 901–904, 2000.

Brown, D., J. Alvarez–Marrón, A. Pérez–Estaún, V. Puchkov, Y. Gorozhanina and P. Ayarza, Structure and evolution of the Magnitogorsk forearc basin: Identifying upper crustal processes during arc-continent collision in the southern Urals, *Tectonics*, 20, 364–375, 2001.

Buschmann, B., P. Jonas, J. Maletz, V. Maslennikov and V. Zaykov, About age relationships and petrochemical peculiarities of hangingwall volcanites of the Blyava massive sulfide deposit (South Urals), in *Metalogeny of Ancient and Modern Oceans*, edited by V. V. Zaykov and H. Belogub, 2001 7th Miass Student's Summer School, Miass, 247–251, 2001.

Chemenda, A., P. Matte and V. Sokolov, A model for Paleozoic obduction and exhumation of high-pressure/low-temperature rocks in the southern Urals, *Tectonophysics*, 276, 217–227, 1997.

Ferstater, G. B., P. Montero, N. S. Borodina, E. V. Pushkarev, V. N. Smirnov and F. Bea, Uralian magmatism: An overview, *Tectonophysics*, 276, 87–102, 1997.

Franklin, J. M., Volcanic-associated Massive Sulphide Deposits, in *Mineral Deposit Modeling*, edited by R. V. Kirkham,

W. D. Sinclair, R. I. Thorpe and J. M. Duke, Geol. Ass. of Canada, Spec. Paper 40, pp. 315–334.

Franklin, J. M., D. M. Sangster and J. W. Lydon, Volcanic-associated massive sulfide deposits, *Econ. Geol. 75th Anniversary Vol.*, 485–627, 1981.

Galley, A. G. and R. A. Koski, Setting and characteristics of ophiolite-hosted volcanogenic massive-sulfide deposits, *Rev. Econ. Geol.*, 8, 221–246, 1999.

Gusev, G. S., A. V. Gushchin, V. V. Zaykov, V. V. Maslennikov, N. V. Mezhelovsky, B. V. Perevozchikov, T. N. Surin, E. I. Filatov and E. P. Shirai, Geology and Metallogeny of Island Arcs (in Russian), in *Geodynamics and Metallogeny: Theory and Implications for Applied Geology*, edited by N. V. Mezhelovsky, Ministry of Natural Resources of the RF and GEOKART Ltd., Moscow, 213–295, 2000.

Hawkesworth, C. J., J. M. Hergt, R. M. Ellam and F. McDermott, Element fluxes associated with subduction related magmatism, *Royal Soc. London, Phil. Trans., A*, 335, 393–405, 1991.

Hawkesworth, C. J., K. Gallagher, J. M. Hergt and F. McDermott, Destructive plate margin magmatism: Geochemistry and melt generation, *Lithos*, 33, 169–188, 1994.

Herrington, R. J., V. V. Zaykov and V. V. Maslennikov, Volcanic-hosted massive sulphide deposits of the Southern Urals, *SGA-IAGOD Field Excursion Guidebook, August 1999*, London, 55, 1999.

Herrington, R. J., V. V. Maslennikov, B. Spiro, V. V. Zay–kov and C. T. S. Little, Ancient vent chimney structures in the Silurian massive sulphides of the Urals, in *Modern Ocean Floor Processes and the Geological Record*, edited by R. A. Mills and K. Harrison, Geol. Soc. London Spec. Publ., 148, 241–257, 1998.

Holland, N. G., Balta Tau — A gold rich VMS deposit in the southern Urals, in *GEODE Workshop "Massive Sulphide Deposits in the Iberian Pyrite Belt: New Advances and Comparison With Equivalent Systems"*, edited by F. Tornos, Abstract Vol. October 2001, Huelva, Spain, 25–27, 2001.

Ivanov, S. N., G. A. Kuritsyna and A. N. Khodalevich, New data of the genesis of massive sulfide deposits in the Urals (in Russian), in *Geneticheskiye Problemy*, edited by D. S. Korzhinskii, Rud, Akad. Nauk SSSR, Moscow, 100–107, 1960.

Ivanov, S. N., V. G. Korinevskii and G. P. Belyanina, Relics of a rift oceanic valley in the Urals (in Russian), *Doklady Acad. Nauk USSR* 211 (4), 939–942, 1973.

Ivanov, S. N. and V. A. Prokin (Eds.), Copper-massive sulphide deposits of Urals: conditions of formation (in Russian), *Sverdlovsk: USC AS of USSR*, 241, 1992.

Ivanov, K. S., Tectonics and geodynamics of the Urals: a development of mobilistic ideas (in Russian), *Abstracts of Papers, Tectonics and Geodynamics: General and Regional Aspects. XXXI Tectonic Conference, 1, Moscow, Geos*, 207–209, 1998.

Juhlin, C., M. Friberg, H. Echtler, A. G. Green, J. Ansorge, T. Hismatulin and A. Rybalka, Crustal structure of the Middle Urals: Results from the (ESRU) Europrobe Seismic Reflection Profiling in the Urals Experiments, *Tectonics*, 17, 710–725, 1998.

Kontar, E. S. and L. E. Libarova, Metallogeny of copper, zinc, lead in the Urals (in Russian), *Ekaterinburg, UralGeoCom*, 233, 1997.

Koroteev, V. A., The ophiolites of the Eastern slope of the South Urals, *Tectonophysics*, 127, 361–369, 1986.

Koroteev, V. A., H. de Boorder, V. M. Netcheukhin and V. N. Sazonov, Geodynamic setting of the mineral deposits of the Urals, *Tectonophysics*, 276, 291–300, 1997.

Kuleyshov, Yu. Y. and V. V. Zaykov, Structure of ore conduits in sulphide hills of the Tash Tau deposit (Maimak ore district south Urals) (in Russian), *Metallogeny of Ancient and Modern Oceans. Ores and Deposits Genesis, Miass, IMin, Ural. Div. RAS*, 84–88, 1998.

Large, R. R., Australian volcanic-hosted massive sulfide deposits: Features, styles and genetic models, *Econ. Geol.*, 87, 471–510, 1992.

Little, C. T. S., R. J. Herrington, V. V. Maslennikov and V. V. Zaykov, The fossil record of hydrothermal vent communities, in *Modern Ocean Floor Processes and the Geological Record*, edited by R. A. Mills and K. Harrison, Geol. Soc. London Spec. Publ., 148, 259–270, 1998.

Little, C. T. S., R. J. Herrington, V. V. Maslennikov, N. J. Morris and V. V. Zaykov, Silurian hydrothermal-vent community from the southern Urals, Russia, *Nature*, 385, 146–148, 1997.

Marcoux, E., Lead isotope systematics of the giant massive sulphide deposits of the Iberian Pyrite Belt, *Min. Deposita*, 33, 31–44, 1998.

Maslennikov, V. V., Sedimentogenesis, halmyrolysis and ecology of massive sulphide-bearing paleohydrothermal fields (after example of the South Ural) (in Russian), *The Scientific Edition, Miass Geotur, ISBN 5-89204-040-2*, 348 pp., 1999

Maslennikov, V. V., V. V. Zaykov and E. V. Zaykova, Paleohydrothermal Fields and Ore Formation Conditions at Massive Sulfide Deposits in the Uralian Paleoocean (in Russian), in *Geodynamics and Metallogeny: Theory and Implications for Applied Geology*, edited by N. V. Mezhelovsky, Ministry of Natural Resources of the RF and GEOKART Ltd., Moscow, 339–357, 2000.

Maslennikov, V. V. and V. V. Zaykov, Massive sulphide-bearing paleohydrothermal fields of Urals marginal oceanic structures (classification, ore facies, formation model) (in Russian), *Miass, IMin, Urals Branch of RAS*, 90, 1998.

Maslov, V. A., Devonian of eastern slope of the south Urals (in Russian), Nauka, Moscow, 224, 1980.

Maslov, V. A., V. L. Cherkasov, V. T. Tischchenko, I. A. Smirnova, O. V. Artyushkova and V. V. Pavlov, On the stratigraphy and correlation of the Middle Paleozoic complexes of the main copper-pyritic areas of the Southern Urals (in Russian), *Ufimian Science Centre, Ufa*, 217, 1993.

Melcher F., W. Grum, T. V. Thalhammer and O. A. R. Thalhammer, The giant chromite deposits at Kempirsai,

Urals: constraints from trace element (PGE, REE) and isotope data, *Mineral. Deposita*, 34, 250–272, 1999.

Mottl, M. J. (Ed.), Proceedings of the Ocean Drilling Program, Scientific Results: College Station, *Texas, Ocean Drilling Program*, 139, 772, 1994.

Netcheukhin, V. M., N. G. Berland and V. N. Puchkov, Deep structure, tectonic, metallogeny of the Urals (in Russian), *Uralskii Nauchnyi Tsentr, Akademiya Nauk SSSR, Sverdlovsk*, 107, 1986.

Nesterenko, V. S., Main features of geological structure and questions of genesis of the Gaiskoye massive copper sulphide deposit in the south Urals (in Russian), *Geologiya Rudnykh Mestorozhdenii*, 3, 24–35, 1978.

Ohmoto, H. and B. J. Skinner (eds.), The Kuroko and related volcanogenic massive sulfide deposits, *Econ. Geol. Monograph*, 5, 604, 1983.

Orgeval, J.-J., S. G. Tesalina, V. V. Zaykov, V. V. Maslen-nikov, C. Guerrot and L. Bailly, Caracteristiques geodina-miques et isotopiques des Amas Sulphures du Sud de I''Oural (Russie), *RST-98, Brest*, 169, 1998.

Ovchinnikov, L. N. and V. D. Baranov, Some aspects of metallogeny of massive sulphide deposits (in Russian), *Regularities in Localisation of Mineral Resources, 12, Moscow, Nauka*, 89–97, 1978.

Petrov G. V. and N. M. Kazakova, Geology of the Uzel'ga Cu-Zn massive sulfide deposit (in Russian), *Geology and Genesis of Ore Deposits, Ufa Inst. Geol., Bashk. Branch, Acad. Sci., USSR*, 54–63, 1978.

Prokin, V. A., Regularities in localization of massive sulphide deposit in the southern Urals (in Russian), *Moscow, Nauka*, 184, 1977.

Prokin, V. A., I. B. Seravkin and F. P. Buslaev, Copper-sulfide deposits of the Urals: Conditions of formation (in Russian), *Russian Academy of Sciences Urals Branch, Ekaterinburg*, 307 pp., 1992.

Prokin, V. A. and F. P. Buslaev, Massive copper-zinc deposits in the Urals, *Ore Geol. Rev.*, 14, 1–69, 1999.

Prokin, V. A., F. P. Buslaev and A. P. Perepetailo, Baimak type sulphide deposits in the Urals: Intermediate between massive sulphide and porphyry copper, in *Mineral Deposits: Processes to Processing*, edited by C. Stanley, Balkema Rotterdam, 567–569, 1999.

Prokin, V. A., O. V. Bogoyavlenskaya, V. V. Maslennikov, Conditions of locations of fossils at massive copper sulphide deposits in the Urals (in Russian), *Geol. Rud. Mestorozh.*, 1, 114–117, 1985.

Prokin, V. A., F. P. Buslaev and A. P. Nasedkin, Types of massive sulphide deposits in the Urals, *Min. Deposita*, 34, 10, 121–126, 1998.

Prokin, V. A. and Yu. A. Poltavets, Geodynamic formation conditions of Urals deposits endogene copper-ore and iron-ore (in Russian), *Annual. Ekaterinburg: IGG.*, 161–165, 1996.

Pshenichniy, G. N., Mineralogy and geochemistry of arsenic and antimony in pres of Uzelga copper-massive sulphide deposit (in Russian), *Ufa: Bashk. Branch of AS of USSR*, 88–110, 1978.

Puchkov, V. N., Structure and geodynamics of the Uralian orogen, in *Orogeny Through Time*, edited by J.-P. Burg and M. Ford, Geol. Soc. Spec. Publ., 121, 201–236, 1997.

Rona, P. A., M. D. Hannington, C. V. Raman, G. Thompson, M. K. Tivey, S. E. Humphris, C. Lalou and S. Petersen, Active and relict seafloor hydrothermal mineralization at the TAG hydrothermal field, Mid-Atlantic Ridge, *Econ. Geol.*, 8, 357–374, 1993.

Rudnitskiy, V. F., Hydrothermal-metasomatical alterations of rocks of massive sulphide deposits of Uzelga ore field (in Russian), *Isv. VUSov, Geol. Serie*, 1, 101–111, 1983.

Sangster, D. and S. D. Scott, Precambrian strata-bound massive Cu-Pb-Zn sulfide ores of North America, in *Handbook of Strata-bound and Stratiform Ore Deposits*, edited by K. H. Wolf, Amsterdam, Elsevier, 120–222, 1976.

Sasaki, A., K. Sato and G. L. Cuming, Isotopic composition of ore lead from the Japanese islands, *Min. Geol.* 32(6), 457–474, 1982.

Savlieva, G. N., A. Y. Sharaskin, A. A. Savliev, P. Spadea and L. Gaggero, Ophiolites of the southern Uralides adjacent to the East European continental margin, *Tectonophysics*, 276, 117–138, 1997.

Sawkins, F. J., Massive sulphide deposits in relation to geotectonics, *Geol. Assoc. of Canada Spec.* Paper 14, 221–240, 1976.

Sazonov, V. N; A. H. van Herk and H. de-Boorder, Spatial and temporal distribution of gold deposits in the Urals, *Econ. Geol.* 96; 685–703.

Seravkin, I. B., Volcanism and massive sulphide deposits of Southern Ural (in Russian), *Nauka, Moscow*, 268, 1986.

Seravkin, I. B, A. M. Kosarev, D. N. Salikhov, S. E. Znamenskii, Z. I. Rodicheva, M. V. Rykus and V. I. Snachev, Volcanism of the South Urals (in Russian), *Nauka, Moscow*, 197, 1992.

Seravkin, I. B, S. E. Znamensky and A. M. Kosarev, Volcanic metallogeny of the southern Urals 8 in Russian), Nauka, Moscow, 152 pp., 1994.

Shatov, V., R. Seltmann and G. Romanovsky, Gold Mineralization map of the southern Urals, scale 1:1,000,000, *London/St. Petersburg: NHM, IAGOD and VSEGEI Publication* 2001.

Smirnov, V. N. (Ed.), Copper-massive sulphide deposits of Urals: Geological placing conditions (in Russian), *Sverdlovsk: USC AS of USSR*, 288, 1988.

Sopko, L. N. (Ed.), Massive sulphide deposits of Baymak ore region (in Russian), Moscow, 224, 1973.

Spadea, P., L. Ya. Kabanova and J. H. Scarrow, Petrology, geochemistry and geodynamic significance of mid-Devonian boninitic rocks from the Baimak–Buribai area (Magnitogorsk Zone, Southern Urals), *Ofioliti*, 23(1), 17–36, 1998.

Sundblad, K., E. Bibikova, E. Kontar, L. Neymark, M. Bechkolmen and V. A. Prokin, Source of lead in sulphide ores in the Urals: *EUROPROBE Uralides Workshop, Granada, Programme with Abstracts*, 12 pp., 1996.

Surin, T. N., Geodynamics and metallogeny of the Uchaly-Alexandrinka Zone (in Russian), *Geodynamics* and *Metallogeny of the Urals, Sverdlovsk, Ural. Div. Of RAS*, USSR, 122–123, 1991.

Surin, T. N., Early Givetian contrasting volcanism of the Uchaly-Alexandrinka Zone, Southern Urals: Petrology, geochemistry and related massive sulphide ore formation (in Russian), unpublished Ph.D Thesis, St-Petersburg St. Univ., 1992.

Surin, T. N., Petrology and geochemistry of volcanic rocks early Givetian rhyolite-basalt association bearing massive sulphide mineralisation (in Russian), *Ufa, Ural. Science Centre of RAS*, 45, 1993.

Taylor, R. N., B. J. Murton and R. W. Nesbitt, Chemical transects across intra-oceanic arcs: Implications for the tectonic setting of ophiolites, in *Ophiolites and Their Modern Oceanic Analogues*, edited by L. M. Parson, B. J. Murton and P. Browning, Geol. Soc. of London Spec. Publ., 60, 117–132, 1992.

Tesalina, S. G., V. V. Maslennikov and T. N. Surin, Alexandrinka copper-zinc massive sulphide deposit (East-Magnitogorsk paleoisland arc, Urals) (in Russian), *Miass, Imin*, 228, 1998.

Tessalina, S. G., V. V. Zaykov, J.-J. Orgeval, T. Auge and P. Omenetto, Mafic-ultramafic hosted massive sulphide deposits in Southern Urals (Russia), in *SEG meeting abstract volume, Krakow, August 2001, Balkema*, 2001.

Tessalina S. G. and J. J. Orgeval, Preliminary lead isotope data from the south Urals VMS deposits (in French), *BRGM Archive Report*, 2001.

Tryggvason, A., D. Brown and A. Perez-Estaun, Crustal architecture of the southern Uralides from true amplitude processing of the URSEIS vibroseis profile, *Tectonics*, 20, 1040–1052, 2001.

Wipfler, E. L., B. Buschmann and V. V. Zaykov, The mafic-ultramafic-hosted massive sulphide deposit of Ishkinino, Southern Urals, in *Processes to processing*, edited by C. Stanley, SGA meeting, London, August 1999, Rotterdam, Balkema, 1999.

Yazeva, R. G. and V. V. Bochkarev, Urals Silurian island arc: Structure, development, geodynamic (in Russian), *Geotectonics*, 6, 32–44, 1995.

Zaykov, V. V., Volcanism and sulphide mounds of paleo-oceanic margins (an example of massive sulphide-bearing zones of Ural and Siberia) (in Russian), Moscow, Nauka, 206, 1991.

Zaykov, V. V., About structure and ore facies of Tash-Tau copper-zink-massive sulphide deposit (Baimak ore region, Ural) (in Russian), *Urals Mineralogical Collects, Miass: IMin Urals Branch of RAS*, 197–215, 1995.

Zaykov, V. V. and V. V. Maslennikov, About seafloor sulphide mounds in Urals massive sulphide deposits (in Russian), *Dokl. AS of USSR*, 293, 1, 181–184, 1987.

Zaykov, V. V., V. V. Maslennikov, E. V. Zaykova and R. J. Herrington, Hydrothermal activity and segmentation in the Magnitogorsk — West Mugodjarian zone on the margins of the Urals paleo-ocean, *Tectonic, Magmatic, Hydrothermal and Biological Segmentation of the Mid Ocean Ridges, London*, 199–210, 1996.

Zaykov, V. V., T. N. Shadlun, V. V. Maslennikov and N. S. Bortnikov, The Yaman Kasy sulfide lode as ruins of ancient black smoker at the floor of Uralian paleo-ocean (in Russian), *Geol. Rudn. Mestorozhd.*, 37, 6, 511–529, 1995.

Zaykov, V. V., E. V. Zaykova and V. V. Maslennikov, Volcanic Complexes and Ore Mineralization in Spreading Basins of the Southern Urals, in *Geodynamics and Metallogeny: Theory and Implications for Applied Geology*, edited by N. V. Mezhelovsk, Ministry of Natural Resources of the RF and GEOKART Ltd., Moscow, 315–337, 2000.

Zierenberg, R. A., J. L. Morton, R. A. Koski and S. L. Ross, Geologic setting of massive sulfide mineralization in Escanaba Trough, *U.S.G.S. Bull.*, 171–197, 1994.

Zonenshain, L. P., M. I. Kuz'min and L. M. Natapov, Tectonics of lithospheric plates at the territory of the USSR, Vol. 1 and 2 (in Russian), Moscow, Nedra, 1990.

Zonenshain, L. P. and M. I. Kuz'min, Paleogeodynamics (in Russian), Nauka, Moscow, 192, 1993.

Zonenshain, L. P., V. G. Korinevsky, V. G. Kazmin, D. M. Pechersky, V. V. Khain and V. V. Matveenkov, Plate tectonic model of the South Urals, *Tectonophysics*, 109, 95–135, 1984.

Zonenshain, L. P., M. I. Kuzmin and L. M. Natapov, Uralian foldbelt, in *Geology of the USSR: A Plate-Tectonic Synthesis*, Geodyn. Ser., vol. 21, edited by B. M. Page, pp. 27–54, AGU, Washington, D.C., 1990.

R. J. Herrington, R. N. Armstrong and S. G. Tessalina, Department of Mineralogy, Natural History Museum, Cromwell Road, London SW7 5BD, England (R.Herrington@nhm.ac.uk); (R.Armstrong@nhm.ac.uk); (S.Tessalina@nhm.ac.uk)

V. V. Zaykov and V. V. Maslennikov, Institute of Mineralogy UB RAS, 456301, Miass, Chelyabinsk District, Russia (mas@ilmeny.ac.ru), (zaykov@ilmeny.ac.ru)

J.-J. Orgeval, BRGM, 3, Avenue Claude Guillemin, BP 6009-45060 Orléans Cedex 2, France. (orgeval@exchange.brgm.fr)

R. N. A. Taylor, School of Ocean and Earth Science, University of Southampton, Southampton Oceanography Centre, European Way, Southampton, SO14 3ZH, UK (rex@soc.soton.ac.uk)

Surface Signals of an Arc-Continent Collision: The Detritus of the Upper Devonian Zilair Formation in the Southern Urals, Russia

A. P. Willner[1], T. Ermolaeva[2,4], Y. N. Gorozhanina[3], V. N. Puchkov[3], M. Arzhavitina[3], V. N. Pazukhin[3], U. Kramm[2] and R. Walter[4]

During collision of the East European Craton continental margin with the Magnitogorsk island arc in the Southern Urals, highly immature sandstones of the Zilair Formation (Famennian to Lower Tournaisian in age) were deposited on both sides of the Main Uralian fault (suture). An integrated study of the light and heavy mineral spectrum, as well as of the chemical composition of detrital minerals in samples from known biostratigraphic positions, yield the following information about the source rock assemblages; (1) metamorphic lithoclasts in addition to detrital quartz, chlorite, some feldspar and most of the heavy minerals (epidote, garnet, tourmaline, Ca- and Na-amphibole, chloritoid, titanite and rutile); (2) volcanic lithoclasts of a calcalkaline source in addition to most feldspar, zircon and some Cr-spinel; (3) few serpentinite and chert lithoclasts as well as abundant Cr-spinel from an ophiolithic source; and (4) mostly intraformational siliciclastic and few carbonate lithoclasts. Such mixed provenance is not properly reflected by conventional light mineral discrimination diagrams. The metamorphic source consisted of low to medium metamorphic grade rocks and high pressure rocks as evidenced by phengites with Si-contents up to 3.45 p.f.u., and rare glaucophane. As the prime source, obducted ophiolites of the Main Uralian fault zone and metamorphic complexes at the rear of the collisional accretionary complex can be envisaged, as well as volcanosedimentary series of the Magnitogorsk arc. The provenance signature remains unchanged throughout the sedimentation period of the Zilair Formation. This time (~376–356 Ma) represents surface uplift of an axial rise between the collisional accretionary prism and the forearc basin during exhumation of high pressure rocks.

[1]Institut für Geologie, Bochum, Germany
[2]Institut für Mineralogie und Lagerstättenlehre, Aachen, Germany
[3]Ufimian Geoscience Centre, Ufa, Russia
[4]Geologisches Institut, Aachen, Germany

Mountain Building in the Uralides: Pangea to the Present
Geophysical Monograph 132
10.1029/132GM10

1. INTRODUCTION

Along with the substantial advances in the understanding of metamorphic complexes during the last decades it becomes a challenge in many orogens to compare and correlate geological processes which occurred contemporaneously at very different crustal depth during a relatively short time period and which are now archived in present surface rocks. Hence, provenance studies of synorogenic detritus could help to understand the concomitant exhumation of high pressure rocks reconstructing the fossil distribution of surface lithology in the source area and its variation with space and time. However, modifications during transport have to be taken into account, but these may also allow rough estimates about former surface conditions. In this respect the southern part of the Uralide orogen represents an exciting possibility to study the crustal-scale evolution of an arc-continent collision, including the exhumation of high pressure rocks [*Chemenda* et al., 1997; *Hetzel*, 1999; *Brown* et al., 1998, 2000, 2001], because it is relatively unaffected by late- to post-collisional processes. Due to preservation of their crustal and lithospheric root even after 300 Ma the Uralides are regarded as an orogen still preserved in its collision stage *Berzin* et al. [1996]. They represent a Late Paleozoic collision between the East European craton (part of the Baltica plate) and the former Kazakhstanian plate with an intervening collage of microcontinental and island arc terranes [*Zonenshain* et al., 1990]. The Main Uralian fault (Figure 1) marks the principal suture zone that can be traced along the entire length of the orogen.

In the Magnitogorsk island arc that collided first with the East European Craton continental margin during Late Devonian the upper crustal evolution from an intraoceanic subduction to a mature arc during collision can be identified due to to its low degree of metamorphic and deformational overprint [*Brown* et al., 2001]. On the other hand contemporaneous subduction towards the east partly to upper mantle depths, metamorphism at very different depths and exhumation of rocks of the colliding East European Craton margin can be traced within two metamorphic complexes to the west of the Main Uralian fault (Figure 1); these are the low grade Suvanyak Complex and the high pressure Maksyutovo Complex [*Hetzel*, 1999; *Brown* et al., 1998, 2000; *Beane and Connelly*, 2000]. The first sediments to be deposited on both sides of the suture zone at the earliest stage of collision during the Uralide orogeny is represented by the synorogenic turbidite succession of the Zilair Formation that occurs between the upper course of the Ufa river in the north and the Or-Ilek watershed in the south (roughly between 49°–56° N). Its major distribution roughly coincides with the surface outcrop of both metamorphic complexes (Figure 1). The sandstones of the Zilair Formation are thought to have been derived from the Magnitogorsk island arc [e.g., *Il'inskaja*, 1980]. *Arzhavitina and Arzhavitin* [1991], however, suggested two sources for it: the volcanic arc in the east and the metamorphic complexes in the west. Following the Late Devonian arc-continent collision, convergence ceased and a carbonate platform developed on both sides of

the Main Uralian fault during Early Carboniferous and Bashkirian time. Renewed E-W convergence started at the end of the Carboniferous (Moscovian) with formation of a west vergent fold and thrust belt [*Puchkov*, 2000].

Hence a detailed study of the Zilair Formation may yield information about surface features during the arc-continent collision that can help to refine current models. In this paper we address the following questions:

– What source lithologies and areas can be defined for the Zilair Formation?
– Can oblique collision between the arc and the continent be proved or disproved?
– What is the time relationship between the exhumation of the high pressure rocks in the Maksyutovo Complex and the deposition of the Zilair Formation? Is erosion a major unroofing process to exhume high pressure rocks?
– How did the morphology of the collision zone look during exhumation of the high pressure rocks?

To address these questions light and heavy mineral analysis as well as mineral chemistry were carried out.

2. GEOLOGICAL SETTING

The Southern Urals can be subdivided into several distinct units according to their different histories and geological processes during the two major stages of the Uralide orogeny [*Puchkov*, 1997; *Brown* et al., 1998, 2000, 2001]. The first stage during the Late Devonian to Early Carboniferous is related to the arc-continent collision. During this time bathyal sediments were accreted and remnants of oceanic crust were obducted onto the subducting East European Craton continental margin. Continental crust was subducted to upper mantle depth and exhumed to lie at the rear of an accretionary complex. Contemporaneously, the Zilair Formation was deposited and sedimentation continued on the former East European Craton continental margin, while volcanism dominated in the eastern part of the Magnitogorsk arc. The second stage of Uralide collision occurred during Late Carboniferous to Early Permian, after a major temporal break, when collisional processes occurred within the East Uralian zone. Several aspects of this stage are discussed by *Brown* et al. [1997] and *Giese* et al. [1999]. Further discussion of this stage of deformation is beyond the scope of this paper.

The units involved in the arc-continent collision are briefly described below (Figure 1) because they are

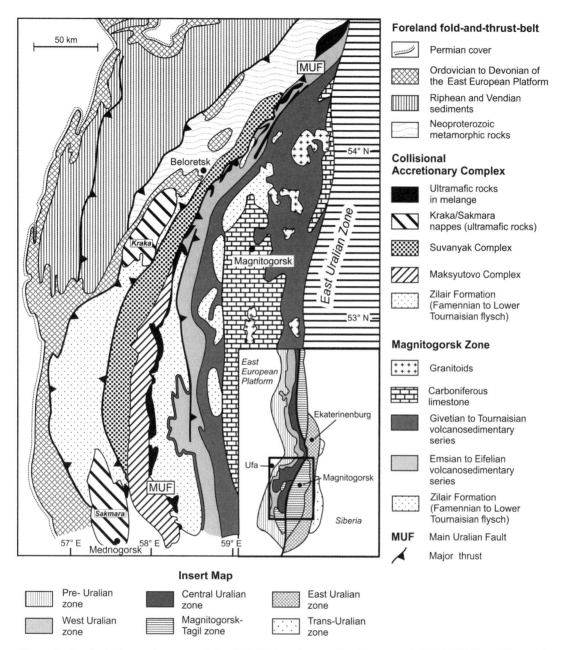

Figure 1. Geological overview map of the SW-Urals redrawn after *Brown* et al. [1998, 2001] and *Lennykh* et al. [1999].

potential sources for the detrital sediments of the Zilair Formation.

The East European Craton margin in the west is dominated by up to 15 km of Riphean strata (1650 to 650 Ma) which were eroded from the basement of the East European platform and deposited in aulocogens. Conformably overlying Vendian sandstones filled a foredeep of a Neoproterozoic orogen that was situated in

the east and included high pressure/low temperature rocks [*Glasmacher* et al., 2001; *Willner* et al., 2001]. During the Early Paleozoic, the area represented a passive continental margin filled with Ordovician to Devonian siliciclastic sediments and limestones that overlie the Neoproterozoic rocks with a marked unconformity [*Puchkov*, 1997]. During the second period of Uralide convergence this former margin was deformed to a west vergent fold and thrust belt [*Brown* et al., 1997; *Giese* et al., 1999].

The Main Uralian fault zone comprises a strongly sheared mélange of mostly serpentinized ultramafic rocks, gabbro, Ordovician volcanics and chert, Silurian graptolite-bearing shale, pillow lava, as well as Devonian volcanic and volcanoclastic rocks (andesites, dacites). Also slices of the Zilair Formation occur. According to *Hetzel* [1999] it acted as a major east dipping detachment zone during exhumation of the high pressure rocks of the Maksyutovo Complex. According to *Brown* et al. [1998] it represents the damage zone that occurred along the backstop during arc-continent collision.

The structural highest position in the accretionary complex west of the Main Uralian fault zone is occupied by the ophiolithic Sakmara and Kraka nappes, which are composed mainly of ultramafic rocks with a mélange base of serpentinites. Basalts and gabbros are rare. Ordovician and Silurian continental slope and rise sedimentary rocks occur at the base of the Kraka nappe as a tectonic sheet [*Puchkov*, 2002].

The Suvanyak Complex comprises a series of low to medium grade metamorphic rocks that were strongly deformed and transposed, thus lacking any stratigraphic continuity. It is composed of quartzites, phyllites and quartz phyllites as well as greenschists and local mylonite zones. Little is known about the PT-evolution of the rocks. Some biostratigraphic evidence (graptolites, conodonts, acritarchs, brachiopods) points to Ordovician to Lower Devonian ages of deposition for some rocks [*Puchkov*, 1997, 2002]. *Brown* et al. [1998] and *Alvarez-Marron* et al. [2000] interpret the Suvanyak Complex to represent shallowly subducted continental shelf and rise material. According to *Hetzel* [1999] the lower boundary to the underlying Maksyutovo Complex is a late normal fault shear zone that dips to the west, although *Brown* et al. [1998] present seismic evidence that the Maksyutovo Complex overlies the Suvanyak Complex.

The Maksyutovo Complex is composed of eclogit facies, high pressure rocks thought to represent sediments of the East European Craton continental margin that were subducted during the continent/arc collision [*Hetzel*, 1999; *Brown* et al., 2000]. A lower unit is comprised mainly of metagreywackes, quartzites and minor mica schists with lenses of graphitic schists, eclogites and blueschists. Determined peak metamorphic conditions in the lower unit range from 15 to 25 kbar and 550° to 700°C [*Karsten* et al., 1994; *Schulte and Blümel*, 1999; *Lennykh and Valizer*, 1999]. Isotope ages of the high pressure rocks of the lower unit scatter from 382 Ma to 357 Ma [Sm-Nd-mineral isochrons; *Shatsky* et al., 1997; *Beane and Connelly*, 2000], 384 Ma to 378 Ma [U/Pb;

rutile; *Beane and Connelly*, 2000] and 387 Ma to 363 Ma [Ar/Ar; white mica; *Matte* et al., 1993; *Lennykh* et al., 1995; *Beane and Connelly*, 2000] dating the early exhumation after peak metamorphic conditions.

The structurally overlying upper unit contains mainly lower grade quartzites and phyllites, metagabbros, marbles, lawsonite-bearing rocks and abundant serpentinites. Peak metamorphic conditions here are considered to be around 450°C/8 kbar [*Hetzel* et al., 1998]. Locally, microfossils of Silurian to Lower Devonian age have been found, suggesting the upper unit may be composed of Paleozoic pelagic sediments mixed with dismembered ophiolitic material. Ar–Ar white mica ages in the upper unit including late retrograde shear zones are 339–332 Ma [*Beane and Connelly*, 2000]. Both units were emplaced against each other along a major retrograde shear zone with normal fault kinematics (top-to-the-ENE) under greenschist facies [*Hetzel*, 1999]. It yielded an Ar–Ar age of 365–355 Ma [*Beane and Connelly*, 2000].

The Magnitogorsk volcanic arc is composed of basalts, andesites, dacites and rhyolites, as well of volcanosedimentary rocks of the same derivation [*Seravkin* et al., 1992]. Conodont biostratigraphy [*Artyushkova and Maslov*, 1998] dates the Magnitogorsk arc as Emsian to Famennian. During two stages it evolved from an incipient intraoceanic subduction to a mature arc [*Brown* et al., 2001]. Arc development starts with the Emsian Baimak-Burubai Formation which occupied the fore-arc position [*Seravkin* et al., 1992; *Brown and Spadea*, 1999] and consists of massive diabase and lava flow basalt interbedded with volcanic breccia and hyaloclastite at the base and high-Mg pyroxene-bearing basalts as well as volcanoclastic rocks in the upper part, which are interbedded with rhyodacites and dacites. The overlying Emsian to Eifelian Irendyk Formation is a typical calc-alkaline volcanic arc sequence, composed of coarse volcanoclastic breccias with some lava flows of andesitic basalts, epiclastic turbidites and debris flow deposits. It is overlain by the Upper Eifelian extrusion of tholeiitic basalts and rhyolites with hyaloclastite and red jasper interbeds of the Karamalytash Formation.

Turbidites with volcanoclastic detritus of the Givetian Ulutau Formation follow upward in the succession. The source rocks for these sediments were andesitic to rhyodacitic volcanoes in the easternmost Magnitogorsk Zone. The Ulutau sediments filled a suture fore-arc basin during Givetian to Lower Frasnian times, while the active volcanic axis of the arc had moved to the east. This is taken as an indication of the arrival of continental crust in the subduction zone [*Brown* et al., 2001]. The Ulutau Formation is overlain by the Frasnian Mukas chert [*Maslov* et al., 1993], which consists of siltstones and dark-grey cherts. During that time the Upper Frasnian volcanic arc complex was also situated in the eastern Magnitogorsk Zone. In the eastern part of the forearc basin the Mukas chert grades into the Koltubanian Formation consisting of poorly-sorted volcanomictic siliceous sandstone. In the Ulutau Formation and the Koltubanian Formation only volcanic and sedimentary clasts were observed. The Mukas chert and the Koltubanian Formation are in turn overlain by the Zilair Formation, which interfingers with coeval volcanic and volcano-sedimentary rocks toward the east.

Lower Carboniferous shallow-water limestones unconformably overlie the weakly folded arc sequence and are themselves slightly deformed and intruded by Carboniferous gabbro and granitoids.

The synorogenic greywackes of the Zilair Formation are found on both sides of the Main Uralian fault, resting on a different basement. In the Magnitogorsk zone the Zilair Formation rests confomably on Upper Devonian sediments and is gently folded, while westward it forms a series of tectonic sheets within the allochthonous Zilair nappe [*Alvarez-Marron* et al., 2000] and tectonic slices within the Main Uralian fault zone. Since the Zilair Formation is the object of this study, it is described in some detail in the following section.

3. THE ZILAIR FORMATION

The Zilair Formation was first described in detail by *Keller* [1949], who subdivided it into five units (Table 1). A somewhat different subdivision was given by *Chibrikova* [1997], who described a lower unit that is locally restricted to the southernmost part of the Zilair Nappe. These six units are well dated biostratigraphically as Famennian to Lower Tournaisian and a summary is given in Table 1. For sake of simplification we refer to three subformations [lower, middle and upper] to which our samples are referenced (Table 1).

In the Zilair nappe the sediments are tightly folded with an overall synformal structure and affected by thrusting. The vergence of folds fans from west vergent on the western limb of the Zilair nappe to east vergent on the eastern [*Bastida* et al., 1997; *Alvarez-Marron* et al., 2000]. The thickness of the series is generally about 2000 m, but according to borehole and seismic data it can be up to 6000 m [*Bastida* et al., 1997; *Alvarez-Marron* et al., 2000; see also *Arzhavitina and Arzhavitin*, 1991]. The degree of metamorphism in the Zilair nappe varies from lowermost greenschist grade in the east to diagenetic conditions in the west [*Bastida* et al., 1997; *Giese* et al., 1999]. A strong cleavage is developed throughout

the Zilair nappe. Southward, in the Sakmara zone, the Zilair Formation in the eastern margin of the Zilair nappe is conformably underlain by the Ibragimovo unit that contains cherts of Upper Frasnian age [*Puchkov*, 1997], while Frasnian shelf limestones mark the base in the western part of the nappe. The pelagic sediments of the Frasnian were deposited on the subducting East European Craton continental margin and represent time equivalents to the shelf limestones further west. The Zilair Formation in the Sakmara zone is facially substituted by a condensed section of Famennian deep water shales, cherts, marls and limestones (Kiya unit) [*Puchkov*, 1997].

In the Main Uralian fault zone the Zilair sediments are more intensely deformed and form blocks of different size (up to several hundred meters) in the serpentinitic melange [*Brown* et al., 2001]. The thickness is not known.

In the Magnitogorsk zone the Zilair Formation has a thickness of up to 2000 m. Olistostrome beds with blocks of limestones, volcanic rocks and siliceous shales were mapped at the base of the formation. They lie on cherts and siltstones of the Mukas chert and sandstones with interbedded tuffaceous chert of the Koltubanian Formation which are dated by Frasnian conodonts [*Maslov* et al., 1993]. On the eastern limb of the Magnitogorsk synform the Zilair Formation interfingers with contemporaneous Famennian to Early Tournaisian calc-alkaline magmatic rocks (andesites, trachyandesites, rhyolite, monzonite, shoshonite, tonalite and volcanoclastic equivalents) providing a possible source for the Zilair detrital components [*Puchkov*, 1997, 2000].

Arzhavitina and Arzhavitin [1991] postulated an emergent area along the Main Uralian fault zone (Figure 2) from which sediments were deposited into two depocentres to the west and east. The basin in the east can be regarded as a suture fore-arc basin in front of a still active Magnitogorsk arc, while toward the west the Zilair sediments were incorporated into the accretionary prism [*Brown* et al., 1998, 2000, 2001].

4. ANALYTICAL METHODS

For light mineral statistics 400 grains from 30 samples of medium to coarse grained sandstones from the three major zones of Zilair sandstones (Table 1) were counted in thin sections using a point counter. Heavy minerals from 36 samples were separated from crushed and sieved 63–180 μm fractions using a LST heavy liquid [lithium heteropolytungstates; density 2.85 g/ccm]. After further sieving fractions of 63–80 μm, 80–125 μm and 125–180 μm were used for heavy mineral statistics. In each fraction 200 transparent grains were counted and

Table 1. Stratigraphy of the Zilair formation.

Samples	Subformation	Member (after 8; in the Zilair zone)	Rock type	Stage	Fossils	Reference	Remarks
47, 48, 50, 1/5, 1/121 (Zilair zone)	Upper	Mazitovian	Siltstone, mica-rich grey-wacke, few limestone	Lower Tournaisian (sulcata – belkai zones)	conodonts, foraminifers	5 7	Only in the Zilair zone
		Yamashlinian	Siltstone, siliceous shale, limestone	Middle-Upper Famennian (postera, expansa, praesulcata zones)	conodonts foraminifers spores, pollen	5 7 6	
U5, U6, U81, U82, U83, U100, Z4, Z11, Z12, Z14, Z15, Z17, Z18, Z19, 97/24, (Zilair zone)	Middle	Avashlinian	Shale, siltstone, greywacke (partly calcareous)	Middle Famennian (trachitera zone)	conodonts, brachiopods, plant remains spores, pollen	5 4 3 6	
Z1, Z2, Z3, U1, U2 (MUF zone) U95, U96, Z30 (Magnitogorsk zone)		Astashian (Ziren'agachian; 6)	Greywacke, siltstone	Lower Famennian (crepida zone)	conodonts spores, pollen	2, 5 9 in Mag. Z. 6	
U91, Z37, Z38 (Zilair zone)	Lower	Yaumbayevian (Vazyamskian; 6)	Shale, siltstone, few lime-stone concretions/beds	Lower Famennian (Upper triangularis zone)	conodonts	1 9 in Mag. Z.	Lowermost unit of the Zilair zone
U27, U102, Z10, Z25, Z26, Z27, Z28, Z29 (MUF zone)					spores, pollen	2 6	
Z8, Z32, Z34, 370 (Magnitogorsk zone)		Dombarkovskian (6)	Greywacke, siltstone, chert intercalations	Upper Frasnian	spores, pollen	6	Only in the Sakmara zone

1 — *Abramova* et al. [1998]; 2 — *Puchkov* et al. [1998]; 3 — *Librovich* [1932]; 4 — *Tyazheva* [1943] in *Keller* [1949]; 5 — *Pazukhin* et al. [1996]; 6 — *Chibrikova* [1997]; 7 — *Sinitsina* et al. [1984]; 8 — *Keller* [1949]; 9 — *Maslov* et al. [1993].

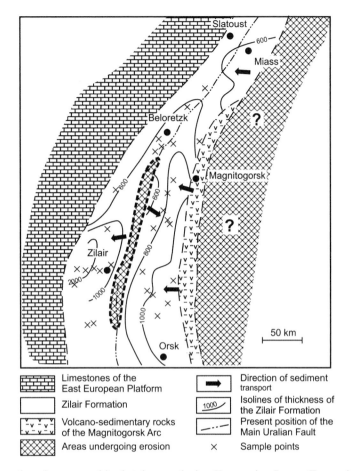

Figure 2. Nonpalinspastic paleogeographic sketch map during Famennian-Lower Tournaisian in the SW-Urals redrawn after *Arzhavitina and Arzhavitin* [1991].

means from 600 grains were calculated. Morphology of the heavy minerals were studied with a binocular microscope and with a scanning electron microscope at Ruhr-Universität Bochum. Back-scattered electron images of tourmaline in polished thin sections were performed with a Jeol JXA-8900 microprobe at Göttingen University.

Quantitative chemical analyses of polished heavy mineral concentrates were performed on a Jeol JXA-8900 microprobe at Aachen University. Operating conditions were an acceleration voltage of 15 kV (for garnet, micas, amphibole, epidote) and 20 kV (for chromspinel), respectively, a beam current of 20 nA (for micas, amphibole, epidote), 25 nA (for chromspinel), 30 nA (for garnet) as well as 10 s counting time per element. When measuring white micas the beam was slightly defocussed (10 μm) in order to avoid loss of alkalies. The following standards were used: jadeite for Na, spinel for Mg and Al, rutile for Ti, plagioclase for Ca and Si, fayalite for Fe, Mn, Si, orthoclase for K, chromite for Cr, Al, Mg, Fe and manganese oxide for Mn. Representative analyses and

structural formulae of the minerals are presented in Table 2. A full set of analytical data, of the light and heavy mineral statistics as well as exact location of the samples is available from the two senior authors.

5. PETROGRAPHICAL CHARACTERISTICS AND LIGHT MINERAL SPECTRUM

The Zilair Formation is characterized by dominant monotonous medium to coarse grained greywackes that alternate with minor siltstones and shales. No detailed facies analysis has yet been done, but grain flow and proximal turbiditic deposits characterize the succession. Typically these rocks have a distinct olive-green colour due to dispersed detrital chlorite or epidote. Lithoclasts (17 to 54 vol%) and feldspar (6 to 29 vol%) are dominant and macroscopically conspicuous. According to the classification of *Folk* [1980] the rocks are litharenite, arkosic litharenite or lithic arkose (Figure 3a). Sorting is very poor. Within these highly immature rocks grains of

Table 2. Representative analyses of detrital minerals 1 rocks (*value calculated).

Oxide compositions (wt %):

| | White Mica | | | | Chlorite | | | | | | Epidote | | Amphibole | | | |
| | Phengite | Muscovite | | | light green | dark green | | fine-grained aggregates | | | | | Ca-Amphibole | | Glaucophane | |
Sample	47	U96	U2	Z19	Z15	U82	50	Z3	U96	Z10	Z3	Z8	U100	U102	Z14	U2
SiO_2	50.76	50.79	45.87	46.86	26.88	29.15	28.86	32.60	30.58	27.76	38.10	37.99	42.22	48.18	56.25	56.26
TiO_2	0.19	0.13	0.98	0.24	0.05	0.01	0.00	0.08	0.53	0.04	0.11	0.06	0.56	0.68	0.10	0.01
Al_2O_3	26.09	26.94	34.67	32.25	19.52	17.02	20.94	14.27	17.75	17.81	23.33	23.69	14.06	6.23	1.84	11.32
FeO / Fe_2O_3	4.78	2.20	1.10	3.40	27.71	24.66	28.03	23.84	24.35	28.88	12.43	11.68	16.15	13.92	7.21	10.28
MnO	0.03	0.04	0.00	0.01	0.45	0.32	0.09	0.35	0.34	0.32	0.13	0.17	0.31	0.28	0.17	0.01
MgO	2.80	3.30	0.74	1.06	13.51	16.66	9.45	16.01	13.05	13.59	0.02	0.06	10.20	15.48	20.18	11.42
CaO	0.04	0.04	0.03	0.00	0.10	0.09	1.00	0.67	1.21	0.14	23.49	23.92	11.69	12.29	13.20	0.65
Na_2O	0.12	0.35	0.73	0.76	0.03	0.04	0.05	0.10	0.10	0.03	0.01	0.02	1.44	0.93	0.16	6.91
K_2O	10.43	10.31	10.66	9.89	0.00	0.02	0.03	0.03	0.29	0.06	0.00	0.00	0.46	0.93	0.04	0.01
H_2O*	4.44	4.45	4.47	4.43	11.32	11.50	11.43	11.61	11.57	11.30	1.92	1.92	2.00	1.95	2.17	2.20
Sum	99.67	98.54	99.24	98.91	99.57	99.47	99.88	99.57	99.75	99.92	99.53	99.50	99.08	100.22	101.32	99.07

Cations — White Mica (O = 11):

	Phengite 47	Muscovite U96	Muscovite U2	Muscovite Z19
Si	3.427	3.425	3.076	3.171
Al^{IV}	0.573	0.575	0.924	0.829
Al^{VI}	1.503	1.565	1.815	1.743
Ti	0.010	0.006	0.049	0.012
Fe	0.270	0.124	0.062	0.192
Mn	0.002	0.002	0.000	0.001
Mg	0.281	0.332	0.074	0.107
Sum	2.066	2.030	2.000	2.054
Ca	0.003	0.003	0.002	0.000
Na	0.016	0.045	0.095	0.100
K	0.898	0.886	0.911	0.854
Sum	0.917	0.934	1.008	0.954
OH	2.000	2.000	2.000	2.000
O	11	11	11	11

Cations — Chlorite (O = 28):

	Z15	U82	50	Z3	U96	Z10
Si	5.695	6.078	6.053	6.737	6.342	5.891
Al^{IV}	2.305	1.922	1.947	1.263	1.658	2.109
Al^{VI}	2.569	2.262	3.230	2.213	2.680	2.345
Ti	0.007	0.001	0.000	0.013	0.082	0.006
Fe	4.909	4.301	4.917	4.120	4.222	5.124
Mn	0.081	0.057	0.016	0.061	0.060	0.058
Mg	4.266	5.179	2.956	4.933	4.036	4.298
Ca	0.023	0.020	0.225	0.148	0.268	0.031
Na	0.011	0.015	0.019	0.042	0.042	0.012
K	0.000	0.004	0.008	0.007	0.076	0.015
Sum	11.867	11.838	11.372	11.537	11.466	11.889
OH	8.000	8.000	8.000	8.000	8.000	8.000
O	28	28	28	28	28	28

Cations — Epidote (O = 12.5):

	Z3	Z8
Si	2.982	2.968
Al	2.152	2.181
Ti	0.006	0.004
Fe^{3+}	0.814	0.763
Mn	0.008	0.011
Mg	0.003	0.007
Sum	2.982	2.965
Ca	1.969	2.002
Na	0.001	0.003
Sum	1.970	2.005
OH	1.000	1.000
O	12.5	12.5

Cations — Amphibole (O = 23):

	Ca-Amphibole U100	Ca-Amphibole U102	Glaucophane Z14	Glaucophane U2
Si	6.257	6.895	7.733	7.677
Al	1.743	1.051	0.267	0.323
Sum	8.000	8.000	8.000	8.000
Al	0.711	0.000	0.030	1.498
Fe^{3+}*	0.560	0.777	0.228	0.802
Ti	0.063	0.073	0.010	0.002
Mg	2.253	3.303	4.316	2.323
Fe^{2+}	1.393	0.798	0.586	0.374
Mn	0.019	0.017	0.010	0.001
Sum	5.000	5.000	5.000	5.000
Fe^{2+}	0.047	0.037	0.014	
Mn	0.019	0.017	0.010	
Ca	1.857	1.885	1.944	0.095
Na	0.076	0.062	0.032	1.828
Sum	2.000	2.000	2.000	1.923
Na	0.336	0.197	0.010	
K	0.087	0.014	0.008	
Sum	0.423	0.211	0.018	
OH	2.000	2.000	2.000	2.000
O	23	23	23	23

Table 2. (continued)

Cr-Spinel

	Z3	U6	U81	Z18
Al_2O_3	46.38	10.17	9.74	13.07
V_2O_3	0.07	0.00	0.00	0.00
ZnO	0.14	0.15	0.07	0.19
Cr_2O_3	21.74	60.19	55.30	43.88
MnO	0.11	0.25	0.21	0.33
MgO	17.37	11.65	12.29	6.80
TiO_2	0.01	0.09	0.45	0.33
NiO	0.21	0.04	0.07	0.05
FeO	12.26	15.59	15.15	23.75
Fe_2O_3*	1.43	2.14	7.42	11.67
Sum	99.72	100.27	100.70	100.08
Al	12.025	3.148	3.054	4.291
V	0.013	0.000	0.000	0.000
Zn	0.022	0.028	0.014	0.040
Cr	3.781	12.495	11.626	9.662
Mn	0.021	0.055	0.047	0.079
Mg	5.697	4.560	4.871	2.822
Ti	0.001	0.018	0.090	0.070
Ni	0.037	0.007	0.015	0.012
Fe^{2+}	2.255	3.423	3.369	5.532
Fe^{3+}*	0.237	0.424	1.484	2.446
O	32	32	32	32

Garnet

	Z1	U1	Z19	U96
SiO_2	36.05	37.52	37.02	36.32
TiO_2	0.08	0.02	0.03	0.17
Al_2O_3	21.25	21.89	21.66	20.12
Cr_2O_3	0.01	0.00	0.04	0.01
Fe_2O_3*	1.07	0.13	2.13	2.58
FeO	35.60	30.93	26.36	21.08
MnO	1.61	2.11	0.52	9.75
MgO	2.14	5.20	3.82	0.65
CaO	1.74	2.06	8.33	9.02
Sum	99.53	99.86	99.93	99.69
Si	5.878	5.943	5.847	5.893
Al^{IV}	0.122	0.057	0.153	0.107
Sum	6.000	6.000	6.000	6.000
Al^{VI}	3.958	4.025	3.876	3.738
Fe^{3+}*	0.131	0.016	0.253	0.315
Ti	0.010	0.002	0.004	0.021
Cr	0.001	0.000	0.005	0.001
Sum	4.100	4.043	4.138	4.074
Fe^{2+}	4.855	4.096	3.482	2.860
Mg	0.519	1.229	0.899	0.157
Mn	0.223	0.283	0.069	1.340
Ca	0.303	0.350	1.410	1.568
Sum	5.900	5.957	5.862	5.926
Alm	82.286	68.758	59.394	48.262
And	3.188	0.388	6.110	7.732
Gross	1.937	5.486	17.809	18.720
Pyrope	8.800	20.626	15.343	2.649
Spess	3.774	4.743	1.184	22.612
Uvaro	0.016	0.000	0.133	0.025
O	24	24	24	24

Clinopyroxene

	Z14
SiO_2	51.51
TiO_2	0.17
Cr_2O_3	1.09
Al_2O_3	3.43
FeO	0.09
MnO	16.81
MgO	24.90
Na_2O	0.22
Sum	98.22
Si	1.908
Al	0.048
Ti	0.005
Fe^{2+}	0.175
Mn	0.003
Mg	0.928
Ca	0.989
Na	0.016
Sum	4.071
O	6

Chloritoid

	Z19	Z19
SiO_2	24.30	24.47
TiO_2	0.00	0.01
Cr_2O_3	0.02	0.00
Al_2O_3	40.63	40.74
FeO	25.05	25.33
MnO	0.58	0.51
MgO	3.12	3.08
H_2O*	7.28	7.31
Sum	100.99	101.47
Si	1.991	1.997
Ti	0.000	0.001
Cr	0.001	0.000
Al	3.924	3.918
Fe	1.717	1.729
Mn	0.040	0.035
Mg	0.382	0.375
OH	4.000	4.000
O	14	14

Tourmaline U100

	monocyclic	recycled	authigenic
SiO_2	36.21	36.23	36.51
B_2O_3*	10.29	10.21	10.41
TiO_2	0.72	0.40	0.73
Cr_2O_3	0.03	0.00	0.04
Al_2O_3	34.49	31.24	33.29
FeO	9.40	10.11	5.67
MnO	0.00	0.01	0.00
MgO	5.49	7.34	9.28
CaO	0.23	0.13	1.24
Na_2O	2.13	2.90	1.91
K_2O	0.01	0.03	0.00
H_2O*	3.55	3.52	3.59
Sum	102.55	102.13	102.66
Si	5.865	5.96	5.844
B	3.000	3.000	3.000
Ti	0.0880	0.049	0.088
Cr	0.0030	0.000	0.005
Al	6.5850	6.043	6.279
Fe	1.2730	1.387	0.759
Mn	0.0000	0.002	0.000
Mg	1.3260	1.796	2.213
Ca	0.0400	0.023	0.213
Na	0.6690	0.922	0.594
K	0.0010	0.006	0.000
OH	4.000	4.000	4.000
O	31	31	31

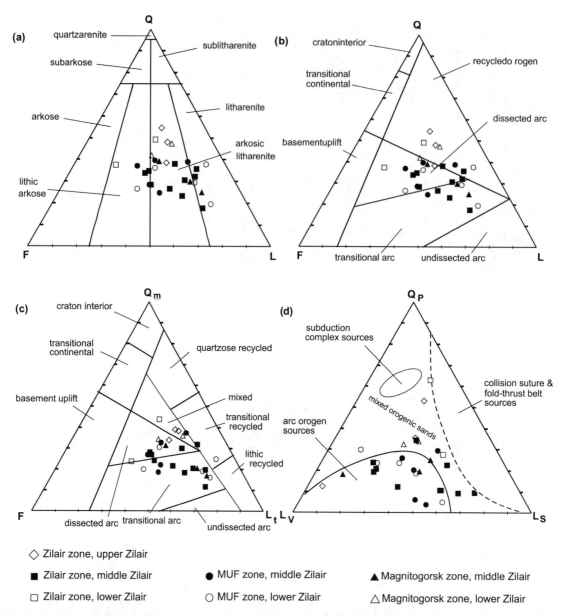

Figure 3. The light mineral spectrum of the Zilair Formation: (a) classification after *Folk* [1980]: Q — quartz, F — feldspar, L — lithoclasts; (b–d) discrimination diagrams after *Dickinson* [1985]: Qm — monocrystalline quartz, Qp — polycrystalline quartz, Lt = L + Qp, Ls — sedimentary lithoclasts, Lv — volcanic lithoclasts.

mono- and polycrystalline quartz and feldspars are angular to subangular. Sericitic white mica (< 20 μm) and some very fine-grained chlorite form a framework that is interpreted as pseudomatrix formed during compaction of small soft lithoclasts. Therefore the amount of cement is low (< 5 %) comprising some authigenic carbonate and chert material.

Although the metamorphic grade may reach lower greenschist facies conditions in the eastern Zilair nappe,

where even some greywackes are strongly cleaved, we analysed only samples from areas where diagenetic and very low grade conditions were determined and quartz was not recrystallised.

5.1. Monomineralic Clasts

5.1.1. Quartz. Quartz grains are predominantly monocrystalline (9 to 27%; Figure 4). Most grains are undulatory, often with deformation lamellae or subgrain

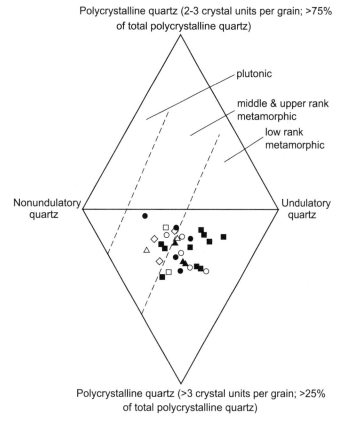

Figure 4. Variation of quartz fabric after *Basu* et al. [1975]; signatures as in Figure 4.

formation. Nonundulatory and weakly undulatory clear grains, but partly with fluid inclusions, are less abundant. Some grains contain solid inclusions of white mica or vermicular chlorite. Some rare monocrystalline quartz grains are elongated and show concave grain boundaries that suggest pressure solution processes in the source area. Polycrystalline quartz grains (up to 11%) are characterized by a polygonal fabric and to a minor degree by strongly sutured grain boundaries between grains of less than 50 µm. The polycrystalline quartz detritus reflects all stages of protomylonite and mylonite formation as shown by the variable degree of recrystallisation and preservation of the grains.

5.1.2. Feldspar. The feldspar component (6 to 29%) is dominated by albite, while partly microperthitic K-feldspar is minor. Many feldspar clasts have an euhedral or subhedral shape which suggests a volcanic source. Rare albite clasts may have an elongate shape with concave/convex grain boundaries presumably due to pressure solution in the source area. Some albite clasts contain abundant white mica inclusions as known from

porphyroblasts in low grade metamorphic rocks. Some plagioclase clasts are partly replaced by calcite, others by sericite.

5.1.3. Phyllosilicate. Large and mostly bent or kinked detrital flakes of phyllosilicate (30 to 100 µm) are comprised predominantly of chlorite (up to 13 vol%), abundant white mica (1 to 9 vol%), white mica and chlorite intergrowth and brown biotite chloritized to variable degree. Furthermore, rare glauconite clasts were observed in two samples (47, 50) from the upper part of the Zilair Formation. Generally more white mica is observed in the upper part of the section. Detrital phyllosilicates can easily be distinguished from those of the pseudomatrix by markedly different grain size.

5.1.4. Carbonate. Three types of sparitic carbonate (up to 13%) have been identified: (1) Many monocrystalline and polycrystalline terrigenous clasts were observed in most samples. Grain shape varies from subangular to subrounded. (2) Cement filling is sometimes difficult to distinguish from clasts. (3) Some carbonate occurs as diagenetic replacement of plagioclase. Generally more carbonate is found in the upper part of the section.

5.2. Lithoclasts

5.2.1. Sedimentary Clasts.
Sedimentary clasts (2 to 18 vol% of all clasts; 6 to 38 vol% of the lithoclasts) include abundant shale and slate, siltstone, rare chert, arenite, and fine-grained greywacke clasts. Shale clasts with sizes up to several millimeters are generally macroscopically detectable. Many of these, but also siltstone clasts, show soft sediment deformation. They may be predominantly intraformational rip-up clasts and form transitions to pseudomatrix. The amount of siliceous extra- and intraclasts cannot be estimated. Furthermore the amount of sedimentary lithoclasts relative to metasedimentary lithoclasts (see below) may be overestimated due to a wide transition and ambiguous grain size criteria [Dickinson, 1985]. Arbitrarily, a limit was set at a grain size of 20 µm and good development of mineral orientation within the clasts.

5.2.2. Volcanic Clasts.
Volcanic clasts (3 to 24 vol% of all clasts; 15 to 72 vol% of the lithoclasts) are highly variable, consisting of (i) altered glass, (ii) acid tuff with pyroclasts of quartz and feldspar within recrystallized quartz, as well as (iii) aggregates containing small euhedral plagioclase phenocrysts within a matrix of either brown glass, pyroxene aggregates, chlorite-opaque aggregates or carbonate of diagenetic origin. Abundant euhedral plagioclase aggregates with trachytoidal texture or lithoclasts with micrographic intergrowth were observed. It is not possible to attribute these clasts to definite rock types due to size and alteration of the clasts, but an entire basic to acid volcanic rock suite appears to be present.

5.2.3. Metamorphic Clasts.
Metamorphic clasts (4 to 25 vol% of all clasts; 15 to 60 vol% of the lithoclasts) include quartz + white mica + chlorite aggregates, plagioclase + white mica aggregates, abundant phyllite occasionally showing a crenulation cleavage, quartzite, rare graphitic quartzite, mylonite, epidosite, epidote + chlorite + albite aggregates or quartz + epidote + albite aggregates. There is a marked transition between this group and many siliciclastic sedimentary fragments.

5.2.4. Plutonic Clasts.
Plutonic clasts (0 to 3 vol% of all clasts; 0 to 10 vol% of the lithoclasts) include quartz + feldspar aggregates and graphic intergrowths of quartz and feldspar, although the first may also represent medium to high grade metamorphic rocks, the latter represents subvolcanic rocks.

5.2.5. Serpentinite Clasts.
Serpentinite clasts (0 to 5 vol% of all clasts; 0 to 24 vol% of the lithoclasts) are very fine-grained yellow-greenish aggregates with fibrous to platy fabric. These clasts may be interpreted as serpentinized and chloritized ultramafic fragments. There appears to be a correlation between the amount of serpentinite clasts and chromian spinel.

5.3. Light Mineral Variation

When light mineral compositions (carbonates excluded) are plotted on the discrimination diagrams of *Dickinson and Suzcek* [1979] and *Dickinson* [1985] (Figure 3) that are based on simplified provenance models, most samples fall into fields assigned to magmatic arcs. This is partly conformable with the abundant volcanic lithoclasts detected in the light mineral spectrum. However, the strong scatter of our samples among several subfields indicates a mixture of materials with entirely different sources. While the quartz/feldspar ratio is fairly narrow (between 0.4 and 0.7), the content of lithic fragments varies strongly. Also, the variation of the clast spectrum is evidently similar in space and time.

According to *Basu* et al. [1975], quantification of the types of quartz clasts provides an indication of the metamorphic grade of the source rocks (Figure 4). The Zilair Formation quartz falls within the fields of middle and low rank metamorphic sources. The latter is clearly consistent with the type of lithoclasts, namely phyllite, quartz-phyllite, quartzite and epidote-bearing rock fragments, as well as chlorite clasts. Mono- and polycrystalline quartz clasts with abundant fluid inclusion trails, chlorite and white mica inclusions are typical for quartz veins which usually develop parallel to the predominant foliation particularly during a greenschist facies metamorphism. The concave grain boundaries of elongated monocrystalline quartz grains are commonly attributed to processes of pressure solution, a deformation mechanism common in greenschist facies metamorphic rocks. Polycrystalline grains with sutured grain boundaries and locally subgrain formation may indicate that low grade shear zones were a source.

A plot of three major types of lithoclasts (Figure 5a) shows that sedimentary, metamorphic and volcanic clasts have similar ranges of composition, which also prevails in space and time. Furthermore, the distribution of the lithoclasts does not conform with any discrimination field established by *Ingersoll and Suczek* [1979]. However, there is a fair correlation between polycrystalline quartz clasts and metamorphic lithoclasts, as well as between feldspar and volcanic lithoclasts.

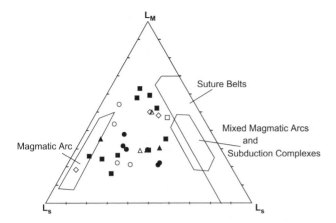

Figure 5. Variation of sedimentary [L$_S$], metamorphic [L$_M$] and volcanic lithoclasts [L$_V$]. Signatures as in Figure 4. Discrimination fields after *Ingersoll and Suczek* [1979].

6. THE HEAVY MINERAL SPECTRUM

6.1. Ophiolitic Source

Angular fragments of brown chromian spinel are the most abundant component among transparent heavy minerals. Complete crystals are euhedral (Figure 6b). Chromian spinel generally comprises 22 to 55% of the total transparent heavy minerals, but in seven samples

it is the dominant heavy mineral with as much as 67 to 87% or it can locally be as low as 12 to 14% (samples Z4, Z14). Such Cr-spinel concentrations suggest an important ophiolitic source for all samples.

6.2. Metamorphic Source

Four types of heavy minerals may be assigned to a metamorphic source only: epidote, garnet, titanite and chloritoid. Most of these are present as irregular, angular fragments of crystals and crystal aggregates. Greenish-yellow epidote is the most abundant metamorphic component, comprising 12 to 52% of the transparent heavy minerals. However, in 12 samples it is the dominating heavy mineral being as high as 56 to 79% (samples Z4, Z14). On the other hand, seven samples are devoid of epidote or contain only up to 4%.

Reddish garnet comprises up to 1 to 9%, locally reaching up to 19% (sample Z18). Rarely, grains may be euhedral or hypidioblastic, most are strongly rounded (Figure 6a). Most garnet crystals contain tiny quartz inclusions. Brownish-yellow titanite, which comprises up to 2.8%, is a minor component. Platy euhedral to hypidioblastic chloritoid of greyish green colour has only been observed in two sample (Z19; Z32). This mineral has also been observed by *Arzhavitina* [1978] and *Arzhavitina and Arzhavitin* [1978] in the Zilair Formation.

Figure 6. Secondary electron images of heavy mineral grains of: (a) edge rounded garnet (sample Z1); (b) euhedral chromian spinel (sample Z14); (c) euhedral tourmaline (sample Z19) and (d) hypidioblastic amphibole fragments (sample U102).

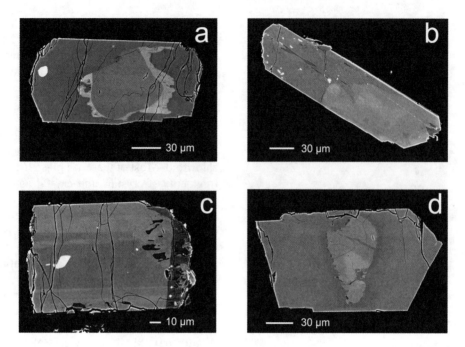

Figure 7. Back scattered electron images of polished tourmaline discontinuously zoned grains: (a) rounded recycled detrital core, irregular overgrowth and broad continuously zoned outer rim (sample U100); (b) detrital recycled core with overgrowth containing oriented quartz inclusions (sample U83); (c) detrital grain with detrital core with oriented quartz inclusions and authigenic overgrowth (sample Z19); (d) recycled, cataclastically deformed detrital core with broad overgrowth (sample U100).

6.3. Metamorphic or Igneous Source

Colourless to slightly orange, generally prismatic crystals of apatite with partly rounded edges and rare angular fragments comprise 0.5 to 7%, but locally even 46% (sample U100). Euhedral, dark long needles of green amphibole may make up to 7% of the heavy minerals (locally 20% in sample Z19; Figure 6d). Partly it occurs as aggregates of fibrous crystals. A rare occurrence of blue amphibole has been described by *Arzhavitina* [1978] in the Zilair detritus and was redetected during this study at least in one sample (U2). Angular fragments of slightly greenish pyroxene (up to 3%) were observed. Pyroxene has also been described by *Arzhavitina and Arzhavitin* [1991].

6.4. Stable Heavy Minerals

This group comprises the heavy minerals zircon, rutile and tourmaline that may be of metamorphic or igneous origin, but due to their stability with respect to weathering and transport they may often be derived from reworked (meta)sediments. The stable heavy mineral content is low in the Zilair Formation, generally less than 10% of the transparent heavy minerals, although zircon comprises 22% in sample U83. Colourless, clear, long as well as short prismatic grains of zircon prevail over rounded

grains and angular fragments, suggesting an overall monocyclic source of igneous origin for this mineral. Similarly, orange to brownish red rutile, which is prismatic with slightly rounded edges, is also interpreted to be derived from a monocyclic metamorphic source.

Tourmaline comprises less than 2% of the transparent heavy minerals in most samples, reaching 7% in sample U83. 36 tourmaline grains were separated from Zilair sandstones and analysed. Tourmaline shape is quite sensitive to transport and the degree of authigenic overgrowth. Tourmaline grains are generally angular fragments, fragments of euhedral prismatic crystals and perfectly euhedral crystals (Figure 6c). Slightly rounded grains are very rare. Their colour is invariably brownish. With back scattered electron images the internal structure of the grains can easily be made visible (Figure 7). The following structural types are observed (abundances are given in brackets; note that a grain may show different types of internal structure):

(i)   Grains with discontinuous zoning (47%; Figure 7a, b, d): The cores of the grains strongly contrast with their broad rims in back scattered electron images, indicating different compositions. The cores are interpreted as older detrital grains with erosive former grain boundaries. 76% of these cores are well rounded, others angular, but strongly edge rounded. 59% of the cores are unzoned, others are slightly concentrically zoned or with parallel lamellae. This indicates recycling of former clastic sedimentary rocks that were metamorphosed up to low and medium metamorphic grade in the source region, when broad rims were produced [*Henry and Dutrow*, 1996]

(ii) Grains with oriented quartz inclusions (53%; Figure 7b, c): Elongate quartz inclusions of 5 to 30 µm size occur parallel to faintly visible lamellae of elongate tourmaline. These are indicative of a low grade metamorphic crystal growth.

(iii) Grains with broad oscillatory lamellae are rare (14%; Figure 7c).

(iv) Grains with double zoned rims (Figure 7a): Two grains show an irregular asymmetrical overgrowth on detrital cores that in turn is overgrown by a broad rim with idiomorphic outer boundaries of the tourmaline grain. According to *Henry and Dutrow* [1996] this is typical of authigenic overgrowth under diagenetic or very low grade conditions overgrown again under low or medium grade metamorphic conditions.

(v) Unzoned grains: Completely unzoned fragments are minor (11%), but 35% of the broad rims on the discontinuously zoned grains do not show evidence of zoning.

(vi) Cataclastically deformed grains: one grain and one inner detrital core show evidence of fractures healed by authigenic tourmaline (Figure 7d).

(vii) Authigenic outer rims: 44% of the studied grains show bipolar authigenic overgrowth with irregular outer boundaries (Figure 7c). However, overgrowth at the analogous (−) pole is not often developed. Such thin (10 to 15 µm) overgrowths are commonly formed under diagenetic and very low grade conditions [*Henry and Dutrow*, 1996].

6.5. Opaque Minerals

This group was not quantified. Generally, angular grains of mostly magnetite, pyrite and more rarely ilmenite were detected. Attribution to particular source rock types is impossible.

6.6. Heavy Mineral Variation

Quantitative plots of heavy mineral distributions in Figures 8 and 9 also show that there is no fundamental variation of the spectra either in space or in time. However, rather a strong local influence of specific source rocks can often be detected, where certain heavy minerals

like chromian spinel, epidote, garnet, zircon or apatite exceed their normal content by far. This often observed phenomenon is also attributed to very short transport, a low degree of mixing of detritus and a very close source region. Figure 9 shows that the heavy mineral spectrum is essentially dominated by varying contributions from metamorphic and ophiolitic rock sources.

Similar observations were made by *Arzhavitina and Arzhavitin* [1991]. However, these authors describe an overall quantitative decrease of total heavy minerals in the Zilair nappe with respect to the Magnitogorsk zone.

7. CHEMICAL COMPOSITION OF DETRITAL SOLID SOLUTION MINERALS

7.1. Spinel

250 microprobe analyses of spinel grains from 13 samples were made (Figure 10; Table 2). These analyses do not show significant compositional variation with respect to lithologic age or sample location. Major end member components are chromite (0.25 to 0.61, locally 0.85 mole%), spinel (0.08 to 0.54, locally 0.63 to 0.71 mole%), Mg-chromite (0.00 to 0.46 mole%) and magnetite (0.00 to 0.19 mole%). Minor end member components are ulvöspinel (0.00 to 0.09 mole%), galaxite (0.00 to 0.01, locally 0.06 to 0.13 mole%) and gahnite (0.00 to 0.01, locally 0.02 mole%). The most important variation, particularly with respect to ultrabasic source rocks, is $X_{Mg} = Mg/(Mg + Fe^{2+})$ versus $X_{Cr} = Cr/(Cr + Al)$. This is due to element partitioning between melts and minerals and is dependent on temperature and oxygen fugacity [*Dick and Bullen*, 1984]. The variation of chromian spinel in peridotites as given in the literature [*Pober and Faupl*, 1988] is shown in Figure 10c. Similar variation is also seen in chromian spinels within peridotite bodies in the Main Uralian fault zone and the Kraka Nappe [*Savelieva* et al., 1997; *Smirnov and Volchenko*, 1992] (Figure 10b). It was shown by *Savelieva* et al. [1997] that over a distance of 300 km peridotite massifs in the Southern Urals are uniform in structure and composition. Chromian spinel of distinct composition is found in harzburgites, lherzolites and chromitite lenses in these massifs. The majority of the detrital chromian spinel in the Zilair Formation (Figure 10a) has a compositional variation equivalent to those in harzburgites (approximately 61%) and are to a minor extent comparable only with those from lherzolites (2.5%) or chromite lenses (10%). Others fall into fields of overlap or cumulates.

Besides ultramafic rocks, chromian spinel also occurs in basaltic rocks. Such chromian spinel, however, has higher contents of TiO_2 (mostly > 0.25 wt%) and higher

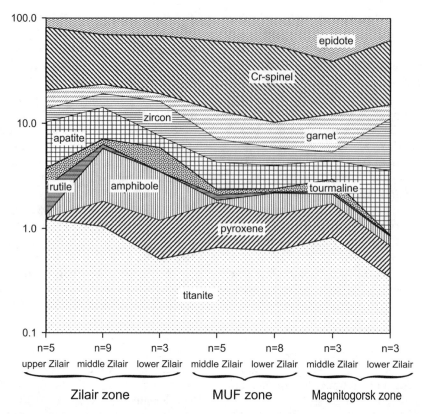

Figure 8. Variation of the transparent heavy minerals of the Zilair Formation according to location and stratigraphic position. Means of samples of similar zones and stratigraphic position are indicated.

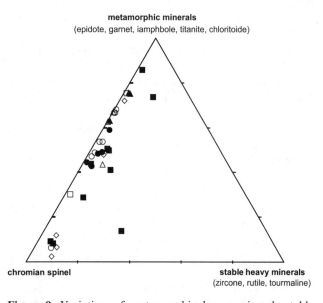

Figure 9. Variation of metamorphic heavy minerals–stable heavy minerals–chromian spinel. Signatures as in Figure 4.

$X_{Fe^{3+}} = Fe^{3+}/(Fe^{3+} + Cr + Al)$ (mostly > 0.05) [*Arai*, 1992; *Allen*, 1992]. Only cumulates have similar TiO_2 and Fe^{3+} contents. Some detrital chromian spinel crystals may actually have been derived from basic volcanic rocks (or corresponding cumulates), because 9.5 wt% of all detrital chromian spinel have TiO_2 contents > 0.25 wt% (up to 0.45, locally 1.2 to 1.54 wt%) and also have an $X_{Fe^{3+}} > 0.05$. Such rocks are associated with ophiolitic massifs in the Southern Urals and occur in tectonic slivers within the Main Uralian fault zone. Figure 11 shows that those Cr-spinels could be assigned to island arc basalts.

7.2. White Mica

Variation of the chemical composition of white mica (Figure 12; Table 2) is similar throughout the Zilair Formation. The Si-content in white mica indicates the degree of tschermak substitution $(SiR^{2+}Al^{IV}_{-1}Al^{VI}_{-1})$ and is roughly an expression of the pressure of formation [*Massonne and Schreyer*, 1987; *Massonne*, 1991]. Minimum contents of 3.02 to 3.17 p.f.u. correspond to typical muscovite, while white mica with maximum contents

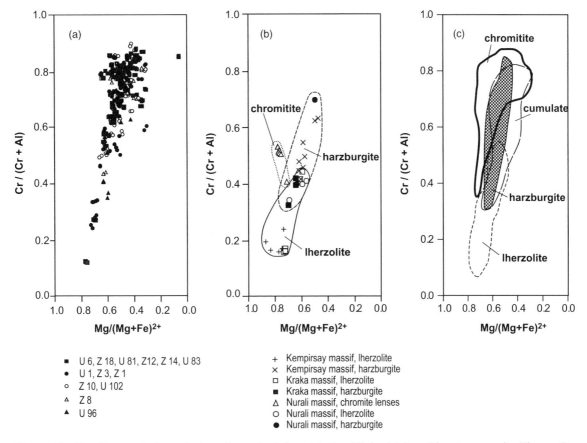

Figure 10. X_{Cr}–X_{Mg} variation of chromian spinel for: (a) the Zilair detritus (signatures as in Figure 4); (b) Southern Uralian ultramafic rocks [*Savelieva* et al., 1997; *Smirnov and Volchenko*, 1992]; (c) of typical ultramafic rocks in ophiolites [*Pober and Faupl*, 1988].

of 3.31 to 3.45 p.f.u. is phengite suggesting elevated pressures of formation. General deviation from the line of ideal tschermak substitution suggests variable Fe^{3+} contents up to about 0.2 p.f.u. (Figure 12a).

X_{Mg} in the Zilair Formation white mica is roughly positively correlated with Si in most samples (Figure 12b). Because this does not correlate with any substitution it should reflect changing temperature and pressure during formation of the source rocks. While the Si content is pressure dependent, X_{Mg} is temperature dependent due to Fe/Mg exchange with neighbouring Fe/Mg-phases. Ti contents show widest variation from nearly devoid to maxima of 0.037 to 0.087 p.f.u. (0.72 to 1.72 wt%). Most white micas have paragonite components below 10%, although values up to 28% and one paragonite were detected. X_{Mg} generally shows a wide range between 0.15 to 0.47 and 0.5–0.79.

The phengite compositions fit partly with those detected in high pressure rocks of the Maksyutovo

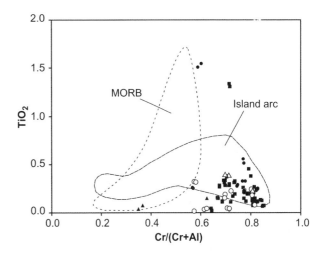

Figure 11. TiO_2-X_{Cr} variation for chromian spinel with $Fe^{3+} \geq 0.05$ p.f.u. Discrimination fields after *Arai* [1992].

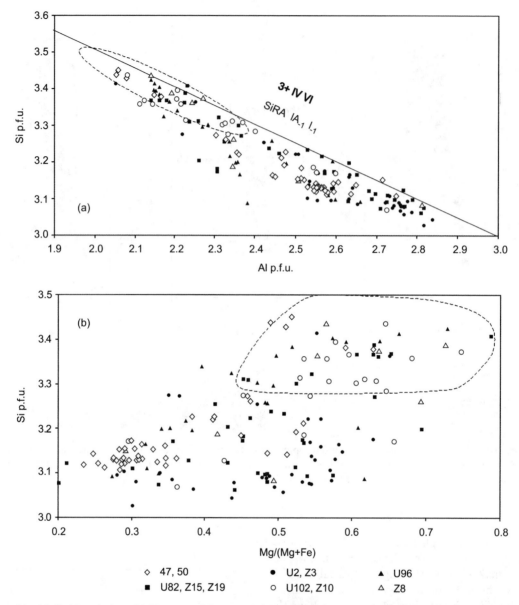

Figure 12. (a) Si-Al-variation, (b) $X_{Mg} = Mg/(Mg + Fe)$- Si variation in detrital white micas. Position of the ideal Tschermak substitution line is marked. Area surrounded by dashed line indicates the position of compositions of white micas from the Maksyutovo Complex [*Schulte and Blümel*, 1999]. Signatures as in Figure 4.

Complex (Figure 12) [*Schulte and Blümel*, 1999], but low Si compositions were yet not analysed in this complex. Unfortunately no white mica compositions are known from the Suvanyak Complex. It is very important to note that high Si-phengite (i.e., a high pressure signature) already occurs in the lowermost stratigraphic levels of the Zilair Formation. It can be concluded that white mica were derived from a metamorphic source which at least partly contained high pressure rocks.

7.3. Garnet

Garnet from ten samples (Figure 13; Table 2) show a similar compositional range of end member contents of almandine (43 to 84 mole%), pyrope (0.5 to 36 mole%), spessartine (0.5 to 27 mole%), grossular (0.5 to 27 mole%), andradite (0.5 to 12 mole%) and uvarovite (0 to 0.3 mole%). There is a significant overlap with the compositional field of garnet from the Maksyutovo Complex covering garnet from eclogite, blueschist

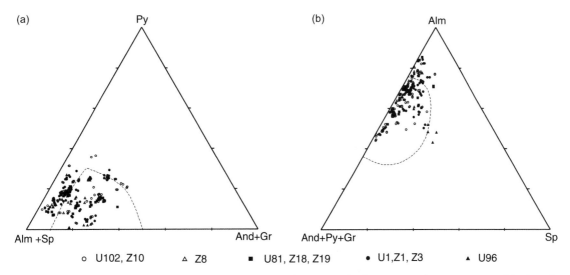

Figure 13. Triangular plots of (a) almandine + spessartine vs. andradite + grossular vs. pyrope; (b) almanine vs. spessartine vs. pyrope + andradite + grossular. Areas surrounded by dashed lines indicate the position of composition of garnet from the Maksyutovo Complex [*Hetzel* et al., 1998; *Schulte and Blümel*, 1999; *Lennykh and Valiser*, 1980]. Signatures as in Figure 4.

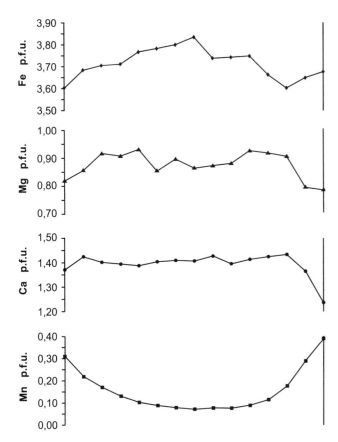

Figure 14. Inverse Mn zonation in garnet clast (sample Z19). Width of grain is 125 μm.

and metapelitic schist [*Hetzel* et al., 1998; *Schulte and Blümel*, 1999; *Lennykh and Valizer*, 1980].

Three types of garnet zonation can be observed (in spite of the fact that mostly fragments were analysed): unzoned grains (7 grains); bell-shaped zonations with Mn, Fe and Mg content decreasing from core to rim and increasing Ca content (1 grain); inverse Mn zonations with Mn content increasing from core to rim compensated by decreasing Mg, Ca and Fe contents (2 grains; Figure 14). The cores of large garnet in eclogites and mica schists from the Maksyutovo Complex are in fact largely unzoned, although weak bell-shaped zoning, particularly of Mn, is observed there [*Schulte and Blümel*, 1999; *Karsten* et al., 1994]. Inverse Mn zonation has only rarely been identified in the Maksyutovo Complex [*Karsten* et al., 1994]. According to *Tracy* [1982] such zoning is due to diffusion effects in unzoned garnet of high grade rocks during slow cooling.

7.4. Amphibole

Nearly all detrital amphibole detected during this study is Ca-amphibole. Ca-Amphibole shows similar composition throughout the Zilair Formation. All are magnesian with a range of $X_{Mg} = Mg/(Mg + Fe) = 0.54$ to 0.99. There is a continuous variation in composition from actinolite towards magnesiohornblende and tschermakite (Figure 15; Table 2). Some grains with $(Na + K)_A > 0.5$ p.f.u. plot into the field of pargasite. Ti content is generally low (0.04 to 0.16 p.f.u.), while Fe^{3+}

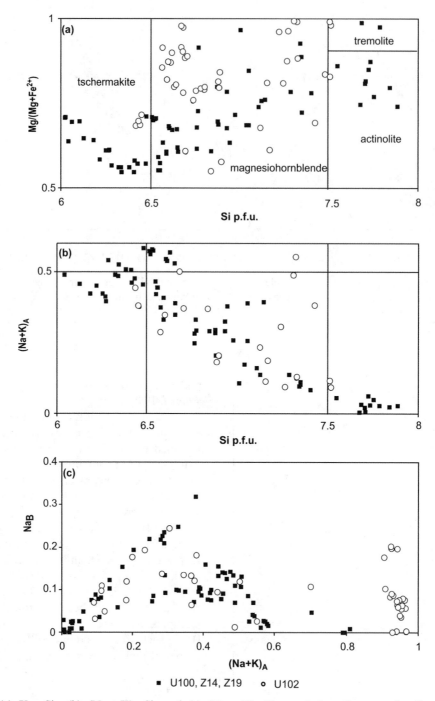

Figure 15. (a) X_{Mg}-Si-, (b) $(Na+K)_A$-Si- and (c) $(Na+K)_A$-Na_B-variation diagrams for Ca amphiboles. Signatures as in Figure 4.

can range from 0.1 to 1.1 p.f.u.. The range of Na_B is 0.00 to 0.32 p.f.u., that of Na_A 0.01 to 0.59 p.f.u.. X_{Mg} correlates with Si as well as $(Na+K)_A$ does. This points to temperature variation during the growth of the crystals in transition from greenschist to amphibolite facies, which is defined by continuous tschermakite

and edenite substitutions. On the other hand, Na_B plotted against $(Na+K)_A$ shows different populations with both parameters correlating positively or negatively. From the correlation pattern shown we suggest a metamorphic source for the detrital amphibole. Alternatively, Ca-amphibole might occur as asbestos veins

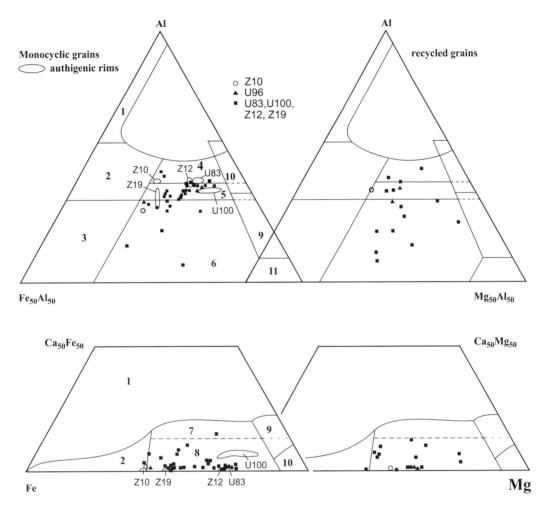

Figure 16. Al-Fe-Mg- and Ca-Fe-Mg-variation diagrams for tourmaline. Discrimination fields for various rock types according to *Henry and Guidotti* [1985] are as follows: 1 — Li-rich granitoids and pegmatites; 2 — Li-poor granitoids and pegmatites; 3 — hydrothermally altered granites; 4 — Al-rich metapelites and –psammites; 5 — Al-poor metapelites and –psammites; 6 — Fe^{3+}-rich metapelites and calcsilicates rocks; 7 — Ca-rich metapelites, –psammites and calcsilicate rocks; 8 — Ca-poor metapelites and –psammites; 9 — metacarbonates; 10 — ultramafics; the mixibility gap between elbaite and dravite is indicated. Signatures as in Figure 4.

within ultrabasic rocks. Aggregates of extremely long needles of Ca-amphibole might indicate such source.

The blue amphibole detected in sample U2 has a distinct glaucophane composition with $Na_B = 1.81$ to 1.90 p.f.u. and $X_{Mg} = 0.82–0.86$ (Table 2).

7.5. Tourmaline

The chemical composition of monocyclic detrital tormaline grains, recycled detrital grains (i.e., cores within the former; see above) and authigenic overgrowth is distinguished in Al-Fe-Mg-Ca-discrimination diagrams of *Henry and Guidotti* [1985] which allow a good assignment to many different source rocks (Figure 16; Table 2). Monocyclic and recycled detrital

tourmaline show the same range in all samples studied. They are all from a metasedimentary source and plot mainly into the fields of Al-poor, Fe^{3+}-rich and Ca-poor metapsammopelites. Few grains derived from Al-rich and Ca-rich metapelitic sources. No tourmaline of clear magmatic origin has been detected. Some monocyclic as well as recycled grains have somewhat elevated contents of Ti (0.09 to 0.3 p.f.u.; 0.7 to 2.6 wt% TiO_2), Cr (0.02 to 0.04 p.f.u.; 0.15 to 0.30 wt% Cr_2O_3) and Mn (0.02 to 0.05 p.f.u.; 0.1 to 0.3 wt% MnO). Composition of authigenic rims of the tourmaline invariably plot into the field of Al-poor metapsammopelites and slightly overlap with the field of Al-rich metapsammopelites.

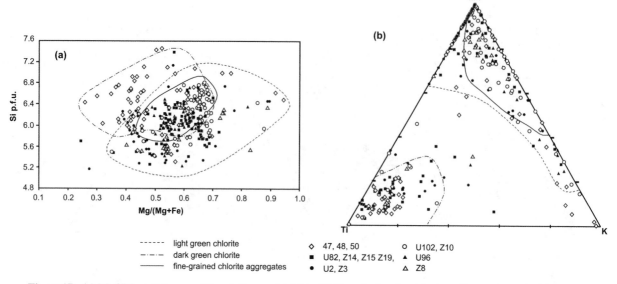

Figure 17. (a) Mg/(Mg + Fe) versus Si variation and (b) Mn-Ti-K variation for chlorites. Signatures as in Figure 4.

Based on the composition and zonation pattern (see above), it is evident that all tourmaline was derived from a low to medium grade metasedimentary source that contained recycled, rounded tourmaline clasts.

7.6. Epidote

The composition of 59 epidote grains from 9 samples shows a uniform pattern (Table 2). The high pistacite components range between 51 and 97 mole%, and the only significant minor elements are Ti (generally <0.02 p.f.u.; maximum 0.10 p.f.u.) and Mn (generally <0.04 p.f.u.; maximum 0.07 p.f.u.). Compositional uniformity suggests a uniform source, such as probably metabasites with similar metamorphic overprint.

7.7. Chloritoid

Chloritoid was detected in two samples only (Table 2). It has a X_{Mg} of 0.18 to 0.24 and low Mn content (0.01 to 0.24 p.f.u.). This fits well with the range known for chloritoids from the Maksyutovo Complex with X_{Mg} of 0.01 to 0.42 and Mn contents of 0.00 to 0.20 p.f.u. [*Schulte and Blümel*, 1999]. According to *Chopin and Schreyer* [1983] such compositions are typical for blueschist facies rocks rather than for low pressure rocks with often enhanced Mn contents.

7.8. Clinopyroxene

Analyses of pyroxene were made only from one sample (Table 2). It has a diopside composition with $X_{Mg} = 0.84$ and low contents of Na (0.01 to 0.03 p.f.u.)

and Al (0.05 to 0.12 p.f.u.). It is compatible with a magmatic origin.

7.9. Chlorite

Detrital chlorite grains are abundant in the Zilair Formation. Three groups of chlorite are optically distinguished; single light green chlorite clasts, dark green chlorite clasts, and fine grained chlorite aggegates. The latter were counted with the serpentinite clasts (see above). All types of chlorite are randomly distributed among samples of different age and zones.

The composition of the three groups of chlorites within 13 investigated samples (Table 2) is characterized as follows. The light green clasts have a wide range of X_{Mg} and Si (Figure 17a), while the composition of the dark green chlorite and the fine-grained chlorite aggregates are more restricted. In a Mn-Ti-K-plot (Figure 17b) the distinct composition of the dark green chlorites becomes more apparent. They are particularly Ti-rich, suggesting a precursor Ti-rich phase and also show a trend towards K-enrichment. In back scattered electron images K-rich domains ≤ 1 μm parallel to (001) planes suggest interlayering of chlorite with a K-bearing phyllosilicate. Similar chlorites, which were interpreted as former biotite, have also been observed in Vendian detritus by *Willner* et al. [2001].

8. DISCUSSION

The strong angularity of the majority of the clasts, the presence of low stability minerals in the detritus such as

K-feldspar, amphibole, pyroxene, biotite (decomposed), garnet or carbonate and the frequent local enrichment of specific heavy minerals point to a very low degree of transport abrasion and therefore a nearby source area. A relatively low quartz/feldspar ratio, as well as the presence of carbonate clasts, point to a low semiquantitative weathering index of 0 resulting from ln (Q/F) = −0.4 to −0.9 versus ln (Q/L− 1.28 to + 0.78) according to the scheme of *Weltje* [1994]. This may indicate a high relief independent of climate. The high amount of intraclasts, the predominance of grain flow and proximal turbiditic deposits and the presence of olistostrome deposits at the base would be in line with this. It is also suggested that the degree of modification of the detritus during weathering and erosion from its source, its transport to the depositional basin, its deposition and its diagenetic interaction with pore waters and neighbouring minerals is relatively low and, therefore, the composition of detritus is mainly defined by its provenance.

No unambiguous variation of the provenance signal in the Zilair Formation was detected, neither in space nor in time. Furthermore, no diachronous evolution of the Zilair Formation is evident from the known biostratigraphical record. On the other hand, geodynamical conditions changed dramatically before and after deposition of this unit. According to the time scale of *Harland* et al. [1982] the biostratigraphical age of the Zilair Formation was equivalent to 367–356 Ma. More recent studies [*Tucker* et al., 1998] place the lower boundary of the Famennian at 376.5 Ma. This relatively short time span overlaps with the above cited record of isotope age range of 387 Ma to 357 Ma in the lower unit of the Maksyutovo Complex, where different minerals were dated using different isotopic systems. According to *Beane and Connelly* [2000] this dates the early exhumation of the high pressure rocks after peak metamorphic conditions around 385 Ma. Hence there is an evident time relation between this exhumation and surface uplift evidenced by the Zilair Formation. This uplift must have stopped with the end of the deposition of the Zilair Formation, when a Lower Carboniferous carbonate platform developed and convergence ceased. According to the isotope ages of *Beane and Connelly* [2000] amalgamation of the lower and upper units of the Maksyutovo Complex within a mid crustal level occurred at that time. However, Ar–Ar ages of 339–332 Ma [*Beane and Connelly*, 2000] and apatite fission track data around 315 Ma [*Leech and Stockli*, 2000] in the upper unit of the Maksyutovo Complex show that the presently exposed rocks of the Maksyutovo Complex were definetely exhumed to the surface much later during, the second period of Uralide convergence. In this respect only the detritus of the Zilair Formation can give indications of the surface lithologies during the first period of convergence. However, we see no indications for a major oblique convergence, which should result in a more pronounced variation of detritus composition and diachronous development.

In contrast to the underlying sandstones in the Magnitogorsk arc, the Zilair Formation contains a substantial amount of clasts other than of volcanic and sedimentary origin. This points to the arrival of a further source other than the magmatic arc. Of particular importance is the ophiolitic detritus. Ultrabasic rocks have a minor representation among light mineral clasts. Analyses of many serpentinite-like clasts turned out to be fine-grained chlorite aggregates, which nevertheless can be derived from altered ultramafic rocks as well as some fibrous Ca-amphibole clasts. However, chromian spinel is one of the dominant single transparent heavy mineral species indicating that abundant ultrabasic rocks must have occurred in the source region. Their chemistry points to the entire spectrum of ultrabasic and minor basic rock types (of island arc derivation). These potential source rocks are present in the Main Uralian fault zone and within the Kraka and Sakmara nappes. Chert lithoclasts may also point to an ophiolithic source. This also shows that the Main Uralian fault zone already formed an axial rise over a backstop system separating the accretionary prism in the west and the fore-arc basin in the east [see *Brown* et al., 1998, 2000, 2001].

Although metamorphic rocks do not dominate among lithoclasts (Figure 5), they seem to have provided the largest volume of the total Zilair Formation. Apart from the metamorphic lithoclasts, some sedimentary lithoclasts may have been derived from very low to low grade metasediments. Metamorphic detrital light minerals include white mica, chlorite, most quartz clasts, potassic feldspar and a minor part of albite. Among transparent heavy minerals the total of metamorphic heavy minerals clearly dominates, i.e., the sum of epidote, garnet, tourmaline, amphibole, titanite and chloritoid (presumably also most or all rutile and apatite). A source region for the Zilair Formation appears to have been dominated by greenschist to amphibolite facies rocks. The presence of phengite with up to 3.4 to 3.45 Si p.f.u. suggests minimum pressures of formation in the range of 5 to 12 kbar at 300 to 600°C [according to the barometer by *Massonne and Schreyer*, 1987 and improved by *Massonne*, 1991] indicating the presence of medium to high pressure rocks in at least some part of the area. This is also in line with the occurrence of glaucophane. Based on tourmaline compositions, Al- and Ca-poor, partly Fe^{3+} rich metapsammopelitic rocks appear to have been a source for the Zilair Formation apart from greenschists, mylonites and epidosite.

Common greenschists and phyllite clasts could be derived from the Suvanyak Complex, which, however, is devoid of garnet, chloritoid and glaucophane. Although the chemical compositions of white mica, garnet and chloritoid partly fit with those determined from the Maksyutovo Complex the minor representation of glaucophane and the abundance of Ca-amphibole make the presently exposed rocks of the Maksyutovo Complex an unlikely source, because blue amphibole is abundant here and Ca-amphibole were not described yet except some retrograde actinolite rims [*Schulte and Blümel*, 1999]. This is also in line with the much later exhumation of the Maksyutovo rocks to the surface [*Beane and Connelly*, 2000; *Leech and Stockli*, 2000]. However, garnet, Ca-amphibole, phengite and chloritoid occurs in the nearby Precambrian metamorphic Beloretzk Complex [*Glasmacher* et al., 2001] which was part of the East European Craton continental margin (Figure 1). Nevertheless, this also seems to be an unlikely source, because its present exposure was still under cover of a carbonate platform during deposition of the Zilair Formation [e.g., *Puchkov*, 1997, 2000]. Most probably high pressure rocks were exhumed to the surface in front of the backstop system which were originally structurally overlying the presently exposed Maksyutovo high pressure rocks and which are eroded now. Similar distribution of phengite and garnet compared with that of chromian spinel throughout the Zilair Formation would suggest exposure of such earlier high pressure rocks within the axial rise parallel to the belt of ultramafic rocks similar to the Maksyutovo Complex.

Evidently exhumation of the Maksyutovo high pressure rocks was relatively slow [*Beane and Connelly*, 2000] and even an exhumation rate of only 0.3 to 1.5 mm/a was estimated by *Leech and Stockli* [2000]. Nevertheless erosion could have played only a minor part as an unroofing process considering even high erosion rates. A simple balancing makes this apparent. Total volume calculated from the paleogeographic map (Figure 2) suggests a volume of about 7.7×10^4 km^3 for the Zilair Formation. Under the assumption that even all was eroded from the present exposure area of the Suvanyak and Maksyutovo metamorphic complexes, we may place this volume on this surface in Figure 1 (2.8×10^4 km^2). The thickness of eroded Zilair material would only be 2.7 km compared to rocks coming from up to 60 km depth in the metamorphic complexes. However, original surface exposure of the metamorphic rocks during erosion might have been considerably smaller.

Volcanic lithoclasts of basic to acid composition are abundant in the Zilair Formation. Most euhedral albite clasts (some partly filled with carbonate) also come from a volcanic source. On the other hand, the presence of lithoclasts derived from plutonic rocks is doubtful and evidence for a magmatic source among heavy minerals is poor. Only euhedral long prismatic zircon could be assigned to a volcanic source as also possibly some apatite needles. Approximately 10% of chromian spinel comes from a basaltic source (or their cumulates). The source of volcanic lithoclasts may have been volcanic units of the Magnitogorsk arc or volcanic rocks that are found as blocks within the Main Uralian fault zone. An uplifting axial rise along the Main Uralian fault zone contributing ultrabasic, volcanic and metamorphic detritus to either side might presumably have shielded any detritus transport from the Magnitogorsk arc toward the west into the Zilair nappe. However, on both sides of the Main Uralian fault the same detritus signature prevails. This shows that either the axial rise was not a continuous high relief feature or the volcanic clasts came essentially from this axial rise and not from the Magnitogorsk arc at all.

Finally, sedimentary clasts dominate among lithoclasts. Unfortunately these cannot be used as clear indicators of provenance. For example, minor carbonate clasts could also have been derived from the carbonate platform in the west or from marbles in the metamorphic complexes, siliceous sedimentary lithoclasts on the other hand from some siliceous sediments in the Main Uralian fault zone or in the Magnitogorsk arc. Probably most of the sedimentary clasts represent intraclasts being refcycled, when the detritus was deposited into the accretionary prsism in the west and the forearc basin in the east.

9. CONCLUSIONS

The time of deposition of the Zilair Formation (376 to 356 Ma) correlates with the time of exhumation of the continental high pressure rocks and related surface uplift in the rear of the Uralide collisional accretionary prism. An exhumation channel within a steep low viscosity zone flanked by the backstop damage zone along the Main Uralian fault developed with continuous buoyancy-driven exhumation of high pressure rocks during concomitant underthrusting by ongoing subduction (*Brown* et al., 2000). The Zilair detritus, which was also mixed with eroded material from the Magnitogorsk arc was distributed to either side of this rise of elevated relief into the accretionary prism in the west and the suture forearc basin in the east. The end of the deposition of the Zilair Formation marks the end of the arc-continent collison, when a carbonate platform developed indicating a general low relief again.

Our studies of light and heavy mineral distribution as well as the variation of the chemical composition of solid solution minerals did not reveal any significant difference during the entire period of deposition of the Zilair Formation to either the east or the west of the Main Uralian fault zone. This may be interpreted to mean that:

(1) there was presumably no change in the type of source region during time. Although it cannot be excluded, the detritus spectrum does not suggest that oblique collision occurred in the Southern Urals.

(2) The accretionary complex, in particular the backstop damage zone containing ultramafic rocks along the Main Uralian fault and metamorphic complexes to the west of it with high pressure rocks, contributed most and approximately equal amounts of detritus throughout the Zilair Formation. This indicates that it had significant exposure.

(3) It appears that ultramafic rocks and low to medium grade metamorphic rocks including high pressure rocks were at the surface from the beginning of the deposition of the Zilair Formation. This could point to at least initially rapid exhumation.

Alvarez-Marron et al. [2000] and *Brown* et al. [2001] suggest that the preservation of the Uralide arc-continent collision system correlates well with recent analogs along the northern Australian continental margin. A similar setting as the Zilair Formation in a forearc position of an active arc-continent collisional system is given by the Erap Complex in eastern Papua New Guinea [*Abbott* et al., 1994], where a complex collision of the Australian plate with terranes including active arcs is currently taking place. Similarly mixed metasedimentary and volcanic detritus is observed. However, recent exhumation of high pressure rocks including eclogites within continental crust was reported in the same system somewhat further south in the D'Entrecastaux Islands [*Hill* et al., 1992]. Here a rapidly uplifting and eroding ridge is currently contributing detritus to either side.

Acknowledgments. Financial support by the Deutsche Forschungsgemeinschaft (grants Wi 875-6/1 and 6/2) is gratefully acknowledged. We thank B. Schulte for further unpublished mineral analyses from the Maksyutovo Complex supplementing his published data. Perfect logistic support by the Ufa crew and invaluable field guidance by V. I. Kozlov and A. Alekseev during the 1999 field leg made this work possible. G. Zuffa, an anonymous reviewer and D. Brown helped to improve the manuscript substantially. We also thank D. Brown for his careful editorial handling and his patience. This paper is a contribution to EUROPROBE [URALIDES], an International Lithosphere Program and sponsored by the European Science Foundation.

REFERENCES

Abbot, L. D., E. A. Silver, P. R. Thompson, M. V. Filewicz, C. Schneider and Abdoerrias, Stratigraphic constraints on the development and timing of arc-continent collision in northern Papua New Guinea, *J. Sediment. Res.*, B64, 169–183, 1994.

Abramova, A. N., V. N. Maslov, O. V. Artyushkova and V. N. Baryshev, About the lower boundary of the Zilair Formation in the section near Yaumbaevo Village (in Russian), in *Yearbook-96 Inst. Geology Ufimian Sci. Centre Ufa*, 32–34, 1998.

Alvarez-Marron, A., D. Brown, A. Pérez-Estaún, V. Puchkov and Y. Gorozhanina, Accretionary complex structure and kinematics during Paleozoic arc-continent collision in the Southern Urals, *Tectonophysics*, 325, 175–191, 2000.

Allen, J. F., Cr-spinel as a petrogenetic indicator: Deducing magma composition from spinels in highly altered basalts from the Japan Sea, sites 794 and 797, *Proc. ODP Sci. Results*, 127/128, 837–847, 1992.

Arai, S., Chemistry of chromian spinel in volcanic rocks as a potential guide to magma chemistry, *Mineral. Mag.*, 56, 173–184, 1992.

Artyushkova, V. and V. Maslov, Paleontologic justification of stratigraphic subdivision of the pre-Famennian volcanic complexes of the Verkhneuralsk and Magnitogorsk regions (in Russian), 156 pp., *Institute of Geology, Ufa*, 1998.

Arzhavitina, M. Y., Mineral composition of the terrigenous rocks of the Zilair-Formation in the Magnitogorsk syncline (in Russian), in *Mineral Chemistry of Volcanogenic and Sedimentary Deposits of the Southern Urals*, Geol. Inst. USSR Akad. Sci. Ufa, 48–55, 1978.

Arzhavitina, M. Y. and P. V. Arzhavitin, Mineral components of the terrigenous flysch in the Magnitogorsk Syncline (in Russian), in *Nappe Tectonics and its Importance for Ore Deposits*, USSR Akad. Science Ufa, 82–83, 1991.

Bastida, F., J. Aller, V. N. Puchkov, C. Juhlin and A. Oslansky, A cross-section through the Zilair Nappe [southern Urals], *Tectonophysics*, 276, 253–264, 1997.

Basu, A., S. W. Young, L. J. Suttner, W. C. James and G. H. Mack, Reevaluation of the use of undulatory extinction and polycrystallinity in detrital quartz for provenance interpretation, *J. Sediment. Petrol.*, 45, 873–882, 1978.

Beane, R. J. and J. N. Connelly, $^{40}Ar/^{39}Ar$, U-Pb and Sm-Nd constraints on the timing of metamorphic events in the Maksyutov Complex, southern Ural Mountains, *J. Geol. Soc. London*, 157, 811–822, 2000.

Berzin, R., O. Onken, J. H. Knapp, A. Pérez-Estaún, T. Hismatulin, A.Yunusov and A. Lipilin, Orogenic evolution of the Ural Mountains: Results from an integrated seismic experiment, *Science*, 274, 220–221, 1996.

Brown, D., J. Alvarez-Marron, A. Perez-Estaun, Y. Gorozhanina, V. Baryshev and V. N. Puchkov, Geometric and kinematic evolution of the foreland thrust and fold belt in the southern Urals, *Tectonics*, 16, 551–562, 1997.

Brown, D., C. Juhlin, J. Alvarez-Marron, A. Perez-Estaun and A. Oslianski, Crustal-scale structure and evolution of an arc-continent collision zone in the southern Urals, Russia. *Tectonics*, 17, 158–171, 1998.

Brown, D. and P. Spadea, Processes of forearc and accretionary complex formation during arc-continent collision in the Southern Ural Mountains, *Geology*, 27, 649–652, 1999.

Brown, D., R. Hetzel and J. H. Scarrow, Tracking arc-continent collision subduction zone processes from high-pressure rocks in the southern Urals, *J. Geol. Soc. London*, 157, 901–904, 2000.

Brown, D., J. Alvarez-Marrón, A. Pérez-Estaún, V. N. Puchkov, Y. Gorozhanina and P. Ayarza, Structure and evolution of the Magnitogorsk forearc basin: Identifying upper crustal processes during arc-continent collision in the southern Urals, *Tectonics*, 20, 364–375, 2001.

Chemenda, A. I., P. Matte and V. Sokolov, A model of Paleozoic obduction and exhumation of high-pressure/low-temperature rocks in the southern Urals, *Tectonophysics*, 276, 217–227, 1997.

Chibrikova, V., Age and stratigraphy of the Zilair deposits of the Urals (in Russian), *Otechestvennaya Geologiya*, 12, 31–35, 1997.

Chopin, C. and W. Schreyer, Magnesiocarpholite and magnesiochloritoid: Two index minerals of pelitic blueschists and their preliminary phase relations in the model system MgO-Al₂O₃-SiO₂-H₂O, *Am. J. Sci.*, 283, 72–96, 1983.

Dickinson, W. R. and C. A. Suczek, Plate tectonics and sandstone compositions. *AAPG Bull.*, 63, 2164–2182, 1979.

Dickinson, W. R., Interpreting provenance relations from detrital modes of sandstones, in *Provenance of arenites*, edited by G. G. Zuffa, *Nato ASI Series C*, 148, 333–361, Reidel Publishing Co., Dordrecht, 1985.

Dick, H. J. B. and T. Bullen, Chromian spinel as a petrogenetic indicator in abyssal and alpine-type peridotites and spatially associated lavas, *Contrib. Mineral. Petrol.*, 86, 54–76, 1984.

Folk, R. L., *Petrology of sedimentary rocks*, 182 pp., Hemphill Publishing Company, Austin/Texas, 1980.

Gebauer, D., H. P. Schertl, M. Brix and W. Schreyer, 35 Ma old ultrahigh-pressure metamorphism and evidence for very rapid exhumation in the Dora Maira Massif, Western Alps, *Lithos*, 41, 5–24, 1997.

Giese, U., U. A. Glasmacher, V. I. Kozlov, I. Matenaar, V. N. Puchkov, L. Stroink, W. Bauer, S. Ladage and R. Walter, Structural framework of the Bashkirian anticlinorium, SW Urals, *Geol. Rundsch.*, 87, 526–544, 1999.

Glasmacher, U. A., W. Bauer, U. Giese, P. Reynolds, B. Kober, V. N. Puchkov, L. Stroink, A. Alekseyev and A. P. Willner, The metamorphic complex of Beloretzk, SW Urals, Russia - a terrane with a polyphase Meso- to Neoproterozioic thermodynamic evolution, *Precam. Res.*, 110, 185–213, 2001.

Harland, W. B., A. V. Cox, P. G. Llewllyn, C. A. G. Pickton, A. G. Smith and R. Walters, *A Geologic Time Scale*, 131 pp., Cambridge University Press, 1982.

Henry, D. J. and C. V. Guidotti, Tourmaline as a petrogenetic indicator mineral: an example from the staurolite-grade metapelites of NW Maine, *Am. Mineral.*, 70, 1–15, 1985.

Henry, D. J. and B. J. Dutrow, Metamorphic tourmaline and its petrologic applications. in *Boron: Mineralogy, Petrology and Geochemistry*, edited by E. S. Grew and L. M. Anovitz, *Rev. Mineral.*, 33, 503–558, Mineral. Soc. Am., 1996.

Hetzel, R., H. P. Echtler, W. Seifert, B. A. Schulte and K. S. Ivanov, Subduction- and exhumation-related fabrics in the Paleozoic high pressure/low-temperature Maksyutov complex, Antingan area, southern Urals, Russia. *Geol. Soc. Am. Bull.*, 110, 916–930, 1998.

Hetzel, R., Geology and geodynamic evolution of the high-P/low-T Maksyutov Complex, southern Urals, Russia, *Geol. Rundsch.*, 87, 577–588, 1999.

Hill, E. J., S. L. Baldwin and G. S. Lister, Unroofing of active metamorphic core complexes in the D'Entrecasteaux Islands, Papua New Guinea, *Geology*, 20, 907–910, 1992.

Ingersoll, R. V. and C. A. Suzcek, Petrology and provenance of Neogene sand from Nicobar and Bengal fans, DSDP Sites 211 and 218, *J. Sediment. Petrol.*, 49, 1217–1228, 1979.

Il'inskaja, M. N., Sandstone composition of the Zilair Formation in the Southern Urals (in Russian), in *Lithology and Mineral Deposits*. USSR. Akad. Science Ufa, 55–66, 1980.

Karsten L. A., K. S. Ivanov and V. N. Puchkov, New data on the geological structure and metamorphism of the Maksyutovo eclogite-glaucophane-schist complex (Southern Urals) (in Russian), *Ezhegodnik-1993*, Institute of Geology and Geochemistry, Ekaterinburg, 20–25, 1994.

Keller, B. M., The Paleozoic flysch formation in the Zilair synclinorium of the Southern Urals and comparable complexes (in Russian), *Geol. Inst. Academy of Sci. USSR*, 104, N34, Moscow, 165 pp., 1949.

Librovich, L. S., On the geology of the southern part of the Bashkirian Urals (in Russian), *The All Union Geological Prospecting Incorporation*, 144, Leningrad-Moscow, 66 pp., 1932.

Leech, M. L. and D. F. Stockli, Exhumation history of the ultrahigh-pressure Maksyutov Complex, south Ural mountains, from new apatite fission-track data. *Tectonics*, 19, 153–167, 2000.

Lennykh V. I. and P. M. Valizer, Garnets from eclogites and glaucophane schists of the Polar and South Urals, in *Garnets in Metamorphic Complexes of the Urals*, Sverdlovsk, UNC AN SSSR, 22–37, 1980.

Lennykh, V. I., P. M. Valizer, R. Beane, M. Leech and W. G. Ernst, Petrotectonic evolution of the Maksyutov Complex, southern Urals, Russia: implications for ultra-high-pressure metamorphism. *Int. Geol. Rev.*, 37, 584–600, 1995.

Lennykh, V. I. and P. M. Valizer, High pressure metamorphic rocks of the Maksyutov Complex (Southern Urals), *Field Guide 4. Internat. Eclogite Field Symp.*, 64 pp., *Russian Academy of Science Novosibirsk*, 1999.

Maslov V. A., V. L. Cherkasov, V. T. Tischenko, A. I. Smirnova, O.V. Artyushkova and V. V. Pavlov, Stratigraphy and correlation of the Middle Paleozoic volcanic complexes of the main copper-sulphide areas on the Southern Urals (in Russian), *Ufimian Sci.Centre RAS, Ufa*, 217 pp., 1993.

Massonne, H.-J. and W. Schreyer, Phengite geobarometry based on the limiting assemblage with K-feldspar, phlogopite and quartz, *Contrib. Mineral. Petrol.*, 96, 212–224, 1987.

Massonne, H.-J., *High-pressure, Low-Temperature Metamorphism of Pelitic and Other Protoliths Based on Experiments in the System K_2O-MgO-Al_2O_3-SiO_2-H_2O*, unpublished thesis of habilitation, Univ. Bochum, Germany, 172 pp., 1991.

Matte, P., H. Maluski, R. Caby, A. Nicholas, P. Kepezhinkskas and S. Sobolev, Geodynamic model and $^{39}Ar/^{40}Ar$ dating for the generation and emplacement of the high pressure metamorphic rocks in the SW Urals, *C. R. Acad. Sci. Paris*, 317, 1667–1674, 1993.

Pazukhin, V. N., V. N. Puchkov and V. N. Baryshev, New data on the stratigraphy of the Zilair series (Southern Urals) (in Russian), in *Year-book '95 Inst. Geol. Ufimian Sci. Centre RAS, Ufa*, 34–41, 1996.

Pober, E. and P. Faupl, The chemistry of detrital chromian spinels and its implications for the geodynamic evolution of the Eastern Alps, *Geol. Rundsch.*, 77, 641–670, 1988.

Puchkov, V. N., Structure and geodynamics of the Uralian orogen, in *Orogeny Through Time*, edited by J.-P. Burg and M. Ford, *Geol. Soc. London Spec. Publ.*, 121, 201–236, 1997.

Puchkov, V. N., V. N. Baryshev and V. N. Pazukhin, New data on the stratigraphy of terrigenious-siliceous Devonian at the western slope of the Bashkirian Urals (in Russian), in: *Year-book '96 Inst. of Geology Ufimian Sci. Centre RAS, Ufa*, 24–31, 1998.

Puchkov, V. N., *Paleogeodynamics of the Southern and Middle Urals* (in Russian), 146 pp., Dauria, Ufa, 2000.

Savelieva, G. N., A. Y. Sharaskin, A. A. Saveliev, P. Spadea and L. Gaggero, Ophiolites of the southern Uralides adjacent to the East European continental margin, *Tectonophysics*, 276, 117–137, 1997.

Schulte, B. A. and P. Blümel, Metamorphic evolution of eclogite and associated garnet-mica schist in the high-pressure metamorphic Maksyutov complex, Ural, Russia, *Geol. Rundsch.*, 87, 561–576, 1999.

Seravkin, I., A. M. Kosarev and D. N. Salikhov, *Volcanism of the Southern Urals* (in Russian), 195 pp., Nauka, Moscow, 1992.

Sinitsina Z. I., I. I. Sinitsin and D. F. Shamov, Brief stratigraphical essay on the Upper Paleozoic of the Southern Urals, in *Guidebook for the Southern Urals Excursion 047*,edited by O. L. Eynor, I. I. Sinitsin and A. Kochetkova, 27th International Geological Congress Moscow, *Nauka*, 9–19, 1984.

Shatsky, V. S., E. Jagoutz and O. A. Koz'menko, Sm-Nd dating of the high-pressure metamorphism of the Maksyutov Complex, southern Urals, *Doklady [Earth Sci. Section]*, 353, 285–288, 1997.

Smirnov S. V. and Y. A. Volchenko, The first find of platinoid mineralization in chromite ores in the Nurali massif on the Southern Urals (in Russian), in *Year-book-1991, Institute of Geology and Geochemistry Uralian Branch RAS, Ekaterinburg*,115–117, 1992.

Tracy, R. J., Compositional zoning and inclusions in metamorphic minerals, in *Characterization of metamorphism through mineral equilibria*, edited by J. M. Ferry, *Reviews in Mineralogy*, 10, 355–393, Mineralogical Society of America, 1982.

Tucker, R. D., D. C. Bradley, C. A. Ver Straeten, A. G. Harris, J. R. Ebert and S. R. McCutcheon, New U-Pb zircon ages and the duration and division of Devonian time, *Earth Planet. Sci. Letters*, 158, 175–186, 1998.

Weltje, G. J., Provenance and dispersal of sand-sized sediments, *Geol. Ultraiectina*, 121, 1–208, 1994.

Willner, A. P., T. Ermolaeva, L. Stroink, U. Giese, U. A. Glasmacher, V. M. Puchkov, V. I. Kozlov and R. Walter, Contrasting provenance signals in Riphean and Vendian sandstones in the SW Urals (Russia): constraints for a change from passive to active continental margin conditions in the Late Precambrian, *Precamb. Res.*, 110, 215–239, 2001.

Zonenshain, L. P., M. I. Kuzmin and L. M. Natapov, Geology of the USSR: a plate-tectonic synthesis, *Geodynamic series*, 21, Amer. Geophys. Union, Washington, D.C., 242 pp., 1990.

A. Willner, Institut für Geologie, Ruhr-Universitat Bochum, D-44780 Bochum, Germany (arne.willner@ruhr-uni-bochum.de)

T. Ermolaeva and U. Kramm, Institut für Mineralogie und Lagerstättenlehre, RWTH Aachen, D-52062 Aachen, Germany

Y. Gorozhanina, V. Puchkov, M. Arzhavitina and V. Pazukhin, Ufimian Geoscience Centre, Russian Academy of Sciences, 16/2 K. Marx, Ufa 450000, Russia

T. Ermolaeva and R. Walter, Geologisches Institut, RWTH Aachen, D-52062 Aachen, Germany

Granitoids of the Uralides: Implications for the Evolution of the Orogen

F. Bea[1], G. B. Fershtater[2] and P. Montero[1]

[1]*University of Granada, Spain*
[2]*Russian Academy of Sciences, Ekaterinburg, Russia*

Uralide granitoids formed at an almost constant rate from 370 Ma to 250 Ma, first in subduction and then in intracontinental settings. Subduction-related granites were produced in two episodes. The first occurred about 370 Ma to 350 Ma in the eastern sector of the East Uralian zone. It was caused by a subduction zone located to the east of the Devonian Magnitogorsk arc, which had already collided with the East European Craton, dipping to the east underneath the Siberia-Kazakhstan continent. It produced the batholiths of the Valerianovsky arc, composed of I-type granitoids with a recognizable continental component. The second episode occurred about 335 Ma to 315 Ma in the western part of the East Uralian zone, between 55° N and 58° N. It was caused by a subduction zone located to the east of the Silurian Tagil arc, already accreted to the East European craton, dipping to the east underneath the Valerianovsky arc. It generated batholiths composed of I- and M-type granitoids with little, if any, continental component. The gabbro-granite plutons of the Magnitogorsk lineament are coeval with Late Carboniferous subduction granites, but they formed, however, in an extensional or passive geodynamic regime, probably from basaltic arc-magmas highly enriched in incompatible elements and volatiles. The magmatic activity directly related to subduction ended after the Carboniferous. The generation of granites, in contrast, continued during the Permian, first in the southern part (292 Ma to 280 Ma) and then in the northern part (270 Ma to 250 Ma) of the East Uralian zone. Permian granites are high in SiO_2, mildly peraluminous, with elevated Rb, Cs, Ba, Th and U, but with an unusually primitive Sr and Nd isotope composition. They derived from recycling of the Valerianovsky arc materials deeply buried after the collision. Permian crustal melting was driven by the combination of three mechanisms: (1) radiogenic heating of an overthickened sialic crust that did not undergo late-orogenic extension (2) local underplating

Mountain Building in the Uralides: Pangea to the Present
Geophysical Monograph 132
10.1029/132GM11

by mafic magmas and (3) local accumulation of heat and fluids related to the oblique, crustal-scale strike-slip shear zones that finally assembled the Middle Urals.

1. INTRODUCTION

From ~370 Ma to ~250 Ma, the Uralides were the locus of sustained granite magmatism related to the closure of the Uralian paleocean and subsequent collision between the East Baltica, Kazakh and Siberian continental plates [*Hamilton*, 1970; *Fershtater*, 1984; *Zonenshain* et al., 1990; *Puchkov*, 1997]. These events produced numerous medium to small-sized batholiths composed of diverse associations of granitoids whose typology and associated ore deposits show a marked spatial and temporal variability, presumably reflecting the transition from subduction to collision [*Fershtater* et al., 1998a]. Due to their remarkable economic importance and scientific interest, the granites of the Uralides have been the object of numerous studies [e.g., *Fershtater* et al., 1994, and references therein]. During the 1990's, research on Uralide granites underwent a new surge owing to the cooperation (within the framework of the **URALIDES-EUROPROBE** project) of the Ekaterinburg Institute of Geology and Geochemistry, Russia, with the University of Granada, Spain. A joint team of granite petrologists, geochemists and geochronologists from both Institutions launched a long term project for studying the geology and petrogenesis of the main Uralide batholiths [*Fershtater and Bea*, 1993; *Fershtater and Bea*, 1996; *Bea* et al., 1997; *Fershtater* et al., 1997; *Montero* et al., 2000; *Gerdes* et al., 2001] whose most important results, including new and still unpublished data on trace elements, Sr and Nd isotopes, and zircon chronology, are summarized here.

The aim of this paper is to give an updated overview on the Uralide granitoids and their evolution in space and time. For this purpose, we have first grouped the granite batholiths according to their age and position. Then, we briefly describe the geology, mineralogy, chemical and isotopic composition, and zircon chronology of twelve selected examples that represent the whole spectrum of granite rocks found in the Uralides. We finally discuss the most important implications of granite studies, especially those derived from timing, nature of the source rock, input of energy for melting and polarity of successive magmatic arcs, for understanding the evolution of the Uralide orogen. We focus on the sector between the latitudes 53° N and 58° N, which comprises the majority of the Uralide granitoids except for some minor bodies in the north, associated with the mafic-ultramafic complexes of the Platinum-Bearing Belt, and the batholith swarm of Neplyuchevsk, Suunduk, etc. in the south, about which detailed information can be found in *Samarkin and Samarkina* [1988].

2. GEOLOGICAL BACKGROUND

The Uralides form a nearly linear N-S-trending upper Paleozoic orogenic belt that extends for more than 2500 km between the eastern European and the westernSiberian-Kazakh cratons [*Zonenshain* et al., 1990; *Matte*, 1995]. The hinterland of the Middle and South Urals is a narrow area bounded by the Main Uralian fault in the west and by the Chelyabinsk fault in the east (Figure 1). From west to east, it consists of: (i) the Suture Sector, (ii) two well-preserved volcanic arcs, Tagil in the north and Magnitogorsk in the south, and (iii) the East Uralian zone, separated from the Tagil and Magnitogorsk arcs by the Serov-Mauk and the Uisk-Kazbakhsk faults respectively. Between 55° 0′ N and 55° 40′ N, the East Uralian zone is directly in contact with the Main Uralian fault. West of the Tagil arc, between 58° N and 64° N, approximately, there is a N-S lineament of concentrically-zoned dunite-clinopyroxenite-gabbro intrusive complexes (Alaskan-type) that form the Platinum-Bearing Belt [*Efimov and Efimova*, 1967; *Fershtater* et al., 1999].

The Suture Sector is a 1 km to 5 km wide melange zone located immediately east of the Main Uralian fault. It is characterized by an intense positive gravimetric anomaly and abundant mafic-ultramafic complexes which, at least in part, represent fragments of Uralide paleoceanic crust. The Silurian Tagil and the Devonian Magnitogorsk arcs are mainly composed of volcanic rocks and volcanoclastic sediments. The Tagil arc does not host Carboniferous or Permian granite batholiths; the Magnitogorsk arc, however, hosts a lineament of Late Carboniferous high level gabbro-granite plutons (the Magnitogorsk lineament) and the Early Carboniferous Akhunovo granite batholith (Figure 1).

The East Uralian zone contains most of the Uralide granitoids, with medium- to small-sized batholiths usually aligned NS, NNE or SSW (Figure 1). It also contains several belts of strongly sheared melange rocks, with abundant serpentinites, which represent minor

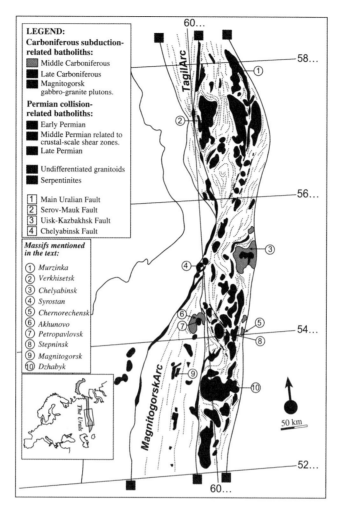

Figure 1. Geological sketch of the Uralides between 58° N and 52° N with the names of the plutonic complexes mentioned in the text except Kytlym and Khabarny, which are outside this area located in the north and the south, respectively.

sutures marking the boundaries between different terranes. The granites are hosted by volcanic rocks, metapelites and gneisses with a variable metamorphism that can locally reach granulite facies. Migmatitic complexes are uncommon. At latitude 57° N, where the East Uralian zone reaches its maximum width, the granites display a strong W-E polarity owing to the decreasing age and increasing continental signatures to the east [*Fershtater* et al., 1994; *Fershtater* et al., 1997] (Figure 1).

3. DESCRIPTION OF GRANITE TYPES

For the following description we have grouped the Uralide granitoids into three groups: Late Silurian to Early Devonian ophiolite-related granitoids, Late Devonian to Carboniferous, mostly subduction-related granitoids, and Permian collision-related granitoids.

3.1. Late Silurian to Early Devonian Ophiolite-Related Granitoids

Two kinds of granite rocks are found spatially related to ophiolitic complexes: oceanic plagiogranites, produced as a consequence of the magmatic evolution of ophiolitic magmas, and anatectic granites, produced by local partial melting of crustal rocks at the base of the ophiolite during its emplacement. Their age is not yet well known. Radiometric dating of the associated ophiolites, especially those in the Suture Sector, has yielded ages clustering around 400 Ma [*Lawrence and Wasserburg*, 1985; *Pushkarev and Kaleganov*, 1993; *Fershtater* et al., 2000; *Scarrow* et al., 1999].

3.1.1. Oceanic Plagiogranites. Uralide ophiolite complexes locally contain small, but significant, volumes of oceanic plagiogranites. Where the initial ophiolitic structure is well preserved, plagiogranites appear at the top of the noncumulate gabbros and diorites, either forming the cement of intrusive breccias, or as dikes and small, irregular intrusive bodies. Petrographically, they consist of leucodiorites and trondhjemites, frequently with a granophyric texture, composed of quartz, plagioclase (An_{50-15}), hornblende, rare biotite and scarce K-feldspar as major minerals. Owing to the high deformation and alteration of the ophiolites in the study area, the chemical and isotopic data we present here come from fresh representative samples from the well preserved Kempirsai-Khabarny complex in the South Urals (50° N–51° N) [*Fershtater and Krivenko*, 1991].

The chemical composition of the Khabarny plagiogranites (Table 1) is characterized by low concentrations of potassium ($K_2O < 1$ wt.%) and the most incompatible trace elements. When the concentration of trace elements normalized to the average Continental Crust [*Wedepohl*, 1995] is plotted in order of increasing compatibility (calculated as the abundance ratio between the continental crust [*Wedepohl*, 1995] and the silicate earth [*McDonough and Sun*, 1995] (Figure 2), the oceanic plagiogranites show a pattern with a positive slope that increases, first steeply from Cs to U, and then, more gently, from Th to the HREE, with a small positive anomaly of U and a marked negative anomaly of Li. Chondrite-normalized REE patterns are almost flat, with $La_N \approx 20$ to $30x$ chondrite, little LREE/HREE fractionation ($La_N/Lu_N \approx 2$) and no Eu anomaly (Figure 2).

The Nd isotope composition of the Khabarny plagiogranites is MORB-like, with $\varepsilon_{Nd}(400$ Ma) in the range of $+6.3$ to $+6.9$, slightly lower than in related mafic and ultramafic rocks ($+7.3$ to $+9.0$) [*Lawrence and Wasserburg*, 1985]. The Sr isotope composition is moderately radiogenic, with initial $^{87}Sr/^{86}Sr$ in the range of 0.7040 to 0.7043 and $\varepsilon_{Sr}(400$ Ma) ≈ 0 to $+4.1$. In the $\varepsilon_{Nd}(t)$ vs $\varepsilon_{Sr}(t)$ diagram, therefore, they plot above the mantle array (Figure 6), a feature characteristic of most of the Uralide granitoids.

3.1.2. Crustally Derived Anatectic Granites. Small intrusive bodies and dikes of peraluminous granites either with gradational contacts to the high-grade host metamorphic rocks or crosscutting the ophiolite itself, occur at the bottom of the ophiolite complexes. The largest body (Banka) has been found in an erosional window of the Khabarny ophiolite. These granites stand out due to their varied mineralogy: aluminous biotite,

garnet, and tourmaline, an assemblage never found in the rest of the Uralide granitoids. They have not yet been well studied. Our provisional data indicate they are the products of local anatexis caused during the emplacement of high temperature ophiolite blocks in the continental crust.

3.2. Late Devonian to Carboniferous Granitoids

All of the Paleozoic subduction related granitoids of the Uralides were generated from 370 Ma to 315 Ma. For descriptive purposes, we have grouped them in the following categories:

- Late Devonian (or early) non-batholithic, calcic (Peacock index > 70) granitoids of the Platinum-Bearing Belt.
- 370 to 350 Ma calc-alkaline (Peacock index ≈ 57 to 59) deep seated batholithic granitoids, preferentially located in the easternmost part of the East Uralian zone near the Chelyabinsk fault, between ~54° N to 55° 30′ N.
- 335 Ma to 315 Ma calc-alkaline (Peacock index ≈ 58 to 61) deep seated batholithic granitoids, preferentially located in the west of the East Uralian zone, closely related to the Main Uralian or Serov-Mauk faults, between ~55° N to 58° N.

Additionally, we describe in this section the 335 to 315 Ma alkaline-calcic (Peacock index ≈ 58 to 51) high level gabbro-granite complexes of the Magnitogorsk arc.

3.2.1. Granitoids of the Platinum-Bearing Belt: Kytlym. The Platinum-Bearing Belt is formed by a lineament of supra subduction, concentrically-zoned dunite-clinopyroxenite-gabbro massifs (Alaskan-type) with chemical and isotopic features of arc magmas (Figure 2), intruded in Paleozoic accreted oceanic terranes [*Ivanov* et al., 1999, and references mentioned therein]. Crosscutting these rocks there are a few leucocratic pods and dikes with a geometry very similar to migmatitic leucosomes [*Fershtater* et al., 1998b]. In the case of Kytlym (59° 30′ N, 59° 50′ E), one of the largest and most complex bodies of the Platinum-Bearing Belt [*Efimov and Efimova*, 1967], the composition of the leucocratic bodies ranges from low-K leucodiorites and trondhjemites to medium- and high-K granites (Figure 2). Low-K series are, by far, the most abundant. They are composed of quartz, plagioclase (An_{40-10}) and hornblende, with rare biotite and accessory K-feldspar. High-K varieties have a similar modal mineralogy, but a greater abundance of biotite and, especially, K-feldspar.

Table 1. Chemical composition of selected samples of Uralide granitoids. Major elements and Zr are XRF data. The rest are ICP-MS data.

	Khabarny		Kytlym low-K	high-K	Akhunovo		Chelyabinsk		Chernorechenks			
n^a	1	2	3	4	5	6	7	8	9	10	11	12
SiO_2	70.62	67.26	62.98	73.93	68.74	72.06	62.17	68.21	75.87	75.87	62.34	67.56
TiO_2	0.48	0.82	0.34	0.04	0.41	0.25	0.78	0.44	0.25	0.17	0.58	0.51
Al_2O_3	13.48	14.3	18.75	14.28	15.84	15.55	16.04	16.54	13.18	13.55	16.13	16.7
FeO^1	4.53	6.11	3.63	1.67	2.77	1.12	5.83	3.12	1.26	1.37	4.81	3.78
MgO	0.63	0.43	0.9	0.11	0.94	0.32	2.89	1.25	0.16	0.25	3.28	1.34
MgO	0.04	0.08	0.07	0.02	0.06	0.02	0.1	0.06	0.06	0.05	0.06	0.07
CaO	2.13	3.31	6.18	1.01	3.48	1.39	4.94	3.09	0.9	0.97	3.41	3.29
Na_2O	5.48	4.26	5.49	4.07	3.82	3.79	3.08	3.27	3.25	3	2.9	3.47
K_2O	0.22	0.33	0.16	3.89	1.86	3.51	2.31	3.38	4.48	4.03	3.18	2.06
P_2O_5	0.13	0.24	0.09	0.01	0.13	0.08	0.25	0.15	0.04	0.06	0.31	0.2
Li	1.2	1.2	0.3	0.8	12	11.2	18.2	18.3	25.6	41.3	35.3	19.3
Rb	1.8	2.8	1	43	46	67	76	103	293	173	81	59
Cs	0.01	0.05	0.06	0.23	1.17	1.4	2.22	5.11	3.09	7.32	3.68	2.14
Be	1.05	0.98	0.45	0.37	1.28	1.25	1.68	2.99	6.61	3.59	2.28	2.57
Sr	211	242	856	186	617	797	572	606	118	192	1203	595
Ba	81	145	105	1069	638	1060	740	1184	296	410	1345	683
Sc	10	15.9	4.2	0.1	5.6	1.6	14	6.4	5.3	3.4	11	7.2
V	12.6	21	31.2	60.4	51	21.4	113	46.4	11.9	13.6	98.5	69.2
Cr	33.5	31.3	62.5	20.9	14.5	7.4	107	37.9	0.8	3	129	34.2
Co	4.3	6.3	6.5	2.1	6	2	16.3	7	1.1	1.6	16.6	9.6
Ni	21.7	7.5	36.4	111	7.6	2.8	46.6	16.2	4.7	3.7	86.5	20.7
Cu	14.2	6.8	25.7	36.7	10	16.5	22.6	5.2	4.7	4.5	23.5	20.2
Zn	12.5	25.4	27.3	8	46.6	30	74.9	52.4	35.5	65.6	83.1	61.3
Ga	14.6	16.8	17.4	12	18.5	17.6	19.5	19.2	22.8	17.8	17.9	20.1
Y	23.9	25.8	4.9	2.4	7.5	2.4	17.8	10.8	14.8	14.1	11.4	8.9
Nb	3.2	3	0.7	0.7	5.2	2	10	7.7	49.5	16	7.8	7.4
Ta	0.2	0.2	0.1	0.1	0.4	0.2	0.7	1.1	5.7	2.6	0.5	0.5
Zr	115	99	46	47	110	102	196	161	130	98	160	160
Hf	3.18	2.64	2.27	2.01	2.25	2.37	3.71	3.83	5.19	3.2	3.93	2.83
Mo	3.61	1.2	5.09	16.21	0.13	0.02	0.26	0.11	2.01	0.13	0.11	0.14
Sn	0.1	0.1	0.42	1.12	4.32	2.42	4.97	5.32	6.37	8.27	2.06	1.36
Tl	2.14	0.07	0.12	0.16	0.2	0.58	0.39	0.55	1.47	0.95	0.81	0.32
Pb	92.6	0.8	1.6	10.4	12.3	32.3	13.8	18.2	38.2	29.4	22.6	19.3
U	0.39	0.33	0.03	0.13	1.18	1.75	2.33	4.42	17.1	3	2.05	1.17
Th	1.03	0.86	0.07	0.16	5.72	6.27	9.22	9.13	36.7	12.1	15.4	8.34
La	6.61	7.49	1.8	1.18	18.7	14.1	38.8	31.5	37.9	15.7	40.4	30
Ce	17.9	17.9	4.53	2.95	37	29	75.4	56.6	68.6	37.2	79.7	59.5
Pr	2.26	2.49	0.7	0.38	3.78	2.64	7.99	5.56	6.11	3.46	8.61	6.41
Nd	9.92	11.6	3.42	1.4	13.2	8.96	29.2	19.1	18.4	12.2	33.3	22.5
Sm	2.91	3.41	0.97	0.39	2.39	1.5	5.46	3.26	2.82	2.8	5.32	3.61
Eu	0.84	1.07	0.48	0.21	0.67	0.42	1.3	0.91	0.41	0.42	1.46	1
Gd	3.12	3.76	0.98	0.33	1.81	0.91	4.29	2.54	2.22	2.47	3.55	2.55
Tb	0.6	0.66	0.16	0.05	0.24	0.1	0.57	0.35	0.33	0.41	0.43	0.34
Dy	4	4.05	0.95	0.33	1.27	0.48	3.2	1.88	1.9	2.28	2.14	1.84
Ho	0.96	1.01	0.19	0.08	0.24	0.08	0.62	0.35	0.43	0.45	0.39	0.33
Er	2.66	2.77	0.5	0.22	0.63	0.21	1.54	0.98	1.25	1.23	0.96	0.84
Tm	0.39	0.42	0.08	0.04	0.09	0.03	0.23	0.15	0.24	0.21	0.15	0.13
Yb	2.52	2.72	0.48	0.3	0.57	0.18	1.4	0.87	1.65	1.37	0.85	0.73
Lu	0.4	0.4	0.07	0.06	0.09	0.03	0.21	0.13	0.29	0.22	0.13	0.11
ASI	1.03	1.07	0.92	1.12	1.08	1.24	0.97	1.13	1.11	1.23	1.12	1.2
Th/U	2.66	2.58	2.5	1.17	4.85	3.57	3.96	2.06	2.15	4.04	7.53	7.16
Nd/Th	9.7	13.5	48.9	9	2.3	1.4	3.2	2.1	0.5	1.1	2.1	2.7
La_N/Lu_N	1.7	2	2.9	2.2	22.2	59.6	19.8	25	13.9	7.7	32.7	28
Eu/Eu*	0.85	0.91	1.51	1.83	0.98	1.09	0.82	0.97	0.49	0.49	1.03	1.01

Table 1. (continued)

n[a]	Syrostan				Verkhisesk older			Younger		
	13	14	15	16	17	18	19	20	21	22
SiO_2	49.89	64.72	67.94	73.57	50.53	57.12	65.49	70.1	72.39	71.01
TiO_2	1.24	0.73	0.48	0.21	1.06	0.55	0.64	0.16	0.36	0.43
Al_2O_3	14.18	17.16	16.42	14.88	15.5	20.00	16.06	17.01	15.27	15.08
FeO^1	9.18	3.67	2.97	1.00	10.37	6.47	4.35	1.42	2.50	2.62
MgO	9.82	1.59	1.36	0.35	5.57	2.32	2.06	0.81	0.89	0.94
MgO	0.15	0.08	0.07	0.03	0.26	0.17	0.08	0.03	0.04	0.03
CaO	7.31	3.65	2.78	1.56	9.33	6.91	4.04	3.64	2.38	2.30
Na_2O	2.78	5.06	4.74	4.52	3.18	4.68	4.76	5.04	2.55	4.35
K_2O	2.97	2.17	2.78	3.68	0.87	1.13	1.97	1.86	2.74	3.13
P_2O_5	0.44	0.26	0.27	0.06	0.48	0.24	0.22	0.09	0.04	0.11
Li	26.5	17.3	18.6	9.3	6.1	24.1	16.5	4.2	23.9	25.7
Rb	54	61	60	62	8.2	26	37	5.3	69	71
Cs	1.54	1.17	1.78	0.51	0.19	1.22	1.05	0.24	2.29	1.98
Be	1.13	2.08	1.95	2.04	1.11	1.19	1.56	1.51	1.56	1.81
Sr	1331	1131	1133	560	869	731	496	567	549	552
Ba	874	1007	1306	1116	282	677	576	650	959	928
Sc	22.3	6.3	4.5	0.6	22.4	10.7	10.1	4.6	2.9	2.8
V	160	62.2	38.2	9.3	276	105	80.4	20.7	26.1	34.9
Cr	342	32.7	7.6	1.1	108	2.7	31.5	8.9	4.7	7.4
Co	50.1	8.8	6.6	1.4	30.3	11.3	9.4	2.3	2.8	3.7
Ni	260	10.8	7.0	1.4	39.1	4.9	25.0	7.5	3.0	4.6
Cu	57.8	11.4	10.4	12.7	82.3	12.9	14.7	23.3	11.7	23.5
Zn	101	86.4	88.9	65.5	102	52.4	64.4	17.5	53.3	51.5
Ga	17	21.7	20.7	17.7	21.9	21.3	16.2	14.5	17.6	18.5
Y	17.9	11.8	13.7	5	24.9	17.2	11.9	3.7	5.5	5
Nb	10.4	14.3	16	8.6	5.2	3.5	7.7	2.9	4.2	4.9
Ta	0.7	1.1	1.2	0.4	0.2	0.2	0.8	0.2	0.3	0.4
Zr	129	201	183	92	109	39	160	71	131	164
Hf	3.86	6.2	4.67	3.02	4.3	0.44	0.47	2.1	3.42	4.39
Mo	0.39	0.32	0.11	0.12	0.43	0.11	0.69	0.44	0.02	0.04
Sn	1.24	1.55	0.38	0	0.45	1.03	2.54	0.49	2.76	2.75
Tl	0.38	0.32	0.67	0.3	0.17	0.16	0.23	0.11	0.83	0.4
Pb	6.9	14.2	12.2	18.5	3.9	7.3	15.7	15.8	23	21.2
U	1.08	2.27	1.54	0.72	0.31	0.51	1.7	1.22	0.81	0.76
Th	3.9	5.16	9.04	6.3	0.95	1.71	4.99	4.26	4.97	8.36
La	41.3	29.6	42.6	19.5	34.1	10.6	17.28	12.28	9.63	19.2
Ce	84.5	57.5	98.9	40.4	84.2	26.0	35.6	22.1	33.5	51.0
Pr	10.2	6.63	8.38	3.75	10.8	3.28	4.37	2.1	1.95	3.8
Nd	38.9	24.4	28.7	12.7	45.7	14.1	15.0	7.1	6.17	11.7
Sm	6.25	3.93	4.82	1.96	9.11	3.38	2.73	1.13	1.22	1.91
Eu	1.72	1.09	1.00	0.36	2.33	1.08	0.95	0.48	0.54	0.67
Gd	4.46	2.84	3.48	1.35	7.57	3.13	2.66	0.90	1.19	1.90
Tb	0.63	0.38	0.50	0.18	0.96	0.5	0.37	0.13	0.17	0.23
Dy	3.44	2.18	2.63	0.96	4.57	2.79	2.09	0.68	0.92	0.97
Ho	0.68	0.42	0.53	0.19	0.88	0.6	0.41	0.15	0.19	0.18
Er	1.77	1.11	1.42	0.5	2.23	1.66	1.15	0.39	0.46	0.45
Tm	0.26	0.17	0.21	0.08	0.36	0.27	0.17	0.06	0.08	0.06
Yb	1.62	0.96	1.35	0.47	2.23	1.74	1.03	0.41	0.53	0.42
Lu	0.24	0.16	0.19	0.08	0.33	0.25	0.16	0.06	0.07	0.06
ASI	0.67	0.99	1.04	1.04	0	0.93	0.93	1.01	1.33	1.02
Th/U	3.62	2.27	5.88	8.75	3.08	3.35	2.94	3.48	6.12	10.9
Nd/Th	10.0	4.7	3.2	2.0	48.0	8.2	3.0	1.7	1.2	1.4
La_N/Lu_N	18.1	19.7	24.1	27.4	11.1	4.54	11.2	20.6	14.1	34.3
Eu/Eu*	0.99	1.00	0.74	0.67	0.86	1.02	1.07	1.47	1.37	1.07

Table 1. (continued)

	Magnitogorsk			Dzhabyk					Stepninsk			Murzinka			
				low-K	Granites		q-monzonites					Vathika		Murzinka	
n^a	23	24	25	26	27	28	29	30	31	32	33	34	35	36	37
SiO$_2$	44.45	57.37	70.62	61.05	71.40	69.10	72.14	68.7	50.89	67.20	62.04	73.41	70.2	72.48	73.38
TiO$_2$	3.50	2.10	0.38	1.15	0.25	0.60	0.30	0.43	1.25	0.63	0.91	0.05	0.26	0.14	0.09
Al$_2$O$_3$	13.67	14.18	15.21	16.38	14.9	14.9	14.24	15.11	18.03	15.72	16.76	15.1	14.7	15.27	13.91
FeO[1]	16.05	8.77	3.20	5.80	1.64	3.58	2.20	3.18	8.64	3.53	5.19	0.87	3.31	1.67	2.62
MgO	5.50	3.53	0.92	2.51	0.54	1.28	0.61	1.11	6.37	1.43	2.17	0.20	0.59	0.29	0.13
MgO	0.22	0.15	0.05	0.18	0.02	0.07	0.02	0.05	0.16	0.05	0.08	0.01	0.02	0.01	0.01
CaO	10.00	5.62	1.41	3.69	1.26	1.85	0.98	1.68	10.73	2.71	4.16	1.60	1.32	1.48	0.74
Na$_2$O	1.2	3.55	4.76	7.2	4.16	4.53	4.00	3.81	2.59	3.14	3.66	3.77	3.67	4.82	3.90
K$_2$O	1.37	1.71	2.56	0.18	4.8	2.95	4.82	4.96	0.33	3.80	3.26	4.72	4.65	3.19	4.15
P$_2$O$_5$	0.22	1.00	0.05	0.27	0.08	0.15	0.16	0.19	0.21	0.30	0.47	0.01	0.13	0.05	0.05
Li	25.8	6.5	6.8	6.1	25	62.3	11.8	18.9	13.1	31	36	8.5	14	118	86
Rb	43	31	54	9.3	119	85	162	101	11	130	94	97	61	200	250
Cs	1.05	0.35	0.49	0.28	2.44	5.18	3.05	2.99	0.53	4.6	2.06	0.45	0.74	5.92	6.83
Be	0.56	2.33	1.5	2.17	2.64	4.56	2.2	2.12	0.73	2.47	3.04	1.66	1.23	2.66	4.15
Sr	219	367	128	317	167	216	292	475	417	562	977	273	181	217	145
Ba	483	320	457	54	841	771	596	912	117	1013	1427	830	506	598	514
Sc	56.8	19.5	4.7	14.4	1.6	5.7	3.2	4.8	29.1	7.9	10.3	0.7	2.9	2	2.5
V	629	161	36	147	16.5	40.4	20.5	45.1	191	72.1	102	2.9	17.8	13.6	8.1
Cr	15.8	22.7	32.9	20.9	3.5	8.8	10.3	14.3	339	46.4	67.1	1.9	10.3	6.2	8.4
Co	44.9	13.5	4.3	13.8	0.1	2.8	3.4	5.4	32.6	8.5	13.6	0.1	2.3	1.4	0.6
Ni	9.2	3.9	7.6	4.4	0	5	7.7	9.8	73	19.3	31.9	0.1	3.6	1.7	1.7
Cu	430	277	348	27.1	12.5	18.1	15.9	11	8.2	17.7	24.9	8.5	16.9	11.6	9.5
Zn	424	213	243	53.9	43.4	91.7	32.8	55.7	69.9	57.4	86.9	25.8	57.2	96.7	36.2
Ga	20.7	20	14	20.2	17.2	22.6	16.8	16.7	17.2	20.7	22.0	17.1	19.6	20.8	19.8
Y	17.4	42.0	26.9	32.7	3.3	6.3	13.2	10.3	23.7	9.7	17.9	1.0	3.0	3.2	7.9
Nb	1.8	9.0	3.7	9.0	5.5	21	14.1	21.4	4.1	13.1	31.4	1.9	9.9	10.1	10.2
Ta	0.6	0.6	0.4	0.6	0.6	1.7	0.9	1.6	0.3	0.9	2.3	0.2	0.3	0.7	1.0
Zr	25	171	128	121	144	246	170	262	111	188	294	64	212	86	82
Hf	0.89	4.29	3.81	3.03	4.3	6.1	5.29	7.2	2.49	4.88	6.65	1.98	3.70	2.83	2.73
Mo	0.63	1.23	0.72	0.98	0.05	0.26	0.42	1.86	0.23	0.35	0.94	0.05	0.48	0.33	0.24
Sn	1.29	3.19	1.45	2.53	0.22	2.73	1.44	1.92	2.25	3.00	3.10	0.10	0.11	2.26	3.07
Tl	0.15	0.19	0.3	0.06	0.91	1.10	0.62	0.60	0.09	1.07	0.52	0.5	0.62	3.80	42.8
Pb	25.3	10.2	13.9	3.3	31.9	27.6	37.9	28.3	5.8	17.1	20.5	30.5	23.9	27.3	31.4
U	0.21	1.59	2.58	1.03	1.96	1.6	3.64	3.94	0.3	4.37	3.24	0.42	0.68	2.44	3.00
Th	0.42	5.25	4.58	3.04	15.86	9.98	23.12	19.53	1.02	19.68	14.86	0.86	10.32	4.83	26.34
La	4.71	26.6	14.8	25	19.7	20.36	39.8	58.1	8.58	40.0	77.7	3.77	25.3	7.62	30.1
Ce	12.7	59.7	36.6	55.4	47.4	44.4	76.2	118	21.0	84.8	160	6.49	52.2	15.6	57.9
Pr	1.93	7.47	3.98	6.34	4.41	4.34	7.26	12.3	2.81	9.57	16.8	0.71	5.85	1.49	5.85
Nd	9.78	31.0	15.7	24.6	14.7	14.2	22.8	41.4	13.2	30.5	56.3	2.31	19.45	5.37	19.5
Sm	3.04	7.46	3.8	5.96	2.44	2.23	3.44	6.03	3.68	4.00	7.21	0.48	3.02	0.96	3.44
Eu	1.03	2.23	0.76	2.37	0.37	0.43	0.65	1.09	1.32	1.09	1.91	0.42	0.43	0.28	0.51
Gd	3.13	7.87	3.65	6.17	1.67	1.70	2.24	3.62	4.04	2.9	5.52	0.38	1.94	0.85	3.00
Tb	0.52	1.28	0.66	1.06	0.18	0.23	0.38	0.48	0.69	0.37	0.67	0.04	0.22	0.11	0.38
Dy	3.2	7.5	4.41	5.70	0.78	1.25	2.3	2.33	4.38	1.94	3.32	0.17	0.96	0.51	1.55
Ho	0.66	1.61	0.95	1.30	0.14	0.25	0.45	0.41	0.90	0.35	0.6	0.03	0.13	0.09	0.27
Er	1.63	4.29	2.8	3.43	0.40	0.70	1.35	1.05	2.40	0.87	1.48	0.07	0.32	0.23	0.65
Tm	0.24	0.64	0.45	0.54	0.06	0.11	0.21	0.15	0.36	0.13	0.22	0.01	0.04	0.04	0.10
Yb	1.40	4.10	3.03	3.20	0.41	0.69	1.32	1.08	2.21	0.85	1.3	0.04	0.24	0.21	0.60
Lu	0.22	0.62	0.46	0.49	0.07	0.11	0.20	0.15	0.33	0.12	0.19	0.01	0.04	0.03	0.10
ASI	0.63	0.79	1.16	0.87	1.04	1.06	1.05	1.03	0.75	1.11	0.98	1.06	1.09	1.09	1.14
Th/U	2.04	3.29	1.78	2.94	8.08	6.22	6.36	4.95	3.37	4.50	4.59	2.03	15.2	1.98	8.77
Nd/Th	23.1	5.9	3.4	8.1	0.9	1.4	1.0	2.1	12.9	1.5	3.8	2.7	1.9	1.1	0.7
La$_N$/Lu$_N$	2.3	4.5	3.4	5.4	31.0	20.3	21.1	40.1	2.7	36.7	44.3	79.6	76.1	24.3	33.1
Eu/Eu*	1.02	0.88	0.62	1.19	0.55	0.67	0.71	0.71	1.04	0.98	0.92	2.96	0.55	0.94	0.48

[1]total iron.

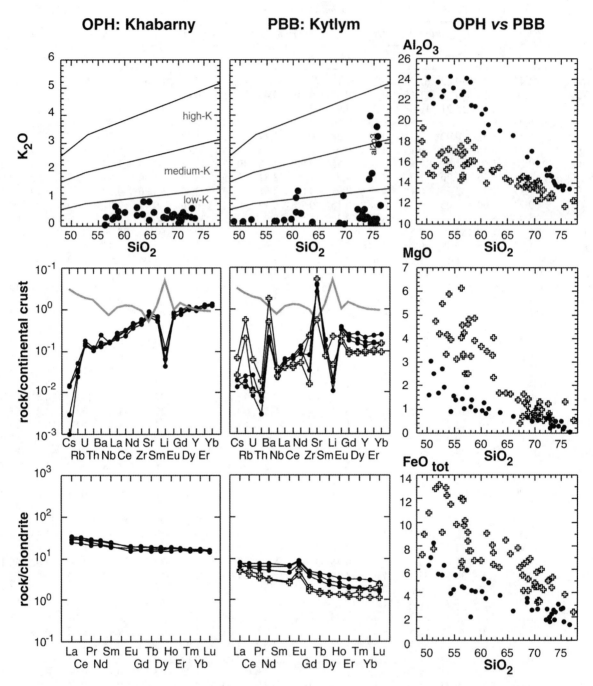

Figure 2. K_2O vs SiO_2, Continental Crust (CC)-normalized trace-element and chondrite-normalized REE plots of trondhjemites (plagiogranites) from the Khabarny ophiolite and the Kytlym concentrically zoned dunite-pyroxenite-gabbro body. In CC-normalized plots, the gray line is the average of anatectic Variscan granites from central Iberia, which represent pure crustal melts with no detectable mantle component [*Bea* et al., 1999]. See text for normalization values. Al_2O_3, MgO and FeO_{total} vs SiO_2 plots comparing the major element composition of ophiolitic plagiogranites (crosses) with low-K granites of the Platinum-Bearing Belt (dots). Note how, in spite of being similarly poor in potassium, they are totally different in major and trace elements. Note also the strong arc magma signature of the Platinum-Bearing Belt plagiogranites.

Both the low-K and normal-K series are characterized by small concentrations of nearly all trace elements except Sr, Ba and, in the case of normal-K series, Rb (Table 1). In the Continental Crust (CC)-normalized trace-element plot, low-K series show patterns with a moderate positive slope from Cs to the HREE, pronounced positive Sr and Ba anomalies, a pronounced negative Li anomaly, and with minor negative anomalies of Nb, Th and U (Figure 2). High-K varieties have similar CC-normalized patterns but with some differences: the appearance of a positive anomaly of Rb and (less intense) Cs and U, an increase in the positive Ba anomaly, and a decrease in the negative Li anomaly. The REE concentration is notably low, especially in high-K varieties. Chondrite-normalized REE patterns are similar in both rock series, with moderate LREE/HREE fractionation ($La_N/Lu_N \approx 2$ to 5) and a positive Eu anomaly.

The trace-element features of Kytlym granitoids are totally different from ophiolitic plagiogranites, but comparable to those of anatectic leucosomes [*Bea* et al., 1994; *Bea and Montero*, 1999]. This feature, together with their migmatite-like appearance, supports an origin by partial melting of gabbros either alone, in the case of low-K varieties, or with some contribution from host metavolcanites and sediments, in the case of high-K varieties [*Fershtater* et al., 1998b].

The age of plutonic complexes of the Platinum-Bearing Belt is problematic. Field relationships and a few U-Pb data (zircon concentrates) point to a Silurian age. However, dating by single-grain techniques (Pb-Pb evaporation and U-Pb ion microprobe) of zircons with an ultimate crustal origin discovered in the dunites of Kytlym [*Bea* et al., in press] has yielded ages of about 360–370 Ma for the diapir formation within the mantle wedge (coeval with the Uralides high-pressure metamorphism [*Glodny* et al., 1999]), and ~330 Ma (337 $\pm$ 22 Ma by Rb-Sr) for the late gabbros. The palingenetic granites of Kytlym have not yet been dated, but considering they are younger than the late gabbros and based on analogous field relationships with well dated batholiths such as Syrostan or Verkhisetsk (see below), we provisionally suggest a Late Carboniferous or Early Permian age.

3.2.2. 370 to 350 Ma Subduction Related Batholiths. Late Devonian to Early Carboniferous subduction related batholiths are common in the Middle Urals between 54° N and 55° 30′ N. So far, we have dated three of them: Chelyabinsk and Chernorechensk, in the East Uralian zone close to the Chelyabinsk fault, and Akhunovo, located inside the Magnitogorsk arc (Figure 1).

a) The Akhunovo batholith. The spatially related Akhunovo-Karagai, Uisk and Petropavlosk intrusions form a medium sized batholith situated in the northern part of the Magnitogorsk arc near the boundary with the East Uralian zone (Figure 1). Akhunovo-Karagai, in the west, and Uisk, in the east, are composed of biotite $\pm$ hornblende granites and granodiorites. Single-zircon Pb-Pb stepwise evaporation and ion microprobe U-Pb dating in Akhunovo-Karagai (unpublished data of the authors) have yielded ages of 365 $\pm$ 3 Ma and 365 $\pm$ 9 Ma, respectively. Petropavlosk forms the core of the batholith, which is composed of gabbros and diorites intruding the Akhunovo granites and has a zircon age of 315 $\pm$ 8 Ma (unpublished data of the authors). As they have the same petrography and similar field relationships, the Akhunovo-Karagai and Uisk plutons are described together under the name of the Akhunovo batholith. The Petropavlosk pluton is described in another section, together with the gabbro-granite complexes of the Magnitogorsk lineament, of which it likely represents the northernmost pluton.

Akhunovo granites are composed of quartz, oligoclase, microcline, biotite and occasional hornblende as major minerals. The fabric is strongly sheared and most minerals show evidence of intense alteration and/or recrystallization. This fact might explain why internal Rb-Sr isochrons, relying heavily on different fractions of K-feldspar, have yielded ages of 305 Ma to 308 Ma [*Bogatov and Kostits'in*, 1999], notably younger than the zircon ages.

Akhunovo granites are calc-alkaline, with $Na_2O > K_2O$. Its trace element composition is characterized by high Sr, Li, Ba and, in most leucocratic facies, U (Table 1). In the CC-normalized trace element plot (Figure 3), Akhunovo granites have a pattern with a moderate negative slope and important positive anomalies of Li, Sr, Ba and U, and negative Nb and Zr anomalies. Chondrite-normalized REE patterns display variable but always intense LREE/HREE fractionation ($La_N/Lu_N \approx 20$ to 60) and a moderate negative, or no, Eu anomaly ($Eu/Eu^* \approx 0.61$ to 1.0).

The chemical composition of Akhunovo granites has features typical of arc magmas, such as the negative Nb anomaly and positive Sr anomaly, together with some characteristics of crustal granites, such as the elevated Li and U contents. The Sr and Nd isotope composition, however, is very primitive, with initial $^{87}Sr/^{86}Sr$ in the range 0.7040 to 0.7043, $\varepsilon_{Sr}(365\ Ma) \approx -1$ to $+2.2$, and $\varepsilon_{Nd}(365\ Ma) \approx +3.9$ to $+4.1$. These values are identical to the Sr isotopes and slightly less radiogenic with respect to the Nd isotopes of the Khabarny ophiolite plagiogranites.

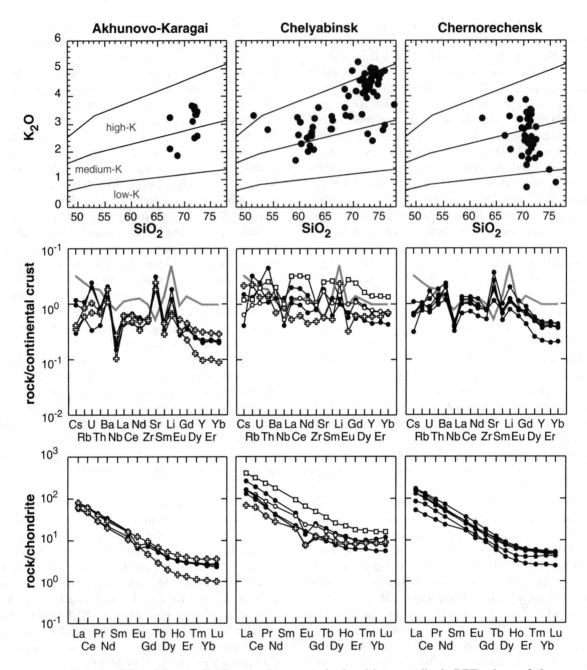

Figure 3. K$_2$O vs SiO$_2$, CC-normalized trace-element and chondrite-normalized REE plots of Lower Carboniferous subduction related granitoids. In the K$_2$O vs SiO$_2$ plot all samples are represented as dots. In the other plots circles represent diorites, dots represent granodiorites, and crosses represent granites.

b) The Chelyabinsk batholith. Chelyabinsk is one of the largest batholiths of the Uralides (Figure 1). It is composed of medium to high-K tonalites, granodiorites and granites forming different intrusive units. Ongoing zircon dating (unpublished data of the authors) reveals a complex age pattern. Zircons from the tonalitic Smolenskiy unit and the Main Granite have yielded uniform

Pb-Pb evaporation ages of 355 ± 7 Ma and 360 ± 9 Ma, respectively. Zircons of the Kazanskiy granodiorite, however, yielded a complex age pattern, from ~360 Ma to ~330 Ma. The youngest granite of this batholith is Kremenkul, a small intrusion of undeformed pinkish leucocratic granites with a biotite K-Ar age close to 280 Ma [*Grabezhev* et al., 1986; *Kaleganov*, pers. comm.].

The primary mineralogy of Chelyabinsk tonalites and granodiorites consists of quartz, plagioclase (An_{45-15}), biotite and green amphibole, with conspicuous, primary looking epidote (XPist ≈ 0.21 to 0.29) and large allanite crystals, usually with spectacular oscillatory zoning. The primary mineralogy is diversely affected by recrystallization processes. The less transformed rocks still have recognizable igneous textures, with the plagioclase variably saussuritized and the primary hornblende partially transformed to tremolite-actinolite. The most transformed rocks have a texture comprising quartz aggregates in a fine-grained groundmass of plagioclase, biotite, K-feldspar and quartz with a clear granoblastic texture; primary amphibole phenocrysts are transformed to fine-grained biotite aggregates.

Compositionally, Chelyabinsk granitoids are high-K calc-alkaline rocks (Figure 3) with elevated concentrations of trace elements that decrease with increasing silica (Table 1). In the CC-normalized plot they show a negative Nb anomaly, a small positive Sr anomaly and, in many cases, positive U and Th anomalies, making them the Uralide granitoids with the highest concentrations of heat producing elements. The Li anomaly is variable, from negative in tonalites to strongly positive in granites. The REE contents are high, especially in tonalites, which have La_N ≈ 200× to 400× chondrite (Figure 3). La_N/Lu_N is in the range of 15 to 30, and the Eu anomaly increases from Eu/Eu* ≈ 1 to 0.9 in tonalites to Eu/Eu* ≈ 0.6 to 0.5 in the most silicic granites.

The Sr and Nd isotope composition of Chelyabinsk granites is the less primitive among the Uralide granitoids, with the single exception of some facies of the Permian Murzinka batholith (see below). Chelyabinsk granites have initial $^{87}Sr/^{86}Sr$ in the range of 0.7046 to 0.7056. $\varepsilon_{Sr}(360\ Ma)$ ≈ +6 to +22, and $\varepsilon_{Nd}(360\ Ma)$ ≈ 0 to −4.1.

c) The Chernorechensk batholith. Chernorechensk is a small N-S trending batholith cropping out immediately west of the Chelyabinsk fault, at latitude 54° N. It is composed of medium-K to high-K granodiorites and granites with a zircon age of 354 ± 7 Ma (unpublished data of the authors). Petrographically, it stands out due to the strong recrystallization displayed by all the rock facies. The main minerals are quartz, plagioclase (An_{43-22}), K-feldspar and biotite, with occasional hornblende and scarce but constant primary looking epidote (XPist ≈ 0.20 to 0.27) and large allanite crystals with marked oscillatory zoning. Textures are nearly identical to the most recrystallized samples of Chelyabinsk (i.e., a fine-grained granoblastic groundmass of quartz, plagioclase, K-feldspar and biotite enclosing coarse-grained aggregates of quartz).

The chemical composition is very similar to that of the Akhunovo granodiorites (Table 1). In the CC-normalized plot they display a pattern with a positive slope from Cs to Ba, a negative Nb anomaly, positive Sr and Li anomalies and a negative slope from Gd onward (Figure 3). Chondrite-normalized REE patterns have a marked negative slope from La to Er, with no Eu anomaly, and then flatten out from Ho to Lu, with La_N/Lu_N ≈ 12 to 35 (Figure 3).

The Sr isotope composition is, as in the case of Chelyabinsk, moderately radiogenic, with initial $^{87}Sr/^{86}Sr$ in the range of 0.7047 to 0.7051 and $\varepsilon_{Sr}(355\ Ma)$ ≈ +8 to +14. The Nd isotope composition, however, is more primitive, with $\varepsilon_{Nd}(355\ Ma)$ ≈ −0.2 to +3.3, intermediate between Chelyabinsk and Akhunovo.

3.2.3. 335 Ma to 315 Ma Subduction-Related Batholiths. During the Late Carboniferous, the plutonic activity of the Uralides was concentrated in the neighborhood of the Main Uralian and Serov-Mauk faults. Between 55° N and 58° N, it produced deep seated gabbrodiorite-tonalite-trondhjemite-granite batholiths. Between 54° 30′ N and 51° N, it produced the Magnitogorsk lineament of high level gabbro-granite plutons. In this section we describe the northernmost group, using the small, but highly complicated, Syrostan pluton and the large Verkhisetsk batholith as examples.

In all cases studied so far Upper Carboniferous subduction related batholiths are composed of older, strongly deformed rocks, and younger undeformed rocks [Fershtater et al., 1994]. The older rocks are the less silicic; they commonly form the periphery of the batholiths and have a subvertical planar shearing fabric roughly parallel to the outer boundary of the batholiths. The younger rocks are the most silicic; they appear as dike swarms, as migmatite-like leucosome veins, or as irregular intrusive bodies, commonly situated in the core of the batholiths. The hiatus between the two rock series is variable, from ~7 Ma in Syrostan to ~30–40 Ma in Verkhisetsk [Montero et al., 2000]. Field relations, together with petrographic, geochemical and isotopic evidence indicate that the younger rocks derived from the older rocks through partial melting [Bea et al., 1997]. They are therefore identical in this respect to the Kytlym palingenetic granites.

a) The Syrostan batholith. Syrostan is a small, but highly complex, body situated within the Main Uralian fault, intruding the serpentinites and the metasediments of the Uralide foreland (Figure 1). It is composed of gabbrodiorites, tonalites, granodiorites, and granites that apparently form two rock series, gabbrodiorites on one hand and tonalites to granites on the other (Figure 4).

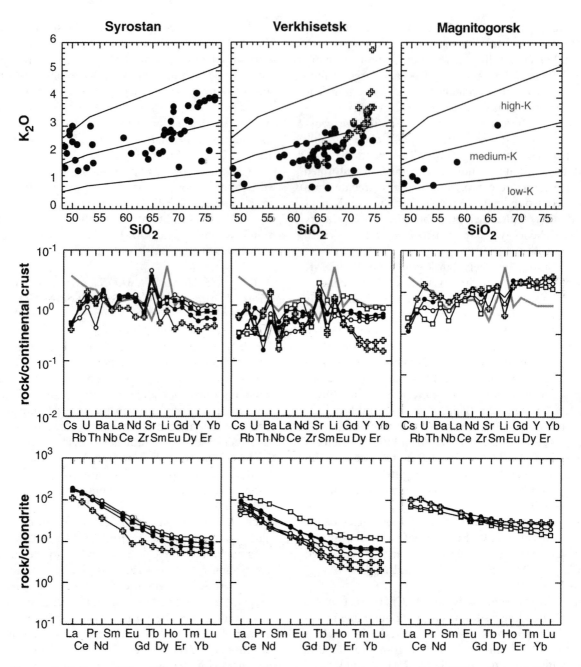

Figure 4. K₂O vs SiO₂, CC-normalized trace-element and chondrite-normalized REE plots of Upper Carboniferous granitoids. In the K₂O vs SiO₂ plot all samples are represented as dots except in the case of Verkhisetsk, where younger granitoids are represented as crosses. In the other plots squares represent gabbros, circles represent diorites, dots represent granodiorites, and crosses represent granites.

The gabbrodiorites are composed of plagioclase (An_{54-28}), Al-rich hornblende, primary looking epidote (up to 10 vol.%) and minor biotite, with abundant titanite, magnetite, allanite, and primary looking carbonates. The tonalites and granodiorites are composed of quartz, plagioclase (An_{40-23}), epidote, biotite, rare

amphibole (limited to the less silicic samples) and accessory titanite and magnetite. The granodiorites are either massive or appear as folded veins and blobs within gabbrodiorites, with a migmatite-like structure. The existence in some places of mafic selvages of hornblende in between granodioritic veins and gabbrodiorites suggests these structures were formed by in situ partial melting. The gabbrodiorites, tonalites and granodiorites are deformed and have a well developed subvertical planar fabric. The granites, however, are undeformed and appear either as dikes crosscutting the other rocks or forming the cement of intrusive breccias. They are composed of quartz, plagioclase (An_{25-20}), microcline, minor biotite and epidote, and rare muscovite. The most conspicuous mineralogical feature of Syrostan is therefore the abundance of primary epidote as a major mineral, present in all facies and locally dominant over biotite.

Zircon dating in Syrostan has yielded an age of 333 ± 3 Ma for the deformed gabbros and granodiorites and an age of 327 ± 2 Ma for the undeformed granites [*Montero* et al., 2000].

The chemical composition of all Syrostan rocks corresponds to high-K calc-alkaline series (Table 1), but, as expected from their field relationships, they do not form a differentiated sequence. Syrostan gabbros are the richest of all the subduction gabbros of the Uralides in LILE and LREE elements, with well-defined positive K and Sr anomalies and negative Nb and Ti anomalies (Figure 4). Their chondrite-normalized REE patterns are smoothly negative from La to Er, with no Eu anomaly, and then flat to Lu, with La_N/Lu_N in the range of 15–20. The tonalites and granodiorites are richer in LILE, have nearly the same LREE, but are poor in HREE, with $La_N/Lu_N \approx 25$ and 45 and $Eu/Eu^* \approx 0.7$ to 1. The granites have similar LILE contents as the granodiorites, but are lower in all other trace elements, including the REE. Their chondrite-normalized REE patterns have moderate LREE/HREE fractionation, with $La_N/Lu_N \approx 3$ to 25, and a higher Eu anomaly, with $Eu/Eu^* \approx 0.5$ to 0.7.

The Sr isotope composition of Syrostan granitoids is the less radiogenic of all the Uralide plutonic rocks, including ophiolites. The above mentioned existence of two rock series is confirmed by the bimodality of Syrostan granitoids with respect to Sr and Nd isotopes. The gabbrodiorites have initial $^{87}Sr/^{86}Sr$ in the range of 0.7029 to 0.7033 and $\varepsilon_{Sr}(330\ Ma) \approx -12$ to -18, with $\varepsilon_{Nd}(330\ Ma) \approx +2.6$ to $+6.6$. The granodiorites and granites are slightly more radiogenic, with initial $^{87}Sr/^{86}Sr$ in the range of 0.7035 to 0.7039, $\varepsilon_{Sr}(330\ Ma) \approx -2$ to -8, and $\varepsilon_{Nd}(330\ Ma) \approx -0.9$ to $+1$.

b) The Verkhisetsk batholith. The Verkhisetsk batholith is the largest subduction-related batholith in the western Uralides [*Fershtater* et al., 1994; *Bea* et al., 1997]. It is placed between the Serov-Mauk (west) and Ekaterinburg (east) shear zones (Figure 1), intruding middle Paleozoic metavolcanic rocks. It comprises an outer envelope of older, coarse-grained, strongly deformed tonalites, trondhjemites and granodiorites with an age of 315–320 Ma, intruded by an inner core of younger, fine-grained, undeformed granodiorites, adamellites and granites dated at 275–285 Ma [*Montero* et al., 2000]. Younger rocks usually have sharp intrusive contacts with older rocks, although gradational contacts and migmatite-like relationships consisting of stromatic structures formed by leucosome veins of younger rocks within a mesosome of older rocks are also common. They also frequently contain rounded and partially assimilated enclaves of older rocks. These facts, together with the similarities in chemical and isotopic composition (see paragraphs below) indicate that the younger rocks derived from older rocks through partial melting [*Fershtater* et al., 1994; *Bea* et al., 1997].

Older tonalites and granodiorites have an assemblage of quartz, plagioclase (An_{40-30}), hornblende, biotite and minor K-feldspar, with abundant accessory epidote and titanite. Closely associated with the older rocks are a few gabbros and diorites appearing either as enclaves or as dikes. They are composed of plagioclase (An_{53-39}) and hornblende, with minor biotite, quartz, titanite, and primary epidote ($XPist \approx 0.21$–0.29). Younger granodiorites have nearly the same modal proportions and mineral chemistry as the older granodiorites, but with a greater abundance of K-feldspar. Most younger granites, however, are peraluminous and have primary phengitic muscovite, always subordinate to aluminous biotite, and contain no calcic amphibole. Remarkably, they also contain large magmatic looking epidote crystals, commonly rimmed by biotite or, more often, by secondary muscovite.

The chemical composition of the older rocks corresponds to medium-K calc-alkaline series with a strong signature of arc magmas (Table 1). Compared with their equivalents in Syrostan, the Verkhisetsk rocks are poorer by a factor of 0.4 to 0.6 in LILE and LREE, and by about 0.9 in HREE (Figure 4). The gabbrodiorites have the lowest LILE contents but are the richest rocks in REE. Their REE chondrite-normalized patterns are smoothly negative from La to Dy and then are flat to Lu, with $La_N/Lu_N \approx 10$ and no Eu anomaly. The tonalites and granodiorites have slightly higher LILE contents than the gabbrodiorites, but are poorer in the other trace elements by a factor of 0.6–0.8, with nearly parallel chondrite-normalized patterns. The composition of the older rocks is consistent with generation by direct melting of the subducted slab; they show pronounced lateral zoning, with increasing HREE, Cr, Ni and #Mg to the east, which reflects the increasing depth of magma generation eastward [*Bea* et al., 1997].

The composition of the younger rocks varies between two extremes, from Na-rich granodiorites to normal or

K-rich granites (Table 1). The less silicic varieties (SiO_2 < 68 wt.%) almost exactly match the composition of older rocks with the same silica contents. The most silicic varieties, however, gradually become peraluminous and are enriched in Cs, Rb, K, Pb, Nb, Ta, Th, and U, but depleted in REE and Sr. Remarkably, no Verkhisetsk rocks show any appreciable negative Eu anomaly.

The Sr and Nd isotope composition of the Verkhisetsk granitoids is less primitive than in Syrostan. The older rocks have initial $^{87}Sr/^{86}Sr$ in the range of 0.7040 to 0.7046 and ε_{Sr}(320 Ma) ≈ −2 to +7, with ε_{Nd}(320 Ma) ≈ +0.4 to +4.8. The younger granites have initial $^{87}Sr/^{86}Sr$ in the range of 0.7043 to 0.7052, ε_{Sr}(320 Ma) ≈ +1.2 to +14.7, and ε_{Nd}(320 Ma) ≈ +0.4 to +4.4, almost exactly the same isotopic composition as the older granites if we take into account the different ages for calculating the isotopic parameters, 320 Ma and 280 Ma, respectively.

3.2.4. 335 Ma to 315 Ma Gabbro-Granite Complexes of the Magnitogorsk Lineament.

The Magnitogorsk gabbro-granite series is represented by small (up to 150 km²) high level plutons cropping out in the eastern part of the Magnitogorsk arc, where they form a meridian lineament. Rb-Sr dating of the Magnitogorsk intrusion has yielded an age of 333 ± 4 Ma [*Ronkin*, 1989]. Dating of associated peralkaline granites by single-zircon stepwise Pb-Pb evaporation yielded an age of 327 ± 8 Ma. Zircon dating of the northernmost body of the lineament, Petropavlosk, yielded 315 ± 8 Ma (unpublished data of the authors). It therefore seems that, despite many differences, the deep seated batholiths of the north and the high level gabbro-granite complexes of the Magnitogorsk arc are coeval.

The Magnitogorsk type plutons are subvolcanic, emplaced at a depth of ~1 km [*Fershtater*, 1987]. They consist of nearly vertical stocks, with thin granitic caps and apophyses, occasionally accompanied by cumulate titanomagnetite and skarn magnetite deposits. Intensive drilling in the Magnitogorsk body has allowed an accurate reconstruction of its internal structure [*Fershtater*, 1987]. From the top downward it consists of three zones: a 50 to 200 m upper layer composed of amphibole bearing pink granites and quartz-syenites; a 20 to 50 m layer of magmatic breccia formed of angular blocks of gabbros and diorites cemented by granite; and a lower layer of gabbros and melanodiorites with a minimum depth of 5 km. Surrounding the main body there are concordant sheets and minor stocks of peralkaline granites.

The chemical composition of the Magnitogorsk gabbros is rather peculiar (Table 1). They are not calc-alkaline, but alkaline calcic, frequently nepheline-normative (CIPW), with higher $Na_2O + K_2O$ and TiO_2 than any subduction gabbro of the Uralides. Nevertheless, they still have a trace element signature characteristic of arc magmas, with marked negative Nb and Zr anomalies and positive Sr and K anomalies (Figure 4). Rocks with SiO_2 > 55 wt.% are bimodal, with high-K (dominant) and low-K (subordinate) varieties (Figure 4). The silicic edge of the high-K series is composed of ($Na_2O + K_2O$)—rich granites that can be slightly peralkaline (agpaitic index ≈ 1), with modal riebeckite ± aegirine. The Magnitogorsk granitoids, especially those from the high-K series, have elevated concentrations of trace elements. In the CC-normalized plot (Figure 4), they show higher than crustal average HREE, a pronounced negative Li anomaly, no Zr anomaly except in gabbros, no Sr anomaly, a small negative Nb anomaly, and elevated contents of Ba, Th and, especially, U, which increase toward the most silicic extreme. Low-K varieties are similar, but with a strong depletion in Ba, Rb and Cs. The REE content is high, with moderate LREE/HREE fractionation (La_N/Lu_N ≈ 3.5 to 5). In contrast with the other Uralides differentiated rock series, the REE concentration increases from gabbros to granites, which develop a small negative Eu anomaly and, remarkably, a small positive Ce anomaly.

The existence of titanomagnetite cumulates, the development of giant skarns, and the abundance of hydrothermal manifestations indicate that Magnitogorsk magmas were extraordinarily rich in volatiles. The peculiar chemistry of the whole rock series, characterized by the enrichment of Zr, Nb, Th, and REE toward the silicic edge, is compatible with this idea and suggests differentiation in the presence of a halogen-rich volatile phase of a basaltic magma initially enriched in incompatible elements.

The Sr and Nd isotope composition of Magnitogorsk gabbros and granites is not well known yet. We shall therefore use available data on the Petropavlosk pluton as an example for Magnitogorsk type rocks. They have initial $^{87}Sr/^{86}Sr$ in the range of 0.7044 to 0.7046 and ε_{Sr}(315 Ma) ≈ +3.9 to +6.7, with ε_{Nd}(315 Ma) ≈ +1.5 to +4.0.

3.2.5. Permian Collision-Related Batholiths.

The ages of three bodies emplaced in the Main Uralian fault, Syrostan (55° 5′ N) at ~335 Ma, Sukhoviaz (56° 01′ N) at ~323 Ma, and Verkhisetsk (56° 49′ N to 57° 22′ N) at ~320 Ma [*Montero* et al., 2000] mark the transition from subduction to continent-continent collision at different latitudes. Afterward, the generation of granites continued and caused both the partial reactivation of older batholiths and the generation of new batholiths. All the subduction granites of the Uralides, especially the largest ones, were reactivated to some extent, so that they always contain a fraction of younger, undeformed silicic rocks. But most of the collisional granites form their own batholiths which, according to their age and field relationships, can be grouped as follows:

- 300 Ma to 290 Ma, apparently unrelated to the main tectonic events
- 285 Ma to 275 Ma, synkinematic with the strike-slip crustal scale shear zones
- 260 Ma to 250 Ma, related to the migmatitic complexes.

The 300 Ma to 290 Ma batholiths appear mostly in the southern Urals, between 55° N and 53° N. The most representative example is the large Dzhabyk batholith (~53° 2′ N to ~53° 20′ N). It lies west of the Chelyabinsk fault and intrudes Paleozoic gneisses, schists and serpentinites [*Fershtater* et al., 1994; *Gerdes* et al., 2001]. It is the only major Uralide batholith with its major axis oriented W-E instead of N-S. The southern part of the Dzhabyk batholith is locally affected by normal shear zones associated with its emplacement; the eastern zone is also affected by a strong regional shear zone, but, in general, the Dzhabyk rocks are undeformed.

The Dzhabyk batholith consists of 6 different units. Velikopetrovsk, Ol'khovsk, and Dzhabyk *s.s.*, form the main part of the batholith and are composed of granites and leucogranites. Rodnichki and Mochagi are composed of quartz syenites and monzonites, and the small Kozubaeevsk is comprised of gabbros to diorites. The Velikopetrovsk, Ol'khovsk, and Dzhabyk granites, hereafter "the granites", have quartz, albite-oligoclase (An_{25-10}), porphyritic K-feldspar, aluminous biotite and, often, primary muscovite. They have been dated at 291 ± 4 Ma by single zircon $^{207}Pb/^{206}Pb$ evaporation analysis [*Montero* et al., 2000]. Rodnichki and Mochagi quartz-monzonites, quartz-syenites and syenogranites, hereafter "the quartz-monzonites", are composed of quartz, K-feldspar, plagioclase (An_{35-25} to An_{10-15}), moderately aluminous biotite and, locally, magnesian hornblende. Rb-Sr data [*Montero* et al., 2000] indicate they have the same age, but slightly different initial $^{87}Sr/^{86}Sr$ (see below), as the main granites. The small Kozubaeevsk pluton consists mainly of K-rich mafic to intermediate rocks; its age and Sr-Nd isotope composition is not yet well known.

Dzhabyk granites are high-silica, peraluminous, and rich in K_2O (Table 1, Figure 5). The Sr and Nb anomalies characteristic of subduction-related granites are either lacking or very minor. Dzhabyk granites, however, show a marked positive Li anomaly, and notable Th, U, and Rb contents, features commonly displayed by the crustal granites of western Europe. Their chondrite-normalized REE patterns show a smooth decrease from La to Sm, a negative Eu anomaly (Eu/Eu* ≈ 0.4 to 0.7), and a flat or positive slope from Gd to Lu, with $La_N/Lu_N \approx 10$ to 50. Dzhabyk quartz-monzonites have a distinctive geochemistry with respect to the granites. They are metaluminous to slightly peraluminous, with lower concentrations of Li, Rb, and Cs, and higher LREE and Sr; in the CC-normalized trace element plot they also lack the features of arc magmas (Figure 5).

Despite the crustal signature with respect to major and trace elements, the Sr and Nd isotope composition of Dzhabyk rocks is nearly indistinguishable from that of the subduction granites. Dzhabyk granites have initial $^{87}Sr/^{86}Sr$ in the range of 0.7038 to 0.7046, ε_{Sr}(290 Ma) ≈ −6 to +7, and ε_{Nd}(290 Ma) ≈ +0.8 to +1.5. In spite of having a less crustal trace element signature than the granites, Dzhabyk quartz-syenites have a more crustal isotope signature, with initial $^{87}Sr/^{86}Sr$ in the range of 0.7045 to 0.7056, ε_{Sr}(290 Ma) ≈ +6 to +20, and ε_{Nd}(290 Ma) ≈ 0 to +0.7.

The 285 Ma to 275 Ma batholiths related to strike-slip crustal scale shear zones are grouped into two lineaments (see Figure 1) of small but complex plutons composed of magnetite-bearing syenogranites and leucogranites that were emplaced in active shear zones between 275 Ma and 285 Ma. In the paragraphs below we shall describe Stepninsk, the largest pluton of this category located in the southern lineament.

Stepninsk (54° 04′ N, 60° 26′ E) is a 18 km × 14 km oval body with its major axis oriented N 15° W. It has a marked zonal, nearly concentric structure that occasionally has been interpreted as a ring complex. Our field studies, nonetheless, reveal that Stepninsk was emplaced in a NW-SE dextral strike-slip shear zone and its zonal structure is due to synmagmatic tectonic rotation. Lithologically it comprises ~15% metamorphic palimpsests, ~45% extremely deformed coarse-grained quartz-monzonites and syenogranites, ~2% gabbrodiorites, intimately associated with the quartz-monzonites, and ~38% undeformed leucogranites. All these materials are disposed in nearly concentric bands, except for a few discordant leucogranite dikes. The metamorphic palimpsests are limestones, metavolcanites and cordierite-bearing metapelites, variously affected by contact metamorphism. Ongoing geobarometric estimates on the metapelites indicate an emplacement pressure of between 3 and 3.7 kbars (unpublished data of the authors).

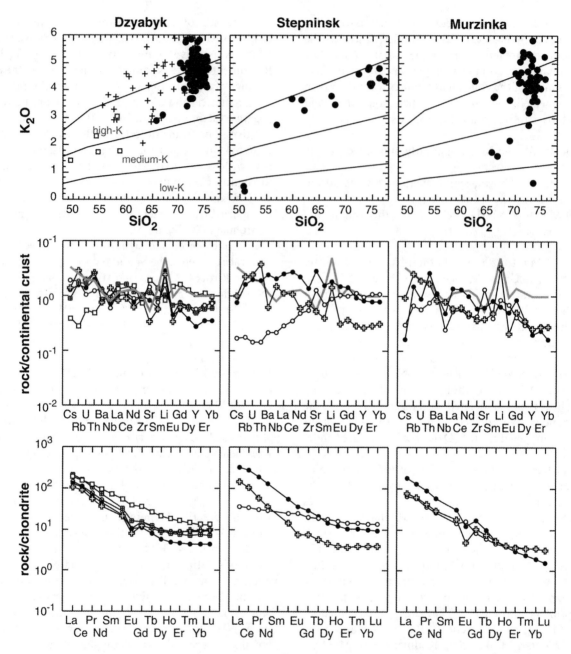

Figure 5. K_2O vs SiO_2, CC-normalized trace-element and chondrite-normalized REE plots of Permian collision related granitoids. In the K_2O vs SiO_2 plot all samples are represented as dots except in the case of Dzhabyk, where Kozubaeevsk gabbrodiorites are represented as squares, quartz-monzonites are represented as crosses and granites are represented as dots. In the other plots squares represent gabbros, circles represent diorites, dots represent granodiorites, and crosses represent granites. For Dzhabyk, additionally, crossed squares and open circles represent Mochagi and Rodnichki quartz-monzonites respectively.

Petrographically, Stepninsk quartz-syenites are composed of K-feldspar, oligoclase (An_{27-23}), quartz, pargasitic amphibole, and biotite, with abundant magnetite and titanomagnetite, and minor titanite. The leucogra-

nites have quartz, K-feldspar and albite, with little biotite and muscovite as major minerals; they also contain abundant Fe-Ti oxides, as in the case of the quartz-syenites. The gabbrodiorites are composed of plagioclase

(An$_{45-40}$, rarely An$_{50-54}$) and hornblende, locally with scarce K-feldspar phenocrysts, and abundant accessory titanite and Fe-Ti oxides.

Geochemically, the quartz-syenites and leucogranites, but not gabbrodiorites, form a continuous series related by a process of magmatic differentiation. Quartz-syenites have elevated contents of most trace elements (Table 1), with a convex CC-normalized pattern having a maximum in Ba, Nb and LREE, a small negative Zr anomaly, and a small positive Sr anomaly. Their chondrite-normalized REE patterns are smoothly negative, with La$_N$/Lu$_N$ ≈ 35 to 40, and no Eu anomaly. The leucogranites have lesser contents of all trace elements except the most incompatible ones, especially Th. Their CC-normalized patterns are steeply positive from Cs to Th, with a marked Ba negative anomaly, caused by feldspar fractionation, and then decrease slightly toward the most compatible elements, with an intense positive Li anomaly (Figure 5). The LREE and HREE contents are lower than in quartz-syenites by a factor of ~0.5, but the MREE are still more depleted (~0.3×) and have a small negative Eu anomaly. Gabbrodiorites have a totally different chemistry, with the lowest contents in trace elements of any similar rocks from the rest of the Uralide batholiths. Their chondrite-normalized REE patterns are almost flat, with La$_N$/Lu$_N$ ≈ 2.5 and a small positive, or no, Eu anomaly (Figure 5).

All the magmatic rocks of Stepninsk, whether deformed or undeformed, have yielded the same age of 283 ± 4 Ma by both single zircon ^{207}Pb/^{206}Pb stepwise analysis and by whole rock Rb-Sr. Quartz-syenites and leucogranites plot in the same isochron, thus supporting the idea that they are cogenetic.

The quartz-syenites and leucogranites have initial ^{87}Sr/^{86}Sr in the range of 0.7046 to 0.7052; with ε$_{Sr}$(285 Ma) ≈ +5.6 to 18.1 and ε$_{Nd}$(285 Ma) ≈ −2.2 to +0.4. The gabbrodiorites are more radiogenic with respect to both Sr and, especially, Nd; they have initial ^{87}Sr/^{86}Sr in the range of 0.7052 to 0.7053, ε$_{Sr}$(285 Ma) ≈ +15 to 17 and ε$_{Nd}$(285 Ma) ≈ +6.0 to +6.4.

The 260 Ma to 250 Ma batholithic granites related to migmatitic complexes appear in the eastern part of the East Uralian zone between latitudes 57° N and 58° N. In contrast to the rest of the Uralide granites, they are spatially related to migmatitic complexes. They are formed of granites and leucogranites, commonly with elevated concentrations of elements such as Li, Rb, Cs, Nb, Ta, Sn, etc., and well developed pegmatites.

The best studied of these batholiths is Murzinka (Figure 1) [*Fershtater*, 1984; *Fershtater* et al., 1986; *Fershtater and Bea*, 1993]. It is a N-S elongated granitic batholith with maximum dimensions of ~75 km N-S and ~25 km E-W, dated with single zircon evaporation analyses at 254 ± 5 Ma [*Montero* et al., 2000]. It lies west of the Alapaeevsk shear zone, intruding gneisses, volcano-sedimentary units and serpentinites primarily of Paleozoic age. The batholith comprises three parts, Yuzhakovsk in the west, Vatikha in the middle, and Murzinka *sensu stricto* in the east.

Yuzhakovsk is a migmatitic complex formed by a network of discordant veins (~0.5–10 m thick), either trondhjemitic (K-poor) or leucogranitic (K-rich), cutting amphibolite grade orthogneiss and paragneiss. Yuzhakovsk is intruded by the massive granites of the Vatikha unit, comprised of medium-grained, sometimes porphyritic, granites composed of quartz, oligoclase (An$_{24-15}$), highly monoclinic orthoclase, and aluminous biotite with scarce, but very characteristic, large magnetite crystals. Vatikha is intruded by the Murzinka granites. They are mildly peraluminous granites and leucogranites formed of quartz, albitic plagioclase (An$_{18-5}$), porphyritic microcline, aluminous biotite, muscovite and, locally, almandine garnet.

Yuzhakovsk K-poor veins have the lowest contents of all trace elements, with positive Eu and Sr but negative Nb anomalies and low initial ^{87}Sr/^{86}Sr (see below). K-rich veins are richer in LIL elements, with positive Eu, Ba and Sr anomalies and high initial ^{87}Sr/^{86}Sr (Figures 5 and 6). Yuzhakovsk veins bear a significant analogy with Kytlym veins and are therefore interpreted in the same way, as partial melts from either a metaigneous protolith, in the case of low-K veins, or a protolith with a significant metasedimentary component, in the case of high-K veins. Vatikha rocks are peraluminous silicic granites with K$_2$O > Na$_2$O (Table 1). They have low Li, Cs, and U, but considerably high LREE and Th (Figure 5). Chondrite-normalized REE patterns are steeply negative, with La$_N$/Lu$_N$ ≈ 70–250 and negative Eu anomalies. Murzinka s.s. granites have a major element composition similar to Vatikha (Table 1). They contain the highest concentrations of the most incompatible elements, with a strong positive anomaly of Li (Figure 5). Chondrite-normalized REE patterns (Figure 5) show moderate LREE/HREE fractionation, with La$_N$/Lu$_N$ ≈ 10−60 and negative Eu anomalies, and flat profiles for the heaviest HREE.

Isotopically, the Murzinka batholith is the most heterogeneous of the Uralides. It is bimodal with respect to Sr isotopes, with two clearly separated populations that plot into parallel isochron lines at 248 ± 6 Ma and 259 ± 8 Ma [*Gerdes* et al., 2002], in good agreement with the single zircon age of 254 ± 5 Ma [*Montero* et al., 2000]. The less radiogenic population has initial ^{87}Sr/^{86}Sr in the range of 0.7041 to 0.7054 and ε$_{Sr}$(250 Ma) ≈ −1 to +19,

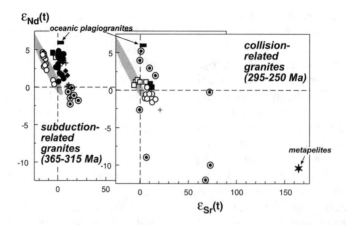

Figure 6. $\varepsilon_{Nd}(t)$ vs $\varepsilon_{Sr}(t)$ of Uralide granitoids. In the case of subduction related granitoids open circles represent Syrostan, dots represent Verkhisetsk, open squares represent Akhunovo, solid squares represent Kytlym gabbros, rombs represent Petropavlosk, crosses represent Chernorechensk, and dotted circles represent Chelyabinsk. Note how the western Early Carboniferous batholiths, Chelyabinsk and Chernorechensk, are significantly less primitive than the eastern Late Carboniferous batholiths, Syrostan or Verkhisetsk, due to the presence of an older crustal component. In the case of collision related granitoids open circles represent Stepninsk, open squares represent Dzhabyk granites, solid squares represent Dzhabyk quartz-monzonites, dotted circles represent Murzinka, and crosses represent Kremenkul Note the anomalously primitive isotope composition of collision granites, except some facies of Murzinka, indicating that no older-than-Silurian crustal material was significantly involved in Permian crustal melting. For the sake of clarity, the younger granites of Verkhisetsk and Syrostan are nor represented; they plot exactly into the same field as the older rocks from which they were generated by in-situ partial melting.

similar to the rest of the Uralide granitoids. The most radiogenic population has initial $^{87}Sr/^{86}Sr$ in the range of 0.7087 to 0.7093 and $\varepsilon_{Sr}(250\ Ma) \approx +65$ to $+75$, being, therefore, the only case among the Uralide granitoids with a Sr isotope composition similar to the crustal granites of western Europe. The distribution of Nd isotopes is also bimodal, but there is no clear correspondence with the populations defined by the Sr isotopes. The less radiogenic population has $\varepsilon_{Nd}(250\ Ma)$ in the range of -5 to $+5$ and is mainly, but not entirely, comprised of low radiogenic Sr samples. The most radiogenic population has $\varepsilon_{Nd}(250\ Ma)$ in the range of -7.5 to -12 and is mainly, but not entirely, comprised of high radiogenic Sr samples. It therefore seems that the Murzinka granitoids come from a hybrid protolith in which two components, one young and metaigneous and the other old and metasedimentary, can be identified. Whereas the first component is similar to the rest of the Uralide Permian granites, the second has so far only been found in Murzinka.

4. DISCUSSION

Uralide granitoids, with the single exception of Murzinka, have a very primitive Sr and Nd isotope composition, more characteristic of mantle than crustal

materials (Figure 6). In the case of subduction related granites, neither $^{87}Sr/^{86}Sr(t)$ nor $^{143}Nd/^{144}Nd(t)$ bear any relation with the age, but depend on the geographical longitude. The Early Carboniferous batholiths in the east, Chelyabinsk and Chernorechensk, are composed of granitoids with significantly higher $^{87}Sr/^{86}Sr(t)$ but lower $^{143}Nd/^{144}Nd(t)$ than similar rocks of the Late Carboniferous batholiths in the west, which have identical (in some cases more primitive) $^{87}Sr/^{86}Sr(t)$ and just slightly lower $^{143}Nd/^{144}Nd(t)$ than oceanic plagiogranites (Figure 6). Moreover, the eastern subduction granites are also enriched in K and trace elements of continental affinity such as Rb, Ba, Th, U, Li, etc. It seems, therefore, that the western protolith of Late Carboniferous granitoids consisted solely of oceanic or very young continental materials, whereas the eastern protolith of Early Carboniferous granitoids was composed of oceanic materials plus a significant fraction of old crustal materials.

The 365 Ma Akhunovo batholith emplaced in the Magnitogorsk volcanic series is isotopically more primitive than the eastern batholiths, with $^{87}Sr/^{86}Sr(t)$ and $^{143}Nd/^{144}Nd(t)$ matching Verkhisetsk or Syrostan. It might be related to the last stages of the eastward subduction of the Uralian paleocean beneath the Magnitogorsk arc [e.g., *Brown* et al., 1998]. Notwithstanding,

Chelyabinsk and Chernorechensk can hardly be related to the same phenomenon. In the Early Carboniferous the Magnitogorsk arc had already collided with the East European Craton, but the Uralian paleocean had not yet closed. Looking at the Uralides palinspastic reconstruction for that time [*Zonenshain* et al., 1990, figure 45], and considering the nature and age of the materials from the Magnitogorsk arc, we must admit that the only potential source of an old crustal component was the Siberian-Kazakhstan continent, at that time separated from the Magnitogorsk arc by 500 km of Uralian ocean. In this situation, therefore, the most probable scenario for the generation of subduction batholiths with a recognizable crustal signature is an Andean-type arc caused by the eastward subduction of the Uralian paleocean crust underneath the Siberia-Kazkhastan continent. This arc existed; it is called the Valerianovsky arc and is currently recognizable stretching between the Uralide fold belt and Central Kazakhstan [*Segedin*, 1981, see also *Zonenshain* et al., 1990].

We therefore suggest that Chelyabinsk and Chernorechensk are remnants of the Valerianovsky arc, accreted to the Uralides when it collided with the Magnitogorsk arc, completely closing the Urals paleocean. In the same way, the occurrence in the Middle Urals of high grade metamorphic complexes whose protolith consisted of subduction granitoids with a crystallization age of around 350 Ma (e.g. the Teliana suite [*Friberg* et al., 2000]) indicates that the Valerianovsky arc occupied a significant part of the East Uralian zone and, as discussed in the paragraphs below, it was probably largely recycled to produce Permian granites.

During the Late Carboniferous, the locus of magmatic activity shifted again to the west and produced two groups of plutonic bodies: the deep seated subduction batholiths of Syrostan, Atlian, Turgoyak, Verkhisetsk, Shartash, etc. on one hand, and the gabbro-granite high level plutons of the Magnitogorsk lineament on the other hand. Late Carboniferous subduction occurred north of the Magnitogorsk arc, between approximately 55° N and 58° N. The subduction zone was located in between the Tagil arc, already welded to the East European Craton, and the Late Devonian to Middle Carboniferous subduction complexes of the Middle Urals that we consider a part of the Valerianovsky arc. The main suture related to this subduction event, the Serov-Mauk Fault (Figure 1), lies in a western position. Furthermore, Verkhisetsk, the largest batholith involved, shows pronounced lateral zoning that reflects the increasing depth of magma generation to the east [*Bea* et al., 1997]. Together, these two circumstances reveal that the Benioff zone dipped to the east, underneath the Valerianovsky arc.

The gabbro-granite plutons of the Magnitogorsk lineament were also produced during the Late Carboniferous, but neither their geodynamic significance nor the origin of their parental melts are well understood yet. On one hand, their field relationships and the occurrence of peralkaline leucocratic fractionates could indicate they were rift magmas [*Fershtater*, 1984; *Fershtater and Bea*, 1993; *Fershtater* et al., 1994]. On the other hand, the most mafic, and presumably the less differentiated rocks of the Magnitogorsk intrusion, have a geochemical signature completely different to that of rift related magmas, more characteristic of arc magmas, though with some peculiarities. At the present, the lack of isotope data does not allow us to state a reasonable genetic model for this conjunction of apparently contradictory features. As a working hypothesis, we provisionally suggest that the Magnitogorsk gabbro-granite series might have consisted of volatile-rich basaltic arc magmas enriched in incompatible elements, which evolved in a passive or locally extensional environment.

The Uralian paleo ocean was already totally closed at ~300 Ma, so that there is no magmatic activity directly related to subduction after the Carboniferous. The numerous Permian granite batholiths found in the East Uralian zone were generated, by all available geological evidence, within the continental crust. Surprisingly, however, their isotopic signature is also very primitive, with $^{87}Sr/^{86}Sr(t)$ and $^{143}Nd/^{144}Nd(t)$ that match subduction granites (Figure 6). This feature completely excludes continental materials older than Silurian, such as those of the Mugodzharia microcontinent [*Zonenshain* et al., 1990], as a possible protolith. On the other hand, the average Permian granite of the Uralides is high-silica, K-rich and mildly peraluminous, with a trace element composition characteristic of continental granites, in which some trace element anomalies characteristic of arc magmas, although attenuated, are still recognizable (see Figures 3 to 5). The only materials able to produce partial melts with this conjunction of mantle like isotope and crust like chemical composition are subduction related rocks with a short crustal residence time of a few tens of Ma. We therefore suggest that the Permian granites were caused by the recycling of early subduction granitoids, either Late Devonian or, preferentially, Carboniferous. The batholiths of the Valerianovsky arc, such as Chelyabinsk and Chernorechensk, are especially good candidates owing to their elevated heat production (1.5 to 2.2 μWm^{-3}).

Our observations indicate that the heat for melting was supplied by one or more of three mechanisms: radiogenic heating of an overthickened sialic crust that did not undergo significant late-orogenic extension

[*Brown* et al., 1998; *Steer* et al., 1998, *Tryggvason* et al., 2001, *Brown and Tryggvason*, 2001], underplating by mafic magmas and, perhaps, local accumulation of heat and fluids related to crustal-scale shear zones, the relative importance of which may vary largely from place to place.

The production of Permian granites in the Uralides happened at an almost constant rate from ~295 Ma until ~255 Ma, but the locus of magmatic activity migrated from the south, where it was preferentially located during the Early Permian, to the north, where it was active mostly during the Late Permian. At each latitude, crustal melting happened about 50 Ma after the onset of the continent-continent collision [*Puchkov*, 1997; *Montero* et al., 2000]. This was approximately the time required for crustal heating to anatexis solely by radiogenic heating of a protolith with a heat production similar to the average Chelyabinsk granitoids. This pattern suggests that radiogenic heating was the main factor responsible for crustal melting, and is consistent with the elevated contents of heat producing elements and the insignificant volumes of coeval mafic rocks associated with the largest Permian batholiths [*Gerdes* et al., 2002].

Radiogenic heating, however, was not the only cause of crustal melting after continent-continent collision. The majority of the pre-Permian batholiths were locally reactivated during the Permian; this phenomenon produced younger granitoids from older rocks by in-situ partial melting, as can be directly observed in Verkhisetsk, Syrostan, Kytlym, etc. [*Fershtater* et al., 1994]. Since the western Late Carboniferous subduction granitoids have a heat production of about 1 μWm^{-3} or less, it is evident that to heat them to anatexis could not be achieved solely by radiogenic heating, but would require an external heat supply. The common occurrence of dike swarms of mafic rocks apparently coeval (but not cogenetic) with palingenetic granitoids suggests that mafic magmas undergoing crystallization deep within the crust could have supplied the heat required for partial melting [*Bea* et al., 1997].

Another mechanism capable of producing crustal melting is crustal scale shearing [*Harrison* et al., 1998]. Between 54° N and 57° N, a definite group of small Permian plutons were synkinematically emplaced in the oblique crustal scale strike-slip shear zones that finally assembled the terranes forming the Middle Urals. At the present, we cannot determine whether the shear zones triggered crustal melting or simply acted as collectors of melt generated by other mechanisms. The fact that coeval (but not cogenetic) mafic rocks frequently appear associated with shear zone granites might support the

latter alternative. Current ongoing work on Stepninsk will hopefully permit us to clarify these aspects.

5. CONCLUSIONS

The granite rocks of the Uralides formed from 365 Ma to 250 Ma, first in subduction and then in intracontinental settings. Neither the Silurian nor the Devonian subduction events, which produced massive accumulations of volcanic rocks in the Tagil and Magnitogorsk arcs, respectively, seem to have generated granite batholiths, except perhaps Akhunovo.

Subduction related granites were produced in two episodes. The oldest occurred between 365 Ma and 350 Ma and generated subduction batholiths with a recognizable older crustal component along the active continental margin of the Siberia-Kazakhstan continent. These batholiths, together with metamorphosed subduction complexes of the same age, such as the Teliana gneisses of the Middle Urals [*Friberg* et al., 2000], were a part of the Valerianovsky arc accreted to the Tagil and Magnitogorsk arcs when the Uralian paleocean closed completely.

The youngest episode occurred between 335 Ma and 315 Ma. It generated the western subduction batholiths such as Syrostan or Verkhisetsk, which are characterized by the lack of any recognizable old crustal component. They are located between 55° N and 58° N, and were generated by an east dipping subduction zone underneath the Valerianovsky arc just before it collided with the Tagil arc.

The gabbro-granite plutons of the Magnitogorsk lineament are coeval with Late Carboniferous subduction granites, but they formed when the Uralian ocean was already closed at that latitude. Their field relationships, mineralogy and geochemistry suggest they consisted of basaltic arc magmas enriched in incompatible elements and volatiles that evolved in an extensional or passive environment until they produced leucocratic peralkaline fractionates. The ultimate source of magmas might have consisted of undepleted lherzolites brought to the melting zone by the dynamics of the mantle wedge beneath the Magnitogorsk arc after collision with the East European craton.

There is no magmatic activity directly related to subduction after the Carboniferous. The generation of granites in the East Uralian Domain, however, continued uninterruptedly during the Permian, first in the south, between 53° N and 55° N, and then in the north, between 56° 30′ N and 58° N. Permian granites were generated within a continental environment, but their extremely primitive Sr and Nd isotope composition in otherwise typical continental granites indicates they were produced by recycling of island arc complexes with a short crustal residence. Owing to their elevated heat production and geological position, we suggest that the main source for Permian granites were the subduction batholiths of the Valerianovsky arc and metasediments derived from them.

The main engine driving Permian crustal melting was radiogenic heating of an overthickened granodioritic crust that did not undergo late orogenic extension. Heat released from mafic magmas crystallizing at depth, however, also contributed. It probably caused the partial melting events that locally reactivated pre-Permian batholiths with low heat production at different moments during the Permian. In the same way, the oblique crustal scale strike-slip shear zones that finally assembled the terranes forming the Middle Urals might have had some relevance. It is not clear whether the shear zones triggered crustal melting or simply acted as collectors of melt generated by other mechanisms, but they certainly controlled the emplacement and magmatic evolution of many small, Middle Permian plutons.

Acknowledgments. We are grateful to V. Smirnov, N. Borodina, Ye. Zinkova, T. Ossipova, G. Shardakova, A. Gerdes, J. Scarrow, V. Puchkov and D. Brown for many discussions in the field that have helped us to clarify our viewpoint about Uralide granitoids. We are also grateful to N. Petford and T. Waight for their helpful revisions, and to Christine Laurin for her help with the English. This work has been supported by the Spanish DGICYT grant PB98-1345 and the Russian Fond for Basic Research (RFFI) grant 01-05-65184.

REFERENCES

Bea, F., G. Fershtater, P. Montero, V. Smirnov and E. Zinkova, Generation and evolution of subduction-related batholiths from the central Urals: Constraints on the P-T history of the Uralian orogen, *Tectonophysics*, 276, 103–116, 1997.

Bea, F., G. B. Fershtater, P. Montero, M. J. Whitehouse, V. Ya. Levin, J. H. Scarrow, H. Austrheim and E. H. Pushkariev, Recycling of continental crust into the mantle as revealed by Kytlym Dunite zircons, Urals Mts. Russia, *Terranova*, in press, 2002.

Bea, F. and P. Montero, Behavior of accessory phases and redistribution of Zr, REE, Y, Th, and U during metamorphism and partial melting of metapelites in the lower crust: An example from the Kinzigite Formation of Ivrea-Verbano, NW Italy, *Geochim. Cosmochim. Acta*, 63, 1133–1153, 1999.

Bea, F., P. Montero and J. F. Molina, Mafic precursors, peraluminous granitoids, and late lamprophyres in the

Avila batholith: A model for the generation of Variscan batholiths in Iberia, *J. Geol.*, 107, 399–419, 1999.

Bea, F., M. D. Pereira and A. Stroh, Mineral/leucosome trace-element partitioning in a peraluminous migmatite (a laser ablation-ICP-MS study), *Chem. Geol.*, 117, 291–312, 1994.

Bogatov, V. I. and Y. L. Kostits'in, Rb-Sr isotopic age and geochemistry of granitoids of noth Magnitogorsk arc, South Urals, *Geokhimiya*, 99–2, 34–41, 1999.

Brown, D., C. Juhlin, J. AlvarezMarron, A. PerezEstaun and A. Oslianski, Crustal-scale structure and evolution of an arc-continent collision zone in the southern Urals, Russia, *Tectonics*, 17, 158–170, 1998.

Brown, D. and A. Tryggvason, Ascent mechanism of the Dzhabyk batholith, southern Urals: Constraints from URSEIS reflection seismic profiling, *J. Geol. Soc. London*, 158, 881–884, 2001.

Efimov, A. A. and L. P. Efimova, *The Kytlym Platiniferous Massif* (in Russian), 336 pp., Nedra, Moscow, 1967.

Fershtater, G. B., *Eugeosynclinal Gabbro-granite Series* (in Russian), 264 pp., Nauka, Moskow, 1984.

Fershtater, G. B., *Petrology of Major Intrusive Associations* (in Russian), 232 pp., Moscow, 1987.

Fershtater, G. B. and F. Bea, Geochemical evidences of two contrasting granite types. The influence of different source materials. Magnitogorsk and Murzinska, the Urals., *Geokhimia*, 93–11, 1579–1599, 1993.

Fershtater, G. B. and F. Bea, Geochemical typification of Ural ophiolites, *Geokhimiya*, 96–3, 195–218, 1996.

Fershtater, G. B., F. Bea, N. S. Borodina and P. Montero, Lateral zonation, evolution, and geodynamic interpretation of magmatism of the Urals: New petrological and geochemical data, *Petrology*, 6, 409–433, 1998a.

Fershtater, G. B., F. Bea, N. S. Borodina and M. P. Montero, Anatexis of basites in a paleosubduction zone and the origin of anorthosite-plagiogranite series of the Ural platinum-bearing belt, *Geokhimiya*, 98–8, 768–781, 1998b.

Fershtater, G. B., F. Bea, E. V. Pushkarev, G. Garuti, P. Montero and F. Zaccarini, Insight into the petrogenesis of the Ural Platinum belt: New geochemical evidence, *Geokhimiya*, 99–4, 352–370, 1999.

Fershtater, G. B., N. S. Borodina, M. S. Rapoport, T. A. Osipova, B. H. Smirnov and M. Y. Levin, *Orogenic Granitoid Magmatism of the Urals* (in Russian), 247 pp., Russian Academy of Sciences. Urals Branch, Miass, 1994.

Fershtater, G. B., V. A. Chashukina and V. A. Vilisov, Geochemical criteria for crystallization sequences in igneous rocks, *Geokhimiya*, 86–6, 771–779, 1986.

Fershtater, G. B., A. B. Kotov, V. S. Smirnov, E. V. Pushkariev, E. B. Salnikova, V. P. Iovach, S. Z. Yakovleva and N. G. Berezhnaya, U-Pb age of the zircon from the diorite of Nurali lherzolite-gabbro massif in the South Urals (in Russian), *Doklady R.A.S.*, 371, 96–100, 2000.

Fershtater, G. B. and A. P. Krivenko, *Petrology of Post-Harzburgite Intrusives of Kempirsay-Khabarny Ophiolite Association (south Urals)* (in Russian), 159 pp., Ints. Geol. Geochem. R.A.S., Sverdlovsk, 1991.

Fershtater, G. B., P. Montero, N. S. Borodina, E. V. Pushkarev, V. N. Smirnov and F. Bea, Uralian magmatism: An overview, *Tectonophysics*, 276, 87–102, 1997.

Friberg, M., A. Larionov, G. A. Petrov and D. G. Gee, Paleozoic amphibolite-granulite facies magmatic complexes in the hinterland of the Uralide Orogen, *Int. J. Earth Sci.*, 89, 21–39, 2000.

Gerdes, A., P. Montero, F. Bea, G. Fershtater, N. Borodina, T. Ossipova, and G. Shardakova, Peraluminous granites with mantle-like isotope compositions: The continental-type Murzinka and Dzhabyk batholiths of the eastern Urals, *Int. J. Earth Sci.*, 91, 1–17, 2002.

Glodny, J., H. Austrheim, B. Bingen, A. Rusin and J. H. Scarrow, New age data for HP rocks and ophiolites along the Main Uralian Fault, Russia: implications for the Uralian Orogeny, *Terranostra*, 99, 89–90, 1999.

Grabezhev, A. I., V. A. Chashchukhina and V. G. Vigorova, *Geochemical Criteria of the Rare Metal ore Potential of the Granitic Rocks* (in Russian), 128 pp., Uralian Branch of Academy of Science USSR, Sverdlovsk, 1986.

Hamilton, W. B., The Uralides and the motion of the Russian and Siberian platforms, *Geol. Soc. Am. Bull.*, 81, 2553–2576.

Harrison, T. M., M. Grove, O. M. Lovera and E. J. Catlos, A Model for the Origin of Himalayan Anatexis and Inverted Metamorphism, *J. Geophys. Res.*, 103, 27017–27032, 1998.

Ivanov, K. S., E. V. Anikina, A. A. Efimov, E. V. Pushkarev, G. B. Fershtater and V. R. Shmelev, *Platiniferous Belt of the Urals* (in Russian), 96 pp., Inst. Geol. and Geochem., Ekaterinburg, 1999.

Lawrence, E. R. and G. J. Wasserburg, The age and emplacement of obducted oceanic cruts in the Urals from Sm-Nd and Rb-Sr systematics, *Earth and Planet. Sci. Lett.*, 72, 389–404, 1985.

Matte, P., Southern Uralides and Variscides: Comparison of their anatomies and evolutions, *Geol Mijnbouw*, 74, 151–166, 1995.

McDonough, W. F. and S. S. Sun, The composition of the Earth, Chem, *Geology*, 120, 223–253, 1995.

Montero, P., F. Bea, A. Gerdes, G. Fershtater, E. Zinkova, N. Borodina, T. Osipova and V. Smirnov, Single-Zircon Evaporation Ages and Rb-Sr Dating of 4 Major Variscan Batholiths of the Urals — A Perspective on the Timing of Deformation and Granite Generation, *Tectonophysics*, 317, 93–108, 2000.

Pushkarev, E. V. and B. A. Kaleganov, K-Ar dating of magmatic complexes of Khabarny gabbro-ultrabasite massif (South Urals). *Dokl. Ross. Akad. Nauk.*, 328, 241–245, 1993.

Puchkov, V. N., Structure and Geodynamics of the Uralian Orogen, in *Orogeny Through Time*, edited by J. P. Burg and M. Ford, Geol. Soc. Spec. Publ., 121, 201–236, 1997.

Ronkin, Y. L., Strontium isotopes - indicators of the evolution of the Uralian magmatism (in Russian), *Yearbook-1988. Inst. Geol. Geochem. R.A.S., Sverdlovsk*, 107–109, 1989.

Samarkin, G. I. and E. Y. Samarkina, *Granitoids of the South Urals* (in Russian), 209 pp., Nauka, Moskow, 1988.

Scarrow, J. H., P. Spadea, J. Glodny, P. Montero, G. N. Savelieva, A. N. Pertsev, L. Cortesogno and L. Gaggero, The Mindyak Palaeozoic Lherzolite Ophiolite, Southern Urals: Geochemistry and Geochronology, *Ofioliti*, 24, 241–248, 1999.

Segedin, R. M., *Explanatory Notes for the Gelogical Map of Kazakhstan SSR, Scale 1:500,000, Turgai-Mugodjary Series* (in Russian), 228 pp., Min. Geol., Alma Ata, 1981.

Steer, D. N., J. H. Knapp, L. D. Brown, H. P. Echtler, D. L. Brown and R. Berzin, Deep structure of the continental lithosphere in an unextended orogen: An explosive-source seismic reflection profile in the Urals (Urals Seismic Experiment and Integrated Studies, URSEIS 1995), *Tectonics*, 17, 143–157, 1998.

Tryggvason, A., D. Brown and A. Perez-Estaun, Crustal architecture of the southern Uralide from true amplitude processing of the Urals Seismic Experiment and Integrated Studies (URSEIS) vibroseis profile, *Tectonics*, 20, 1040–1052, 2001.

Wedepohl, K. H., The composition of the continental crust, *Geochim. Cosmochim. Acta*, 59, 1217–1232, 1995.

Zonenshain, L. P., M. I. Kuzmin and L. M. Natapov, Geology of the USSR: A Plate-Tectonics Synthesis, in *Geology of the USSR: A Plate-Tectonics Synthesis*, edited by B. M. Page, pp. 27–54, Geodynamic Series, V. 21, American Geophysical Union, Washington, D.C., 1990.

Fernando Bea and Pilar Montero, Department of Mineralogy and Petrology. University of Granada. Campus Fuentenueva, 18002 Granada, Spain; E-mail: fbea@goliat.ugr.es

German B. Fershtater, Institute of Geology and Geochemistry, Russian Academy of Sciences, Pochtovi per. 7, 620219 Ekaterinburg, Russia; E-mail: gerfer@online.ural.ru

Four Decades of Geochronological Work in the Southern and Middle Urals: A Review

J. H. Scarrow[1], R. Hetzel[2], V. M. Gorozhanin[3], M. Dinn[4], J. Glodny[2], A. Gerdes[5], C. Ayala[6] and P. Montero[1]

The Uralide Orogen, the geographic and geologic divide between Europe and Asia, has been the subject of geochronological study for more than 40 years. This compilation summarizes age data from the Southern and Middle Urals beginning with Archean to Proterozoic dates from the East European Craton in the west and advancing eastward where progressively younger geological events are recorded. Archean to Proterozoic basement rocks crop out throughout the East European Craton. Neoproterozoic gneisses provide evidence of a Pre-Uralian orogeny that affected large parts of the eastern margin of the East European Craton. Early Paleozoic magmatism related to rifting of the East European Craton in the Middle Urals is recorded by geographically restricted nepheline syenite massifs. The most westerly, and oldest, material accreted during the Paleozoic orogeny is Middle Ordovician to Late Devonian ophiolites and ultramafic/mafic massifs generally associated with the principal suture of the orogen, the Main Uralian fault. Closure of the Uralian paleo-ocean basin led to eastward subduction of the leading edge of the thinned East European Craton generating Late Devonian high pressure complexes now exposed in the Main Uralian fault footwall. East of the Main Uralian fault, Silurian to Early Carboniferous ocean volcanogenic complexes crop out. These accreted island arc terranes are intruded by abundant Late Devonian to Late Carboniferous plutonic complexes with subduction-related characterisitcs. Further to the east, Permian granitoids

[1]Department of Mineralogy and Petrology, University of Granada, Spain
[2]GeoForschungsZentrum, Potsdam, Germany
[3]Geological Institute, Russian Academy of Sciences, Ufa, Russia
[4]British Antarctic Survey, Cambridge, U.K.
[5]NERC Isotope Geosciences Laboratory, Keyworth, Nottingham, U.K.
[6]Institute of Earth Sciences, Barcelona, Spain

Mountain Building in the Uralides: Pangea to the Present
Geophysical Monograph 132
Copyright 2002 by the American Geophysical Union
10.1029/132GM12

were generated by melting of orogenically thickened crust. In general, post-Paleozoic magmatism is sparse throughout the Urals.

1. INTRODUCTION

This paper presents a compilation of the last 40 years of geochronological work in the Southern and Middle Urals (Figure 1). It attempts to summarise all the absolute age determinations, contained in over one hundred papers, available for the area. Much of this work is published in the Russian literature and is not accessible to western geologists. As a result, we synthesise and review this work to make this valuable information generally available. We offer our apologies to researchers who find their data absent from the present compilation and would appreciate being informed of any oversight; almost certainly some data have escaped our attention and have consequently been omitted.

The Uralides are a 2500 km long, north to south-trending orogenic belt (Figure 1) that extends from 68° to 48° N, from the Arctic Ocean to just north of the Caspian and Aral seas. They mark the geographic and geologic divide between Europe and Asia. The Main Uralian fault is the principal suture of the Uralides, marking the junction between Paleozoic microcontinents, island arcs, and oceanic terranes to the east, and the Paleozoic to Mesozoic sediment-covered Archean to Proterozoic East European continental margin, to the west [*Savelieva*, 1987; *Savelieva and Nesbitt*, 1996] (Figure 1). This principal tectonic boundary is a several kilometre wide tectonic mélange containing abundant serpentinite, basalts, and gabbros [*Ivanov* et al., 1975; *Zonenshain* et al., 1984]. The Uralide orogen has traditionally been divided into six longitudinal megazones that extend parallel to the former margin of the East European Craton (Figure 1) [*Ivanov* et al., 1975]. Two of these, the Pre-Uralian foredeep and the Western Uralian zone, consist of autochthonous and para-autochthonous units of the East European Craton. The Central Uralian zone comprises Precambrian sedimentary, metamorphic, and magmatic rocks in the footwall of the Main Uralian fault. In contrast, the Magnitogorsk-Tagil zone in the hanging wall is made up of Silurian (Tagil) and Devonian (Magnitogorsk) island arc sequences [*Zonenshain* et al., 1984]. Further to the east, the East and Trans-Uralian zones comprise microcontinental and island arc terranes of undefined origin and evolution, which are intruded by Late Paleozoic granites. A review of the structure and geodynamics of the Uralide orogen has recently been published by *Puchkov* [1997] and the

reader is referred to that work for a more detailed description, in particular of the sedimentology and paleontological dating of the area.

Here, the geochronological data are presented on 5 maps (Figures 2 to 6) that cover the area from 50° to 60° N and 56° to 62° E; the information is summarised broadly from west to east, that is, generally from oldest to youngest. The latest version of the IUGS-UNESCO International Stratigraphic Chart [*IUGS*, 2000] is used throughout the paper, but with the timing and duration of the Devonian period from *Tucker* et al. [1998]. On the maps, information about whether dates were obtained from whole rocks or minerals is represented by different symbols. Dates from the same complex are grouped in rectangles and for each location available K/Ar and Ar/Ar dates are presented first, followed by Rb/Sr and Sm/Nd dates, and finally U/Pb and Pb/Pb dates. The date and error are placed next to the symbol and are followed by a number referring to the accompanying table. The tables give further details regarding the data, including: geographical sample location; analytical method; rock type description; interpretation; and, the reference in which the dates have been published. Figure 2 and Table 1 present Mesoproterozoic dates from the East European Craton. Figure 3 and Table 2 shows Neoproterozoic dates from the East European Craton. Figure 4 and Table 3 include Paleozoic nepheline syenite, ophiolite, ultramafic, and mafic massifs dates. Figure 5 and Table 4 include data from Paleozoic metamorphic complexes, both the older high pressure eclogitic units associated with the Main Uralian fault and the somewhat younger gneisses associated with granitoid intrusions located mainly east of the Main Uralian fault. Figure 6 and Table 5 present dates from Late Paleozoic granitoid batholiths. Following the presentation of the data a broad comparison is made between earlier geochronological work and new dates produced in recent years, and a synthesis of the dates and tectonic significance of the main Uralian units is given.

2. GEOLOGICAL BACKGROUND

The Paleozoic Uralide orogenic cycle began in the Late Cambrian to Early Ordovician when the Archean to Proterozoic East European Craton's eastern margin (Figures 2 and 3), which had been affected by a Neoproterozoic orogenic event, rifted producing sparse

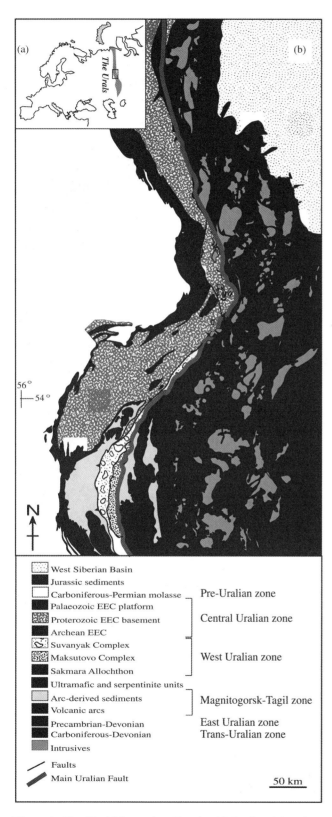

Figure 1. The Ural Mountains, Russia, (a) Regional location, and (b) Geological setting.

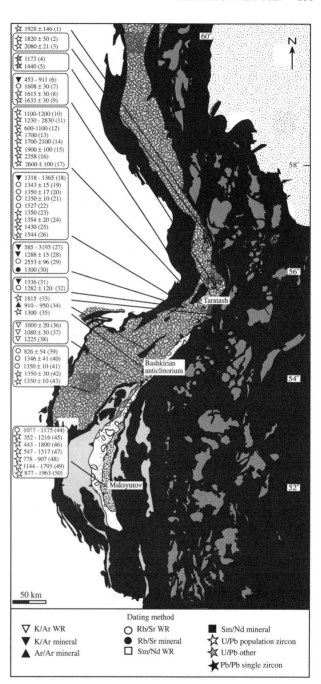

Figure 2. East European Craton Mesoproterozoic ages. Legend as in Figure 1.

nepheline syenite massifs (Figure 4), then developed into a passive continental margin [*McKerrow*, 1994; *Puchkov*, 1997; *Dalziel*, 1997; *Smethurst* et al., 1998]. During the Early Paleozoic the Pre-Uralian ocean opened and oceanic crust was formed. This extension was followed by intra-oceanic convergence, resulting in formation

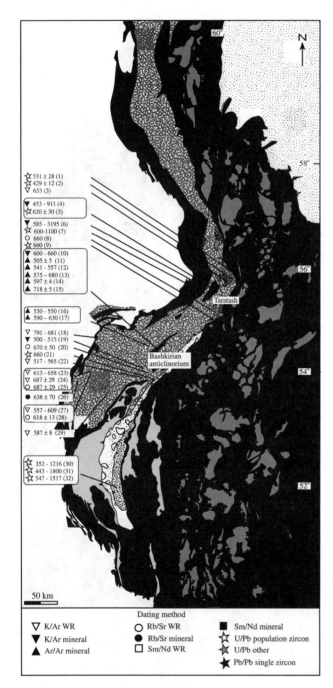

Figure 3. East European Craton Neoproterozoic ages. Legend as in Figure 1.

Figure 4. Accreted terranes Early to Late Paleozoic nepheline syenite, ophiolite, ultramafic, mafic ages. Legend as in Figure 1.

of volcanic arcs and back arc basin-type lithologies. Remnants of this oceanic evolutionary stage are preserved as island arc volcanics, ophiolites and ultramafic and mafic massifs (Figure 4). Subsequent to this, in the Late Paleozoic, Siberian-Kazakhian terrane assemblages accreted to the East European Craton [*Hamilton*, 1970;

Zonenshain et al., 1984; *Puchkov*, 1997]. A main arc-continent collision event occurred in the Southern Urals during the Middle to Late Devonian when continental lithosphere of the East European Craton margin was subducted beneath the Magnitogorsk arc as recorded by the high pressure metamorphic Maksyutov complex

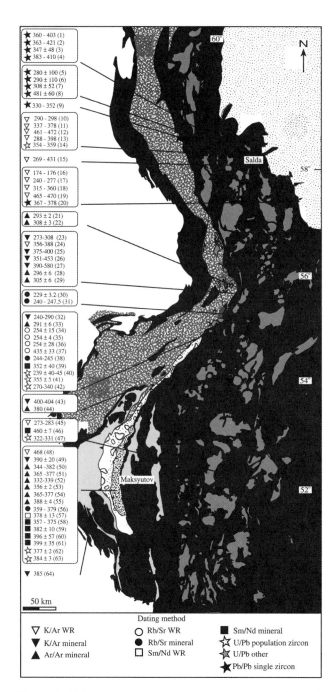

Figure 5. Paleozoic ages from metamorphic complexes in accreted terranes. Legend as in Figure 1.

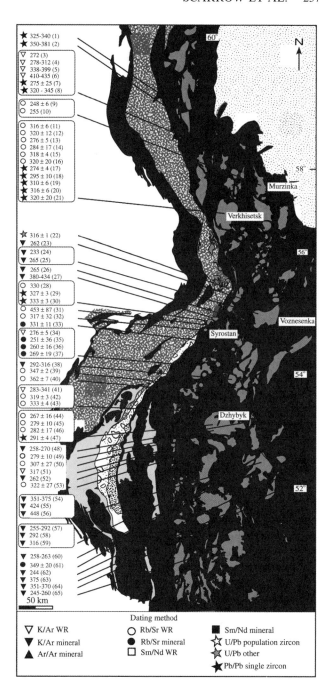

Figure 6. Late Paleozoic ages from intrusions in accreted terranes. Legend as in Figure 1.

(Figure 5) [*Brown* et al., 1998; *Brown and Spadea*, 1999; *Hetzel*, 1999]. In Late Carboniferous to Permian times there was a further collision, between the East European Craton and the Siberian-Kazakhian terrane assemblage, and subsequent generation of granitoid plutons (Figure 6) [*Zonenshain* et al., 1984; *Matte*, 1995; *Otto and Bailey*, 1995; *Puchkov*, 1997].

3. INTERPRETATION OF THE GEOCHRONOLOGICAL DATA

Because of their relative inaccesability the older Russian dates, published prior to use of modern decay constants, have been recalculated using data from *Steiger and Jäger* [1977] and are quoted at 2σ.

Table 1. East European Craton Mesoproterozoic dates presented in Figure 2.

#	Location	Date (Ma)	Method	Rock type	Interpretation	Reference
1	Chelyabinsk	1928 ± 146	U/Pb zrn	gneiss	ampite metam	*Krasnobaev* et al. [1995a]
2	Sysert Ilmenogorsk	1820 ± 50	U/Pb zrn	gneiss	metam	*Bibikova* et al. [1973]
3	Sysert Ilmenogorsk	2080 ± 21	U/Pb zrn	px hbl schist	retro metam	*Krasnobaev* et al. [1998]
4	Ufaley	1173	U/Pb ep	pegmatite		*Mineev* [1959]
5	Ufaley	1440	U/Pb ep	pegmatite		*Mineev* [1959]
6	Navysh	453–911	K/Ar WR, pl	trachybasalt	k metasom	*Harris* [1977]
7	Navysh	1608 ± 30	Rb/Sr WR	basalt	xlln	*Krasnobaev* et al. [1992]
8	Navysh	1615 ± 30	U/Pb	basalt	xlln	*Krasnobaev* et al. [1992]
9	Navysh	1635 ± 30	U/Pb zrn	basalt	xlln	*Krasnobaev* et al. [1992]
10	Taratash	1100–1200	U/Pb zrn		metam	*Lennykh & Krasnobaev* [1978]
11	Taratash	1230–2830	U/Pb zrn			*Ivanov* et al. [1986]
12	Taratash	600–1100	U/Pb zrn	ampite	ampite metam	*Ivanov* et al. [1986]
13	Taratash	1700	U/Pb zrn	migmatite		*Lennykh & Krasnobaev* [1978]
14	Taratash	1700–2100	U/Pb zrn	gneiss		*Lennykh & Krasnobaev* [1978]
15	Taratash	1900 ± 100	U/Pb zrn	granite	anatexis	*Krasnobaev* [1986]
16	Taratash	2358	U/Pb zrn	gneiss	gran metam	*Tugarinov* et al. [1970]
17	Taratash	2600 ± 100	U/Pb zrn	gneiss	gran metam	*Krasnobaev* [1986]
18	Berdyaush	1318–1365	K/Ar WR, amp bt, pl	gabbro rapakivi granite	xlln	*Harris* [1964]
19	Berdyaush	1343 ± 15	Rb/Sr WR	gabbro rapakivi granite	xlln	*Krasnobaev* et al. [1981]
20	Berdyaush	1350 ± 17	Rb/Sr WR	nph syenite	xlln	*Krasnobaev* et al. [1981]
21	Berdyaush	1350 ± 10	Rb/Sr	rapakivi granite	xlln	*Koroteev* et al. [1997]
22	Berdyaush	1527	Rb/Sr WR	rapakivi granite	xlln	*Salop & Murina* [1970]
23	Berdyaush	1350	U/Pb	rapakivi granite	xlln	*Koroteev* et al. [1997]
24	Berdyaush	1354 ± 20	U/Pb zrn	rapakivi granite	xlln	*Krasnobaev* et al. [1984]
25	Berdyaush	1430	U/Pb zrn	rapakivi granite	xlln	*Turgarinov* et al. [1970]
26	Berdyaush	1544	U/Pb zrn	rapakivi granite	xlln	*Salop & Murina* [1970]
27	Kusa-Kopanka	585–3195	K/Ar WR, amp, pl, bt	layered gabbro	xlln & metam	*Dunaev* et al. [1973]
28	Kusa-Kopanka	1288 ± 15	K/Ar ms	ct alt wallrock	pluton intrsn	*Gorozhanin* et al. [1995]
29	Kusa-Kopanka	2553 ± 96	Rb/Sr WR	gabbro-granite	wallrock assmln	*Gorozhanin* et al. [1995]
30	Kusa-Kopanka	1300	Rb/Sr ms, ap	ct alt wallrock	pluton intrsn	*Gorozhanin* et al. [1995]
31	Gubensky	1336	K/Ar bt	bt grano-gneiss	metasom	*Harris* [1977]
32	Gubensky	1282 ± 120	Rb/Sr WR	bt grano-gneiss	xlln	*Dunayev & Durneva* [1966]
33	Selyankino	1815	U/Pb zrn			*Krasnobaev* et al. [1974]
34	Inzer west fold/thrust belt	910–950	Ar/Ar kfs	conglomerate	cooling	*Glasmacher* et al. [1999]
35	Ryabinovsk	1300	U/Pb zrn	granite	xlln	*Turgarinov* et al. [1970]
36	Avzyan	1000 ± 20	K/Ar	gabbro dyke	xlln	*Keller & Chumakov* [1983]
37	Avzyan	1080 ± 30	K/Ar	gabbro dyke	xlln	*Keller & Chumakov* [1983]
38	Avzyan	1225	K/Ar	glauconite	sedmtn	*Polevaya & Kazakov* [1962]
39	Mashak	826 ± 54	Rb/Sr WR	alt rhyolite	metam	*Krasnobaev* et al. [1985]
40	Mashak	1346 ± 41	Rb/Sr WR	alt rhyolite	xlln	*Krasnobaev* et al. [1985]
41	Mashak	1350 ± 10	Rb/Sr WR	rhyolite	xlln	*Koroteev* et al. [1997]

Table 1. (continued)

#	Location	Date (Ma)	Method	Rock type	Interpretation	Reference
42	Mashak	1350 ± 30	U/Pb zrn	rhyolite	xlln	*Krasnobaev* et al. [1985]
43	Mashak	1350 ± 10	U/Pb			*Koroteev* et al. [1997]
44	Maksyutov	1077–1175	Rb/Sr	eclogite	protolith xlln	*Dobretsov* [1974]
45	Maksyutov	352–1216	U/Pb zrn		protolith xlln	*Krasnobaev* et al. [1995b]
46	Maksyutov	443–1800	U/Pb zrn		protolith xlln	*Krasnobaev* et al. [1995b]
47	Maksyutov	547–1517	U/Pb zrn		protolith xlln	*Krasnobaev* et al. [1995b]
48	Maksyutov	778–907	U/Pb zrn	metavolcanics	protolith xlln	*Dobretsov* [1974]
49	Maksyutov	1144–1795	U/Pb zrn	metavolcanics	protolith xlln	*Dobretsov* [1974]
50	Maksyutov	877–1963	U/Pb zrn	terrig material	parent province	*Kozlov* [1982]

Abbreviations used in the tables: act — actinolite, alt — altered, amp — amphibole, ampite — amphibolite, ap — apatite, assmln — assimilation, bt — biotite, chl — chlorite, cpx — clinopyroxene, crb — carbonate, ct — contact, defmn — deformation, diag — diagenesis, fmn — formation, gran — granulite, grph — graphite, grt — garnet, hbl — hornblende, ep — epidote, grano — granitic, il — ilmenite, intermed — intermediate, intrsn — intrusion, k — potassic, kfs — alkali feldspar, ky — kyanite, m — mica, metam — metamorphism, metasom — metasomatism, mfic — mafic, min — mineral, ms — muscovite, nph — nepheline, parent — parental, phn — phengite, pl — plagioclase, px — pyroxene, qtz — quartz, retro — retrograde, rt — rutile, sedmtn — sedimentation, subalk — subalkaline, terrig — terrigenous, tlc — talc, ttn — titanite, wm — white mica, WR — whole-rock, xlln — crystallisation, zo — zoisite, zrn — zircon.

Summarising the geochronological work in the Uralides provides the possibility for a comment on the change in emphasis of dating techniques used over the last 40 years. As is shown by the references for the data presented in the maps (Figures 2 to 6 and Tables 1 to 5) the methods used in early days were predominantly K/Ar on whole rock, biotite and hornblende, Rb/Sr whole rock, and multigrain zircon U/Pb analysis. More recent work, although still including these methods, has focused on Ar/Ar mineral, Pb/Pb single zircon, and Sm/Nd and Rb/Sr mineral dating.

Geochronological data has classically been interpreted in terms of either crystallisation ages or the ages of temperature dependant closure of isotopic systems. However, over the last few years it has become clear that there are additional controls on isotopic systems such as deformation, recrystallisation, fluid availability, and the mineralogical composition of a rock. The term "age" can, therefore, refer to a number of different processes. Only thorough understanding of the petrological history of a rock or mineral can determine whether an isotopic date reflects the age of crystallisation, cooling, mineral reactions, deformation, a fluid-rock interaction event, or another geological process. Although, notably, the coincidence of two or more different geochronological methods giving the same date can usually be taken as good evidence for a date of a geological event, the ad-hoc interpretation of isotopic dates as cooling or crystallisation ages is obsolete.

In a compilation work we can only refer to the original papers for the precise interpretation and the

geological inferences of each isotopic date. Nevertheless, in the tables accompanying the maps, we have, where possible, extracted from the original papers both petrographic information on the dated rock and the proposed interpretation of the isotopic data. Such information should be taken as good evidence for an age of a geological event.

4. SYNTHESIS OF THE GEOCHRONOLOGICAL DATA

The Uralide geochronological data are discussed from oldest to youngest, which more or less corresponds to a west to east progression across the mountain belt (Figures 1 to 6, Tables 1 to 5).

4.1. Precambrian

4.1.1. East European Craton. Metamorphic rocks thought to be typical of the Archean and Proterozoic basement of the East European Craton are exposed in the Central Uralian Zone, in the Taratash complex [*Puchkov*, 1997] (Figure 2). This 45 km long by 15 km wide complex comprises granulite facies rocks including two-pyroxene schists and gneisses containing garnet, sillimanite, cordierite, and hypersthene [*Ivanov* et al., 1986]. Zircons from the variably deformed gneisses have been dated at 2.83 Ga to 1.23 Ga by U/Pb [*Lennykh and Krasnobaev*, 1978; *Ivanov* et al., 1986; *Krasnobaev*, 1986] (Table 1). More specifically, granulite facies metamorphism has been dated at 2.6 Ga to 2.36 Ga by U/Pb on zircons [*Krasnobaev*, 1986; *Tugarinov* et al., 1970].

Table 2. East European Craton Neoproterozoic dates presented in Figure 3.

#	Location	Date (Ma)	Method	Rock type	Interpretation	Reference
1	Chelyabinsk	531 ± 28	U/Pb	gneiss	ampite metam	*Krasnobaev* et al. [1995]
2	Sysert Illmenogorsk	429 ± 12	U/Pb zrn	px hbl schist	plaiogranite fmn	*Krasnobaev* et al. [1998]
3	Taganai Iremel	633	K/Ar WR	phyllite	metam	*Harris* [1964]
4	Navysh	453–911	K/Ar WR, pl	trachybasalt	k metasom	*Harris* [1977]
5	Navysh	620 ± 30	U/Pb zrn	basalt	k metasom	*Krasnobaev* et al. [1992]
6	Kusa-Kopanka	585–3195	K/Ar WR, amp, pl, bt	layered gabbro	metam	*Dunaev* et al. [1973]
7	Taratash	600–1100	U/Pb zrn	ampite	ampite metam	*Ivanov* et al. [1986]
8	Arsha	660	Rb/Sr	glauconite		*Ancygin* et al. [1994]
9	Barangul Ural-Tau	660	U/Pb	granite	xlln	*Zykov* et al. [1985]
10	Beloretsk	600–660	K/Ar ms	quartzite	metam	*Lennykh* [1968]
11	Beloretsk	505 ± 5	Ar/Ar ms		metam	*Glasmacher* et al. [1998]
12	Beloretsk	541 ± 3	Ar/Ar ms	mica schist	metam	*Glasmacher* et al. [1999]
12	Beloretsk	543 ± 4	Ar/Ar ms	mica schist	metam	*Glasmacher* et al. [1999]
12	Beloretsk	543 ± 4	Ar/Ar ms	mica schist	metam	*Glasmacher* et al. [1999]
12	Beloretsk	550 ± 3	Ar/Ar ms	mica schist	metam	*Glasmacher* et al. [1999]
12	Beloretsk	557 ± 4	Ar/Ar ms	mica schist	metam	*Glasmacher* et al. [1999]
13	Beloretsk	575–680	Ar/Ar ms	mica schist	metam	*Glasmacher* et al. [1999]
14	Beloretsk	597 ± 4	Ar/Ar ms	mica schist	metam	*Glasmacher* et al. [1999]
15	Beloretsk	718 ± 5	Ar/Ar amp	ampite	metam	*Glasmacher* et al. [1999]
16	Inzer west fold/ thrust belt	530–550	Ar/Ar kfs	conglomerate	cooling	*Glasmacher* et al. [1999]
17	Inzer west fold/ thrust belt	590–630	Ar/Ar kfs	conglomerate	cooling	*Glasmacher* et al. [1999]
18	Inzer	791–681	K/Ar WR	glauconite	diagen	*Harris & Keller* [1983]
19	Achermovo	500–515	K/Ar bt, ms	granite gneiss	metam	*Lennykh* [1968]
20	Arsha Tirlyan Mulden	670 ± 50	Rb/Sr WR	basalt andesite	xlln	*Gorozhanin* [1995]
21	Barangul Ural Tau	660	U/Pb zrn	granite	xlln	*Zykov* et al. [1985]
22	Mayardak	517–565	K/Ar WR	mica schist	metam	*Harris* [1977]
23	Uk	613–658	K/Ar WR	glauconite	diagen	*Harris & Keller* [1983]
24	Uk	687 ± 29	K/Ar WR	glauconite	diagen	*Gorozhanin* [1995]
25	Uk	687 ± 29	Rb/Sr WR	glauconite	diagen	*Gorozhanin* [1995]
26	Suirovo	638 ± 70	Rb/Sr il		diagen	*Gorozhanin* [1995]
27	Bakeevo	557–609	K/Ar	glauconite	diagen	*Harris & Keller* [1983]
28	Bakeevo	618 ± 13	Rb/Sr WR	glauconite	diagen	*Gorozhanin* [1995]
29	Avashlya	587 ± 8	K/Ar WR	syenite	xlln	*Alexeev & Alexeeva* [1982]
30	Maksyutov	352–1216	U/Pb zrn	eclogite	retro metam	*Krasnobaev* et al. [1995]
31	Maksyutov	443–1800	U/Pb zrn	eclogite	retro metam	*Krasnobaev* et al. [1995]
32	Maksyutov	547–1517	U/Pb zrn	eclogite	retro metam	*Krasnobaev* et al. [1995]

Abbreviations used in the table: see footnote to Table 1.

The lack of detailed field and petrographic descriptions, however, makes these dates difficult to interpret. Retrogression of the granulites, dated at 1.1 Ga to 600 Ma by U/Pb population zircon [*Ivanov* et al., 1986], to amphibolite and greenschist facies has occurred in shear zones along the eastern margin of the complex. *Lennykh and Krasnobaev* [1978] dated mylonites from the shear zones at 1.37 Ga to 370 Ma, by Rb/Sr and K/Ar. In both cases, such a large spread of dates makes it difficult to define which geological event they represent.

Paleontological data indicate Mesoproterozoic (1.6–1.0 Ga) to Neoproterozoic (1.0–0.54 Ga) ages for

Table 3. Accreted terranes Early to Late Palaeozoic ophiolite, ultramafic, mafic, and nepheline syenite massif dates presented in Figure 4.

#	Location	Date (Ma)	Method	Rock type	Interpretation	Reference
1	Kumba	428 ± 7	Pb/Pb zrn	gabbro	xlln	*Bosch* et al. [1997]
2	Kytylm	300	Rb/Sr WR	gabbro	xlln	*Bea* et al. [in press]
3	Kytylm	300	Pb/Pb zrn	dunite	xlln	*Bea* et al. [in press]
4	Imennya	428 ± 7	Pb/Pb zrn	gabbro	xlln	*Friberg* et al. [2000]
5	Sysert Ilmenogorsk	446 ± 13	Rb/Sr WR	nph syenite	xlln	*Kramm* et al. [1993]
6	Vishnevogorsk Ilmenogorsk	478 ± 55	Rb/Sr WR	nph syenite	xlln	*Kramm* et al. [1993]
7	Sysert Ilmenogorsk	434 ± 10–15	U/Pb	nph syenite	xlln	*Kramm* et al. [1993]
8	Syrostan	330	Rb/Sr WR	gabbro	xlln	*Montero* et al. [2000]
9	Syrostan	333 ± 3	Pb/Pb zrn	gabbro	xlln	*Montero* et al. [2000]
10	Nurali	399	U/Pb zrn	gabbro-diorite	xlln	*Smirnov* [1995]
11	Mindyak	414 ± 4	Sm/Nd WR-grt	metagabbro	metam	*Scarrow* et al. [1999]
12	Mindyak	410 ± 5	U/Pb zrn	metagabbro	metam	*Saveliev* et al. [2001]
13	Mindyak	411 ± 4	Pb/Pb zrn	metagabbro	metam	*Scarrow* et al. [1999]
14	Mindyak	483	Pb/Pb zrn	metagabbro	protolith xlln	*Scarrow* et al. [1999]
15	Petropavlovsk	315 ± 10	Pb/Pb zrn	gabbro	xlln	*Bea & Montero* [unpublished]
16	Baimak–Buribai	420 ± 16	Rb/Sr WR	basalt-rhyolite		*Ronkin* [1989]
17	Baimak–Buribai	374 ± 39	Rb/Sr WR,cpx			*Gorozhanin* [1995]
18	Kharbany	378 ± 9	K/Ar bt	px-ampite	cooling	*Pushkarev & Kaleganov* [1993]
18	Kharbany	381 ± 10	K/Ar bt	ampite	cooling	*Pushkarev & Kaleganov* [1993]
18	Kharbany	396 ± 8	K/Ar amp	px-ampite	cooling	*Pushkarev & Kaleganov* [1993]
18	Kharbany	411 ± 9	K/Ar amp	ampite	cooling	*Pushkarev & Kaleganov* [1993]
18	Kharbany	412 ± 11	K/Ar amp	ampite	cooling	*Pushkarev & Kaleganov* [1993]
18	Kharbany	414 ± 9	K/Ar bt	ampite	cooling	*Pushkarev & Kaleganov* [1993]
19	Kharbany Molostov	355 ± 12	K/Ar amp	gabbro	cooling	*Pushkarev & Kaleganov* [1993]
20	Kharbany Molostov	370 ± 16	K/Ar amp	gabbro	cooling	*Pushkarev & Kaleganov* [1993]
20	Kharbany Molostov	370 ± 11	K/Ar bt	granodiorite	cooling	*Pushkarev & Kaleganov* [1993]
20	Kharbany Molostov	376 ± 16	K/Ar amp	gabbro	cooling	*Pushkarev & Kaleganov* [1993]
20	Kharbany Molostov	378 ± 10	K/Ar bt	gabbro	cooling	*Pushkarev & Kaleganov* [1993]
20	Kharbany Molostov	386 ± 8	K/Ar bt	pyroxenite	cooling	*Pushkarev & Kaleganov* [1993]
20	Kharbany Molostov	386 ± 12	K/Ar bt	gabbro	cooling	*Pushkarev & Kaleganov* [1993]
20	Kharbany Molostov	391 ± 20	K/Ar amp	gabbro	cooling	*Pushkarev & Kaleganov* [1993]
20	Kharbany Molostov	399 ± 13	K/Ar amp	gabbro	cooling	*Pushkarev & Kaleganov* [1993]
20	Kharbany Molostov	402 ± 10	K/Ar bt	granodiorite	cooling	*Pushkarev & Kaleganov* [1993]
21	Kharbany Molostov	432 ± 13	K/Ar amp	gabbro	cooling	*Pushkarev & Kaleganov* [1993]
22	Kharbany Molostov	454 ± 10	K/Ar amp	px -ampite	cooling	*Pushkarev & Kaleganov* [1993]
23	Kharbany Bostochno	376 ± 10	K/Ar amp	gabbro	cooling	*Pushkarev & Kaleganov* [1993]
23	Kharbany Bostochno	381 ± 13	K/Ar amp	pyroxenite	cooling	*Pushkarev & Kaleganov* [1993]
23	Kharbany Bostochno	387 ± 10	K/Ar amp	hornblendite	cooling	*Pushkarev & Kaleganov* [1993]
23	Kharbany Bostochno	389 ± 9	K/Ar amp	hornblendite	cooling	*Pushkarev & Kaleganov* [1993]
23	Kharbany Bostochno	389 ± 10	K/Ar bt	pyroxenite	cooling	*Pushkarev & Kaleganov* [1993]
23	Kharbany Bostochno	390 ± 15	K/Ar bt	pyroxenite	cooling	*Pushkarev & Kaleganov* [1993]
23	Kharbany Bostochno	399 ± 9	K/Ar bt	pyroxenite	cooling	*Pushkarev & Kaleganov* [1993]
23	Kharbany Bostochno	405 ± 10	K/Ar amp	pyroxenite	cooling	*Pushkarev & Kaleganov* [1993]
24	Kharbany Bostochno	434 ± 13	K/Ar amp	hornblendite	cooling	*Pushkarev & Kaleganov* [1993]
25	Kharbany Akermanovsky	357 ± 35	K/Ar amp	diorite	cooling	*Pushkarev & Kaleganov* [1993]

Table 3. (continued)

#	Location	Date (Ma)	Method	Rock type	Interpretation	Reference
25	Kharbany Akermanovsky	358 ± 65	K/Ar amp	gabbro	cooling	*Pushkarev & Kaleganov* [1993]
25	Kharbany Akermanovsky	366 ± 38	K/Ar amp	gabbro	cooling	*Pushkarev & Kaleganov* [1993]
26	Kharbany Akermanovsky	390 ± 19	K/Ar amp	plagiogranite	cooling	*Pushkarev & Kaleganov* [1993]
26	Kharbany Akermanovsky	392 ± 25	K/Ar amp	gabbro	cooling	*Pushkarev & Kaleganov* [1993]
26	Kharbany Akermanovsky	395 ± 60	K/Ar amp	gabbro	cooling	*Pushkarev & Kaleganov* [1993]
26	Kharbany Akermanovsky	403 ± 32	K/Ar amp	gabbro	cooling	*Pushkarev & Kaleganov* [1993]
26	Kharbany Akermanovsky	412 ± 18	K/Ar amp	gabbro	cooling	*Pushkarev & Kaleganov* [1993]
26	Kharbany Akermanovsky	424 ± 80	K/Ar amp	gabbro	cooling	*Pushkarev & Kaleganov* [1993]
27	Kempersay	370	K/Ar WR	tonalite	cooling	*Bogdanov & Savelyev* [1979]
28	Kempersay	390	K/Ar	ampite	metam	*Pavlov & Chuprynina* [1968]
29	Kempersay	400–470	K/Ar WR	gabbro	cooling	*Knipper & Perfiliev* [1979]
30	Kempersay	400 ± 10	Rb/Sr min	plagiogranite	xlln	*Buyakayte* et al. [1983]
31	Kempersay	396 ± 33	Sm/Nd	gabbro	xlln	*Edwards & Wasserburg* [1985]
32	Kempersay	397 ± 20	Sm/Nd	gabbro	xlln	*Edwards & Wasserburg* [1985]

Abbreviations used in the table: see footnote to Table 1.

sedimentary and volcanic rocks deposited unconformably upon the metamorphic cover rocks of the East European Craton [*Puchkov*, 1997]. The most complete Precambrian section is exposed in the Bashkirian Anticlinorium in the Southern Urals (Figures 2 and 3). These sediments, with minor, intercalated volcanic rocks, have a thickness of about 12 to 15 km and range in age from 1.65 to 0.65 Ga [*Maslov* et al., 1997 and references therein]. The older Mesoproterozoic metasediments were deposited in intracratonic aulacogens [*Maslov* et al., 1997].

4.1.2. The Pre-Uralide Orogeny.
The Late Neoproterozoic molasse-like sediments of the Bashkirian Anticlinorium, including in the west metamorphic clasts, have been interpreted as being derived from erosion of a Pre-Uralide orogen [*Puchkov*, 1997; *Glasmacher* et al., 1999]. It is known that a Pre-Uralide orogeny of Neoproterozoic age affected large parts of the eastern margin of the East European Craton, particularly in the Northern and Polar Urals [*Puchkov*, 1993 and references therein; *Pease and Gee*, 1999] (Figure 1). Some evidence of this event is also present in the Southern Urals (Figures 2 and 3). In the eastern Bashkirian Anticlinorium, Mesoproterozoic metasediments (unconformably overlain by

Ordovician sandstones and conglomerates) near the town of Beloretsk are medium- to high-grade gneisses that are locally migmatised and intruded by granites. This metamorphic complex experienced two phases of Late Proterozoic ductile deformation, associated with the Pre-Uralide orogeny, for which ~ 550 Ma white mica cooling ages have been obtained [*Matte* et al., 1993; *Glasmacher* et al., 1999].

4.2. Paleozoic

4.2.1. Nepheline Syenites.
An episode of Early Paleozoic magmatism in the Middle Urals is constrained by two Rb/Sr whole-rock and U/Pb zircon dates [*Kramm* et al., 1983, 1993] (Figure 4). These were obtained from the Ilmenogorsk-Vishnevogorsk nepheline syenites located in the Sysert-Ilmenogorsk anticlinorium. The Rb/Sr whole rock dates of 446 ± 13 Ma and 478 ± 55 Ma and the U/Pb zircon date of 434 ± 15–10 Ma, which are within error of each other, were interpreted as the age of crystallisation of the syenites, the generation of which may have been related to rifting of the East European Craton.

4.2.2. Ophiolites, Ultramafic, and Mafic Massifs.
The ophiolites and ultramafic massifs of the Uralides have

Table 4. Paleozoic dates from metamorphic complex in accreted terranes presented in Figure 5.

#	Location	Date (Ma)	Method	Rock type	Interpretation	Reference
1	Brodovo	360 ± 12	Pb/Pb zrn	intermed gneiss	protolith xlln	*Friberg* et al. [2000]
1	Brodovo	362 ± 14	Pb/Pb zrn	intermed gneiss	protolith xlln	*Friberg* et al. [2000]
1	Brodovo	376 ± 13	Pb/Pb zrn	intermed gneiss	protolith xlln	*Friberg* et al. [2000]
1	Brodovo	387 ± 15	Pb/Pb zrn	intermed gneiss	protolith xlln	*Friberg* et al. [2000]
1	Brodovo	403 ± 23	Pb/Pb zrn	intermed gneiss	protolith xlln	*Friberg* et al. [2000]
2	Brodovo	363 ± 54	Pb/Pb zrn	felsic gneiss	protolith xlln	*Friberg* et al. [2000]
2	Brodovo	390 ± 28	Pb/Pb zrn	felsic gneiss	protolith xlln	*Friberg* et al. [2000]
2	Brodovo	397 ± 28	Pb/Pb zrn	felsic gneiss	protolith xlln	*Friberg* et al. [2000]
2	Brodovo	402 ± 12	Pb/Pb zrn	felsic gneiss	protolith xlln	*Friberg* et al. [2000]
2	Brodovo	413 ± 25	Pb/Pb zrn	felsic gneiss	protolith xlln	*Friberg* et al. [2000]
2	Brodovo	421 ± 26	Pb/Pb zrn	felsic gneiss	protolith xlln	*Friberg* et al. [2000]
3	Brodovo	347 ± 48	Pb/Pb zrn	felsic gneiss	protolith xlln	*Friberg* et al. [2000]
4	Brodovo	383 ± 7	Pb/Pb zrn	felsic gneiss	protolith xlln	*Friberg* et al. [2000]
4	Brodovo	383 ± 16	Pb/Pb zrn	felsic gneiss	protolith xlln	*Friberg* et al. [2000]
4	Brodovo	392 ± 16	Pb/Pb zrn	felsic gneiss	protolith xlln	*Friberg* et al. [2000]
4	Brodovo	410 ± 15	Pb/Pb zrn	felsic gneiss	protolith xlln	*Friberg* et al. [2000]
5	Emekh	280 ± 100	Pb/Pb zrn	mafic gneiss	metam	*Friberg* et al. [2000]
6	Emekh	290 ± 110	Pb/Pb zrn	mafic gneiss	metam	*Friberg* et al. [2000]
7	Emekh	308 ± 52	Pb/Pb zrn	mafic gneiss	metam	*Friberg* et al. [2000]
8	Emekh	481 ± 60	Pb/Pb zrn	mafic gneiss	metam	*Friberg* et al. [2000]
9	Teliana	330 ± 13	Pb/Pb zrn	mafic gneiss	protolith xlln	*Friberg* et al. [2000]
9	Teliana	346 ± 17	Pb/Pb zrn	mafic gneiss	protolith xlln	*Friberg* et al. [2000]
9	Teliana	348 ± 25	Pb/Pb zrn	mafic gneiss	protolith xlln	*Friberg* et al. [2000]
9	Teliana	352 ± 18	Pb/Pb zrn	mafic gneiss	protolith xlln	*Friberg* et al. [2000]
10	Salda central dome	290 ± 13	K/Ar WR	amp bt gneiss	metam	*Friberg* et al. [2000]
10	Salda central dome	298 ± 28	K/Ar WR	amp bt gneiss	metam	*Friberg* et al. [2000]
11	Salda central dome	337 ± 30	K/Ar WR	ampite	metam	*Friberg* et al. [2000]
11	Salda central dome	378 ± 35	K/Ar WR	amp bt gneiss	metam	*Friberg* et al. [2000]
12	Salda central dome	461 ± 7	K/Ar WR	grt amp bt gneiss	metam	*Friberg* et al. [2000]
12	Salda central dome	469 ± 15	K/Ar WR	ampite gneiss	metam	*Friberg* et al. [2000]
12	Salda central dome	472 ± 37	K/Ar WR	ampite	metam	*Friberg* et al. [2000]
13	Salda rim dome	288	K/Ar WR	gneiss	metam	*Friberg* et al. [2000]
13	Salda rim dome	292	K/Ar bt	gneiss	metam	*Friberg* et al. [2000]
13	Salda rim dome	314 ± 6	K/Ar WR	grtbt gneiss	metam	*Friberg* et al. [2000]
13	Salda rim dome	314 ± 10	K/Ar WR	bt amp gneiss	metam	*Friberg* et al. [2000]
13	Salda rim dome	316 ± 2	K/Ar WR	pl ampite	metam	*Friberg* et al. [2000]
13	Salda rim dome	339 ± 16	K/Ar WR	bt gneiss	metam	*Friberg* et al. [2000]
13	Salda rim dome	345	K/Ar WR	gneiss	metam	*Friberg* et al. [2000]
13	Salda rim dome	348 ± 25	K/Ar WR	2 m gneiss	metam	*Friberg* et al. [2000]
13	Salda rim dome	355 ± 31	K/Ar WR	ep pl ampite	metam	*Friberg* et al. [2000]
13	Salda rim dome	363 ± 37	K/Ar WR	bt amp gneiss	metam	*Friberg* et al. [2000]
13	Salda rim dome	364 ± 12	K/Ar WR	gneiss	metam	*Friberg* et al. [2000]
13	Salda rim dome	368 ± 28	K/Ar WR	grt bt amp gneiss	metam	*Friberg* et al. [2000]
13	Salda rim dome	373	K/Ar WR	grt 2 m gneiss	metam	*Friberg* et al. [2000]
13	Salda rim dome	386 ± 5	K/Ar WR	amp gneiss	metam	*Friberg* et al. [2000]
13	Salda rim dome	386 ± 12	K/Ar WR	grt bt gneiss	metam	*Friberg* et al. [2000]
13	Salda rim dome	387 ± 18	K/Ar WR	qtz ms schist	metam	*Friberg* et al. [2000]
13	Salda rim dome	393	K/Ar WR	bt ampite	metam	*Friberg* et al. [2000]
13	Salda rim dome	396 ± 33	K/Ar WR	kfs ampite	metam	*Friberg* et al. [2000]
13	Salda rim dome	398	K/Ar WR	ampite gneiss	metam	*Friberg* et al. [2000]
14	Salda	354 ± 3	U/Pb	mafic granulite	metam	*Friberg* et al. [2000]
14	Salda	359 ± 3	U/Pb	mafic granulite	metam	*Friberg* et al. [2000]

Table 4. (continued)

#	Location	Date (Ma)	Method	Rock type	Interpretation	Reference
15	Istok	269 ± 6	K/Ar WR	chl schist	metam	*Friberg* et al. [2000]
15	Istok	285 ± 11	K/Ar WR	act bt schist	metam	*Friberg* et al. [2000]
15	Istok	285 ± 28	K/Ar WR	kfs amp schist	metam	*Friberg* et al. [2000]
15	Istok	294 ± 14	K/Ar WR	qtz ms chl schist	metam	*Friberg* et al. [2000]
15	Istok	298 ± 3	K/Ar WR	m kfs schist	metam	*Friberg* et al. [2000]
15	Istok	307	K/Ar bt	m schist	metam	*Friberg* et al. [2000]
15	Istok	309 ± 22	K/Ar WR	ampite microgneiss	metam	*Friberg* et al. [2000]
15	Istok	314	K/Ar WR	schist	metam	*Friberg* et al. [2000]
15	Istok	332 ± 2	K/Ar WR	ampite schist	metam	*Friberg* et al. [2000]
15	Istok	339 ± 4	K/Ar WR	ms ky schist	metam	*Friberg* et al. [2000]
15	Istok	351 ± 36	K/Ar WR	bt qtz schist	metam	*Friberg* et al. [2000]
15	Istok	351	K/Ar WR	2m microgneiss	metam	*Friberg* et al. [2000]
15	Istok	357	K/Ar ms	ms microgneiss	metam	*Friberg* et al. [2000]
15	Istok	372 ± 6	K/Ar WR	micro-ampite	metam	*Friberg* et al. [2000]
15	Istok	374 ± 40	K/Ar WR	grph m qtz schist	metam	*Friberg* et al. [2000]
15	Istok	376 ± 24	K/Ar WR	tlc crb m schist	metam	*Friberg* et al. [2000]
15	Istok	377 ± 4	K/Ar WR	ms quartzite	metam	*Friberg* et al. [2000]
15	Istok	387 ± 10	K/Ar WR	grph kfs m schist	metam	*Friberg* et al. [2000]
15	Istok	392 ± 11	K/Ar WR	ep m chl amp schist	metam	*Friberg* et al. [2000]
15	Istok	395 ± 1	K/Ar WR	grph m qtz schist	metam	*Friberg* et al. [2000]
15	Istok	407 ± 11	K/Ar WR	2 m schist	metam	*Friberg* et al. [2000]
15	Istok	431 ± 29	K/Ar WR	act chl schist	metam	*Friberg* et al. [2000]
16	Murzinka Adui	174–176	K/Ar WR	m gneiss	metam	*Friberg* et al. [2000]
17	Murzinka Adui	240–277	K/Ar WR	bt gneiss	metam	*Friberg* et al. [2000]
18	Murzinka Adui	315–360	K/Ar WR	bt gneiss	metam	*Friberg* et al. [2000]
19	Murzinka Adui	465–470	K/Ar WR	bt gneiss	metam	*Friberg* et al. [2000]
20	Murzinka Adui	367 ± 6	Pb/Pb zrn	gneiss	metam	*Friberg* et al. [2000]
20	Murzinka Adui	378 ± 19	Pb/Pb zrn	gneiss	metam	*Friberg* et al. [1999]
21	Kurtinsky	293 ± 2	Ar/Ar ms	grt ms bt schist		*Glasmacher* et al. [1999]
22	Kurtinsky	308 ± 3	Ar/Ar ms	grt ms bt schist		*Glasmacher* et al. [1999]
23	Ufaley	273–308	K/Ar ms, pl, WR	metam rocks	metam	*Harris* [1964]
24	Ufaley	356–388	K/Ar	aplite dyke	metam	*Harris* [1977]
25	Ufaley	375–400	K/Ar	gneiss	metam	*Harris* [1977]
26	Ufaley	351–453	K/Ar	crystal schist	metam	*Harris* [1977]
27	Ufaley	390–580	K/Ar	amphibolites	metam	*Harris* [1977]
28	Ufaley	296 ± 6	Ar/Ar wm		cooling	*Eide* et al. [1997]
29	Ufaley	305 ± 6	Ar/Ar wm		cooling	*Eide* et al. [1997]
30	Kyshtym	299.1 ± 3.2	Rb/Sr bt, pl, ap	bt metagranitic mylonite	defmn	*Hetzel & Glodny* [2002]
31	Kyshtym	240.0 ± 1.4	Rb/Sr ms, ap	ms metagranitic mylonite	defmn	*Hetzel & Glodny* [2002]
31	Kyshtym	240.4 ± 2.3	Rb/Sr ms, kfs	ms metagranitic mylonite	defmn	*Hetzel & Glodny* [2002]
31	Kyshtym	244.5 ± 6.5	Rb/Sr ms, ap, pl	ms metagranitic mylonite	defmn	*Hetzel & Glodny* [2002]
31	Kyshtym	247.5 ± 2.9	Rb/Sr ms, kfs, ap	ms metagranitic mylonite	defmn	*Hetzel & Glodny* [2002]
32	Sysert	240–290	K/Ar m, kfs, amp		cooling	*Echtler* et al. [1997]
33	Sysert	291 ± 6	Ar/Ar amp		cooling	*Eide* et al. [1997]
34	Sysert	254 ± 15	Rb/Sr WR	granitic gneiss	metam	*Echtler* et al. [1997]

Table 4. (continued)

#	Location	Date (Ma)	Method	Rock type	Interpretation	Reference
35	Sysert	254 ± 4	Rb/Sr	gneiss		*Ronkin* et al. [1993]
36	Ilmenogorsk	254 ± 28	Rb/Sr WR	gneiss		*Krasnobaev* et al. [1978]
37	Sysert	435 ± 33	Rb/Sr	gneiss		*Echtler* et al. [1997]
38	Vishnevogorsk Ilmenogorsk	244 ± 8	Rb/Sr min	nph syenite	cooling	*Kramm* et al. [1993]
38	Vishnevogorsk Ilmenogorsk	245 ± 24	Rb/Sr min	nph syenite	defmn	*Kramm* et al. [1993]
39	Sysert	352 ± 40	Sm/Nd min	amphibolite	metam	*Echtler* et al. [1997]
40	Vishnevogorsk Ilmenogorsk	239 ± 40–45	U/Pb	nph syenite	defmn	*Kramm* et al. [1993]
41	Sysert	355 ± 5	U/Pb zrn	gneiss	metam	*Echtler* et al. [1997]
42	Sysert	270–340	U/Pb zrn	gneiss	metam	*Echtler* et al. [1997]
43	Mayardak	400–404	K/Ar ms			*Ovchinnikov* et al. [1969]
44	Mayardak	380	Ar/Ar phn			*Matte* et al. [1993]
45	Suunduk	273–283	K/Ar WR	gneiss	metam	*Ovchinnikov* et al. [1969]
46	Suunduk	460 ± 7	Sm/Nd min	gneiss	metam	*Vinogradov* et al. [1998]
47	Suunduk	322–331	U/Pb zrn	gneiss		*Ovchinnikov* et al. [1969]
48	Maksyutov	468	K/Ar WR	diabase		*Alekseyev* [1976]
49	Maksyutov	390 ± 20	K/Ar ms, WR	metam rocks	retro metam	*Lennykh* [1963]
50	Maksyutov	344 ± 7	Ar/Ar wm	qtz mylonite	retro defmn	*Hetzel & Romer* [2000]
50	Maksyutov	352 ± 6	Ar/Ar wm	qtz mylonite	retro defmn	*Hetzel & Romer* [2000]
50	Maksyutov	356 ± 7	Ar/Ar wm	qtz mylonite	retro defmn	*Hetzel & Romer* [2000]
50	Maksyutov	364 ± 7	Ar/Ar wm	qtz mylonite	retro defmn	*Hetzel & Romer* [2000]
50	Maksyutov	370 ± 7	Ar/Ar wm	qtz mylonite	retro defmn	*Hetzel & Romer* [2000]
50	Maksyutov	382 ± 7	Ar/Ar wm	mfic blueschist	metam	*Hetzel & Romer* [2000]
51	Maksyutov Karayanovo	365 ± 2	Ar/Ar	eclogite	cooling	*Lennykh* et al. [1995]
51	Maksyutov	365.5 ± 1.7	Ar/Ar	quartzite	cooling	*Lennykh* et al. [1995]
51	Maksyutov	375 ± 2	Ar/Ar		cooling	*Lennykh* et al. [1995]
51	Maksyutov Karayanovo	375.4 ± 1.7	Ar/Ar wm	eclogite	cooling	*Lennykh* et al. [1995]
51	Maksyutov Karayanovo	372.9 ± 3.8	Ar/Ar wm	eclogite	metam	*Matte* et al. [1993]
51	Maksyutov Karayanovo	377.7 ± 3.8	Ar/Ar wm	blueschist	metam	*Matte* et al. [1993]
52	Maksyutov Karyanova Upper unit	332 ± 3	Ar/Ar ms	metabasalt	metam	*Beane & Connelly* [2000]
52	Maksyutov Novapakrova Upper unit	332 ± 3	Ar/Ar ms	metabasalt	metam	*Beane & Connelly* [2000]
52	Maksyutov Karyanova Upper unit	333 ± 2	Ar/Ar wm	quartzite	metam	*Beane & Connelly* [2000]
52	Maksyutov Karyanova Upper unit	339 ± 2	Ar/Ar ms	metabasalt	metam	*Beane & Connelly* [2000]
53	Maksyutov Karyanova Middle unit	356 ± 2	Ar/Ar wm	graphite quartzite	retro defmn	*Beane & Connelly* [2000]

Table 4. (continued)

#	Location	Date (Ma)	Method	Rock type	Interpretation	Reference
54	Maksyutov Karayanovo Lower unit	365 ± 2	Ar/Ar phn	quartzite	retro defmn	*Beane & Connelly* [2000]
54	Maksyutov Karayanovo Lower unit	374 ± 4	Ar/Ar wm	eclogite	retro defmn	*Beane & Connelly* [2000]
54	Maksyutov Karayanovo Lower unit	375 ± 4	Ar/Ar phn	eclogite	cooling	*Beane & Connelly* [2000]
54	Maksyutov Karayanovo Lower unit	377 ± 2	Ar/Ar phn	graphite schist	cooling	*Beane & Connelly* [2000]
54	Maksyutov Maksyutova Lower unit	372 ± 2	Ar/Ar wm	mica schist	cooling	*Beane & Connelly* [2000]
54	Maksyutov Shubino Lower unit	374 ± 3	Ar/Ar phn	eclogite	cooling	*Beane & Connelly* [2000]
55	Maksyutov Karayanovo	387.9 ± 4	Ar/Ar wm	blueschist	metam	*Matte* et al. [1993]
56	Maksyutov	359 ± 4	Rb/Sr WR wm	quartz mylonite	retro defmn	*Hetzel & Romer* [2000]
56	Maksyutov	363 ± 10	Rb/Sr WR wm	quartz mylonite	retro defmn	*Hetzel & Romer* [2000]
56	Maksyutov	364 ± 4	Rb/Sr WR wm	qrtz mylonite	retro defmn	*Hetzel & Romer* [2000]
56	Maksyutov	366 ± 12	Rb/Sr WR wm	qrtz mylonite	retro defmn	*Hetzel & Romer* [2000]
56	Maksyutov	379 ± 10	Rb/Sr WR wm	mfic blueschist	metam	*Hetzel & Romer* [2000]
57	Maksyutov	378 ± 13	Sm/Nd WR	eclogite	metam	*Shatsky* et al. [1997]
58	Maksyutov Karayanovo	357 ± 15	Sm/Nd grt, amp, zo		metam	*Shatsky* et al. [1997]
58	Maksyutov Karayanovo	366 ± 7	Sm/Nd grt, cpx, wm	eclogite	metam	*Shaksky* et al. [1997]
58	Maksyutov Karayanovo	375 ± 3	Sm/Nd grt, cpx	eclogite	metam	*Shatsky* et al. [1997]
59	Maksyutov Karayanovo Lower unit	382 ± 10	Sm/Nd grt, rt, cpx, ap	eclogite	metam	*Beane & Connelly* [2000]
60	Maksyutov Karayanovo	396 ± 57	Sm/Nd grt, cpx		protolith xlln	*Shatsky* et al. [1997]
61	Maksyutov Karayanovo Lower unit	399 ± 35	Sm/Nd ap, rt	eclogite		*Beane & Connelly* [2000]
62	Maksyutov Karayanovo Lower unit	377 ± 2	U/Pb rt	eclogite	metam	*Beane & Connelly* [2000]
63	Maksyutov Karayanovo Lower unit	384 ± 3	U/Pb rt	eclogite	metam	*Beane & Connelly* [2000]
64	Suvanyak	385	K/Ar ms	quartzite	metam	*Lennykh* [1963]

Abbreviations used in the table: see footnote to Table 1.

Table 5. Late Palaeozoic dates from intrusion in accreted terranes presented in Figure 6.

#	Location	Date (Ma)	Method	Rock type	Interpretation	Reference
1	Emekh	325 ± 7	Pb/Pb zrn	granite	xlln	*Friberg* et al. [2000]
1	Emekh	326 ± 13	Pb/Pb zrn	granite	xlln	*Friberg* et al. [2000]
1	Emekh	335 ± 7	Pb/Pb zrn	granite	xlln	*Friberg* et al. [2000]
1	Emekh	340 ± 6	Pb/Pb zrn	granite	xlln	*Friberg* et al. [2000]
2	Teliana	350 ± 23	Pb/Pb zrn	granodiorite	xlln	*Friberg* et al. [2000]
2	Teliana	351 ± 8	Pb/Pb zrn	granodiorite	xlln	*Friberg* et al. [2000]
2	Teliana	359 ± 11	Pb/Pb zrn	granodiorite	xlln	*Friberg* et al. [2000]
2	Teliana	360 ± 10	Pb/Pb zrn	granodiorite	xlln	*Friberg* et al. [2000]
2	Teliana	362 ± 15	Pb/Pb zrn	granodiorite	xlln	*Friberg* et al. [2000]
2	Teliana	373 ± 29	Pb/Pb zrn	granodiorite	xlln	*Friberg* et al. [2000]
2	Teliana	381 ± 17	Pb/Pb zrn	granodiorite	xlln	*Friberg* et al. [2000]
3	Salda	272	K/Ar WR	bt granite	cooling	*Friberg* et al. [2000]
4	Salda	278 ± 10	K/Ar WR	granite	cooling	*Friberg* et al. [2000]
4	Salda	296	K/Ar WR	granite	cooling	*Friberg* et al. [2000]
4	Salda	300 ± 4	K/Ar WR	granite	cooling	*Friberg* et al. [2000]
4	Salda	300 ± 11	K/Ar WR	bt granite	cooling	*Friberg* et al. [2000]
4	Salda	311	K/Ar WR	bt granite	cooling	*Friberg* et al. [2000]
4	Salda	312	K/Ar WR	2 m granite	cooling	*Friberg* et al. [2000]
4	Salda	312	K/Ar WR	granite	cooling	*Friberg* et al. [2000]
5	Salda	338	K/Ar WR	granite	cooling	*Friberg* et al. [2000]
5	Salda	348	K/Ar WR	ms granite	cooling	*Friberg* et al. [2000]
5	Salda	355 ± 1	K/Ar WR	andesite-dacite porphyry	cooling	*Friberg* et al. [2000]
5	Salda	355	K/Ar WR	qtz pl porphyry	cooling	*Friberg* et al. [2000]
5	Salda	358	K/Ar WR	porphyritic granite	cooling	*Friberg* et al. [2000]
5	Salda	367 ± 8	K/Ar WR	granodiorite	cooling	*Friberg* et al. [2000]
5	Salda	367 ± 29	K/Ar WR	gabbro ampite	cooling	*Friberg* et al. [2000]
5	Salda	368 ± 14	K/Ar WR	gabbro ampite	cooling	*Friberg* et al. [2000]
5	Salda	374 ± 26	K/Ar WR	diorite	cooling	*Friberg* et al. [2000]
5	Salda	375 ± 30	K/Ar WR	bt pl granite	cooling	*Friberg* et al. [2000]
5	Salda	381	K/Ar WR	plagiogranite	cooling	*Friberg* et al. [2000]
5	Salda	393 ± 17	K/Ar WR	granite	cooling	*Friberg* et al. [2000]
5	Salda	399 ± 7	K/Ar WR	qtz diorite	cooling	*Friberg* et al. [2000]
6	Salda	410 ± 11	K/Ar WR	granodiorite	cooling	*Friberg* et al. [2000]
6	Salda	413 ± 10	K/Ar WR	amp gabbro	cooling	*Friberg* et al. [2000]
6	Salda	425	K/Ar WR	diorite	cooling	*Friberg* et al. [2000]
6	Salda	435 ± 34	K/Ar WR	gabbro-norite	cooling	*Friberg* et al. [2000]
7	Salda	275 ± 25	Pb/Pb zrn	granite	xlln	*Friberg* et al. [2000]
8	Salda	320	Pb/Pb zrn	plagiogranite	xlln	*Friberg* et al. [2000]
8	Salda	333 ± 5	Pb/Pb zrn	monzonite	xlln	*Friberg* et al. [2000]
8	Salda	337 ± 5	Pb/Pb zrn	subalk gabbro	xlln	*Friberg* et al. [2000]
8	Salda	340	Pb/Pb zrn	plagiogranite	xlln	*Friberg* et al. [2000]
8	Salda	345 ± 30	Pb/Pb zrn	plagiogranite	xlln	*Friberg* et al. [2000]
9	Murzinka	248 ± 6	Rb/Sr WR	granite	xlln	*Gerdes* et al. [1999]
10	Murzinka	255	Rb/Sr WR	granite	xlln	*Montero* et al. [2000]
11	Verkhisetsk	316 ± 6	Rb/Sr WR	granite	xlln	*Bea* et al. [1997]
12	Verkhisetsk	320 ± 12	Rb/Sr WR	granite	xlln	*Bea* et al. [1997]
13	Verkhisetsk	276 ± 5	Rb/Sr WR	granite	xlln	*Bea* et al. [1997]
14	Verkhisetsk	284 ± 17	Rb/Sr WR	granite	xlln	*Bea* et al. [1997]
15	Verkhisetsk	318 ± 4	Pb/Pb zrn	granite	xlln	*Montero* et al. [2000]
16	Verkhisetsk	320 ± 20	Rb/Sr WR	granite	xlln	*Montero* et al. [2000]
17	Verkhisetsk	274 ± 4	Pb/Pb zrn	granite	xlln	*Montero* et al. [2000]
18	Verkhisetsk	295 ± 10	Pb/Pb zrn	granite	xlln	*Montero* et al. [2000]

Table 5. (continued)

#	Location	Date (Ma)	Method	Rock type	Interpretation	Reference
19	Verkhisetsk	310 ± 6	Pb/Pb zrn	granite	xlln	*Montero et al.* [2000]
20	Verkhisetsk	316 ± 6	Pb/Pb zrn	granite	xlln	*Montero et al.* [2000]
21	Verkhisetsk	320 ± 20	Pb/Pb zrn	granite	xlln	*Montero et al.* [2000]
22	Verknij Ufaley	316 ± 1	U/Pb ttn		xlln	*Hetzel & Romer* [1999]
23	Uvildy	262	K/Ar ms	granite	xlln	*Bushlyakov et al.* [1994]
24	Kisegach	233	K/Ar bt	granite	xlln	*Bushlyakov et al.* [1994]
25	Kisegach	265	K/Ar ms	granite	xlln	*Bushlyakov et al.* [1994]
26	Argasy	265	K/Ar ms	granite	xlln	*Bushlyakov et al.* [1994]
27	Zlatoust	380–434	K/Ar ms, pl	pegmatie	metam	*Lennykh* [1968]
28	Syrostan	330	Rb/Sr WR	granite	xlln	*Montero et al.* [2000]
29	Syrostan	327 ± 3	Pb/Pb zrn	granite	xlln	*Montero et al.* [2000]
30	Syrostan	333 ± 3	Pb/Pb zrn	granodiorite	xlln	*Montero et al.* [2000]
31	Karagai Kulsakmaro Voznesenka	453 ± 87	Rb/Sr WR	granodiorite	xlln	*Gorozhanin* [1995]
32	Balbukovo	317 ± 32	Rb/Sr WR	granodiorite	xlln	*Gorozhanin* [1995]
33	Varlamovka	331 ± 11	Rb/Sr min	granite migmatite	defmn	*Gorozhanin* [1995]
34	Chelyabinsk	276 ± 5	K/Ar WR			*Krasnobaev et al.* [1995]
35	Chelyabinsk	251 ± 36	Rb/Sr min	granodiorite	altn	*Gorozhanin* [1995]
36	Chelyabinsk	260 ± 16	Rb/Sr min	bt granite	xlln	*Gorozhanin* [1995]
37	Chelyabinsk	269 ± 19	Rb/Sr min	pl migmatite	altn	*Gorozhanin* [1995]
38	Stepninsky	292–316	K/Ar bt, amp	granite	xlln	*Lozovaya et al.* [1972]
39	Zamatocha	347 ± 2	Rb/Sr WR	granodiorite	xlln	*Gorozhanin* [1995]
40	Verchne–Ural	362 ± 7	Rb/Sr WR	gabbro-syenite	xlln	*Gorozhanin* [1995]
41	Magnitogorsk	283–341	K/Ar WR	gabbro-granite	xlln	*Lozovaya et al.* [1972]
42	Magnitogorsk	319 ± 3	Rb/Sr WR	phonolite	xlln	*Ronkin* [1989]
43	Magnitogorsk	333 ± 4	Rb/Sr WR	phonolite	xlln	*Ronkin* [1989]
44	Dzhabyk–Karagai	267 ± 16	Rb/Sr WR	adamellite	xlln	*Ronkin* [1989]
45	Dzhabyk	279 ± 10	Rb/Sr WR	granite	xlln	*Ivanov et al.* [1995a]
46	Dzhabyk	282 ± 17	Rb/Sr WR	granite	xlln	*Montero et al.* [2000]
47	Dzhabyk–Olhovsk	291 ± 4	Pb/Pb zrn	granite	xlln	*Montero et al.* [2000]
48	Varna	258	K/Ar WR	granite-aplite	xlln	*Ovchinnikov* [1963]
48	Varna	270	K/Ar pl	bt granite	alt	*Harris* [1963]
49	Mochagin	279 ± 10	Rb/Sr WR	monzodiorite-syenite	xlln	*Ivanov et al.* [1995b]
50	Kizil	307 ± 27	Rb/Sr WR	rhyolite	xlln	*Gorozhanin* [1995]
51	Kaczbach	317	K/Ar WR	granodiorite	xlln	*Ovchinnikov et al.* [1969]
52	Uvildy	262	K/Ar ms	granite	xlln	*Bushlyakov et al.* [1994]
53	Ershovo–Cheka	322 ± 27	Rb/Sr WR	rhyolite	xlln	*Gorozhanin* [1995]
54	Marinian	351	K/Ar bt	hornblendite	xlln	*Ovchinnikov et al.* [1969]
54	Marinian	360	K/Ar ms	pegmatite	xlln	*Lozovaya and Popov* [1966]
54	Marinian	361	K/Ar ms	pegmatite	xlln	*Harris* [1964]
54	Marinian	361	K/Ar bt	diorite	alt	*Ovchinnikov et al.* [1969]
54	Marinian	365	K/Ar WR	granite	xlln	*Ovchinnikov et al.* [1969]
54	Marinian	375	K/Ar bt	plagiogranite	xlln	*Harris* [1964]
55	Marinian	424	K/Ar ms	pegmatite	xlln	*Ovchinnikov et al.* [1969]
56	Marinian	448	K/Ar hbl	diorite	xlln	*Ovchinnikov et al.* [1969]
57	Suunduk	255–292	K/Ar bt, pl, ms	granite	xlln	*Lozovaya and Popov* [1966]
58	Suunduk	292	K/Ar bt	granite	xlln	*Lozovaya and Popov* [1966]
59	Suunduk	316	K/Ar bt	granodiorite	xlln	*Lozovaya and Popov* [1966]
60	Naslednitskoe	258	K/Ar kfs	pegmatite	xlln	*Lozovaya and Popov* [1966]
60	Naslednitskoe	258	K/Ar bt	pegmatite	xlln	*Lozovaya and Popov* [1966]
60	Naslednitskoe	263	K/Ar pl	pegmatite	xlln	*Lozovaya and Popov* [1966]

Table 5. (continued)

#	Location	Date (Ma)	Method	Rock type	Interpretation	Reference
61	Kaindinsky	349 ± 20	Rb/Sr min	granodiorite	xlln	*Gorozhanin* [1995]
62	Jetygara	244	K/Ar bt	granite	xlln	*Ovchinnikov* et al. [1969]
63	Mechet	375	K/Ar bt	granite	xlln	*Ovchinnikov* et al. [1969]
63	Mechet	375	K/Ar bt	granite	xlln	*Ovchinnikov* et al. [1969]
64	Milyutino	351	K/Ar bt	qtz diorite	xlln	*Lozovaya and Popov* [1966]
64	Milyutino	370	K/Ar bt	diorite	xlln	*Lozovaya and Popov* [1966]
65	Kumak	245	K/Ar WR	granite	xlln	*Lozovaya and Popov* [1966]
65	Kumak	254	K/Ar bt	pegmatite	xlln	*Lozovaya and Popov* [1966]
65	Kumak	260	K/Ar ms	pegmatite	xlln	*Lozovaya and Popov* [1966]

Abbreviations used in the table: see footnote to Table 1.

been divided into four main types [*Savelieva and Nesbitt*, 1996]: i. Lherzolitic massifs thrust onto the East European Craton (e.g., Kraka); ii. Harzburgitic massifs thrust onto the East European Craton (e.g., Voykar and Kempersay); iii. Lherzolitic complexes within the Main Uralian fault suture of the orogen (e.g., Nurali and Mindyak); and, iv. Harzburgitic complexes located east of the Main Uralian fault within more minor easterly sutures of the orogen (e.g., Alapaevsky and Bazhenovsky) (Figure 4). Precise dating of mafic to ultramafic rocks is a notoriously difficult task because the overall abundance of radioactive elements (U, K, Rb, Sm) is usually very low.

To date, isotopic age data have not been obtained for the type i. lherzolitic complexes.

The harzburgitic, type ii. ophiolites preserved as klippen to the west of the Main Uralian fault are the most extensively dated. Plagiogranites from Kempersay in the Southern Urals (Figure 4) have been dated as Devonian (370 Ma) by K/Ar whole rock [*Bogdanov and Savelyev*, 1979] and by Rb/Sr mineral date of 400 ± 10 Ma [*Buyakayte* et al., 1983]. Rather than representing the age of formation of the ophiolite crust protolith these dates probably record, respectively, cooling during obduction of the ophiolite and the age of melt generation associated with ophiolite obduction. By contrast, the ophiolite protolith oceanic crust crystallisation has been dated as Early Devonian to Middle Ordovician (about 390 Ma to 470 Ma) by Sm/Nd mineral and whole rock dating on gabbros from Kempersay [*Edwards and Wasserburg*, 1985] and by K/Ar whole rock and mineral ages from Kharbany massif [*Pushkarev and Kaleganov*, 1993] (Figure 4).

The lherzolitic massifs closely associated with the Main Uralian fault are poorly dated. Mindyak massif garnet metagabbros have been variously dated as crystallising at the Silurian-Devonian boundary (~ 415 Ma) [*Scarrow* et al., 1999; *Saveliev* et al., 2001]. Accordingly,

a U/Pb zircon date of 399 Ma has been obtained for crystallisation of a gabbro diorite from Nurali, about 50 km to the north [*Smirnov*, 1995].

Few dates exist for the accreted oceanic terranes to the east of the Main Uralian fault (Figure 4). The age of the volcanogenic complexes was defined by Rb/Sr whole rock and mineral method. The dates obtained range from Silurian (420 ± 16 Ma) [*Ronkin*, 1989] to Early Carboniferous (374 ± 39 Ma) [*Gorozhanin*, 1995] for the Baimak-Buribai Formation and 347 ± 20 Ma [*Gorozhanin*, 1995] for the Karamalytash Formation. Similarly, geochronological constraints on the mafic massifs of the Platinum Belt (Figure 1) are few. Gabbros of the Kumba Massif in the Tagil volcanogenic zone have been dated as Silurian, 428 ± 7 Ma by Pb/Pb on zircons [*Bosch* et al., 1997]. In addition, a Pb/Pb date of 428 ± 7 Ma was obtained on a mafic body of the Imennya Formation, Tagil arc (*Friberg* et al., 2000). Notably, considerably younger dates [300 to 330 Ma] have been obtained from subduction-related mafic rocks from intrusive massifs such as Kytlym and Syrostan [*Montero* et al., 2000; *Bea* et al., in press].

4.2.3. Metamorphic Complexes. Closure of the Uralian paleo-ocean by eastward subduction beneath Silurian-Devonian oceanic island arcs ultimately led to subduction of the leading edge of the thinned East European Craton continental margin. This caused the generation of high pressure complexes now exposed in the footwall of the Main Uralian fault from the Southern to the Polar Urals [*Lennykh*, 1977; *Dobretsov and Sobolev*, 1984] (Figure 5).

The best studied of these high pressure units, the Maksyutov Complex in the Southern Urals is bordered to the east by the Main Uralian fault, and is situated between the Devonian Magnitogorsk island arc and greenschist facies sediments of the East European Craton

[*Lennykh*, 1977; *Hetzel* et al., 1998]. The elongate north-south trending Maksyutov Complex is made up of two contrasting tectono-metamorphic units [*Lennykh* et al., 1995]. The lower unit consists predominantly of Proterozoic clastic sediments and minor amounts of mafic eclogites and blueschists the minerals from which indicate peak metamorphic conditions of 20 ± 3 kbar and $550 \pm 50°$C [*Hetzel* et al., 1998; *Schulte and Blümel*, 1999; *Beane and Connelly*, 2000]. The age of the high pressure metamorphism has been well constrained to 375 Ma to 380 Ma from U/Pb, Sm/Nd, Rb/Sr, Ar/Ar and K/Ar dating [*Lennykh*, 1961; *Matte* et al., 1993; *Shatsky* et al., 1997; *Glodny* et al., 1999; *Beane and Connelly*, 2000]. This lower unit of the complex has been interpreted to represent the passive margin of the East European Craton. In contrast, the upper unit is composed of Paleozoic sediments, volcanic rocks, and a serpentinite mélange containing rodingites; it is interpreted as a remnant of the Uralian paleo-ocean [*Hetzel*, 1999].

Somewhat younger metamorphic rocks genetically associated with Late Paleozoic granitoid batholith intrusion (included here to avoid overcrowding in Figure 6) such as the Salda Complex gneisses, are widespread in the Middle Urals [*Friberg* et al., 2000] (Figure 5).

4.2.4. Granitoid Batholiths. East of the Main Uralian fault numerous Late Paleozoic plutonic complexes crop out over a nearly 200 km wide area in the deeply eroded accreted terrane collage (Figure 6). These plutons constitute a major component of the Uralide crust that formed as a result of, progressively: i. subduction of oceanic and continental crust below island arc complexes and microcontinents, ii. accretion of such terranes along the East European Craton continental margin, and iii. during the final collision between the East European Craton and the Siberian–Kazakhian terrane assemblage. In general, the batholiths are complex with multiphase evolution, e.g., Early Permian late orogenic melting of Early Carboniferous subduction-related plutons [*Fershtater* et al., 1994; *Bea* et al., 1997].

In the East Uralian zone the plutons crop out in a 200 to 500 km long linear array parallel to the main axis of the orogen [*Fershtater* et al., 1997; *Kimbell* et al., 2002]. Rare Early Paleozoic, pre-Middle Devonian, gabbro-tonalite series such as Voznesenka, are interpreted to be relics of early island arcs (Figure 6). Late Devonian to Late Carboniferous, subduction-related, intermediate-mafic intrusions, such as Verkhisetsk, dated at 318 ± 4 Ma by Rb/Sr whole rock analysis [*Montero* et al., 2000] are widespread in the East Uralian zone. These tonalite-granodiorite complexes probably formed

at microcontinental margins [*Samarkin and Samarkina*, 1988; *Yazeva and Bochkarev*, 1996; *Bea* et al., 1997; *Puchkov*, 1997]. This subduction-related granite generation was broadly contemporaneous in the Southern and Middle Urals. Middle Carboniferous deformed granitoids and post-kinematic granites intruded at the periphery of the Main Uralian fault, e.g., Syrostan 333 ± 3 Ma (deformed) and 327 ± 3 Ma (undeformed) Pb/Pb single zircon data supported by comparable Rb/Sr whole-rock dates [*Montero* et al., 2000], have been interpreted to postdate the final closure of the Uralian ocean [*Hetzel and Romer*, 1999]. The east of the East Uralian zone is typified by younger late orogenic Upper Carboniferous to Permian batholiths mainly comprising peraluminous K-rich granitoids such as Murzinka (254 ± 5 Ma Pb/Pb single zircon) and Dzhyabyk (291 ± 4 Ma Pb/Pb single zircon) [*Fershtater* et al., 1994; *Fershtater* et al., 1997; *Montero* et al., 2000]. These youngest batholiths were produced diachronously, 35 Ma earlier in the Southern Urals than in the Middle Urals, by post-ocean closure partial melting of collision-thickened Proterozoic to Paleozoic microcontinental terranes [*Gerdes* et al., 2002].

Further east, in the Trans-Uralian zone, geochronological data are sparse but the age distribution and petrogenetic interpretation of the plutons are broadly comparable to the East Uralian zone granitoids (Figure 6, Table 5). Subduction-related intrusion K/Ar whole rock ages are Middle Ordovician (460 Ma) and Middle to Late Devonian, (360–375 Ma) [*Harris*, 1964; *Ovchinnkov* et al., 1966; *Puchkov* et al., 1986; *Yazeva* et al., 1989]. The common Permian granites (270 to 250 Ma) have been related to magmatism associated with the final Uralide collision [*Lozovaya and Popov*, 1966].

4.3. Mesozoic

Post-Paleozoic magmatism is rare in the Uralides. In the Sysertsk-Ilmenogorsk Complex late stage granitic dykes range in age from Early to Middle Jurassic (181 ± 7 Ma to 172 ± 7 Ma) [*Bushlyakov* et al., 1994]. Some Late Triassic to Early Jurassic (221 ± 14 Ma to 198 ± 4 Ma) lamproites crop out in Magnitogorsk [*Gorozhanin*, 1995]. Comparable, but somewhat broader ranging, Early Triassic [257 ± 2 Ma] to Middle Jurassic [162 ± 7 Ma] dates have been obtained for rhyolites from the Trans-Uralian zone by K/Ar mineral and Rb/Sr whole rock analysis [*Rupasova*, 1966; *Rublev* et al., 2001].

A preliminary fission-track study in the Southern Urals by *Seward* et al. [1997] was undertaken to decipher the late low temperature thermotectonic history of the mountain belt. Fission-track work gives insight into post-collisional cooling and exhumation within a region.

Annealing temperatures for apatite fission-tracks are in the range 60°C to 120°C whereas for zircon a critical temperature of 225 ± 25°C has been proposed [*Hurford*, 1986; *Hurford*, 1991]. *Seward* et al., [1997] concluded that the grouping of ~ 250 Ma dates of zircons from either side of the Main Uralian fault indicated that it has not been reactivated by major vertical movement since the Triassic. *Leech and Stockli* [2000] recorded apatite-fission track dates of 260 Ma to 280 Ma in the Maksyutov high pressure complex adjacent to the Main Uralian fault. In more detail, they noted, apatite fission-track analysis reveals variable denudation rates, the highest of which was in the Permian to Triassic. Following this was a period of burial and then post-Cretaceous final cooling and exhumation. *Seward* et al. [2002] present a more up to date summary of post-collisional cooling in the Southern Urals in the present volume.

5. SUMMARY

Summarizing the existing geochronological data from the Southern and Middle Urals broadly defines the tectonic evolution of the Uralian orogeny from a Mesoproterozoic starting point, through Paleozoic orogenesis, to the final post-orogenic uplift and cooling of the mountain belt. We hope that this review, in addition to being a useful reference work, will highlight targets for further geochronological work in Southern and Middle Urals and stimulate further research in the region.

Acknowledgments. J. H. S., J. G., A. G. and C. A. acknowledge support from European TMR grant URO ERBFMRXCT960009. J. H. S. and P. M. also acknowledge Spanish grant CICYT PB-13435. R. H. and J. G. thank the Deutsche Forschungsgemeinschaft and the GeoForschungs-Zentrum [GFZ] Potsdam for financial support. V. Pease and P. Matte are thanked for detailed and constructive reviews, and D. Brown for editorial comments.

REFERENCES

Alekseyev, A. A., Magmatic Complexes of the Uralian-ridge (zone) (in Russian), *Nauka Moscow*, 1–170, 1976.

Alexeev, A. A. and G. V. Alexeeva, Vendian essecsite-monconite-syenite potassium alcalic gabbroid seria of the Bashkirian Meganticlinorium (in Russian), in *The Questions of the Magmatism and the Metamorphism of the Southern Urals*, Bashkirian R.A.S., pp. 79–84, Ufa, 1982.

Ancygin, N. Y., K. K. Zoloev, M. L. Kluzhina, B. A. Nasedkina, B. A. Popov, B. I. Chuvashev, M. V. Shurygina, O. A. Sherbakov and V. M. Yakushov, *Explanatory Notes to the Stratigraphic Schemes of the Urals (PreCambrian,*

Paleozoic) (in Russian), Roscomnedra, IGG, Uralian R.A.S., 158 pp., Ekaterinburg, 1994.

Bea, F., G. Fershtater, P. Montero, V. Smirnov and E. Zin'kova, Generation and evolution of subduction-related batholiths from the central Urals: constraints on the P-T history of the Uralian orogen, *Tectonophysics*, 276, 103–116, 1997.

Bea, F., G. Fershtater, P. Montero, M. Whitehouse, V. Y. Levin, J. H. Scarrow, H. Austrheim and E. V. Pushkariev, Recycling of continental crust into the mantle as revealed by Kytlym dunite zircons, *Terra Nova.*, in press.

Beane, R. and J. N. Connelly, $^{40}Ar/^{39}Ar$, U-Pb and Sm.-Nd constraints on the timing of metamorphic events in the Maksyutov Complex, southern Ural Mountains, *J. Geol. Soc. Lond.*, 157, 811–822, 2000.

Bibikova E. V., T. V. Gracheva and A. A. Krasnobaev, On the Belomorian phase of metamorphism in the Ilmenogorsky complex (in Russian), *Dokl. Akad. Nauk USSR*, 208, 1165–1167, 1973.

Bogdanov, N. A. and A. A. Savelyev, The ophiolites of the Polar Urals, *Geotectonics*, 13, 169–171, 1979.

Bosch, D., A. A. Krasnobaev, A. A. Efimov, G. N. Savelieva and F. Boudier, Early Silurian ages for the gabbroic section of the mafic-ultramafic zone from the platinum belt, *Terra Abstracts*, 9, 122, 1997.

Brown, D., C. Juhlin, J. Alvarez-Marrón, A. Pérez-Estaún and A. Oslianski, Crustal-scale structure and evolution of an arc-continent collision zone in the southern Urals, Russia, *Tectonics*, 17, 158–171, 1998.

Brown, D. and P. Spadea, Processes of forearc and accretionary complex formation during arc-continent collision in the southern Urals, *Geology*, 27, 649–652, 1999.

Bushlyakov, I. N., B. A. Kaleganov and A. A. Kraesnobaev, New data of the isotopical dating of the Ilmenogorsk complex granitoids and metamorphites (in Russian), *Yearbook 1993 Uralian R.A.S.*, pp. 107–110, Ekaterinburg, 1994.

Buyakayte, M. I., V. I. Vinogradov, V. V. Kuleshov, B. G. Pokrovskiy, A. A. Saveliev and G. N. Savelieva, Isotope geochemistry in ophiolites of the Polar Urals (in Russian), *Nauka* 376, 179 pp., Moscow, 1983.

Dalziel, I., Neoproterozoic–Paleozoic geography and tectonics: Review, hypothesis, environmental speculation, *Geol. Soc. Am. Bull.*, 109, 16–42, 1997.

Dobretsov, N. L., The glaucophane-schist and eclogite-glaucophane-schist complexes of the USSR (in Russian), *Nauka*, 429 pp., Novosibirsk, 1974.

Dobretsov, N. L. and N. V. Sobolev, Glaucophane schists and eclogites in the folded systems of Northern Asia, *Ofioliti*, 9, 401–424, 1984.

Dunayev, V. A. and N. N. Durneva, About age of the rocks from the Kusa-Kopanka basic intrusion region of rubidium–strontium method data (in Russian), in *Absolute Dating of the Tectono-Magmatical Cycles and the Stages of the Ore-Formation on Data*, Nauka, pp. 131–139, Moscow, 1966.

Dunaev, V. A., A. I. Stepanov and M. V. Panova, The age of the Kopanka-Kusa intrusion rocks and the time of their metamorphism (Southern Urals) (in Russian), in *Geological-Radiological Interpretation of the Discordant Ages*, Nauka, pp. 238–247, Moscow, 1973.

Echtler, H. P., K. S. Ivanov, Y. L. Ronkin, L. A. Karsten and A. G. Noskov, The tectono-metamorphic evolution of gneiss complexes in the Middle Urals, Russia: A reappraisal, *Tectonophysics*, 276, 229–251, 1997.

Edwards, L. R. and G. J. Wasserburg, The age and emplacement of obducted oceanic crust in the Urals from Sm-Nd and Rb-Sr systematics, *Earth. Planet. Sci. Lett.*, 72, 389–404, 1985.

Eide, E. A., H. P. Echtler, R. Hetzel and K. S. Ivanov, Cooling age diachroneity and Paleozoic orogenic processes in the Middle and Southern Urals, *Terra Abstracts*, 9, 119, 1997.

Fershtater, G. B., N. S. Borodina and M. S. Rapoport, The orogenical granite magmatism of the Urals (in Russian), *Uralian R.A.S.*, 20 pp., Miass, 1994.

Fershtater, G. B., P. Montero, N. S. Borodina, E. V. Puskarev, V. N. Smirnov and F. Bea, Uralian magmatism: an overview, *Tectonophysics*, 276, 87–102, 1997.

Friberg, M., A. Larinov, G. A. Petrov and D. G. Gee, Paleozoic amphibolite-granulite facies magmatic complexes in the hinterland of the Uralide Orogen, *Int. J. Earth. Sci.*, 89, 21–39, 2000.

Gerdes, A., P. Montero, F. Bea, T. Osipova, N. Borodina and G. Fershtater, Late-orogenic continental-type granites of the Urals: composition, age and petrogenetic implications, *J. Conf. Abstracts*, 4, 6473, 1999.

Gerdes, A., P. Montero, F. Bea, G. B. Fershtater, T. Borodina, Ossipova and G. Shardakova, Peraluminous granites with mantle-like isotope compositions: The continental-type Murzinka and Dzhabyk batholiths of the eastern Urals, *Int. J. Earth Sci.*, 91, 1–17, 2002.

Glasmacher, U. A., U. Giese, L. Stroink, A. A. Alekseev, P. Reynolds, V. Puchkov and R. Walter, A Cadomian terrane at the Eastern Margin of Baltica — implications for the late Proterozoic paleogeography and for the structural evolution of the southwestern Urals, Russia, *Europrobe Workshop Uralides Moscow Abs. Vol.*, 191, 1998.

Glasmacher, U. A., P. Reynolds, A. A. Alekseyev, V. N. Puchkov, K. Taylor, V. Gorozhanin and R. Walter, [40]Ar/[39]Ar Thermochronology west of the Main Uralian fault, southern Urals, Russia, *Geol. Rundsch.*, 87, 515–525, 1999.

Glodny, J., H. Austrheim, B. Bingen, A. Rusin and J. H. Scarrow, New age data for HP rocks and ophiolites along the main Uralian fault, Urals, Russia: Implications for Uralian Orogeny, *Terra Nostra*, 99, 89–90, 1999.

Gorozhanin, V. M., A rubidium-strontium isotopic method in solving of the problems of geology of the Southern Urals (in Russian), *Diss. Inst. Geol. Uralian R.A.S.*, 23 pp., Ekaterinburg, 1995.

Gorozhanin, V. M., A. A. Alexeev and B. A. Kaleganov, New data on geochronology of the Kusa-Kopanka complex (in Russian), *Yearbook 1994, Bashkirian R.A.S.*, pp. 70–73, Ufa, 1995.

Hamilton, W., The Uralides and the motion of the Russian and Siberian platforms, *Bull. Geol. Soc. Am.*, 81(9), 2553–2576, 1970.

Harris, M. A., The main age complexes of magmatical and metamorphical rocks of the Southern Urals and Mugojares (the potassium-argon method data) (in Russian), in *Trans. 1st Uralian Petrogr. Meet.* 1, pp. 83–98, Sverdlovsk, 1963.

Harris, M. A., Geochronological scale of the Urals and their main development stages in Precambrian and Paleozoic (in Russian), *Nauka*, 128–156, 1964.

Harris, M. A., The stages of magmatism and metamorphism in the Prejurassic history of the Urals and Pre-Urals (in Russian), *Nauka*, 266 pp., Moscow, 1977.

Harris, M. A. and B. M. Keller, The stratotype of Riphean Stratigraphy, Geochronology (in Russian), *Trans, Geol. Inst.*, 377, *Nauka*, 184 pp., Moscow, 1983.

Hetzel, R., H. P. Echtler, W. Seifert, A. B. Schulte and K. S. Ivanov, Subduction- and exhumation-related fabrics in the Paleozoic high-P/low-T Maksyutov Complex, Antingan area, Southern Urals, Russia, *Geol. Soc. Am. Bull.*, 110, 916–930, 1998.

Hetzel, R., Geology and geodynamic evolution of the high-P/low-T Maksyutov Complex, Southern Urals, Russia, *Geol. Rundsch.*, 87, 577–588, 1999.

Hetzel, R. and R. L. Romer, U-Pb dating of the Verkniy Ufaley intrusion, Middle Urals, Russia: A minimum age for subduction and amphibolite facies overprint of the East European continental margin, *Geol. Mag.*, 136, 593–597, 1999.

Hetzel, R. and R. L. Romer, A moderate exhumation rate for the high pressure Maksyutov Complex, southern Urals, Russia, *Geol. J.*, 35, 327–344, 2000.

Hetzel, R. and J. Glodny, A crustal-scale, orogen-parallel strike-slip fault in the Middle Urals: Age, magnitude of displacement, and geodynamic significance, *Int. J. Earth Sci.*, in press.

Hurford, A., Cooling and uplift pattern in the Lepontine Alps, South Central Switzerland and an age of vertical movement on the Insubric fault line, *Contrib. Min. Pet.*, 92, 413–427, 1986.

Hurford, A. J., Uplift and cooling pathways derived from fission track analysis and mica dating: A review, *Geol. Rundsch.*, 80, 349–368, 1991.

IUGS *International Commission on Stratigraphy International Stratigraphic Chart*, edited by J. Remane, M. B. Cita, J. Dercourt, P. Bouysse, F. L. Repetto and A. Faure–Muret, courtesy of the division of Earth Sciences UNESCO, 2000.

Ivanov, K. S., P. Bankwitz, Y. L. Ronkin and E. Bankwitz, The model of Jabyk granite pluton (Southern Urals) formation, in *Magmatism and Geodynamic Trans. 1st Petrogr. Meet., Bashkirian R.A.S.*, pp. 69–71, Ufa, 1995a.

Ivanov, K. S., S. N. Ivanov and Y. L. Ronkin, On some of the study of orogenic granite magmatism of the Urals (in Russian), *Yearbook 1994 Uralian R.A.S.*, 171 pp., Ekaterinburg, 1995b.

Ivanov, S., A. Perfiliev, A. Efimov, G. Smirnov, V. Necheukin and G. Fershtater, Fundamental features in the structure and evolution of the Urals, *Am. J. Sci.*, 275, 107–130, 1975.

Ivanov, S. N., A. A. Krasnobaev and A. I. Rusin, Geodynamics regimes in the Precambrian of the Urals, *Precambrian Res.*, 33, 189–208, 1986.

Keller, B. M. and N. M. Chumakov, Stratotype of Riphean Stratigraphy, Geochronology (in Russian), *Nauka*, 184 pp., Moscow, 1983.

Kimbell G. S., C. A. Ayala, A. Gerdes, M. K. Kaban, V. A. Shapiro and Y. P. Menshikov, Insights into the architecture and evolution of the Southern and Middle Urals from gravity and magnetic data, this volume.

Knipper, A. L. and A. S. Perfiliev, Ophiolite belt of the Urals. International Atlas of Ophiolites, IGCP Project 39, *GSA Special Pub.*, map scale 1:2,500,000, 9–11, 1979.

Koroteev, V. A., K. S. Ivanov, H. P. Echtler and Y. L. Ronkin, Evolution of the Urals: Complete geodynamic cycle during 1.6–0.2 Ga, *Terra Abstracts*, 9, 118, 1997.

Kozlov, V. I., Upper Riphean and Vendian of the Southern Urals (in Russian), *Nauka*, pp. 128, Moscow, 1982.

Kramm, U., A. B. Blaxland, V. A. Kononova and B. Grauert, Origins of the Ilmenogorsk–Vishnevogorsk nepheline syenites, Urals, USSR, and their time of emplacement during the history of the Ural fold belt: A Rb-Sr study, *J. Geol.*, 91, 427–435, 1983.

Kramm, U., I. V. Chernyshev, B. Grauert and V. A. Kononova, Zircon typology and U-Pb systematics: A case study of zircons from nepheline syenite from the Il'meny Mountains, Urals, *Petrology*, 1, 474–485, 1993.

Krasnobaev, A. A., Zircons as indicator of geological processes (in Russian), *Nauka*, 148 pp., Moscow, 1986.

Krasnobaev, A. A., E. V. Bibikova and T. V. Gracheva, Belomorian metamorphism of the gneisses from Sysertsk–Ilmenogorsk anticlinorium (in Russian), *Abstr. of the 3rd Ural. Petrogr. Meet. Sverdlovsk*, 1974.

Krasnobaev, A. A., Y. L. Ronkin and A. I. Stepanov, About granitisation age and substrat's nature of the Sysertsk–Ilmenogorsk complex gneisses (in Russian), *Yearbook 1977 Uralian R.A.S.*, pp. 3–6, Sverdlovsk, 1978.

Krasnobaev, A. A., G. B. Feshtater and A. I. Stepanov, Petrology and rubidium-strontium geochronology of the Berdyaush rapakivi massif (Southern Urals) (in Russian), *Izvestia Acad. Nauk, SSSR, Ser. Geol.*, 1, 21–38, 1981.

Krasnobaev, A. A., E. V. Bibikova, A. I. Stepanov and Y. L. Ronkin, Geochronology and genesis of the Berdyaush massif (in Russian), *Izvestia Acad. Nauk, SSSR, Ser. Geol.*, 3, 2–23, 1984.

Krasnobaev, A. A., E. V. Bibikova and A. I. Stepanov, The age of Mashak formation's effusives and the problem of the Lower-Riphean radiological boundary (in Russian), in *Isotopic Dating of the Volcanism and Sedimentation Processes, Nauka*, pp. 118–124, Moscow, 1985.

Krasnobaev, A. A., E. V. Bibikova, Y. L. Ronkin and V. I. Kozlov, Geochronology of the Ai formation's volcanites

and the isotopical age of the Riphean lower boundary (in Russian), *Izvestia Acad. Nauk, SSSR, Ser. Geol.*, 6, 25–40, 1992.

Krasnobaev, A. A., G. P. Kuznetcov, V. A. Davydov, E. P. Shulkin and N. V. Cherednichenko, Uranium–lead age of the zircons from the Chelyabinsk gneiss complex (in Russian), *Yearbook 1994 Uralian R.A.S.*, pp. 34–36, Ekaterinburg, 1995a.

Krasnobaev, A. A., B. A. Davudov, V. I. Lennykh, N. B. Tscherednitschencko and W. J. Kosolov, The ages of zircons and rutiles from the Maksyutov Complex (preliminary data) (in Russian), *Yearbook 1995 Uralian R.A.S.*, pp. 13–16, Ekaterinburg, 1995b.

Krasnobaev, A. A., V. M. Necheukhin, V. A. Davydov and V. V. Sokolov, Zircon geochronology and terrane problem of the Uralian accretionary-folded system (in Russian), *Uralian Mineralogical Book*, 8, 196–206, 1998.

Leech, M. L. and D. F. Stockli, The late exhumation history of the ultrahigh pressure Maksyutov Conplex, south Ural Mountains, from new apatite fisson track data, *Tectonics*, 19, 153–167, 2000.

Lennykh, V. I., Caledonian metamorphism occurence in the rocks of the Maksyutovo complex Ural-Tau Ridge (in Russian), in *The Questions of the Geochronology and Geochemistry in Precambrian and Paleozoic of the Southern Urals and Eastern Part of the Russian Platform*, Institute of Geology, pp. 16–22, Ufa, 1961.

Lennykh, V. I., About the ages of the Ural-Tau zone metamorphic rock on potassium-argon method data (in Russian), *Trans. Absol. Dating Commission, 11th meeting, Moscow, Acad. Sci. USSR*. 1963.

Lennykh, V. I., The regional metamorphism of the Precambrian deposits of the Urals western slope and Ural-Tau Ridge (in Russian), *Uralian R.A.S.*, 67 pp., Sverdlovsk, 1968.

Lennykh, V. I., Eclogite-glaucophane schists belt of the Southern Urals (in Russian), *Nauka*, 160 pp., Moscow, 1977.

Lennykh, V. I. and A. A. Krasnobaev, Petrology and the iron ore deposits of the Taratash Complex (in Russian), *Acad. Nauka*, Sverdlorsk, 1978.

Lennykh, V. I., P. M. Valizer, R. Beane, M. Leech and W. G. Ernst, Petrotectonic evolution of the Maksyutovo Complex, Southern Urals, Russia. Implications for ultrahigh pressure metamorphism, *Int. Geol. Rev.*, 37, 584–600, 1995.

Lozovaya, L. S. and Y. D. Popov, Geochronoagmatical metasomatites from the Southern Ural's eastern slope (in Russian), in *Absolute Dating of the Tectonic-Magmatical Cycles and the Stages of the Ore-Formation on Data, Nauka*, pp. 195–204, Moscow, 1966.

Lozovaya, L. S., M. A. Harris and A. P. Grevtcova, The Hercynian cycle of magmatism and metamorphism in the Urals (in Russian), in *The Isotopical Geochronology's Questions of the Urals and Eastern Part of the Russian Platform, Bashkirian R.A.S.*, pp. 98–114, Ufa, 1972.

Maslov, A. V., B. D. Erdtmann, K. S. Ivanov, S. N. Ivanov and M. T. Krupenin, The main tectonic events, depositional history, and the palaeogeography of the southern

Urals during the Riphean-early Paleozoic, *Tectonophysics*, 276, 313–335, 1997.

Matte, P., H. Maluski, R. Caby, A. Nicolas, P. Kepeshinskas and S. Sobolev, Geodynamic model and ^{40}Ar/^{39}Ar dating for the generation and emplacement of the High Pressure [HP] metamorphic rocks in SW Urals, *C.R. Academie-des-Sciences, Series* 2, 317, 1667–1674, 1993.

Matte, P., Southern Uralides and Variscides: Comparison of their anatomies and evolutions, *Geolische Mijnbouw*, 74, 151–166, 1995.

McKerrow, W. S., Terrane assembly in the Variscan belt of Europe, *Europrobe News*, 5, 4–5, 1994.

Mineev, D. A., RE-bearing epidote from pegmatites of the Middle Urals (in Russian), *Rep. USSR Acad. Sci.* 127, 4, 865–868. 1959.

Montero, P., F. Bea, A. Gerdes, G. Ferstater, E. Zin'kova, N. Borodina, T. Osipova and V. Smirnov, Single zircon stepwise evaporation ^{207}Pb/^{206}Pb and Rb-Sr dating of four major Uralian batholiths. A perspective on the timing of deformation and granite generation, *Tectonophysics*, 317, 93–108, 2000.

Otto, S. C. and R. J. Bailey, Tectonic evolution of the northern Ural Orogen, *J. Geol. Soc. London*, 152, 903–906, 1995.

Ovchinnikov, L. N., The review of absolute age data for geological units of the Urals (in Russian), in *Trans 1st Uralian Petrogr. Meet.*, pp. 111–121, Sverdlovsk, 1963.

Ovchinnikov L. N., R. G. Podlesova, M. V. Panova and F. L. Shangareev, The results of absolute age determination for geological units of the Urals, received by potassium-argon method in 1963–1964 years (in Russian), in *Absolute Dating of the Tectono-Magmatical Cycles and Stages of the Ore-Formation on Data*, pp. 111–121, Moscow, 1966.

Ovchinnikov, L. N., A. I. Stepanov, A. A. Krasnobaev and V. A. Dunaev, Review of the absolute ages data for geological formations of the Urals (in Russian), in *Trans. 2nd Ural. Petrogr. Meet.*, pp. 173–204, Sverdlovsk, 1969.

Pavlov, N. V. and I. I. Chuprynina, The composition of chrome-spinellids and genetic types of chromite mineralisation in the Kempirsay plutonic body, *Geochem. Int.*, 4, 214–227, 1968.

Pease, V. and D. Gee, Baikalian-age complexes along the north-eastern margin of Baltica, *J. Conf. Abstracts*, 4, 109, 1999.

Polevaya, N. I. and G. A. Kazakov, The age subdivision and correlation of ancient sediments on ratio ^{40}Ar/^{40}K in glauconites (in Russian), *Trans. Laborat. Geol. Precambr. Leningrad*, 12, 1962.

Puchkov, V. N., M. S. Rapoport, G. B. Fershtater and E. M. Ananéva, The tectonic control of the Paleozoic granitoid magmatism at the Eastern slope of the Urals (in Russian), in *Investigation in Petrology and Metallogeny of the Urals*, pp. 85–95, Sverdlovsk, 1986.

Puchkov, V. N., Palaeo-oceanic structures of the Ural Mountains, *Geotectonics*, 27, 190–196, 1993.

Puchkov, V. N., Structure and geodynamics of the Uralian orogen, in *Orogeny Through Time*, edited by J. P. Burg and M. Ford, *Geol. Soc. Spec. Public.*, 121, 201–236, 1997.

Pushkarev, E. V. and B. A. Kaleganov, K-Ar dating of magmatic complexes from the Khabarny mafic-ultra-mafic massif (South Urals) (in Russian), *Rpt Acad. Sci., Ekaterinburg*, 328, 241–245, 1993.

Ronkin, Y. L., Strontium isotopes — the indicators of the evolution of the Ural's magmatism (in Russian), *Yearbook 1988 Uralian R.A.S.*, pp. 107–110, Sverdlovsk, 1989.

Ronkin, Y. D., A. G. Noskov and D. Z. Zhuravlev, Sm-Nd isotope system of the Sysert Gneiss-migmatic complex (in Russian), in *Yearbook 1992 Inst. Geol. and Geochem.*, pp. 135–139, Ekaterinburg, 1993.

Rupasova, Z. G., Absolute age for granite intrusion and kainotypical effusives of the Transuralian Zone (in Russian), in *Absolute Dating of the Tectonomagmatic Cycles and the Stages of the Ore-Formation*, Nauka,131–139, Moscow, 1966.

Rubalev, A. G., E. C. Bogomolov, E. E. Poroshin and Y. P. Sheragina, Isotopic age for Triassic Turinian series rhyolites (Kush-Murun graben, Turgai depression) (in Russian), *Strat. Geol. Correl.*, 9, 31–45, 2001.

Salop, L. I. and G. A. Murina, The age of the Berdyaush rapakivi granite pluton and the geochronological boundaries of the Lower Riphean problem (in Russian), *Sovetckaya Geologia*, 6, 15–27, 1970.

Samarkin, G. I. and Y. E. Samarkina, Granitoids from Southern Urals and the origin of granitic belts in folded regions (in Russian), *Nauka*, pp. 209, Moscow, 1988.

Saveliev, A. A., E. V. Bibikova, G. N. Savelieva, P. Spadea, J. H. Scarrow, A. N. Pertsev and T. I. Kirnozova, Garnet pyroxenite of the Mindyak Massif Southern Urals: Age and environment of formation (in Russian), *Bull. Moscow Soc. Nat. Geol*, 76, 23–30, 2001.

Savelieva, G. N., Gabbro-ultrabasic complexes of the Ural ophiolites and their analogues in modern ocean crust (in Russian), *Nauka*, 246 pp., Moscow, 1987.

Savelieva, G. N. and R. W. Nesbitt, A synthesis of the stratigraphy and tectonic seting of the Uralian ophiolites, *J. Geol. Soc. London*, 153, 525–537, 1996.

Scarrow, J. H., G. N. Savelieva, J. Glodny, P. Montero, A. N. Pertsev, L. Cortesogno and L. Gaggero, The Mindyak Palaeozoic lherzolite ophiolite, Southern Urals: Geochemistry and geochronology, *Ofioliti*, 24, 239–246, 1999.

Schulte, B. A. and P. Blümel, Metamorphic evolution of eclogite and associated garnet-mica schist in the high pressure metamorphic Maksyutov Complex, Urals, Russia, *Geol. Rundsch.*, 87, 561–576, 1999.

Seward, D., A. Perez–Estaun and V. Puchkov, Preliminary fission track results from the southern Urals-Sterlitamak to Magnitogorsk, *Tectonophysics*, 276, 281–290, 1997.

Seward, D., D. Brown, R. Hetzel, M. Friberg, A. Gerdes, A. Petrov and A. Perez-Estaun, The syn- to post-orogenic low temperature events of the Southern and Middle Uralides: evidence from fission-track analysis, this volume.

Shatsky, V. S., E. Jagoutz and O. A. Koz'menko, Sm-Nd dating of the high pressure metamorphism of the Maksyutov Complex, Southern Urals (in Russian), *Trans. R.A.S.*, 353, 285–288, 1997.

Smethurst, M. A., A. N. Khramov and T. H. Torsvik, The Neoproterozoic and Palaeozoic palaeomagnetic data for the Siberian Platform: From Rodinia to Pangea, *Earth Sci. Rev.*, 43, 1–24, 1998.

Smirnov, S. V., Petrology of Wehrlite-Clinopyroxenite-Gabbro assosciation of Nurali Ultrabasic Massif and Related Platinum Ore (in Russian), *Cand. Sci. Geol., Dissertation, Inst. Geol. and Geochem.*, 18 pp., Ekaterinburg, 1995.

Steiger, R. H. and E. Jäger, Subcommission on Geochronology: Convention on the use of Decay Constants in Geo- and Cosmochronology, *Earth. Planet. Sci. Lett.*, 36, 359–362, 1977.

Tucker, R. D., D. C. Bradley, C. A. Ver Straeten, A. G. Harris, J. R. Ebert and S. R. McCutcheon, New U–Pb zircon ages and the duration and division of Devonian time, *Earth Planet. Sci. Lett.*, 158, 175–186, 1998.

Tugarinov, A. I., E. V. Bibikova, A. A. Krasnobaev and V. A. Makarov, Geochronology of the Ural Precambrian (in Russian), *Geochem.*, 4, *Nauka*, 501–509, Moscow, 1970.

Vinogradov, V. I., V. M. Gorozhanin and V. I. Murav'ev, Rb-Sr dating of the postsedimentationalteration stage for Lower Riphean deposits in the Southern Urals (in Russian), in *Abstracts, The Sedimentary Formations of the Precambrian and their Mineral Resources*, pp. 11–12, St. Petersburg, 1998.

Yazeva, R. G., V. N. Puchkov and V. V. Bochkarev, The relics of active continental margin in the Urals structure, *Geotectonika*, 3, 76–85, 1989.

Yazeva, R. G. and V. V. Bochkarev, Silurian island arc of the Urals: Structure, evolution and geodynamics, *Geotectonics*, 29, 478–489, 1996.

Zonenshain, L. P., V. G. Korinevskiy, V. G. Kazmin, D. M. Pechersky, V. V. Khain and V. V. Matveenkov, Plate tectonic model of the South Urals, *Tectonophysics*, 109, 95–135, 1984.

Zykov, S. I., N. I. Stupnikova, A. A. Krasnobaev, V. I. Kozlov and V. I. Myasnikova, Lead isotope age of the Ilmenogorian formation and Barangul granite massif, Southern Urals. The problems of the metamorphism and metasomatism processes's isotopic dating (in Russian), *Abstracts, Geochim. Acad. Sci.*, pp. 97–99, Moscow, 1985.

J. H. Scarrow and P. Montero, Department of Mineralogy and Petrology, University of Granada, Campus Fuentenueva, 18002 Granada, Spain; E-mail: jscarrow@ugr.es

R. Hetzel and J. Glodny, GeoForschungsZentrum Potsdam (GFZ), Telegrafenberg, D-14473 Potsdam, Germany

V. M. Gorozhanin, Ufimian Geoscience Center, Russian Academy of Sciences, ul. Karl Marx 16/2, Ufa 45 000 Bashkiria, Russia

M. Dinn, British Antarctic Survey, High Cross, Madingley Road, CB3 0ET, Cambridge, U.K.

A. Gerdes, NERC Isotope Geosciences Laboratory, British Geological Survey, Keyworth, Nottingham NG12 5GG, U.K.

C. Ayala, Instituto de Ciencias de la Tierra "Jaume Almera", c/ Lluis Sole i Sabaris s/n, 08028 Barcelona, Spain.
[#]Previously at; British Geological Survey, Nottingham, U.K.

The Syn- and Post-Orogenic Low Temperature Events in the Southern and Middle Urals: Evidence From Fission-Track Analysis

D. Seward[1], D. Brown[2], R. Hetzel[3], M. Friberg[4], A. Gerdes[5], G. A. Petrov[6] and A. Perez-Estaun[2]

The Uralide orogen is a rare example of a Paleozoic orogenic belt that is preserved, relatively intact, within a tectonic plate (Eurasia); part of this belt still has areas of significant topography. We have utilized the fission-track technique in an attempt to investigate the syn- and post-orogenic development. Zircon ages, throughout the Southern and Middle Urals range from 604 ± 100 Ma to 189 ± 23 Ma. Apatite ages range from 349 ± 27 Ma to 149 ± 31 Ma. Zircon ages in the foreland thrust and fold belt young in the direction of the Main Uralian fault, implying that burial, and subsequent exhumation, was deeper towards the suture. Despite the large range in ages, a simple pattern of cooling affected the orogen as a whole through time. The fission-track analyses, combined with other radiometric ages, suggest rapid cooling ($\geqslant 3$ to $20°C/Ma$) until approximately the end of the Permian, slowing to $\leq 1°C/Ma$ during the Triassic and then $\ll 1°C/Ma$ until today. The age-altitude relationship in the foreland thrust and fold belt, west of the Main Uralian fault is disturbed, suggesting variable post Jurassic exhumation. The correlation of the disturbance in the age-altitude relationship with thrusts suggests that tectonic reactivation has taken place along pre-existing Uralide faults. This is supported by the strong correlation of topography with these faults. The present relief of the Ural Mountains, which in general coincides with the foreland thrust and fold belt, is interpreted to be the product of a post-Uralide event and not of the Uralide orogeny itself.

[1]Geological Institute, ETH, Zürich, Switzerland

[2]Instituto de Ciencias de la Tierra, Barcelona, Spain

[3]Geoforschungs Zentrum, Potsdam, Germany

[4]Dept. of Earth Sciences, Uppsala University, Sweden

[5]NERC Isotope Geosciences Laboratory, Keyworth, UK

[6]Urals Geological Survey Expedition, Ekaterinburg, Russia

Mountain Building in the Uralides: Pangea to the Present
Geophysical Monograph 132
10.1029/132GM13

1. INTRODUCTION

The Uralide orogen of central Russia (Figure 1) is a rare example of a Paleozoic orogen preserved within a tectonic plate (Eurasia), providing a natural laboratory for studying syn- to post-tectonic orogenic processes. The general tectonic evolution of the Uralides is well known [e.g., *Ivanov* et al., 1975; *Zonenshain* et al., 1984, 1990; *Puchkov*, 1997], and recent geological, geophysical, geochemical, and geochronological studies have added significantly to this understanding, providing new ideas on the architecture of the orogen and the tectonic

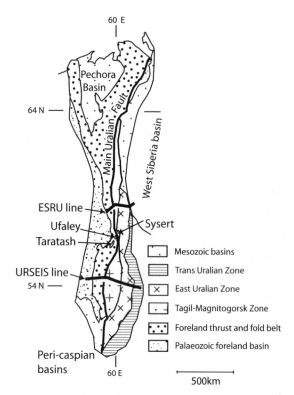

Figure 1. Tectonic subdivisions of the Uralides showing main section lines from which samples were taken.

processes that went into its formation [e.g., *Berzin* et al., 1996; *Perez-Estaun* et al., 1997; *Juhlin* et al., 1998; *Bea* et al., 1997; *Hetzel* et al., 1998; *Brown and Spadea*, 1999]. However, much of the late- to post-orogenic evolution of Uralides, in particular the low temperature evolution, is still not well understood.

There is widespread evidence that the Southern and Middle Urals were eroded and peneplained by the Late Triassic to Jurassic, and the process may have continued until the Early Cretaceous [*Puchkov*, 1997]. Despite having been peneplained recent geophysical experiments show the Uralide crust in the Middle and Southern Urals to be thick, preserving a crustal root of c. 52 km along the central axis of the orogen [*Carbonell* et al., 1998; *Juhlin* et al., 1998]. Low temperature thermochronological studies [*Seward* et al., 1997; *Leech* et al., 2000] have further suggested that the Uralides have been relatively stable since the Jurassic, and that since that time they have experienced very little cooling or erosion (i.e., a low exhumation rate).

Perhaps even more enigmatic is the present topography of the Ural Mountains, which locally reaches c. 1900 m. The topography is almost entirely associated with one tectonic unit of the Uralides (the foreland thrust

and fold belt, west of the Main Uralian fault), and is offset to the west from the deepest part of the crustal root. The geomorphology of the Southern and Middle Ural Mountains is generally mature, being dominated by a system of smooth, north-south trending ridges. Some younger features, such as deeply incised river valleys and elevated river terraces, hint at recent uplift. Being in the middle of a large tectonic plate, and far from any recently or currently active plate boundary, the question of why or when the current topography of the Ural Mountains formed is not fully understood.

This paper uses zircon and apatite fission-track analyses to investigate the syn- to post-orogenic low temperature thermochronological history of the Uralide orogen with the aim of understanding the possible evolution of the topography of the Ural Mountains. It forms an extension of a preliminary study carried out by *Seward* et al. [1997]. Throughout this paper the term Uralides is used for the Paleozoic orogenic belt and Ural Mountains is used for the topographic feature.

2. TECTONIC FRAMEWORK OF THE URALIDES

A generally accepted division of the Uralides consists of a number of longitudinal zones, or megazones [*Ivanov* et al., 1975; *Puchkov*, 1991] that are largely based on dominant rock types, stratigraphic ages and paleoge-ography. In this paper we use a modified subdivision based on the recognition of tectonic units. This sub-division consists of a foreland thrust and fold belt composed predominately of rocks of the East European Craton, the accreted Magnitogorsk and Tagil island arcs, the East Uralian Zone, and the Trans Uralian Zone (Figure 1). These tectonic units are separated by major faults.

The early tectonic evolution of the Uralide orogen involved an arc-continent collision as the former East European Craton continental crust subducted eastward (current coordinates) beneath a chain of Silurian to Devonian age intra-oceanic island arcs (Tagil and Magnitogorsk arcs) during the Late Devonian and Early Carboniferous [e.g., *Brown and Spadea*, 1999]. Arc-continent collision was followed by closure of the Uralian ocean basin to the east with further collision from the Late Carboniferous through to the Late Permian-Early Triassic [*Zonenshain* et al., 1984, 1990; *Puchkov*, 1997]. This latter event was accompanied by westward thrusting of the East European Craton base-ment and its Paleozoic platform cover to form the foreland thrust and fold belt and the foreland basin. During the late stages of the collision, extensive strike-slip, or transpressive faulting appears to have dominated along the central axis of the orogen, fragmenting the Tagil arc and juxtaposing metamorphic terranes within the East and Trans Uralian Zones [*Echtler* et al., 1997; *Ayarza* et al., 2000; *Friberg* et al., 2000; *Hetzel and Glodny*, 2002]. Throughout the Permian, this stage of the development of the orogen was accompanied by widespread melting and emplacement of granitic magmas in the east of the accreted terranes [*Fershtater* et al., 1997; *Bea* et al., 1997; *Montero* et al., 2000; *Gerdes* et al., 2002].

By the end of the Triassic the Uralides were eroded and a peneplain had developed over much of the orogen [*Borisevich*, 1992; *Puchkov*, 1997]. This peneplain is recorded in the Southern Urals, where it is locally overlain by Jurassic and Lower Cretaceous marine and continental sediments (Plate 1), and on the eastern slopes of the Northern Urals. Upper Cretaceous to Eocene marine sediments indicate at least three regression — transgression cycles [*Borisevich*, 1992]. Paleogene-age marine sediments now appear at altitudes ranging from 165 m to 340 m in the Middle and Southern Urals west of the Main Uralian fault, implying that there has been surface uplift of this amount since that time. Post-Cretaceous faulting with up to kilometer-scale displacement has been identified from boreholes in the Middle Urals [*Puchkov*, 1997]. *Puchkov* [1997] also suggested that a renewed phase of intra-continental deformation has taken place since the Late Oligocene.

3. THE URAL MOUNTAINS

The Ural Mountains form the only topographic expression currently associated with the Uralides. In the Southern and Middle Urals, the mountains coincide almost exclusively with the foreland thrust and fold belt, except for the Irendik range in the Magnitogorsk arc. The Ural Mountains are comprised of a series of broad, roughly north-south trending ridges that, in the Southern and Middle Urals, parallel the Uralide structural grain. This is particularly true in the Southern Urals where the relief is stronger. These north-south trending ridges and valleys in the Southern Urals are crosscut by deeply incised river valleys as the generally north-south flowing rivers turn abruptly west to cut through the mountains. Raised post-Paleocene river terraces associated with several of the larger rivers suggest that surface uplift in this area has occurred since the Paleogene [*Borisevich*, 1992]. This uplift was concentrated in the Southern Urals, but decreased eastward. Timing of river downcutting and terrace formation are nowhere well constrained. Doming and surface uplift of the Southern Urals was identified by *Piwowar* [1997] who recognized a broad Neogene uplifted arch that is oblique to the Uralide structural grain and suggested that this arch extends southeastward to the Caucasus and Tien Shan mountains.

4. ANALYTICAL METHODS

For this study samples were collected along two transects in the Southern and Middle Urals that roughly coincide with the URSEIS and ESRU reflection seismic profiles, as well as from the Taratash, Ufaley and Sysert complexes (Figures 1 and 2).

Sample preparation followed the routine technique described by *Seward* [1989]. Etching of apatite grains was done with 7% HNO_3 at 21°C for 55 seconds. Zircon grains were etched in a eutectic mixture of KOH and NaOH at 220°C for between 20 and 32 hours. Titanite was etched in a mixture of $1HF:2HNO_3:3HCl:6H_2O$ for 20 minutes at room temperature [*Naeser and McKee*, 1970]. Irradiation was carried out at the ANSTO facility, Lucas Heights, Australia.

Microscopic analysis was completed using an optical microscope with a computer driven stage ("Langstage" software from *Dumitru* [1995]). All ages were determined using the zeta approach [*Hurford and Green*, 1983] with a zeta value of 360 ± 5 for CN5 — apatite, 120 ± 5 for CN1 — zircon (DS) and 410 ± 10 for CN5 — titanite. They are reported as central ages [*Galbraith and Laslett*, 1993] with a 2σ error (Table 1). Where possible, 20 crystals of each sample were counted for an age determination. The magnification used was ×1250 for apatite and titanite and ×1600 (oil) for zircon. Horizontal confined track lengths were measured at ×1250.

Modeling of the apatite age and track length data was completed with the Monte Trax program of *Gallagher* [1995], using an initial track length of 15.5 μm. In this study we assume an effective closure temperature for apatite of $110 \pm 10°C$ with a partial annealing zone from 110–60°C [*Green* et al., 1989; *Corrigan*, 1993]. For zircon the closure temperature is taken as $250°C \pm 50°C$ with a lower limit of 200°C taken for initiation of annealing [*Yamada* et al., 1995]. The titanite closure lies between 265 and 310°C, the estimated partial annealing zone [*Coyle and Wagner*, 1998].

5. RESULTS

In general zircons were difficult to analyze. They were often metamict and dark, and had a very fast surface etch rate, which hindered the identification of fission-tracks. All zircon ages fall within the range 604 ± 100 Ma to 189 ± 23 Ma (Table 1 and Plate 1). Zircons in

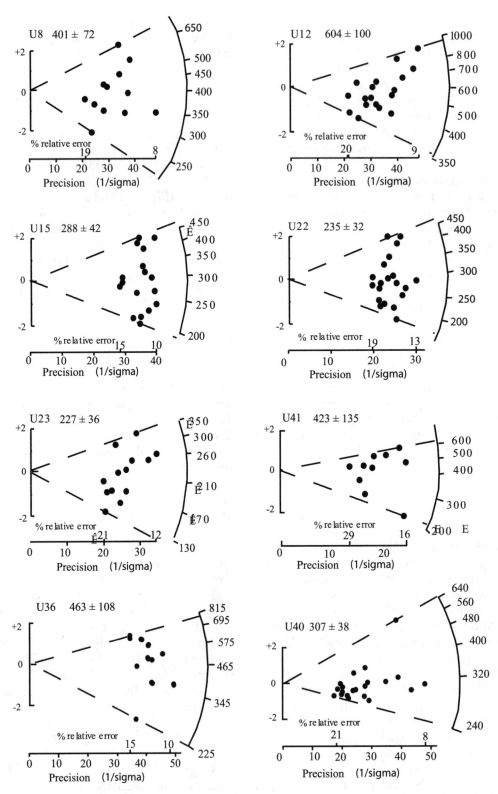

Figure 2. Radial plots [*Galbraith*, 1988] of single crystal zircon ages extracted from sediments from the foreland thrust and fold belt.

the foreland thrust and fold belt range from 604 ± 100 Ma to 226 ± 36 Ma, with the younger ages appearing along its eastern margin towards the suture line, the Main Uralian fault. All ages are thought to have been at least partially reset. Two zircon ages (263 ± 60 and 189 ± 23 Ma) were obtained from the Magnitogorsk zone, and one (225 ± 26 Ma) in the Tagil zone. Five zircon ages were obtained from the East Uralian zone: 273 ± 35 Ma, 292 ± 49 Ma and 223 ± 26 Ma in the Middle Urals, and 205 ± 54 Ma and 245 ± 77 Ma in the Southern Urals.

All apatite ages fall within the range 349 ± 27 to 149 ± 31 Ma (Table 1 and Plate 1), with mean track lengths ranging from 14.33 μm to 12.16 μm. The mean weighted apatite ages from north to south in the foreland thrust and fold belt (Plate 1), are within the 2σ error of each other within the Middle Urals, 220 ± 14 and 226 ± 8 Ma (ESRU section and Taratash and Ufaley complexes, respectively). In the Southern Urals, however, the foreland thrust and fold belt apatites are younger, with a weighted mean age of 206 ± 9 Ma. Mean apatite ages in the Tagil zone are 207 ± 11 Ma, and in the Magnitogorsk zone are 262 ± 5 Ma. There is no statistical difference longitudinally in the mean apatite ages from the East Uralian zone, with mean ages ranging from 216 ± 8 Ma in the north, through 208 ± 16 Ma to 214 ± 11 Ma in the south.

6. INTERPRETATION OF FISSION-TRACK AGES

6.1. Foreland Thrust and Fold Belt

Zircon ages of the Southern Urals foreland thrust and fold belt are generally older in the central and western parts of this zone, where they are considered to represent partially to fully reset ages. All ages are from Precambrian or Paleozoic sediments and most have a large spread (Figure 2). However, there is a younging trend of central ages from the western to the eastern margin of this belt. Interpreting such ages is not always easy because of unanswerable questions regarding the original age of the detrital zirons at deposition, the possibility of recycling, the lack of knowledge about the annealing kinetics of zircons with various chemistries, and the fact that the Precambrian sequences have been exposed to at least two orogenies — the Vendian (630–570 Ma) [*Ivanov* et al., 1986] and the Uralide (380–240 Ma) [*Puchkov*, 1997]. For example, sample U12 (Middle Riphean, Table 1 and Figures 2 and 3) has a central zircon fission-track age of 604 ± 100 Ma, which is younger than the stratigraphic age, but older than ages associated with the Uralide orogeny. However, the spread of single grain ages is from 930 to

357 Ma. This sample may have been partially reset during the Vendian because some ages are still older than 630 Ma. On the other hand, since most ages are less than 570 Ma there is an implication that they underwent heating after this time (i.e., during the Uralide orogeny). Another sample (U41, Figures 2 and 3) has an Upper Devonian to Early Carboniferous stratigraphic age (approximately 360 Ma), but a mean zircon fission-track age of 423 ± 135 Ma, with a range in single grain ages from 643 to 220 Ma. Because of the younger limit, this again implies partial resetting during the Uralide orogeny. Other examples include U8 and U36 (Table 1 and Figures 2 and 3), both with youngest ages of 218 and 225 Ma, respectively, but with older ages of 806 and 813 Ma. These results imply that the samples were buried after sedimentation to depths within the partial annealing zone where the temperature was probably less than 250°C. A depth cannot be estimated as there is no substantial way of determining the paleogeothermal gradient. In contrast, eastward, towards the Main Uralian fault, sample U23 of Middle Riphean stratigraphic age (similar to U12), has a mean zircon fission-track age of 227 ± 36 Ma with an oldest single grain age of 353 Ma and a youngest of 123 Ma (Figure 2). The zircons have clearly been exposed to a higher temperature regime than those discussed above. Other zircon fission-track ages in the eastern part of the foreland thrust and fold belt have similar ages (U22, 235 ± 32 and 97–5, 283 ± 54 Ma; Plates 1 and 3), implying that the amount of denudation increases towards the east. The fact that the zircons range from partially (?) to fully reset from west to east is in agreement with the findings of *Matenaar* et al. [1999] that the metamorphic grade increases from diagenetic to anchizonal in the west to anchizonal to epizonal in the east in the foreland thrust and fold belt about 80 km to the north of the URSEIS survey.

The weighted mean apatite fission-track age in the southern foreland thrust and fold belt is 206 ± 9 Ma ($N = 12$). All apatite ages from sediments in this area are younger than their stratigraphic age. The thermal histories of sedimentary sequences determined from the genetic algorithm modeling of the apatite data (Figure 3) all (except 97–8) point to burial at temperatures $>110°C$, with a return into the partial annealing zone between 300 and 250 Ma. One should note here that the age of sedimentation was initially included in this modeling; in five of the six samples modeled way it may not have been necessary because the samples were buried to temperatures above the apatite annealing temperature and the pre-annealing history within the crystals was lost. However, this was not known before the modeling was

Table 1. Fission-track analysis. A = apatite, Z = zircon, T = titanite, ρ_d, ρ_s and ρ_i represent the standard, sample spontaneous and induced track densities respectively. $P(\chi^2)$ is the probability of χ^2 for v degrees of freedom where v = no. of crystals − 1. All ages are central ages [Galbraith, 1981]. $\lambda_D = 1.55125 \times 10^{-10}$. A geometry factor of 0.5 was used. Zeta = 360 ± 5 for CN5/apatite, 120 ± 8 for CN1/zircon and 410 ± 10 for CN5/titanite. Irradiations were performed at the ANSTO facility, Lucas Heights, Australia.

Sample number	Stratigraphic age and/or rock type	Alt. (m)	Min.	No. grains	$\rho_d \times 10^4$ cm^{-2} (counted)	$\rho_s \times 10^4$ cm^{-2} (counted)	$\rho_i \times 10^4$ cm^{-2} (counted)	$P\chi^2$ (%. variation)	Mean track length (μm)	Standard deviation (μm) (measured)	Central age ± 2σ (Ma)
Foreland thrust and fold belt											
U2	Emsian	440	A	19	109 (3671)	168 (832)	119 (589)	<5 (32)	12.69 ± 0.45	2.12 (22)	262 ± 52
U5	Vendian	640	A	20	103 (3671)	72 (552)	77 (594)	9 (20)	12.81 ± 0.23	1.82 (64)	184 ± 30
U6	Vendian	600	A	22	98.8 (3671)	113 (507)	110 (495)	98 (<1)	12.16 ± 0.45	2.18 (23)	180 ± 23
U7	U. Vendian	550	A	20	672 (3229)	38.9 (258)	218 (1447)	9 (10)			214 ± 32
U9	Emsian	410	A	20	95.7 (3671)	182 (1146)	146 (920)	42 (4)	12.83 ± 0.24	2.06 (71)	212 ± 20
U14	M. Riphean	1100	A	20	91.2 (3671)	51 (176)	56 (192)	99 (0)			149 ± 31
U17	M. Ord	500	A	12	89.6 (3671)	227 (420)	197 (365)	78 (<1)			183 ± 26
U40	U. Devonian	500	A	17	160 (2593)	247 (1042)	283 (1196)	52 (4)			243 ± 24
U41	U. Devonian	450	A	20	137 (2593)	202 (1064)	193 (1017)	27 (7)			251 ± 26
U44	Granite	800	A	10	158 (2593)	225 (1407)	314 (1958)	89 (2)			199 ± 16
CU3	Vendian	190	A	7	138 (3902)	235 (311)	281 (372)	<1 (31)	13.70 ± 0.46	1.52 (11)	192 ± 59
CU4	Vendian	240	A	5	652 (3229)	124 (348)	630 (1769)	90 (0)			227 ± 28
95R12	Ufaley/orthogneiss	400	A	20	128 (2593)	15.7 (196)	12.3 (154)	50 (5)			290 ± 70
95R15	Ufaley gneiss	340	A	10	122 (2593)	8.13 (249)	114 (349)	6 (28)			193 ± 62
97R72	Taratash greenschist	520	A	20	125 (4068)	53.7 (289)	58.2 (313)	81 (10)			200 ± 36
97R95	Taratash gneiss	650	A	20	128 (4068)	123 (1214)	127 (1254)	17 (10)	13.93 ± 0.21	1.28 (37)	215 ± 22
97R70	Taratash gneiss	490	A	14	159 (3860)	272 (895)	336 (1104)	69 (1)	13.58 ± 0.14	1.19 (75)	225 ± 22
97R64	Taratash gneiss	460	A	20	147 (3860)	119 (588)	139 (686)	85 (<1)			220 ± 26
97R111	Taratash gneiss	450	A	10	163 (3860)	135 (288)	134 (286)	11 (14)	14.07 ± 0.17	1.15 (47)	288 ± 60
97R87	Alexandr granite	310	A	20	188 (3860)	236 (1691)	342 (2452)	31 (1)	13.59 ± 0.13	1.14 (81)	226 ± 16
97R89	Alexandr granite	370	A	20	180 (3860)	152 (334)	161 (353)	100 (0)	13.74 ± 0.27	1.15 (18)	295 ± 46
97–7	U. Devonian	500	A	16	151 (3860)	103 (301)	123 (359)	<1 (35)	13.38 ± 0.25	1.62 (42)	228 ± 55
97–8	U. Devonian	560	A	13	129 (4417)	140 (422)	131 (397)	74 (<1)	13.51 ± 0.49	2.83 (33)	239 ± 34
97–27	Granite	440	A	20	167 (3860)	75.7 (560)	90.2 (667)	96 (<1)	13.51 ± 0.18	1.29 (53)	244 ± 30
97–5	Gneiss	600	Z	20	40.81 (2133)	1772 (1479)	150 (125)	69 (1)			283 ± 54
U8	U. Vendian	500	Z	12	39.9 (2311)	218 (3169)	126 (183)	14 (14)			401 ± 72
U12	M. Riphean	550	Z	19	39.2 (2311)	2678 (4663)	120 (209)	<1 (2)			604 ± 100
U15	M. Riphean	1000	Z	17	37.4 (2311)	3223 (4485)	253 (352)	5 (17)			288 ± 42
U22	U. Devonian	720	Z	20	38.1 (2311)	1494 (2963)	143 (283)	38 (7)			235 ± 32
U23	Mylonitic sandstone (age?)	760	Z	20	37.4 (2311)	1688 (1947)	171 (197)	23 (7)			227 ± 36
U36	Devonian	500	Z	20	40.35 (2376)	2752 (2457)	139 (124)	6 (21)			463 ± 108
U40	Up Dev-Lo Carb.	500	Z	20	39.8 (2376)	2490 (4401)	189 (334)	77 (5)			307 ± 38
U41	Up Dev-Lo Carb.	450	Z	20	40.89 (2376)	1447 (940)	8.16 (53)	31 (4)			423 ± 135

Sample	Description										Age
Magnitogorsk											
U31	Carboniferous	570	A	18	120 (2844)	255 (824)	205 (662)	38 (5)			264 ± 32
U32	Visean–rhyolite	440	A	20	130 (1633)	39 (472)	26 (308)	98 (<1)	14.13 ± 0.14	1.35 (100)	349 ± 27
U26	U. Devonian	500	A	10	632 (3229)	154 (198)	732 (942)	24 (<1)	14.33 ± 0.41	1.47 (13)	235 ± 36
97–21	M. Devonian	550	A	4	184 (3860)	25.0 (29)	30.2 (35)	88 (1)			265 ± 133
97–23	M. Devonian	500	A	10	120 (4417)	40.7 (89)	30.2 (66)	77 (<1)			280 ± 91
U26	U. Devonian	500	Z	20	39.6 (2311)	1792 (1130)	159 (100)	18 (11)			263 ± 60
97–30	Granite undeformed	420	Z	20	44.69 (2133)	2841 (2501)	397 (349)	66 (1)			189 ± 23
U31	Carboniferous	570	T	10	119 (2844)	1167 (1689)	843 (1220)	58 (<1)			328 ± 32
Tagil											
CU9	Lower Silurian	285	A	10	136 (3902)	170 (626)	171 (630)	7 (15)			243 ± 37
CU10	Upper Silurian	270	A	3	134 (3902)	94.1 (235)	97.7 (244)	39 (<1)	14.30 ± 0.51	1.25 (6)	225 ± 42
CU11	Silurian-Lo Dev	255	A	15	132 (3902)	373 (2251)	431 (2604)	30 (3)	13.48 ± 0.16	1.33 (69)	199 ± 14
CU9	Lower Silurian	150	Z	20	44.75 (2574)	3017 (6729)	356 (794)	1 (16)			225 ± 26
East Uralian											
U29	Lo Carb-Lo Perm	420	A	20	86.6 (3671)	164 (1204)	137 (1001)	20 (11)	13.07 ± 0.16	1.61 (100)	185 ± 20
U30	Lo Carb-Lo Perm	340	A	10	123 (2844)	142 (764)	150 (805)	36 (6)	13.30 ± 0.36	1.36 (14)	206 ± 24
U42	Granite	400	A	20	134 (2593)	96.9 (412)	103 (437)	84 (1)			221 ± 32
CU15	Upper Devo-Carb	155	A	12	130 (3902)	60.5 (336)	62.2 (340)	43 (19)	12.86 ± 0.57	1.68 (8)	217 ± 52
CU17	Upper Dev-Carb	150	A	8	128 (3902)	385 (857)	398 (886)	2 (16)	13.48 ± 1.18	1.67 (2)	212 ± 37
CU18	Upper Dev-Carb	150	A	11	126 (3902)	93.5 (290)	87.0 (270)	70 (0)	14.62 ± 0.13	0.53 (18)	236 ± 40
CU20	Lower Devonian	135	A	20	122 (3902)	190 (2004)	190 (1999)	2 (13)	14.01 ± 0.11	1.00 (100)	214 ± 20
CU23	Upper Dev-Carb	135	A	20	645 (3229)	17.6 (175)	77.7 (771)	81 (<1)			259 ± 44
CU27	Lower-Middle Perm	195	A	20	120 (3902)	23.1 (288)	25.8 (322)	100 (0)	12.59 ± 0.30	0.84 (8)	188 ± 32
CU32	Siberia Traps–P/T boundary	138	A	20	116 (3902)	93.2 (830)	111 (986)	81 (0)			174 ± 18
95R3	Sysert/granite	340	A	20	152 (2593)	159 (855)	202 (1107)	48 (10)	13.56 ± 0.16	1.58 (100)	206 ± 20
UR1	Sysert gneiss	280	A	16	155 (2593)	135 (669)	147 (730)	0 (24)	13.42 ± 0.14	1.42 (100)	251 ± 42
98R125	Orthogneiss (mylonite)	320	A	20	104 (2921)	306 (955)	298 (930)	18 (12)	13.38 ± 0.21	1.25 (34)	187 ± 21
98R127	Orthogneiss (mylonite)	300	A	10	101 (2921)	208 (451)	131 (284)	66 (1)	14.09 ± 0.22	0.75 (12)	278 ± 44
97–28	Granite	420	A	5	135 (4417)	172 (225)	150 (196)	41 (3)			270 ± 54
97–29	Ordovician?	420	A	17	132 (4417)	98.7 (363)	96.3 (354)	2 (27)			245 ± 50
MRS27	Granite	150	Z	20	170 (2332)	93.0 (745)	122 (974)	41 (9)			223 ± 26
CU15	Upper Dev-Carb	150	Z	20	46.63 (2574)	1447 (2963)	145 (297)	100 (0)			273 ± 35
CU22	Ordovician-Silurian	150	Z	12	43.35 (2574)	1405 (1827)	122 (159)	100 (0)			292 ± 49
97–27	Granite undeformed	440	Z	12	39.70 (2133)	1832 (1359)	193 (143)	1 (30)			205 ± 54
97R98	Gneiss	450	Z	20	40.25 (2133)	1894 (1912)	176 (178)	99 (0)			254 ± 42
DZ-1	Granite	400	Z	4	40.68 (2747)	2475 (460)	242 (45)	63 (15)			245 ± 77

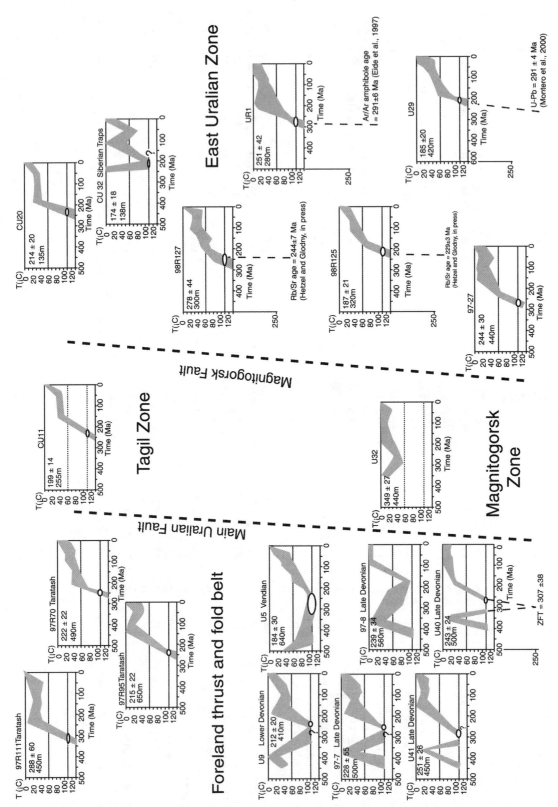

Figure 3. Monte Trax modeling [*Gallagher*, 1995] of some of the data sets positioned according to the tectonic zone from which they were taken and with a general regard to the geographic location. The shaded region represents the best 50 runs. Extension of the cooling paths by a dashed line to temperatures above that of the apatite annealing zone are drawn in towards the approximate closure temperatures of the method. The black ovals represent the time at which the rock passed through the 110°C isotherm.

carried out. Further, when the maximum temperature reached during burial is greater than 110°C the temperature cannot be estimated using the modeling procedure alone because it has constraints only for apatite parameters. This is well illustrated by sample U40 which has a zircon age of 307 ± 38 Ma (Figure 3). On the assumption that this is almost fully reset, (see radial plot, Figure 2) and when added to the apatite model of this same sample at the position of about 250°C, it is clearly seen that burial could have been to greater depths/temperatures than predicted by the apatite modeling alone. This heating event may have been due to burial by sedimentation, but burial by thrust stacking cannot be ruled out in all cases. Burial was followed by a long period of exhumation at a cooling rate of about 3°C/Ma during the Permian with a change to <1°C/Ma afterward. From the Cretaceous onward there is less confidence in the modeled data sets because the rocks had by this time passed through the lower temperature limit (60°C) of the partial annealing zone. However, *Seward* et al. [1997] and *Leech and Stockli* [2000] suggest that the cooling pattern followed in this lower temperature range compliments the known geological history of minor sedimentation during the Jurassic and Cretaceous, followed by later exhumation to the surface in the Tertiary. The fact that marine facies Cretaceous rocks now outcrop at altitudes of up to 500 m (Plate 1) indicates that surface uplift as well as exhumation has occurred since this time west of the Main Uralian fault. The Paleogene-age marine sediments at altitudes ranging from 165 m to 340 m in the Middle and Southern Urals west of the Main Uralian fault further support the timing of uplift.

Northward, in the Taratash and Ufaley complexes, and along the ESRU section, basement rocks of the foreland thrust and fold belt yield mean apatite fission-track ages of 226 ± 8 Ma and 220 ± 14 Ma (N = 8 and 2), respectively (Plate 1). These are essentially indistinguishable from each other. Modeling reveals that these samples passed through the 110°C isotherm at the same time as did the sediments of the Southern Urals and cooled during the Triassic at a rate of <1°C/Ma, to temperatures below 60°C by the latest early Cretaceous (Figure 3). This is similar to the south and we suggest that it represents the long, slow denudation leading in the end to the development of the Late Triassic to Cretaceous peneplain. From the Jurassic until today the cooling rate has been extremely low, much less than 1°C/Ma.

For any undisturbed sequence the oldest ages should occur at the highest topographic sites because these were the first to pass through the partial annealing zone during exhumation. It is very clear that this age-altitude

relationship does not exist in the foreland thrust and fold belt (Figure 4a). In fact, there is a negative trend in which the highest altitudes record the youngest ages for both apatite and zircon. This implies either that there have been disruptions to the pattern after the fission-track ages were set, or that the ages were already within different tectonic blocks that were undergoing different thermal histories. However, since all samples have had similar thermal histories during their pre-Jurassic evolution, then the lack of an age-altitude correlation is interpreted to be the result of variable exhumation after the cooling associated with the Uralide orogeny. A clear example is sample U14 (149 Ma, Table 1 and Plate 1), which was taken from a topographic high in the hanging wall of a thrust, yet it has the youngest age in the region. Two zircon ages from the same region, U15 and U12 (288 and 604 Ma, respectively), are separated by this same thrust. The age obtained from the hanging wall is clearly fully reset while that in the footwall is not, implying greater exhumation in the hanging wall. *Glasmacher* et al. [2000], working in a section north of the URSEIS survey, observed some younger apatite (Cretaceous) ages and suggested also that they indicated a Cretaceous tectonic reactivation. Because of the higher topography, the Southern Urals show this negative age-altitude correlation stronger, but it is also developed in the Taratash and Ufaley complexes. Farther north, along the ESRU survey, it is difficult to determine any relationship, since variation in topography is small and only two samples were analyzed.

The fission-track dataset of the foreland thrust and fold belt reveal a time temperature history of the Uralide orogeny. The Precambrian and lower Paleozoic sediments underwent heating, most likely due to burial, and were brought back up through the 110°C isotherm in the time span 300–250 Ma (i.e., approximately the same time that the Taratash complex passed through this isotherm). Only from one sample was the estimated cooling rate 3°C/Ma during the Permian. From this time on the cooling rate everywhere was slow, on the order of 1°C/Ma, marking the end of the Uralide orogeny. Post-Uralide activity is seen not directly through the ages themselves, but by secondary relationships such as the disturbance of the expected age-altitude pattern.

6.2. Magnitogorsk Zone

The mean apatite age for the Magnitogorsk zone is 262 ± 5 Ma (N = 4). A Carboniferous rhyolite (U32, 349 ± 27 Ma, not included in the mean regional age because of its extrusive nature) records an approximate extrusive age (Visean) which when modeled shows that

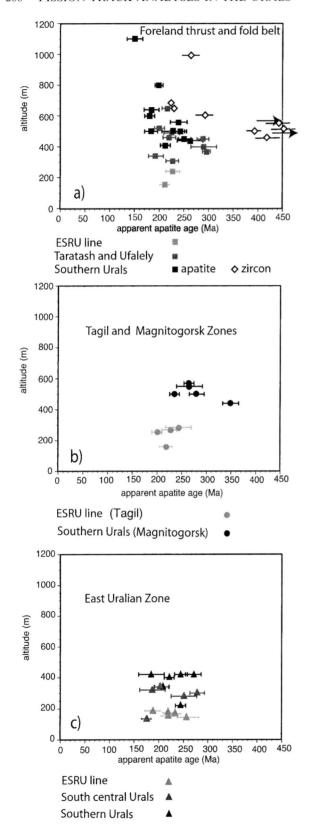

Figure 4. Age-altitude relationships according to the tectonic region. All are apatite data except for 5a which also incorporates zircon.

it was buried to a temperature of less than 60°C (Figure 3). This indicates that there has been little burial, erosion or exhumation since the Carboniferous. A single zircon (U26) yielded an age of 263 Ma, and a titanite (U31) an age of 328 ± 32 Ma (Plate 1 and Table 1). The age-altitude relationship for the Magnitogorsk zone (within error limits) is not strongly disturbed (Figure 4b). However, with a small data set and low relief it is not possible to accurately determine the age-altitude relationships. The extremely low to non-existent metamorphic grade of this zone [*Brown* et al., 2001] corroborates the interpretation of very low exhumation determined through the fission-track analysis.

6.3. Tagil Zone

The mean apatite age in the Tagil zone is 207 ± 11 Ma (N = 3). Sample CU11 (Plate 1 and Table 1) has an apparent age of 199 Ma and modeling (Figure 3) reveals a cooling rate of < 1°C/Ma during the Triassic, with a change to considerably less than 1°C/Ma to bring the rocks to the surface. A single zircon age of 225 ± 26 Ma (CU9) was also obtained from the Tagil zone. The modeled pattern for the Tagil zone is similar to most of the modeled samples from the foreland thrust and fold belt. The younger fission-track age obtained for the Tagil zone samples suggests exhumation continued there, whereas the Magnitogorsk zone remained relatively stable.

Any possible disruption to the age-altitude relationship in the Magnitogorsk and Tagil zones cannot be detected due to the low topography (Figure 4b). All ages statistically overlap.

6.4. East Uralian Zone

Fission-track ages from the East Uralian zone were obtained predominantly from gneisses and the granites which had intruded them. Zircon fission-track ages are in general older in the Middle Urals (292 ± 49, 273 ± 35, 254 ± 42, 223 ± 26 Ma) than in the Southern Urals (245 ± 77 and 205 ± 54 Ma). The 245 Ma (DZ1) and 223 Ma (MRS27) ages are from late-orogenic granites in the Southern and Middle Urals that were emplaced around 291 and 254 Ma, respectively [*Montero* et al., 2000]. Both granites were emplaced at a depth of approximately 12–15 km [*Fershtater* et al., 1994] and if one only considers these crustal levels an average cooling rate of 2–4°C/Ma is determined for the Permian and Middle Triassic. If one takes the crystallization temperature (650°C) then this cooling rate is increased to >9°C/Ma. However, this would involve most of the cooling within the first few million years at most, hence reducing the overall rate towards the 4°C/Ma as calculated above. Rb/Sr mineral isochron ages and K-Ar ages ranging from 280 Ma to 250 Ma obtained from migmatites and granites in the Southern Urals [*Krasnobaev* et al., 1995; *Gorozhanin*, 1995] further indicate a Permian cooling event to below 350°C, which when combined with the zircon fission-track yields again an estimated rate of 2.5°C/Ma.

From north to south, the weighted mean apatite ages are 216 ± 8, 208 ± 16, and 214 ± 11 Ma (Plate 1). Rb/Sr ages of 229 ± 3 and 244 ± 6 Ma have also been determined for two samples (98R125, 98R127) [*Hetzel and Glodny*, 2002]. When added to the modeled apatite fission-track ages, which overlap the Rb/Sr ages (Figure 3) it appears that exhumation in this area occurred at a rate of >20°C/Ma during the Permian to Early Triassic. An ^{39}Ar/^{40}Ar amphibole age of 29 ± 6 Ma from the Sysert Complex [*Eide* et al., 1997] combined with a zircon fission-track age of 251 ± 42 (Figure 3) further indicates that there was rapid cooling in the western half of the East Uralian zone until about 270 Ma. The modeled fission-track ages indicate that once these samples had passed through the 110°C isotherm there was a change to a cooling rate of approximately 1°C/Ma from the Middle Triassic onwards. This slow cooling rate is identical to the results of the other regions. The age-altitude relationship of the East Uralian zone (Figure 4c) does not show any significant disturbance, although this may be strongly biased by the lack of any significant relief.

One sample (CU32; Figures 2 and 4) was from the trap basalts at the Permian-Triassic boundary [*Renne and Basu*, 1991]. The modeling of this sample shows that there was heating perhaps until about 220 Ma. It is possible that a heating event could have been caused either by a tectonic overburden or it was simply by burial, either by more basalts or by sediments, or a combination of both. At 220 Ma the basalt was cooled again slowly, this time at approximately 1°C/Ma throughout the Triassic and Jurassic; the same pattern as recorded elsewhere. A second phase of heating (burial?) beginning at about 120 Ma is clearly recorded. This is similar to the phase of heating only tentatively recognized at other sites. Final cooling then began at about 60 Ma (Paleogene).

7. DISCUSSION

Although there appears to be some continuity and similarity in the Uralide thermal history, one must emphasize that it is important to remember that 2σ errors on fission-track ages are of the order of 10%. This amount does not allow discrimination of events of the order of 20 Ma to be identified in samples with ages of 200 Ma. In Tertiary orogens 20 million years is of major importance in identifying different events. For example, both the Himalayan and Alpine belts contain recorded events within this time span. In the present study this sort of detail is inevitably missing.

The modeled fission-track data, together with other isotopic systems indicate that there is a somewhat similar cooling pattern throughout the Southern and Middle Urals. There is no obvious difference in the timing of the inflexion points of the cooling curves between tectonic units. This suggests that the entire Southern and Middle Urals were affected by the same overall events, which is remarkable when one considers the size of the orogen. This does not necessarily imply that they were affected everywhere to the same degree.

All apatite ages in the Uralides, whether from Precambrian or Paleozoic sedimentary units or extrusive volcanics, are younger than their stratigraphic or eruption ages, implying annealing or partial annealing of the tracks since deposition. This phase of heating could be the result of burial due to thrusting or to sedimentation. If it was due to thrusting then different units should display different thermal effects depending on their structural position. Although the data set is small, this does not seem to be the case (Figures 2 and 3). It would also imply that thrusting began before collision. The final closure of the Paleo-Uralide ocean, which is thought have begun during the Late Carboniferous (approximately 300 Ma, [*Puchkov*, 1997]) is related to the exhumation and cooling which is seen in the modeled data. This was associated with a phase of enhanced exhumation which from our data set is during the Permian and sometimes into the Early Triassic. It was associated with erosion of the mountain belt and deposition of large volumes of synorogenic sediments in the western foreland basin region and in large intermontane depressions on the eastern margin.

Our data suggest that immediately after the collision there was rapid exhumation, but generally from the Triassic onwards, cooling was ≤1°C/Ma (Figure 3). From the modeling one can cautiously comment that in many regions the Uralides were brought close to the surface during the Triassic after a long and continuous cooling period covering one hundred million years, which was time enough for a peneplain to become well developed. Although the Monte Trax modeling is restricted to temperatures greater than 60°C, simple monotonic cooling from 60°C requires a change to an even slower rate ($\ll 1$°C/Ma) in order to bring the rocks to the surface today. With some caution of over-interpretation there may be a small heating event identi-fied in the Jurassic to Cretaceous which we have only managed to positively identify in one sample (CU32). Both continental and marine Jurassic and Cretaceous sediments were deposited during this time; the extent is unknown, but some river valleys are reported to contain Cretaceous sediments [*Borisevich*, 1992] and some sedi-ments of this age are still outcropping today in the south (Plate 1).

Evidence from the zircon fission-track ages suggest that the eastern sector of the foreland thrust and fold belt was buried more deeply than the western margins and has thus undergone greater exhumation/denudation. Temperatures reached during burial in the eastern sector of this region were greater than 250°C. This is in agree-ment with metamorphic studies of *Matenaar* et al. [1999] in the Southern Urals.

The weighted mean apatite ages for the Tagil and Magnitogorsk zones are statistically different and sup-port the suggestion that they have undergone different exhumation histories during and after collision. The Magnitogorsk zone, where the metamorphic grade is negligible to subgreenschist, has remained very stable since the Carboniferous as evidenced by the impercep-tibly small resetting of the rhyolite (sample U32). The mean apatite ages are 262 ± 5 Ma. Although no other samples were suitable for modeling, this age is the time when slow cooling began elsewhere. We suggest that there has been very little disturbance in this region since this time. On the other hand, the mean weighted apatite age for the Tagil zone is younger (207 ± 11 Ma) and in places a greenschist metamorphic grade has been reached. These two facts combined point to later and deeper exhumation than in the Magnitogorsk arc to the south.

Evidence for later tectonic reactivation in the fore-land thrust and fold belt is realized through the dis-turbance of the age-altitude relationship (Figure 4a). Perhaps because the relief in the foreland thrust and fold belt is greatest in the Southern Urals and the southern Middle Urals, the negative relationship with age is seen more strongly in this region. Further, the ages of the southern foreland thrust and fold belt are the youngest, implying later exhumation. This is interpre-ted to be the result of post Uralian reactivation. In the Middle Urals, any disturbance in the age-altitude rela-

tionship is not so clear because the altitude varia-tions between samples are much smaller. In the Tagil, Magnitogorsk and East Uralian zones the variation is not readily seen and we suggest that there has been little reactivation there.

Later reactivation in the Uralides seems to have been dominant in the foreland thrust and fold belt of the Southern and southern Middle Urals forming the topographic expression of the Ural Mountains. There appears to be a correlation between the geomorphol-ogy and the thrust faults within the Ural Mountains (Plate 2) which leads us to suggest that reactivation occurred along the existing Paleozoic faults and that this is the reason that there is a mountain range there today. Surface uplift is recorded by the occur-rence of Cretaceous marine sediments at up to 500 m altitude. On the basis of river terrace studies, *Borisevich* [1992] developed the notion of an uplifted dome structure increasing in amplitude from east to the west (Plate 2). Further evidence for surface uplift comes from the geomorphological study of *Piwowar* [1997] who identified Neogene doming extending southeastwards to Tien Shan. His idea is supported by the work of *Hendrix* et al. [1994], who recorded Late Oligocene to Miocene enhanced exhumation for Tien Shan.

The primary cause of the uplift is more enigmatic. The geographical and tectonic location of the region of higher exhumation suggests that the Ural Mountains may be part of the isostatic shoulder, flexural uplift perhaps related to the downwarping of the West Siber-ian Basin. This conclusion fits well with the estimated uplift and subsidence patterns (Plate 2) calculated by *Borisevich* [1992]. The reason for the localized doming in the Southern Urals foreland thrust and fold belt is that in this region there were suitable structures whereby strain could be accommodated, whereas in the more easterly regions there are fewer lines of pre-existing weakness, except the thrust related Irendik Mountains which are also a topographic high today (Figure 4).

Brown et al. [1997] suggested that in the Southern Urals there had been only 20 km of shortening in an east-west direction during the Permo-Triassic collision. This is very small on an orogenic scale. Further support is given by *Artyushkov* et al. [2000] who concluded that the Uralide orogeny did not produce high mountains, hence the erosion level reached would not have been deep. In order to determine the absolute amount of erosion the paleogeothermal gradient must be known. Without this data it is unwise to extrapolate so far backwards.

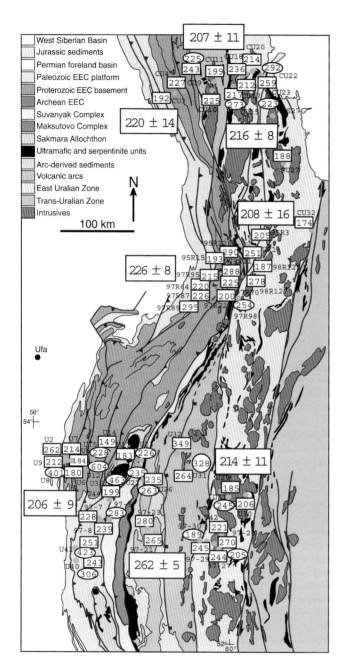

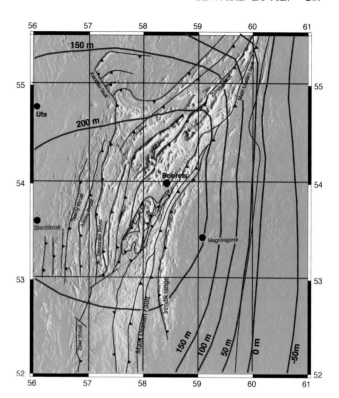

Plate 2. Topographic map of the Southern Urals together with the major thrust faults. The surface uplift proposed by *Borisevich* [1992] is superimposed.

Plate 1. Regional geology, sample numbers and apparent fission-track ages of the South and Middle Uralides. Small rectangles enclose apatite ages, ovals the zircon ages and a circle the titanite age. Weighted mean apatite ages (with 2σ error) for separate regions are in the large rectangles.

8. CONCLUSIONS

The goal of this study was to analyze the late stage history of the Uralides. The results are broadly consistent from place to place — details within a range of less than 20 Ma cannot be discerned due to errors on the ages and in the modeling processes.

The new data set shows:

(1) With the addition of Rb/Sr, U/Pb, Ar/Ar and zircon fission-track ages to the higher temperature part of the apatite cooling curve it is clear that there was, in general, a period of rapid exhumation from temperatures $\geq 250°$ C to 110°C at a rate of up to or greater than 20°C/Ma before approximately 250 Ma. It was at this time that the sediments derived from the mountain belt were being deposited in the western foreland basin.

(2) At about 250 ± 20 Ma there was a slowing down in the cooling rate for approximately the duration of the Triassic. This cooling was slow, generally $\leq 1°$C/Ma. The conclusion is that it represents an erosion phase of the mountain chain that finally ended in the formation of a peneplain [*Puchkov*, 1997]. This represents the end of the Uralide orogeny.

(3) A tentative inflexion in this cooling curve occurs at variable times from the Late Triassic through to the Cretaceous. Although only tightly confirmed in one sample, this represents a period of non-erosion or perhaps minor deposition which is supported by the known regional geology. A monotonic cooling from 60° at about 200 Ma to zero today resulted in a further decrease in the cooling rate to much less than 1°C/Ma.

(4) Post-Paleogene reactivation is recognized through the disturbance of the apatite age-altitude relationship. Other geological studies have suggested that there was surface uplift in the region of the Southern Urals. This represents smoothed regional data (river terrace correlation, etc.), whereas the apatite fission-track system records exhumation and is more localized. The correlation of the present topographic highs in the Southern Urals with the Uralide thrust faults support the idea that the renewed activity took place along these lines of inherent weakness.

We thus conclude that the Ural Mountain chain of today, which is geologically the foreland thrust and fold belt, is the result not of the Uralide orogeny, but of reactivation of old structures that represent inherited zones of weakness such as the Uralide faults. A further, more detailed fission-track study over the thrust systems, particularly of the foreland thrust and fold belt, would yield a more constrained thermal history and would yield more tectonic information. A future effort to determine more precisely the timing of the last cooling/exhumation and associated tectonic and erosion periods would be better constrained through use of the U/Th-He age dating technique as it is more sensitive to lower temperatures (as low as 45°C [*Zeitler* et al., 1987; *Wolfe* et al., 1996]).

Acknowledgments. This work involved the support of many people particularly those in Russia who provided help in the field both physically and intellectually. Victor Puchkov is thanked for his able organization throughout. Reviewers are thanked for their comments, which have helped to improve the manuscript. D. Seward thanks the ETH Zürich for financial support through grant number 0-20-991-95.

REFERENCES

Artyushkov, E. V., M. A. Baer, P. A. Chekhovich and N.-A. Mörner, The Southern Urals. Decoupled evolution of the thrust belt and its foreland: A consequence of metamorphism and lithospheric weakening, *Tectonophysics*, 320, 271–310, 2000.

Ayarza, P., D. Brown, J. Alvarez-Marron and C. Juhlin, Contrasting tectonic history of the arc-continent suture in the Southern and Middle Urals: Implications for the evolution of the orogen, *J. Geol. Soc. London*, 157, 1065–1076, 2000.

Bea, F., G. Fershtater, P. Montero, V. Smirnov and E. Zin'kova, Generation and evolution of subduction related batholiths from the central Urals: Constraints on the P-T history of the Uralian orogen, *Tectonphysics*, 276, 103–116, 1997.

Berzin, R., O. Oncken, J. H. Knapp, A. Perez-Estaun, T. Hismatulin, N. Yunosov and A. Lipilin, Orogenic evolution of the Ural Mountains: Results from an integrated seismic experiment, *Science*, 274, 220–221, 1996.

Borisevich, D. V., Neotectonics of the Urals, *Geotectonics*, 26, 41–47, 1992.

Brown, D., J. Alvarez-Marron, A. Perez-Estaun, V. Puchkov, Y. Gorozhanina and P. Ayarza, Structure and evolution of the Magnitogorsk forearc basin: Identifying upper crustal processes during arc-continent collision in the Southern Urals, *Tectonics*, 20, 364–375, 2001.

Brown, D. and P. Spadea, Processes of forearc and accretionary complex formation during arc-continent collision in the Southern Ural Mountains, *Geology*, 27, 649–652, 1999.

Brown, D., J. Alvarez-Marron, A. Perez-Estaun, Y. Gorozhanina, V. Baryshev and V. Puchkov, Geometric and kinematic evolution of the foreland thrust and fold belt in the Southern Urals, *Tectonics*, 16, 551–562, 1997.

Carbonell, R., D. Lecerf, M. Itzin, J. Gallart and D. Brown, Mapping the Moho beneath the Southern Urals, *Geophys. Res. Lett.*, 25, 4229–4233, 1998.

Corrigan, J. D., Apatite fission-track analysis of Oligocene strata in South Texas, U.S.A. Testing annealing models, *Chem. Geol.*, 104, 227–249, 1993.

Coyle, D. A. and G. A. Wagner, Positioning the titanite fission-track partial annealing zone, *Chem. Geol.*, 149, 117–125, 1998.

Dumitru, T. A., A new computer automated microscope stage system for fission-track analysis, *Nuclear Tracks and Radiation Measurements*, 21, 575–580, 1995.

Echtler, H. P., K. S. Ivanov, Y. L. Ronkin, L. A. Karsten, R. Hetzel and A. G. Noskov, The tectono-metamorphic evolution of gneiss complexes in the Middle Urals, Russia: A reappraisal, *Tectonophysics*, 276, 229–251, 1997.

Echtler, H. P., M. Stiller, F. Steinhoff, C. Krawczyk, A. Suleimanov, V. Spiridonov, J. H. Knapp, Y. Menshikov, J. Alvarez-Marron and N. Yunusov, Preserved collisional structure of the Southern Urals revealed by vibroseis profiling, *Science*, 274, 224–226, 1996.

Eide, E. A., H. P.Echtler, R. Hetzel and K. S. Ivanov, Cooling age diachroneity and Paleozoic orogenic processes in the Middle and Southern Urals, Abstract supplement No. 1, *Terra Nova*, 9, 119, EUG 9, Strasbourg, 1997.

Fershtater, G. B., P. Montero, N. S. Boridona, E. V. Pushkarev, V. Smirnov, E. Zin'kova and F. Bea, Uralian magmatism: An overview, *Tectonophysics*, 276, 87–102, 1997.

Fershtater, G. B., N. S. Borodina, M. S. Rapoport, T. A. Osipova, B. H. Smirnov and M. Y. Levin, Orogenic granitoid magmatism of the Urals (in Russian), *Miacc, Russian Academy of Sciences*, Urals Branch, 247 pp., 1994.

Friberg, M., C. Juhlin, A. G. Green, H. Horstmeyer, J. Roth, A. Rybalka and M. Bliznetsov, Europrobe seismic reflection across the eastern Middle Urals and West Siberian Basin, *Terra Nova*, 12, 252–257, 2000.

Gallagher, K., Evolving temperature histories from apatite fission-track data, *Earth Plan. Sci. Lett.*, 136, 421–435, 1995.

Galbraith, R. F., Graphical display of estimates having differing standard errors, *Technometrics*, 30, 271–281.

Galbraith, R. F. and G. M. Laslett, Statistical models for mixed fission-track ages, *Nuclear Tracks and Radiation Measurements*, 21, 459–470, 1993.

Gerdes, A., P. Montero, F. Bea, G. Fershtater, N. Borodina, T. Osipova and G. Shardakova, Peraluminous granites frequently with mantle-like isotope compositions: The continental-type Murzinka and Dzhabyk batholiths of the east Urals, *Int. J. Earth Sci.*, 91, 1–17, 2002.

Glasmacher, U., G. A. Wagner and V. N. Puchkov, Thermo-tectonic evolution of the western fold and thrust belt, southern Urals, Russia, in *9th International Conference on Fission-track Dating and Thermochronology, Geological Society of Australia Abstracts*, edited by W. P. Noble, P. B. O'Sullivan and R. W. Brown, 58, 129–130.

Gorozhanin, V. M., A rubidium-strontium isotopic method in solving of the problems of geology of the Southern Urals (in Russian), *Autoref. Diss. Inst. Geol.* Uralian branch of the Russ. Ac. Sci. Ekaterinburg, 23 pp., 1995.

Green, P. F. and I. R. Duddy, Some comments on paleo-temperature estimation from apatite fission-track analysis, *J. Pet. Geo.*, 12, 111–114, 1989.

Hetzel, R. and J. Glodny, A crustal-scale, orogen-parallel strike-slip fault in the Middle Urals: age, magnitude of displacement, and geodynamic significance, *Int. J. Earth Sci.*, 91, 231–245, 2002.

Hetzel, R., H. P. Helmut, W. Seifert, B. A. Schulte and S. I. Kirill, Subduction- and exhumation-related fabrics in the Paleozoic high-pressure-low temperature Maksyutov Complex, Antingan area, southern Urals, Russia, *Geol. Soc. Am. Bull.*, 110, 916–930, 1998.

Hendrix, M. S., T. A. Dumitru and S. A. Graham, Late Oliogocene-Early Miocene unroofing in the Chinese Tian Shan: An early effect of the India-Asia collision, *Geology*, 22, 487–490, 1994.

Hurford, A. J. and P. F. Green, The zeta age calibration of fission-track dating, *Chem. Geol.*, 41, 285–317, 1983.

Ivanov, S. N., A. S. Perfiliev, A. A. Efimov, G. A. Smirnov, V. M. Necheukhin and G. B. Fershtater, Fundamental features in the structure and evolution of the Urals, *Am. J. Sci.*, 275, 107–130, 1975.

Ivanov, S. N., A. A. Krasnobayev and A. I. Rusin, Geodynamic regimes in the Precambrian of the Urals, *Precambrian Res.*, 33, 189–208, 1986.

Juhlin, C., M. Friberg, H. P. Echtler, T. Hismatulin, A. Rybalka, A. G. Green and J. Ansorge, Crustal structure of the Middle Urals: Results from seismic reflection profiling in the Urals experiments, *Tectonics*, 17, 710–725, 1998.

Knapp, J. H., C. C. Diaconescu, M. A. Bader, V. B. Sokolov, S. N. Kashubin and A. V. Rybalka, Seismic reflection fabrics of continental collision and post orogenic extension in the Middle Urals, central Russia, *Tectonophysics*, 228, 115–126, 1998.

Krasnobaev, A. A., G. P. Kuznetcov, V. A. Davydov, E. P. Shulkin and N. V. Cherednichenko, Uranium-lead age of the zircons from the Chelyabinsk gneiss complex (in Russian), Yezhegodnik-1994, Inst. Geol. Uralian Branch of the Russ. Ac. of Sci., Ekaterinburg, 34–36, 1995.

Leech, M. L. and D. F. Stockli, The late exhumation history of the ultrahigh-pressure Maksyutov Complex, south Ural Mountains, from new apatite fission-track data, *Tectonics*, 19, 153–167, 2000.

Matenaar, I., U. A. Glasmacher, W. Pickel, U. Giese, V. N. Pazukhin, V. I. Kozlov, V. N. Puchkov, L. Stroink and R. Walter, Incipient metamorphism between Ufa and Beloretzk, western fold and thrust belt, Southern Urals, Russia, *Geol. Rundschau*, 87, 545–560, 1999.

Montero, P., F. Bea, A. Gerdes, G. Fershtater, N. Zin'kova, T. Osipova and V. Smirnov, Single-zircon evaporation ages and Rb/Sr dating of four major Variscan batholiths of the Urals: A perspective on the timing of deformation and granite generation, *Tectonophysics*, 317, 93–108, 2000.

Naeser, C. and E. H. McKee, Fission-track and K-Ar ages of Tertiary ash-flow tuffs, north-central Nevada, *Geol. Soc. Amer. Bull.*, 81, 3375–3384, 1970.

Perez-Estaun, A., J. Alvarez-Marron, D. Brown, V. Puchkov, Y. Gorozhanin and V. Baryshev, Along-strike structural variations in the foreland thrust and fold belt of the Southern Urals, *Tectonophysics*, 276, 265–280, 1997.

Piwowar, T. J., Long-wavelength Neogene flexural uplift of the Southern Urals and central Eurasia, unpublished M.Sc., Cornell University, U.S.A., 74 pp., 1997.

Puchkov, V. N., Paleozoic of the Urals-Mongolian foldbelt, *Occas. Publ., Environ. Syst. Res. Inst., Uni. of S. C., Columbia*, Nov. Ser. II, 69 pp., 1991.

Puchkov, V. N., Structure and geodynamics of the Uralian orogen, in *Orogeny Through Time*, edited by J.-P. Burg and M. Ford, Geol. Soc. London, Spec. Publ. 121, 201–236, London, 1997.

Renne P. R. and A. R. Basu, Rapid eruption of the Siberian trap flood basalts at the Permo-Triassic boundary, *Science*, 253, 176–179, 1991.

Seward, D., Cenozoic basin histories determined by fission-track dating of basement granites, South Island, New Zealand, *Chem. Geol.*, 79, 31–48, 1989.

Seward, D., A. Perez-Estaun and V. Puchkov, Preliminary fission-track results from the Southern Urals: Sterlitamak to Magnitogorsk, *Tectonophysics*, 276, 281–290, 1997.

Wolfe, R. A., K. A. Farley and T. L. Silver, Helium diffusion and low-temperature thermochronometry of apatite, *Geochim. Cosmochim. Acta*, 60, 4231–4240, 1996.

Yamada, R., T. Tagami, S. Nishimura and H. Ito, Annealing kinetics of fission-tracks in zircon: An experimental study, *Chem. Geol.*, 122, 249–248, 1995.

Zeitler, P. K., A. L. Herczeg, I. McDougall and M. Honda, U-Th–He dating of apatite: A potential thermochronometer, *Geochim. Comochim. Acta*, 51, 2865–2868, 1987.

Zonenshain, L. P., V. G. Korinevsky, V. G. Kazmin, D. M. Pechersky, V. V. Khain and V. V. Matveenkov, Plate tectonic model of the South Urals development, *Tectonophysics*, 109, 95–135, 1984.

Zonenshain, L. P., M. I. Kuzmin and L. M. Napatov, Uralian foldbelt, in: *Geology of the USSR; A Plate Tectonic Synthesis*, edited by B. M. Page, AGU, Geodynamics Series, 21, 27–54, 1990.

D. Brown, Instituto de Ciencias de la Tierra "Jaume Almera", CSIC, Barcelona, Spain

M. Friberg, Dept. of Earth Sciences, Uppsala University, Villavägen 16, S-752 36 Uppsala, Sweden

A. Gerdes, NERC Isotope Geosciences Laboratory, Keyworth, Nottingham NG12 5GG, UK

R. Hetzel, Geoforschungs Zentrum, Potsdam, Telegrafenberg C2, 14473 Potsdam, Germany

G. A. Petrov, Urals Geological Survey Expedition UGSE, 55 Ol. Veinara, Ekaterinburg, Sverdlovsk Oblast, 620014 Russia

A. Perez-Estaun, Instituto de Ciencias de la Tierra "Jaume Almera", CSIC, Barcelona, Spain

D. Seward, Geological Institute, Sonneggstrasse 5, ETH Zurich, 8092 Zurich, Switzerland

Constraints on the Neogene–Quaternary Geodynamics of the Southern Urals: Comparative Study of Neotectonic Data and Results of Strength and Strain Modeling Along the URSEIS Profile

V. O. Mikhailov[1], A. V. Tevelev[2], R.G. Berzin[3], E.A. Kiseleva[1], E. I. Smolyaninova[1], A. K. Suleimanov[3] and E. P. Timoshkina[1]

[1] *United Institute of Physics of the Earth, RAS, Moscow, Russia*
[2] *Moscow State University, Moscow, Russia*
[3] *"Spetsgeofizika", Moscow, Russia*

Much geological and geomorphologic evidence indicates that the modern topography of the Southern Urals has been formed during the Neogene–Quaternary due to superposition of, (1) NW–SE compression and asymmetric uplift of the area as a whole, and (2) more vigorous transpressive uplift of the Central and Western Uralian blocks in the Late Pliocene–Quaternary. Strength modeling based on data on the deep structure, temperature and composition of the crust revealed that the Western and Central Uralian blocks are characterized by low total strength. Numerical strain modeling showed that vertical Neogene–Quaternary movements of the area can be a consequence of intraplate compression, maximum deformation being concentrated in weak blocks. The model predicts different deformational style in the upper and lower parts of inhomogeneous Uralide crust: the zone of maximum deformation at the top of the crust occurs in the Main Uralian fault zone while, in the lower crust, it is shifted 70 km to the west, where a vertical Moho offset (the so-called Makarovo fault) is located. Thus, this fault could have developed (or at least been sufficiently renewed) during Neogene–Quaternary.

1. INTRODUCTION

Much new data have been recently acquired for the Southern Urals, especially along the URSEIS profile [*Berzin* et al., 1996; *Carbonell* et al., 1996; *Echtler* et al., 1996; *Knapp* et al., 1996; *Brown* et al., 1997; *Diaconescu* et al., 1998; *Steer* et al., 1998; *Tevelev* et al., 1998]. In addition to models on the deep structure [e.g., *Berzin* et al., 1996; *Carbonell* et al., 1996; *Echtler* et al., 1996; *Knapp* et al., 1996] new thermal [*Kukkonen* et al., 1997] and gravity [*Döring* et al., 1997] models as well as new fission track data [*Seward* et al., 1997, 2002] have been published. These data allow a better understanding of the present structure and evolution of the Southern Urals. Nevertheless, some results obtained by different geological and geophysical methods seem to contradict each other. This relates, for example, to data on the age

Mountain Building in the Uralides: Pangea to the Present
Geophysical Monograph 132

of modern topography and Neogene–Quaternary geo-dynamics of the Southern Urals.

Much evidence points to the conclusion that the Ural Mountain topography is comparatively young [e.g., *Trifonov*, 1960; *Lider*, 1976; *Borisevich*, 1992; *Stephanovsky* et al., 1997; *Puchkov*, 1997; *Tevelev* et al., 1998]. Based on different geologic and geomorphologic data, these authors suggested that the topography of the Ural Mountains has been developed during the Neo-gene–Quaternary. On the other hand, recent prelimi-nary fission-track results obtained for the west-east transect of the Southern Urals [*Seward* et al., 1997] showed zircon and apatite fission-track ages of 200 Ma and older. A number of investigators explained these data as evidence that the present day topography of the Southern Urals is a remnant of relief resulting from Paleozoic deformation [e.g., *Diaconescu and Knapp*, 2002]. On the other hand, additional study of zircon and apatite fission track ages led *Seward* et al. [2002] to the conclusion that the present relief of the Ural Mountains is likely a product of post-Uralide events and not of the Uralide orogeny itself.

The goal of the present paper is to study the mechanism and temporal development of the modern relief of the Southern Urals. To do this we considered geological and geomorphologic data and correlate them to data on the structure and lithospheric strength of different tectonic zones in the vicinity of the URSEIS profile. We demonstrate that uplifts occurred in blocks characterized by lower present day strength. Using results of strength calculations we carried out numerical modeling of deformation of the southern Uralide litho-sphere. For this we employed a simple mechanical model that simulates the main structural features of the Southern Urals along the URSEIS profile. This modeling showed that topography similar to the modern topo-graphy of the Southern Urals can result from horizontal compression of the model by far-field forces alone. Thus, using neotectonic data on Neogene–Quaternary uplift of the region it appears possible to obtain a selfconsistent model of modern structure and geodynamics of the Southern Urals.

2. NEOTECTONIC MOVEMENTS IN THE SOUTHERN URALS

The main topographic features of the Southern Urals and the location of the URSEIS profile are presented in Figure 1. Many researchers of the geology of the Uralides [*Trifonov*, 1960; *Schults*, 1969; *Lider*, 1976; *Borisevich*, 1992; *Puchkov*, 1997; *Stefanovsky*, 1997] have considered the Southern Urals to have been tectonically active

during the Cenozoic. *Schults* [1969] for the first time compared the Southern Urals with such active and well known regions as Tien-Shan, and found clear similarities in their neotectonic development. According to *Lider* [1976], rapid uplift of the Ural Mountains took place at the beginning of the Quaternary, when several hundred meters of coarse gravel was accumulated on the western margin of the West Siberia Depression. These Early Mindel sediments are recognized as molasse of Early Pleistocene orogeny. At the Likhvin time (Mindel–Riss) the lower edge of the East Uralian plateau could be two or three times higher than at present. During the Middle Pleistocene and, in particular, Late Pleistocene, neotec-tonic movements were strongly inhomogeneous [*Borise-vich*, 1992; *Tevelev*, et al., 1998]. Relatively uplifted areas were cut by a series of flat erosional surfaces, correlated with alluvial terraces of the main rivers of the Urals. The actual ages of these geomorphologic complexes in the vicinity of the URSEIS profile was for a long time in dispute, because they were inferred only from morpho-logical and lithological correlations [*Tevelev*, et al., 2002]. Our dating based on floral remnants confirmed the Quaternary age of these complexes. Thus, in general, the current relief of the Southern Urals is more likely newly formed than inherited. Published evaluations on the amplitude and rate of Quaternary uplift in the region are diverse. We consider the estimates given by *Trifonov* [1960] as the most reliable. Based on repeated geodetic measurements he estimated the rate of uplift of the Southern Urals at the latitude of the town of Miass to be 0.45 cm/year and the total Quaternary uplift to be about 1 km [see also, *Borisevich*, 1992].

Evidence of recent tectonic activity in the Southern Urals is found in the western periphery of the region, nearly at the western end of the URSEIS profile. Here, in the Belaya river valley there is a zone of sedimentation similar to that of a foredeep, which started to develop in front of the Southern Urals during the Late Miocene [*Shilts*, 1969]. The coal-bearing Miocene sediments are folded, and it seems probable that the evolution of the main features of the modern topography started with these deformation events. The Pliocene and Quaternary sediments in this depression seem to be composed of both proximal and distal components. The latter are represented almost exclusively by of several series of Caspian transgressions [*Tevelev* et al., 2002]. The well-studied Apsheron (early Quaternary) and Akchagyl (pre-Quaternary) transgressions definitely reached this zone, but spatial configuration of the corresponding basins has not coincided with modern valleys of the Belaya, Ural and Volga rivers. This suggests a young age for these river valleys, which have probably been developed

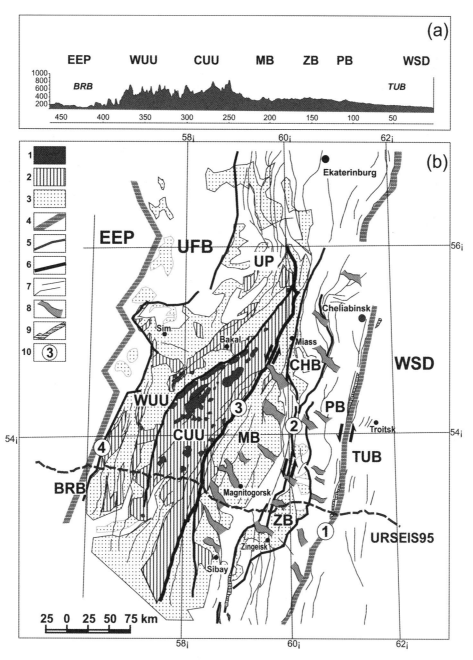

Figure 1. Geomorphologic scheme of the Southern Urals. (a) topography along the URSEIS profile (meters vs. kilometers). (b) geomorphologic sketch map of the Southern Urals: 1. maximally uplifted blocks and massifs of the Central Uralian uplift; 2. comparatively high uplifted blocks and massifs; 3. blocks and massifs characterized by medium uplift; 4. boundaries of the modern orogen; 5. boundaries of structural geomorphologic units of the Southern Urals including: 6. boundary of the Central Uralian uplift and Eastern Uralian Plateau; 7. Late Paleozoic and Early Mesozoic faults; 8. recent depressions; 9. remnants of Early Mesozoic rift basins; 10. reactivated sutures of the eastern part of the Southern Urals (1 — Kopeisk fault zone, 2 — Uysk-Brient fault zone, 3 — Main Uralian Fault Zone, 4 — Shikhan fault zone). Letters on the scheme mark: EEP — Eastern European platform; UFB — uplift of foredeep basin; BRB — Belaya-river block; WUU — Western Uralian Uplift; UP — Ufimian plateau; CUU — Central Uralian Uplift; TUB — Trans-Uralian block; WSD — Western Siberian depression. Blocks of the Eastern Uralian plateau: MB — Magnitogorsk Block, CHB — Chebarkul Block, ZB — Zingeika Block, PB — Plast Block.

during the Pleistocene when relief in the area was finally reshaped.

In fact, there are no data supporting the existence of Late Jurassic relief in the region. According to *Puchkov* [1997], since the end of the Jurassic and during the Cretaceous the Uralides were a lowland partly covered by a sea. In the Early Cretaceous, and in part of the Late Cretaceous, a sea strait between the Polar Urals and the Mediterranean Sea basins occupied the area of the Ufimian plateau situated to the north of the Southern Urals; thus, no subaerial relief existed here. In the Late Cretaceous a wide sea basin occupied the eastern margin of the East European platform. It included the boundary zone of the Uralides, and in the Paleogene (at least in the Paleocene) the orientation of sea basins did not coincide with the strike of the Uralides. Thus, the Uralides probably did not control a geomorphologic structure of the region at that time. The eastern slope of the South Urals is in general exhumed from beneath Eocene sediments and has also Neogene to Quaternary geomorphologic age.

Interpretation of known geological and geomorphologic data [*Sigov*, 1969; *Borisevich*, 1992; *Tevelev* et al., 1998; *Makarova* et al., 2000; *Tevelev* et al., 2002] allow us to suggest that deformation of the Southern Urals in the Neogene–Quaternary time has been complex. Deformation evolved under oblique (NW–SE) compression and consisted, (1) of regional asymmetric arch-shaped uplift, involving the whole area from the Kopeysk fault zone on the east to the Shikhan fault zone on the west (with the maximum uplift being on the order of several hundreds meters), and (2) of more vigorous (of order of thousand meters) superimposed transpressive uplift of the Central Uralian and Western Uralian zones, pushed up along weak marginal zones inherited from old strike-slip faults. Left-lateral transpression, uplift and development of mountain structures in this region correlates with the main stage of Alpine orogeny in surrounding orogens [*Shults*, 1969; *Tevelev and Tevelev*, 1996; *Tevelev* et al., 1998]. Thus, most likely the Ural Mountains represent a part of the collisional collage caused by the India and Eurasia interaction. Deformation started at the end of the Oligocene, was especially active in the Late Pliocene and Late Quaternary time, and is still active now. Deformation is manifested in structure-dependent distribution of the Late Pleistocene – Holocene subsiding and elevated areas on the East Uralian Plateau [*Tevelev* et al., 1998, *Tevelev* et al., 2002, see also Figure 1b], as well as in data of geodetic and horizontal stress measurements [*Puchkov*, 1997; *Zubkov and Lipin*, 1997].

Based on this deformational model we estimated a rate of vertical neotectonic movement along the URSEIS

profile. We used a standard procedure of comparing present day topographic height of pre-orogenic geomorphologic surfaces (see Figure 1a) with the proposed height of their formation. The most reliable estimates can be obtained for the eastern and western ends of the profile, situated within (or next to) the areas of Oligocene to Miocene marine shallow water sedimentation. We suppose that sea level in these basins was close to the modern one, so the present-day altitude of the basins is close to the post-Miocene elevation. In the central part of the profile (in the regions of the East Uralian Plateau, Central and Western Uralian Uplifts) the altitudinal position of the pre-orogenic surface is inferred from analysis of spatial and temporal relations between continental flat erosional surfaces of the Pliocene to Late Pleistocene age. We took into consideration the general curvature of these surfaces, and amplitude of sequential erosional cuttings, and interactions with Cenozoic depositional complexes.

At the eastern termination of the URSEIS profile, Oligocene marine sediments, which are about 20 to 30 meters thick, are now situated at a height of 230 to 250 meters. Thus, a reasonable value of uplift here for the Neogene–Quaternary is about 250 to 300 m. Amplitude of uplift increases smoothly up to 500 m at the western boundary of the Zingeika block, where the altitude of the flat erosional surface exhumed from beneath the Paleogene deposits reaches up to 450 m.

To the west, in the Magnitogorsk block, recent relief has developed over very thick island arc complexes. Amplitudes of the Neogene–Quaternary uplift vary from 400 to 450 m near the east margin of the Magnitogorsk block (Gumbeika river valley) where continental deposits of pre-Neogene age are situated at an altitude of 350 to 400 m, to 800 m near the Main Uralian fault, in the vicinity of which the Early Cenozoic complexes are denuded and the altitude of the Pliocene topographic surfaces is about 600 to 700 m. The Main Uralian fault separates the Magnitogorsk block from the Central transpressive block; the latter includes the Central Uralian and Western Uralian uplifts. Relief within the block is completely erosional; the denudation of pre-deformational surface is roughly estimated at 200 m, so maximum elevation of the Central Uralian uplift along the profile is about 1300 m, and elevation of the Western Uralian uplifts is 1200 to 1000 m from east to west. The western boundary of the block is also represented by a set of strike-slip and reverse faults, and marked by systems of surficial scarps and young sedimentary deposits. The structural pattern of the neotectonic units suggests that deformation of the Central Uralian uplift was governed by diagonal nearly NW–SE compression, which is consistent with instrumental present day stress measurements [*Zubkov and Lipin*, 1997].

The western end of the profile crosses the Urals foredeep, which is partly involved in the recent uplift of the region. Total vertical deformation there consists of a regional component (uplift of the Southern Urals as a whole) and local subsidence during the recent activity. Minimum amplitudes (150 to 200 m in accumulated topography) are in the Belaya River area. The amplitude of uplift increases to the west up to 400 to 450 m, toward the East European platform marginal anticline structures.

The latest results of zircon and apatite fission-track analyses [*Seward* et al., 2002] support the young age of Urals relief. According to these authors disturbance of the apatite age-altitude relationship shows that the Ural Mountains are not the result of the Uralide orogeny, but of reactivation of old structures that determines distribution of present day weak zones.

3. STRENGTH OF THE LITHOSPHERE ALONG THE URSEIS PROFILE

To estimate the strength of the Uralide lithosphere we calculated yield strength profiles using data on the structure, temperature and composition of the lithosphere in the vicinity of the URSEIS profile. We first consider the geothermal data.

Heat flow in the Uralides has been determined from several hundred measurements of temperature and thermal parameters of rocks taken from deep boreholes. The most complete catalogue of heat flow data was collected by *Golovanova* [1994]. The characteristic features of heat flow in the Southern Urals are as follows.

Heat flow in the eastern part of the Eastern European platform ranges between 35 to 45 mW/m^2. Nearly the same heat flow values are found at the western slope of the Urals. In the Magnitogorsk block the heat flow values are considerably lower, about 25 mW/m^2. The zone of comparatively low values is relatively narrow but it stretches for at least 1500 km from 48° N to 60° N. To the east of the Magnitogorsk block heat flow increases to more than 40 mW/m^2. The heat flow decrease in the Magnitogorsk block can be attributed to a number of factors including: (a) high permeability of the rocks, which permit heat transfer from uplifted areas to surrounding foredeeps by underground waters; (b) paleo-climate; (c) low heat generation in rocks of the Magnitogorsk block; (d) low heat flow from the mantle below the Central Urals. Detailed analysis of the possible contribution of each of these factors led *Kukkonen* et al.

[1997] to the conclusion that the main factor responsible for the heat flow minimum is low heat generation in rocks of the Magnitogorsk block.

It is worth noting that the distribution of temperature in the upper lithosphere does not differ substantially in different thermal models [*Bulashevich and Shapov*, 1983; *Salnikov*, 1984; *Khutorskoy*, 1985; *Kukkonen* et al., 1997; *Khachay and Druzhinin*, 1998], as all the models have the same heat flow and temperature at the Earth surface. Thus, estimates of the strength of the upper lithosphere should be similar. We used the model of lithosphere structure, composition, and temperature distribution suggested by *Kukkonen* et al. [1997]. This very detailed thermal model was suggested for the Troitsk DSS profile situated close to the URSEIS profile and crossed the same tectonic units. We used this model to characterize relative variation of the strength of the lithosphere in the main blocks of the Southern Urals. The model of the deep structure used by *Kukkonen* et al. [1997] was slightly modified in the Magnitogorsk block to be closer to data of URSEIS profile (Figure 2a).

To estimate mechanical properties of lithospheric rocks we calculated yield strength profiles using the conventional approach [*Ranalli and Murphy*, 1987; *Kohlstedt* et al., 1995; *Evans and Kohlstedt*, 1995]

$$\sigma = \min\{\alpha \cdot \rho \cdot g \cdot z(1 - \lambda); (\dot{\varepsilon}/A_d)^{1/n}$$
$$\exp(E/n \cdot R \cdot T)\}, \qquad (1)$$

where $\sigma = \sigma_{max} - \sigma_{min}$ is the difference between maximum and minimum principal stresses, α is a coefficient depending upon the type of fault ($\alpha = 3.0$, 1.2 and 0.75 for thrust, transcurrent and normal faulting [*Ranalli*, 1997]); $\rho g z$ is the overburden pressure (product of density, gravitational acceleration and depth); λ is the ratio of fluid pore pressure to lithostatic pressure; $\dot{\varepsilon}$ is a strain rate; A_d and n are the Dorn constants, studied by laboratory experiments for different types of rocks; E is an activation energy; R is the universal gas constant; T is the rock temperature (°K).

Strength diagrams calculated for eight different tectonic units crossed by the URSEIS profile are presented in Figure 3 (location of these units is shown in Figure 2a). It is worth noting that creep parameters have been measured for a limited number of different rock types. As a result, it is difficult to find a close analog for every petrologic unit found in outcrops and boreholes or suggested for deep layers of the Southern Urals. Diagrams presented in Figure 3 were calculated using creep parameters of marble and quartzite as strength characteristics for the rocks marked by number 1 on Figure 2a; granite and quartz-diorite data for

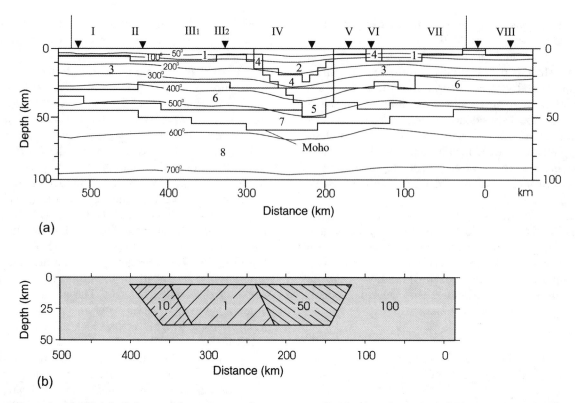

Figure 2. (a) Model of the crustal structure and temperature distribution along the URSEIS profile (modified from [*Kukkonen* et al., 1997]) used in the strength modeling. Temperature distribution is shown by isotherms. The horizontal axis is numbered according to URSEIS profile. Numbers correspond to different lithologies (explanations are in the text). Vertical arrows at the top of the model mark part of the profile for which stress calculations have been done. Black triangles indicate the location of the strength profiles shown on Figure 3. Roman numerals mark the following blocks: I-Eastern European Platform, II-Urals foredeep; III₁-Western Uralian block, III₂-Central Uralian block, IV-Magnitogorsk block; V-Zingeika block; VI-Plast block; VII-Trans-Uralian block; VIII-Western Siberian depression. (b) Distribution of viscosity (in 10^{22} Pa·s) used for numerical modeling.

the layers 2, 3, and 4; anorthosite and quartz-diorite data for the layers 5 and 6; and olivine data for the layers marked 7 and 8. As usual, when a layer consists of several rock types, the minimum yield strength value was adopted for the strength profile. It appeared that for the thermal conditions of the Southern Urals, yield strength of the upper layer (marked 1 and 2) was mainly determined by brittle failure and only in its lower part (diagrams II, IV, VI) by the power law creep of quartzite, so changes of composition of this layer can not change the strength diagrams. Strength of the layers from 2 to 6 was controlled by brittle failure or power law creep of quartz-diorite. Thus, even though in strength calculations we used many rock types, the final strength profiles were determined by only three: quartzite, quartz-diorite and olivine. Their creep parameters are given in Table 1. The similar strength distribution was obtained replacing quartz-diorite by mafic granulite

(Table 1). When replacing olivine rheology in layer 7 by the rheology of dry peridotite (Table 1), the relative distribution of the total strength along the profile remains the same, even the strength of layer 7 reduces considerably above the Moho discontinuity. Thus, stress estimates and especially estimates of the total strength appeared to be stable with regard to variations of rock composition.

The coefficient $\alpha(1-\lambda)$ was assigned a value of 2, averaging α values for the thrust and transcurrent faulting, and also taking into account the possible influence of the pore pressure (when pressure is hydrostatic coefficient λ is equal to 0.38 [*Kohlstedt* et al., 1995]). Strain rate was assumed to be equal to $\dot{\varepsilon} = 10^{-16}c^{-1}$. This value was obtained from numerical modeling of neotectonic movements, which yields maximum $\dot{\varepsilon}$ for the Neogene–Quaternary in the range 4–6 $10^{-16}c^{-1}$ (see section 4.2 and Figure 4 for more details).

Pressure (MPa)

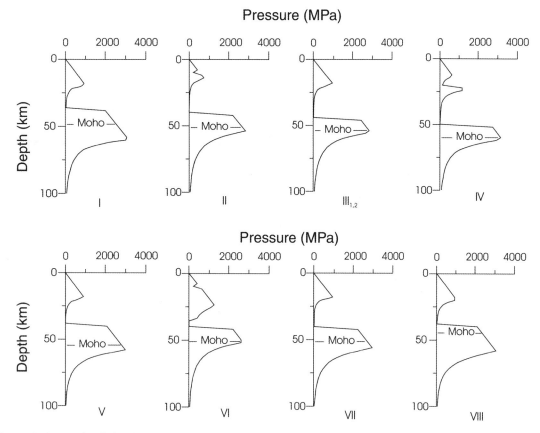

Figure 3. Strength of the lithosphere for different blocks of the URSEIS profile. Position of blocks (black triangles) is shown on Figure 2a.

There are two main brittle layers on the strength diagrams for the Southern Urals (Figure 3). The first one is a brittle layer in the upper crust (which can be subdivided into two sub-layers as on diagrams II, IV, VI) and the second one is a brittle layer incorporating an area above the Moho and continuing into the upper mantle. The existence of the brittle layer above and below the Moho indicates that faults traced on seismic profiles can disrupt the Moho discontinuity (Makarovo fault, for example, see *Diaconescu* et al. [1998]). It is also important to note that not all of the profiles contain the brittle layer below the Moho (see diagrams III, IV, VI).

Absence of brittle layers in the middle crust indicates that faults in the upper crust should not extend directly to deeper parts of the crust.

An important characteristic of the lithospheric blocks appears to be the total strength, which is the result of integration of the strength profiles over depth. The highest total strength has been obtained for the Eastern European platform ($9.3 \cdot 10^4$ MPa·km), the Western Siberian basin ($9.0 \cdot 10^4$ MPa·km), and the Zingeika block ($8.6 \cdot 10^4$ MPa·km). Moderate values of total strength have been found for the Plast ($7.3 \cdot 10^4$ MPa·km) and the Trans-Uralian blocks ($7.5 \cdot 10^4$ MPa·km). Relatively low

Table 1. Creep parameters of rocks used for yield strength calculations.

Material	$A_d \mathrm{Mpa}^{-n}\,\mathrm{s}^{-1}$	n	E kJ mol^{-1}	Reference
Quartzite	$6.7 \cdot 10^{-6}$	2.4	156	*Ranalli*, 1997
Quartz diorite	$1.3 \cdot 10^{-3}$	2.4	219	*Kirby and Kronenberg*, 1987
Mafic granulite	$1.4 \cdot 10^{4}$	4.2	445	*Ranalli*, 1997
Peridotite	$2.5 \cdot 10^{4}$	3.5	532	*Ranalli*, 1997
Olivine	$4.0 \cdot 10^{6}$	3.0	540	*Evans and Kohlstedt*, 1995

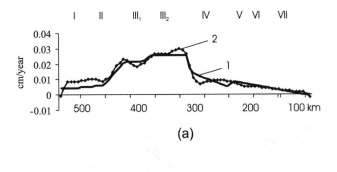

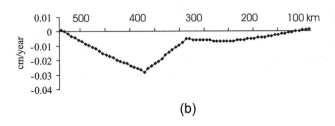

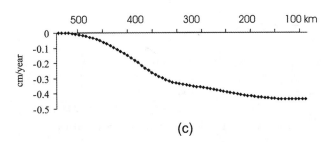

Figure 4. Components of velocity vectors at the surface and base of the model. Notice offset of the maximum of the vertical component of velocity vector at the top (a) and at the base (b) of the model. a. 1-average rate of vertical movement during the Neogene–Quaternary according to geomorphologic data; 2-calculated vertical component of velocity vector at the top of the model; b. Vertical component of velocity at the base of the model; c. Horizontal component of velocity at the base of the model.

total strength is found for the Western Uralian and Central Uralian uplifts ($6.4 \cdot 10^4$ MPa·km) as well as the Urals foredeep ($6.7 \cdot 10^4$ MPa·km) and the Magnitogorsk block ($7.0 \cdot 10^4$ MPa·km). These results are in good agreement with the location of zones of maximum deformation and high rates of neotectonic movements: maximum deformation has occurred in the zones of low total strength.

4. NUMERICAL MODELING

4.1. Background

In this section we consider the results of numerical modeling of deformation for a simple mechanical model simulating the main features of the Southern Urals structure along the URSEIS profile.

For modeling of slow, long-term (several My) deformation at the neotectonic stage we used a simple 2D model of a viscous Newtonian medium. It is possible to use more sophisticated models, although any model will only be a rough approximation of the very complicated structure that is the lithosphere, the southern Uralide lithosphere in particular. The style of deformation in the simple model described below mainly depends on the nonhomogeneous distribution of mechanical properties. Similar styles of deformation can be obtained using other models, in particular the pure elastic one (see below).

In our model the lithosphere comprises three layers: the upper "rigid" 5 km thick layer having apparent viscosity 10^{24} Pa·s, the "ductile" (having low horizontally inhomogeneous viscosity) 35 km thick layer and a "rigid" 10 km thick layer having apparent viscosity 10^{24} Pa·s situated above and below the Moho (see the diagrams in Figure 3). Horizontal inhomogeneity of the crust was simulated by variations of effective viscosity of the middle "ductile" layer over two orders of magnitude (Figure 2b). Based on strength estimations the smallest effective viscosity value was assigned to the middle crust below the Central and Western Uralian uplifts: 10^{22} Pa·s; a higher value of 10^{23} Pa·s was prescribed for the Urals foredeep; the value $5 \cdot 10^{23}$ Pa·s was used for the Magnitogorsk block. Other blocks had the same apparent viscosity values as the viscosity of the upper and the lower "rigid" layers, i.e. 10^{24} Pa·s.

The boundaries between the Urals foredeep and the Western Uralian uplift as well as between the Central Uralian uplift and the Magnitogorsk block (Figure 2b) were assumed dipping to the east [*Tryggvason* et al., 2001]. The boundary between the Magnitogorsk and Trans-Uralian blocks was inferred to dip to the west. The density distribution along the profile was given in accordance with *Döring* et al. [1997]. It is necessary to point out that in the majority of tectonically active areas (including the Southern Urals, as will be shown in our calculations), regional deformation is driven almost exclusively by externally imposed intraplate forces [e.g., *Zoback*, 1992], so local density variations do not strongly influence the results of the modeling.

Using the rheological law of Newtonian viscosity, the velocity field within the model is determined by solving the Stokes equations. We assumed material incompressibility and an absence of body forces other than those of gravity. The upper boundary is assumed to be stress free.

The processes that have produced the present day topography of the Southern Urals are not completely understood. Topography might be caused by horizontal compression by far field forces and by processes in the mantle below the Uralides (e.g., convection, phase transitions). Thus, we have decided not to assign any boundary conditions based on existing geodynamical hypotheses, but to find them by solving the following inverse problem: to find the velocity fields on the side boundaries and at the base of the model so that the top of the model moves in accordance with the rates of neotectonic movements discussed previously. We used a similar approach to estimate the state of stress in different regions [*Kolpakov* et al., 1991; *Smolyaninova* et al., 1996; *Stephenson and Smolyaninova*, 1999].

The problem of independently finding all components of the velocity vector (in the 2D case they are the horizontal U and the vertical W components) has no unique solution. It is clear, for example, from the fact that any distribution of the vertical component of the velocity vector at the top of the model can be obtained as a result of pure vertical movements at the base of the model with horizontal component equal to zero. To avoid this problem we link horizontal and vertical components of the velocity vector at the two vertical side boundaries and the base of the model by the equation:

$$W(x,z) = -(z - z_0)\partial U(x)/\partial x \qquad (2)$$

where x and z are horizontal and upward vertical coordinates respectively and z_0 is a constant. This equation has been extensively used in sedimentary basin modeling (for references see *Cloetingh* et al. [1995]). This equation is based on the following assumption put forward by *Braun and Beaumont* [1989]: when the lithosphere is extended (or compressed) by intraplate forces, there is a horizontal level z_0 which stays horizontal during the process of deformation in the absence of gravity. The same equation was obtained by [*Myasnikov and Savushkin*, 1978] in their consideration of the interaction of the lithosphere, asthenosphere and upper mantle. Using a model based on Newtonian viscosity they found that for regional structures which formed during time periods of several My and more, z_0 coincides with the so-called free mantle or floating level. By definition, free mantle or floating level, z_{fm}, is an equilibrium level to which an inviscid mantle substratum could rise in a well crossing through the crust to the depth of the mantle. An equation relating z_0 to the distribution of rheological properties of the lithosphere assuming a linear relationship of stress and strain (i.e., for an inhomogeneous effective elastic plate) was obtained by *Mikhailov* [1999].

When z_0 does not coincide with the free mantle level, intraplate extension or compression disturbs local isostatic equilibrium. For small deformations of a thin plate this disturbance (referred to as a load) can be found from the equation [e.g., *Mikhailov*, 1999]:

$$q(x) = -(z_0 - z_{fm})\partial U/\partial x/(1 + \partial U/\partial x) \qquad (3)$$

Corresponding vertical isostatic movements can be found using the model of a thin elastic plate [*Braun and Beaumont*, 1989]. In this paper we assigned $z_0 = z_{fm}$, neglecting the role of flexural rigidity of the rocks in the process of the Uralides foredeep formation, as shortening and deformation in the Uralides foredeep in Neogene–Quaternary were small, especially in comparison to the Paleozoic ones [*Brown* et al., 1997].

Thus, we have solved the inverse problem to find $W_0(x)$, the vertical component of the velocity vector at the base of the model, under the condition that it provides a best fit to the rates obtained from the geomorphologic data. To solve this problem numerically, the function $W_0(x)$ was presented as an expansion in a series of elementary functions. The number of functions in this expansion should be less than the number of points where neotectonic data are assigned. Strictly speaking, this inverse problem is unstable; one of the possible ways to arrive to a stable solution is by the reduction of the number of elementary functions. Using the finite element method [e.g., *Zienkevich and Taylor*, 1989] the problem of deformation of the viscous medium was solved for every elementary function of the unit amplitude. After that, amplitude of all the elementary functions was found under the condition of the mean square deviation of calculated vertical component of the velocity vector at the top of the model and geomorphologic data. For a Newtonian medium this problem is linear (see *Smolyaninova* et al. [1996] for more details). For the present case the length of the profile was supposed to be 504 km, depth of the model was 50 km, and the size of mesh was 73 km × 19 km (horizontal × vertical).

It should be stressed that the bottom boundary of the model need not coincide with any physically interpretable boundary because the horizontal component of the velocity vector in the lithosphere below the bottom of the model is supposed to be independent of z (see equation (2)). The only point is that it has to be deeper than the main structural inhomogeneities affecting the stress field, while the ratio of horizontal to vertical dimensions of the model must be much greater than unity (as, strictly speaking, equation (2) is valid only for the thin-sheet model). For more details, see *Stephenson and Smolyaninova* [1999].

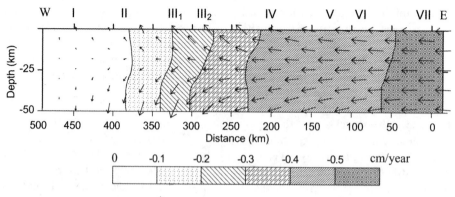

Figure 5. Distribution of horizontal component of velocity (isolines) and velocity vectors (arrows) relative to the left side of the model. Notice decollement of the upper and lower "rigid" plates in the central part of the profile.

When the boundary conditions are determined, components of the stress tensor within the model and other parameters characterizing the deformation of rocks can be calculated [e.g., *Smolyaninova* et al., 1996]. In our study we used the octahedral shear stress and the ratio of the octahedral shear stress to the mean pressure (so-called damage parameter). This parameter can be used to characterize deformation by rock failure or frictional sliding. In areas where this parameter is large, development of faults or movement along pre-existing faults is more likely [*Byerlee*, 1968].

4.2. Results of Numerical Modeling

Results of the inverse problem solution are shown in Figures 4 to 6. Figure 4a demonstrates the quality of the inversion: both the calculated and observed curves of the vertical movements at the surface of the model are very similar. Calculated vertical and horizontal components of the velocity vector at the base of the model are shown in Figure 4 b and c.

The horizontal component of the velocity vector at the base of the model (Figure 4c) looks very simple. It is nearly constant near the left and right sides of the model where deformation is small and has two almost linear intervals which coincide with areas of compression in weak zones of the model. Such a distribution corresponds to simple intraplate compression of the model by forces applied at its side boundaries. Thus, no tectonic forces acting at the bottom of the model are necessary to explain neotectonic movements of the Uralides. Generally, the obtained solution may have been much more complicated. For example, to explain rather complex neotectonic movements along the profile Crimea-Black Sea [*Smolyaninova* et al., 1996], it appeared necessary to

include some mantle induced movements at the bottom of the model.

It can be seen from Figures 4a and 4b that the compression is localized mostly in areas which correspond to the Western and Central Uralian uplifts. Considerably less deformation took place in the Magnitogorsk block and the Urals foredeep. An important peculiarity of the velocity distribution is an offset of the maximum of the vertical velocity at the base of the model in relation to its maximum at the surface. At the surface of the model (Figure 4a) the maximum of the vertical velocity is between 280 and 310 km (i.e., in the Central Uralian uplift area), while at the bottom of the model (Figure 4b) the vertical velocity reaches a maximum between 380 and 350 km, which corresponds to the Western Uralian uplift. Absolute values of the vertical velocity at the surface and at the Moho are comparatively small. If the magnitude of the velocities in the Southern Urals region are of the same order, then the total vertical displacement over the last 10 My would be about several kilometers. Taking into account the low thermal gradient with depth (Figure 2), which appears not to have changed considerably during the last 10 My, one can conclude that these movements were not able to exhume rocks from depths of 15–20 km corresponding to the 300° isotherm. Thus, fission track data do not contradict the suggestion that present day topography has been formed during the Neogene–Quaternary.

Horizontal offset of the area of maximum deformation at the top of the model in relation to its bottom is well demonstrated by Figure 5, where the distribution of the horizontal component of velocity is shown by isolines. Arrows show the direction of movement, their length correspond to absolute values of the velocity vector in

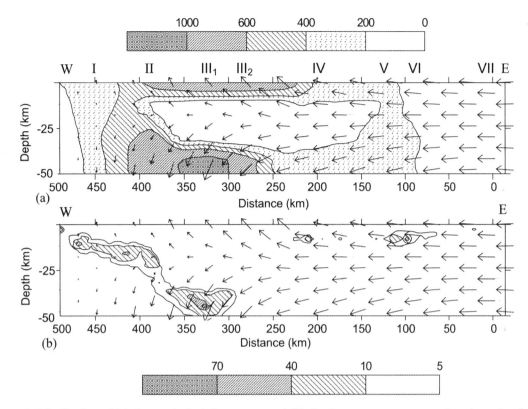

Figure 6. Distribution of (a) the octahedral shear stress and (b) the damage parameter equal to the ratio of the octahedral shear stress to the mean pressure in the crust of the Southern Urals along the URSEIS profile. Absolute values of stress depend on adopted viscosity values. If apparent viscosity of the lithosphere in the both "rigid" layers is equal to 10^{24} Pa·s, then stress in Figure b is in MPa. The Makarovo fault, a distinct 5 km vertical offset of the Moho, which does not disrupt the overlying Riphean sediments, is situated at the point $x = 390$ km [*Diaconescu* et al., 1998], where both the damage parameter and the general shear stress have maximum values.

relation to the left side of the model. To make variations of the vertical component of velocity in Figure 5 visible, the scale for the vertical component was set 10 times greater than for the horizontal one.

It can be seen from Figure 5 that the "rigid" plate at the top of the crust and the other one near the Moho deform in different ways. The area of maximum horizontal shortening (where the vertical component of the velocity vector reaches the maximum value) in the upper layer is offset 70 km eastward in comparison to the area of maximum horizontal shortening of the lower "rigid" layer. This offset is a result of the geometry of the low viscosity zone in the middle part of the model (Figure 2b). When side boundaries of the low viscosity zone are vertical there is no horizontal offset of zones of maximum horizontal shortening in the upper and lower layers. This offset was accompanied by detachment of the lower plate in relation to the upper one along the low viscosity zone in the middle crust in the central part of the model.

Distribution of the octahedral shear stress is presented on Figure 6a. As boundary conditions of the mechanical problem were not boundary forces, but components of the velocity vector (neotectonic movements at the top as well as horizontal components of velocity vector at the bottom and the side boundaries), the relative distribution of stress does not depend on the absolute values of the viscosity. The absolute values of stress are linearly proportional to the viscosity. If apparent viscosity of the lithosphere in both "rigid" layers is equal to 10^{24} Pa·s, then stress units in the Figure 6b are MPa. Figure 6a demonstrates that the model, in general, is characterized by low values of the octahedral shear stress. There are two zones of maximum values which coincide with "rigid" (according to diagrams on Figure 3, brittle) layers. The first one is at the surface of the model, embracing the area of the eastern part of the Urals foredeep and the Western and Central Uralian uplifts. Faulting with thrusting of rigid blocks from east to west can develop there. The second zone of maximum shear stress is situated at the base of the crust below the Western Uralian and western part of the Central Uralian uplifts. Brittle deformation can also take place in this zone (diagram III in Figure 3) because it is characterized by high values of the damage parameter (Figure 6b). This zone coincides with the so-called Makarovo fault [e.g., *Diaconescu* et al., 1998], an apparent 5 km vertical offset of the Moho discontinuity. Thus, this fault could have formed (or at least sufficiently rejuvenated) in the Neogene–Quaternary, contemporaneously with the formation of the modern topography. A distinctive feature of the Makarovo fault is that it has no analog in the upper crustal layers above it, which implies that this fault has a very old age [*Diaconescu* et al., 1998]. Another possible explanation is based on the strength profiles and the octahedral shear stress distribution. They demonstrate that faults at the base of the crust cannot extend to the surface because in the middle crust brittle failure should be replaced by ductile behavior. The slow flow pattern (Figures 5 and 6) shows that shortening of the lower rigid plate between 350 and 380 km is accommodated by shortening of the upper rigid plate not immediately above this zone, but at 70 km to the east in the Main Uralian fault zone.

There are two other zones of high damage parameter values. The first one is between 170 and 120 km, which corresponds to the Suhteli blanket, a thrust zone which is characterized mostly by horizontal displacements (see arrows in Figure 6). The second one is under the Urals foredeep where multiple reflectors are observed in the middle of the crust [*Berzin* et al., 1996; *Tryggvason* et al., 2001]. The direction of velocity vectors shows that displacements here should be nearly horizontal and small.

5. CONCLUSIONS

(1) Geological and geomorphologic data suggest that the present topography of the Southern Urals has been formed during the Neogene–Quaternary. We consider complex deformation of the Southern Urals in this period being due to a superposition of, (a) diagonal NW–SE compression and asymmetric uplift of the Southern Urals area as a whole with maximum uplift of several hundreds meters, and (b) more vigorous superimposed transpressive uplift of the Central and Western Uralian blocks in the Late Pliocene and Quaternary of about one thousand meters.

(2) Modeling of lithospheric strength along the URSEIS profile revealed highest crustal strength in the Western Siberian basin, the East European platform,

and the Zingeika block. Moderate strength values were obtained for the Plast and Trans-Uralian blocks. The strength of the crust below the Urals foredeep and the Magnitogorsk block, and especially below the Western and Central Uralian uplifts, is considerably lower (25% less). This is in a good agreement with the location of the zones of maximum deformation and rates of neotectonic movement in the Southern Urals.

(3) We considered the deformation of a simple model of the lithosphere that includes the main features of the structure of the Southern Urals. It appeared from solution of the inverse problem with the use of geomorphologic data that movements at the Southern Urals during the Neogene and Quaternary can be governed by intraplate compression. The model predicts horizontal offset of the zones of maximum deformation at the top and the base of the crust. The maximum deformation near the surface took place in the area corresponding to the Central Uralian block while in the lower crust maximum deformation was offset to the west and occurred below the area corresponding to the Western Uralian block. This area coincides with an offset in depth of the Moho discontinuity (the Makarovo fault). Thus, this fault could have developed or rejuvenated at the neotectonic stage of the Southern Urals formation.

Acknowledgments. We thank Alexander A. Savelyev (deceased), who attracted our attention to numerical modeling of neotectonic movements of the Southern Urals. We are grateful to two anonymous reviewers for very carefully reading the manuscript, constructive reviews and numerous suggestions, and language corrections. We also thank Robert Simpson (USGS) for language corrections. Attention and help of Dennis Brown, the volume editor, is greatly acknowledged.

REFERENCES

Berzin, R., O. Oncken, J. H. Knapp, A. Pérez-Estaún, T. Hismatulin, N. Yunusov and A. Lipilin, Orogenic evolution of the Uralian mountains: results from an integrated seismic experiment, *Science*, 274, 220–221. 1996.

Byerlee, J. D., Brittle-ductile transition in rocks, *J. Geophys. Res.*, 73, 4741–4750, 1968.

Borisevich, D. V., Neotectonics of the Urals, *Geotectonics*, 26, 41–47, 1992.

Braun, J. and C. Beaumont, A physical explanation of the relationship between flank uplifts and the breakup unconformity at rifted continental margins, *Geology*, 17, 760–764, 1989.

Brown, D., J. Alvarez-Marrón, A. Pérez-Estaún, Y. Gorozhanina, V. Baryshev and V. Puchkov, Geometric and kinematic evolution of the foreland thrust and fold belt in the Southern Urals, *Tectonics*, 16, 551–562, 1997.

Bulashevich, Yu. P. and V. A. Schapov, Geothermal characteristics of the Urals, in *Application of Geothermal Data to Regional Study and Geological Prospecting* (in Russian), edited by Yu. V. Bulashevich, B. P. Dyakonov and Yu. V. Khachay, Acad. Sci. USSR, Uralian branch, Sverdlovsk, 3–17, 1983 (in Russian).

Carbonell, R., A. Pérez-Estaún, J. Gallart, J. Diaz, S. Kashubin, J. Mechie, R. Stadtlander, A. Schulze, J. H. Knapp and A. Morozov, Crustal root beneath the Urals: Wide-angle seismic evidence, *Science*, 274, 222–224, 1996.

Cloetingh, S., J. D. van Wees, P. A. van der Beek and G. Spadini, Role of pre-rift rheology in kinematics of extensional basin formation: Constraints from thermo-mechanical models of Mediterranean and intracratonic basins, *Mar. and Petrol. Geol.*, 12, 793–807, 1995.

Diaconescu, C., J. H. Knapp, L. D. Brown, D. N. Steer and M. Stiller, Precambrian Moho offset and tectonic stability of the East European platform from the URSEIS deep seismic profile, *Geology*, 26, 211–214, 1998.

Diaconescu, C. C. and J. H. Knapp, Role of a phase-change Moho in stabilization and preservation of the Southern Uralian orogen, Russia, this volume.

Döring, J., H.-J. Götze and M. K. Kaban, Preliminary study of the gravity field of the Southern Urals along the URSEIS'95 seismic profile, *Tectonophysics*, 276, 49–62, 1997.

Echtler, H. P., M. Stiller, F. Steinhoff, C. Krawczyk, A. Suleimanov, V. Spiridonov, J. H. Knapp, Y. Menshikov, J. Alvarez-Marrón and N. Yunusov, Preserved collisional crustal structure of the Southern Urals revealed by vibroseis profiling, *Science*, 274, 224–226, 1996.

Evans, B. and D. L. Kohlstedt, Rheology of rocks, in *Rock Physics and Phase Relations. A Handbook of Physical Constants*. AGU Reference Shelf, 3, 148–165, 1995.

Golovanova, I. V., *Catalog of the Uralian Heat Flow Data* (in Russian), Russian Acad. Sci., Ufa Scientific Center, 1994.

Khachay, Yu. V. and V. S. Druzhinin, Geothermal cross-section of the Uralian lithosphere along latitudinal DSS profiles, *Izvestiya, Phys. Solid Earth*, 34, 59–62, 1998.

Khutorskoy, M. D., Heat flux and a model of structure and evolution of the lithosphere in the South Urals and Central Kazakhstan, *Geotectonics*, 19, 215–223, 1985.

Kirby, S. H. and A. K. Kronenberg, Correction to Rheology of the lithosphere: selected topics, *Rev. of Geophys.*, 25, 1680–1681, 1987.

Knapp, J. H., D. N. Steer, L. D. Brown, R. Berzin, A. Suleimanov, M. Stiller, E. Lüschen, D. L. Brown, R. Bulgakov, S. N. Kashubin and A. V. Rybalka, Lithosphere-scale seismic image of the Southern Urals from explosion-source reflection profiling, *Science*, 274, 2226–2228, 1996.

Kohlstedt, D. L., B. Evans and S. J. Mackwell, Strength of the lithosphere: Constraints imposed by laboratory experiments, *J. Geophys. Res.* 100, 17,587–17,602, 1995.

Kolpakov, N. I., V. A. Lyakhovsky, M. V. Mints, E. I. Smolyaninova and Ye. Ye. Shenkman, Geodynamic nature of relief-forming processes on the Kola Peninsula, *Geotectonics*, 25, 161–166, 1991.

Kukkonen, I. T., I. V. Golovanova, Yu. V. Khachay, V. S. Druzhinin, A. M. Kasarev and V. A. Schapov, Low geothermal heat flow of the Urals fold belt — implication of low heat production, fluid circulation or paleoclimate? *Tectonophysics*, 276, 63–85, 1997.

Lider, V. A., *Quaternary Deposits of the Urals* (in Russian), Moscow, Nedra, 1976.

Makarova, N. V., N. I. Korchuganova and V. I. Makarov, Morphological types of orogens as indicators of geodynamical settings of their formation, *Geomorphology*, 1, 14–28, 2000.

Mikhailov, V. O., Modeling of extension and compression of the lithosphere by intraplate forces, *Izvestya, Physics of the Solid Earth*, 35, 228–238, 1999.

Myasnikov, V. P. and V. D. Savushkin, Small parameter method in the hydrodynamic model of the Earth's evolution, *Trans. (Doklady) Russian Acad. Sci./Earth Sc. Sec.*, 238, 1083–1086, 1978.

Puchkov, V. N., Structure and geodynamics of the Uralian orogen, in *Orogeny Through Time*, edited by J.-B. Burg and M. Ford, Geol. Soc. London, Spec. Publ. 121, 201–236, 1997.

Ranalli, G. and D. C. Murphy, Rheological stratification of the lithosphere, *Tectonophysics*, 132, 281–295, 1987.

Ranalli, G., Rheology of the lithosphere in space and time, in: *Orogeny through time*, edited by J. -B. Burg and M. Ford, Geol. Soc. London, Spec. Publ. 121, 19–37, 1997.

Salnikov, V. E., *Geothermal Regime of the Southern Urals* (in Russian), Moscow, Nauka, 1984.

Seward, D., A. Pérez-Estaún and V. Puchkov, Preliminary fission-track results from the southern Urals — Sterlitamak to Magnitogorsk, *Tectonophysics*, 276, 281–290, 1997.

Seward, D., D. Brown, R. Hetzel, M. Friberg, A. Gerdes, G. A. Petrov and A. Pérez-Estaún, The syn- and post-orogenic low temperature events of the Southern and Middle Uralides: evidence from fission-track analysis, this volume.

Schults, S. S., On the newest tectonics of the Urals (in Russian), in: *Materials on Geomorphology and Neotectonics of the Urals and the Volga Basin*, edited by V. P. Sigov, Ufa, 45–59, 1969.

Sigov, A. P., *Metallogeny of the Mesozoic and Cenozoic deposits of the Urals* (in Russian), Moscow, Nedra, 296 pp., 1969.

Smolyaninova, E. I., V. O. Mikhailov and V. A. Lyakhovsky, Numerical modeling of regional neotectonic movements in the Northern Black Sea, *Tectonophysics*, 266, 221–231, 1996.

Steer, D. N., J. H. Knapp, L. D. Brown, H. P. Echtler, D. L. Brown and R. Berzin, Deep structure of the continental lithosphere in an extended orogen: An explosive-source seismic reflection profile in the Urals (Urals

Seismic Experiment and Integrated Studies (URSEIS 1995)), *Tectonics*, 17, 143–157, 1998.

Stefanovsky, V. V., Stratigraphy Scheme of the quaternary deposits of the Urals (in Russian), in *Materials to Stratigraphic Scheme of the Urals (Mz, Kz)*, edited by V. V. Stefanovsky, Ekaterinburgh, 93–139, 1997.

Stephenson, R. A. and E. I. Smolyaninova, Neotectonics and seismicity in the south eastern Beaufort Sea, polar continental margin of the north-western Canada, *Geodynamics*, 27, 175–190, 1999.

Tevelev, Al. V., Arc. V. Tevelev, I. A. Kosheleva and E. F. Burstein, *Explanations to Geological Map of Russian Federation at the 1:200,000 scale. The South Uralian Series* (in Russian), *Sheet N-41-XIX*, S-Petersburg, 2002.

Tevelev, Arc. V. and Al. V. Tevelev, Modern pull-apart basins of passive continental regions, in *Abstracts of the Lomonosov Conference*, Moscow State University, 27–28, 1996.

Tevelev, Arc. V. and Al. V. Tevelev, Mode of recent evolution of the Eastern Urals, *6th Zonenshain Conference on Plate tectonics and Europrobe Workshop on Uralides. Abstracts.* Moscow, Geomar, 203–204, 1998.

Trifunov, V. P., Main features of the Uralian neotectonics, in *Geomorphology and Modern Tectonics of Volga-Urals and Southern Urals Regions* (in Russian), edited by V. V. Stefanovsky, Ufa, 293–300, 1960.

Tryggvason, A., D. Brown and A. Pérez-Estaún, Crustal architecture of the southern Uralides from true amplitude processing of the URSEIS vibroseis profile, *Tectonics*, 2001 (in press).

Zienkiewicz, O. C. and R. L. Taylor, *The Finite Element Method*, 4th Ed., McGraw-Hill, New York, 1989.

Zoback, M. L., First and second order patterns of stress in the lithosphere: The world stress map project, *J. Geophys. Res.*, 97, 11,703–11,728, 1992.

Zubkov, A. V. and Ya. J. Lipin, Stressed state of the Uralian uppermost crust, *Trans. (Doklady) of Russian Acad. Sci.* 336, 792–793, 1997.

R. G. Berzin and A. K. Suleimanov, "Spetsgeofizika", Povarovka, Moscow District, Russia

E. A. Kiseleva, V. O. Mikhailov, E. I. Smolyaninova and E. P. Timoshkina, United Institute of Physics of the Earth, Russian Academy of Sciences, B. Gruzinskaya, 10, Moscow, 123810, Russia

A. V. Tevelev, Moscow State University, Geological Faculty, Vorobievy Gory, Moscow, Russia